Contents

Preface *ix*

I. Life History and Brain Evolution *1*

1. Life History and Cognitive Evolution in Primates 5
CAREL P. VAN SCHAIK AND ROBERT O. DEANER

Case Study 1A. Sociality and Disease Risk: A Comparative Study of Leukocyte Counts in Primates *26*
CHARLES L. NUNN

2. Dolphin Social Complexity: Lessons from Long-Term Study and Life History *32*
RANDALL S. WELLS

3. Sources of Social Complexity in the Three Elephant Species *57*
KATY PAYNE

II. Evolution of Cooperative Strategies *87*

4. Complex Cooperation among Taï Chimpanzees *93*
CHRISTOPHE BOESCH

Case Study 4A. Coalitionary Aggression in White-Faced Capuchins *111*
SUSAN PERRY

Case Study 4B. Levels and Patterns in Dolphin Alliance Formation 115
RICHARD C. CONNOR AND MICHAEL KRÜTZEN

5. The Social Complexity of Spotted Hyenas 121
CHRISTINE M. DREA AND LAURENCE G. FRANK

Case Study 5A. Maternal Rank "Inheritance" in the Spotted Hyena 149
ANNE ENGH AND KAY E. HOLEKAMP

6. Is Social Stress a Consequence of Subordination or a Cost of Dominance? 153
SCOTT CREEL AND JENNIFER L. SANDS

Case Study 6A. Sperm Whale Social Structure: Why It Takes a Village to Raise a Child 170
SARAH L. MESNICK, KAREN EVANS, BARBARA L. TAYLOR, JOHN HYDE, SERGIO ESCORZA-TREVIÑO, AND ANDREW E. DIZON

III. Social Cognition 175

7. Equivalence Classification as an Approach to Social Knowledge: From Sea Lions to Simians 179
RONALD J. SCHUSTERMAN, COLLEEN REICHMUTH KASTAK, AND DAVID KASTAK

8. The Structure of Social Knowledge in Monkeys 207
ROBERT M. SEYFARTH AND DOROTHY L. CHENEY

9. Social Syntax: The If-Then Structure of Social Problem Solving 230
FRANS B. M. DE WAAL

Case Study 9A. Conflict Resolution in the Spotted Hyena 249
SOFIA A. WAHAJ AND KAY E. HOLEKAMP

IV. Communication 255

10. Laughter and Smiling: The Intertwining of Nature and Culture 260
JAN A. R. A. M. VAN HOOFF AND SIGNE PREUSCHOFT

Case Study 10A. Emotional Recognition by Chimpanzees *288*
LISA A. PARR

11. Vocal Communication in Wild Parrots *293*
JACK W. BRADBURY

Case Study 11A. Representational Vocal Signaling in the Chimpanzee *317*
KAREN I. HALLBERG, DOUGLAS A. NELSON, AND SARAH T. BOYSEN

12. Social and Vocal Complexity in Bats *322*
GERALD S. WILKINSON

13. Dolphins Communicate about Individual-Specific Social Relationships *342*
PETER L. TYACK

Case Study 13A. Natural Semanticity in Wild Primates *362*
KLAUS ZUBERBÜHLER

V. Cultural Transmission *369*

14. Koshima Monkeys and Bossou Chimpanzees: Long-Term Research on Culture in Nonhuman Primates *374*
TETSURO MATSUZAWA

Case Study 14A. Movement Imitation in Monkeys *388*
BERNHARD VOELKL AND LUDWIG HUBER

15. Individuality and Flexibility of Cultural Behavior Patterns in Chimpanzees *392*
TOSHISADA NISHIDA

Case Study 15A. Sex Differences in Termite Fishing among Gombe Chimpanzees *414*
STEPHANIE S. PANDOLFI, CAREL P. VAN SCHAIK, AND ANNE E. PUSEY

16. Ten Dispatches from the Chimpanzee Culture Wars *419*
W. C. MCGREW

Case Study 16A. Spontaneous Use of Tools by Semifree-ranging Capuchin Monkeys 440
EDUARDO B. OTTONI AND MASSIMO MANNU

17. Society and Culture in the Deep and Open Ocean: The Sperm Whale and Other Cetaceans 444
HAL WHITEHEAD

Case Study 17A. Do Killer Whales Have Culture? 465
HARALD YURK

18. Discovering Culture in Birds: The Role of Learning and Development 470
MEREDITH J. WEST, ANDREW P. KING, AND DAVID J. WHITE

References 495

Acknowledgments 594

Contributors 601

Index 607

Animal Social Complexity

Intelligence, Culture, and Individualized Societies

EDITED BY

Frans B. M. de Waal
and Peter L. Tyack

HARVARD UNIVERSITY PRESS
Cambridge, Massachusetts
London, England · 2003

Printed in the United States of America

Library of Congress Cataloging-in-Publication Data

Animal social complexity : intelligence, culture, and individualized societies / edited by Frans B.M. de Waal and Peter L. Tyack.
p. cm.
Includes bibliographical references (p.) and index.
ISBN 0-674-00929-0 (cloth : alk. paper)
1. Mammals—Behavior. 2. Birds—Behavior.
3. Social behavior in animals. 4. Learning in animals.
I. Waal, F. B. M. de (Frans B. M.), 1948– II. Tyack, Peter L.

QL739.3 .A56 2003
599.156—dc21 2002027505

Preface

Are we guilty of hubris in editing a book on animal social complexity and intelligence? What could be more difficult than comparing the social complexity of different species, or comparing the intelligence of animals adapted to totally different environments? Even for our own species, for which we understand these issues in far greater detail, researchers cannot agree on how intelligence should be measured and how it may have been shaped by ancestral milieus. Our goal is not to champion one way to measure animal intelligence nor to rank animal species by intelligence. Rather our goal is to explore the many different forms social complexity and intelligence take in a diversity of animals.

The past few decades have seen rapid growth of research on the societies, social cognition, and emotions of long-lived, large-brained animals such as monkeys, apes, whales, elephants, and lions. Unfortunately, but perhaps inevitably, this expansion of research has led to increasing specialization and separation between scientists working on different topics, such as vocal communication versus social cognition, or on different taxonomic groups, such as nonhuman primates versus social carnivores. Each specialized field tends to have its own professional society, its own annual conference, and its own scientific journals, so that the younger generation, educated after all of this fragmentation, hardly realizes the power of generalizations across taxonomic groups.

At the same time, however, we have seen studies on a variety of vertebrates using a variety of methods converge on the insight that many animals rely for survival on special social relationships throughout their life span. The societies of particular interest to us share the following characteristics: they are a) individualized (i.e., members recognize each other individually and form variable relationships built on histories of interaction), and b) longitudinally stable (i.e., species with long life spans have long-term or multigenerational relationships, such as those between grandparents and grand-offspring or friendships among adults going back to youth). Moreover, there are reasons to conclude that c) social behavior and survival strategies are subject to strong learning effects, so much so that we can speak of acquired social skills and cultural transmission of habits and knowledge. The latter seems to make the traditional dichotomy between nature and nurture a rather artificial one.

Even though the societies of social insects, such as ants and honey bees, are far from simple, they don't share this set of characteristics. Societies in which the members recognize one another, remember the outcomes of earlier encounters, and modify how they relate to one another based upon histories of interaction can develop unusual complexity because they are individualized at every level. Needless to say, this sort of sociality is easy for us to relate to since it also characterizes our own societies. For over a quarter of a century, primatologists have recognized the complexity of individualized societies, and have in addition speculated that high intelligence evolved as an adaptation to the social milieu of friendships and rivalries. Now, a much broader array of scientists seems to be reaching similar conclusions with regard to a wider range of species.

To bring such a diverse group of scientists together, a two-tiered conference was organized with the enthusiastic support of Dr. Paul G. Heltne of the Chicago Academy of Sciences from 23–26 August 2000 in the Peggy Notebaert Nature Museum, Chicago, Illinois. The proceedings began with a three-day conference on animal social complexity and intelligence. Second, a special day was organized by the Jane Goodall Institute to celebrate 40 years of chimpanzee studies at Gombe National Park, Tanzania. During these four days, 23 of the most prominent scientists in their respective fields came together from all over the world for an electrifying exchange. An afternoon session with presentations by junior scientists added to the excitement.

What made the conference and what makes the present volume special is the mixture of detailed documentation of the social lives of some of the most fascinating "mega fauna" on earth and recent theoretical advances that help explain the evolution of social systems. We were moved to hold this conference in part to celebrate the enormous efforts of so many dedicated fieldworkers studying animals in habitats all over the world—from the jungle to the desert, from the sea to the air. It was obvious during the conference how much delight the speakers take in watching their study animals, and this pleasure was immediately contagious for the audience. Two remarkable heroes of this tradition, Jane Goodall with her path-breaking fieldwork on wild chimpanzees, and Kenneth Norris, a pioneer in the behavioral study of dolphins and whales, acted as catalysts for the conference that led to this book.

The story of how Jane Goodall initiated one of the first behavioral studies of wild chimpanzees at Gombe against all odds is well known. The enormous time investment to habituate shy wildlife to human presence, the decision to recognize and name individual apes, the effort to follow them on their excursions through the forest, and the formation of a dedicated staff who track them throughout their life cycles have made the Gombe studies a great success that, through popularization in both writing and photography, has reached millions. Major discoveries have been made with regard to tool use, territorial violence, hunting and meat-eating, and—thanks to recent advances in DNA analysis—genetic relationships within a chimpanzee community. The 40 years of fieldwork at Gombe is exactly the kind of longitudinal research that many young researchers now apply to all sorts of animals, not only primates. Let Jane Goodall stand as a symbol of the critical need in this field for long-term studies of identified animals living in the wild, and of the success that can come from an unyielding devotion to fieldwork and to the study animals.

Ken Norris, who inspired an entire generation of marine biologists with prescient insights, helped enthusiastically with the conference planning, which began in 1995. Sadly, he didn't live to see its realization. He was a polymath of natural history, equally at home studying circadian rhythms of desert snakes or effects of temperature on intertidal fish, teaching students about the different ecosystems that made up his native California, and developing new methods for a decisive experiment to show that dolphins can echolocate. When Norris picked up a dolphin jawbone during a walk on the beach, the ideas he had in the first 30 seconds could fuel decades of research. Ken was driven by ideas: he developed radically different studies for different questions. To study dolphin groups in the wild, he devised novel semi-submersible boats for underwater viewing. For studying sound production, he and his students took X-ray film of a dolphin vocalizing in captivity. Let Ken Norris stand as a symbol of the benefits of combining a sharp focus on a scientific question with an open-minded and creative search for the best methods to answer the question.

Contributions to our volume are limited to work on mammals and birds, because that is where we found the most relevant and interesting research in this area. Yet, if a book like this were to be written twenty years from now, it might well include an even broader taxonomic range. We decided to include contributors who use a variety of methods so that we could approach the topic of social complexity from many different angles. We include one chapter, for example, that applies a classic method for comparing animal intelligence on a broad scale—an analysis of factors that correlate with an anatomical indicator of intelligence, the ratio of brain mass to body mass. The use of such an indirect proxy for intelligence facilitates broad comparisons, but poses the risk that the proxy may not always measure what we hope and expect it to. Other contributions compare social behavior and ecological conditions for several populations of one species or several closely related species. At the finest scale of analysis, some contributors have spent years studying a single group of animals of a single species, documenting how their animals develop social relationships through a series of minute-by-minute interactions. We hope that this combination of studies on the ecology of sociality with the behavioral, communicative, and cognitive mechanisms that mediate social behavior will help test assumptions underlying each kind of study, and will also help the reader think about the evolution of complex animal societies and intelligence.

There is a strong tradition of analyzing intelligence in terms of how well animals perform in tests devised by psychologists. While we recognize the importance of this research, our bias is to focus on evidence about how animals solve problems posed by their own social lives rather than how they solve problems posed by a human. Our interest in the evolutionary origins of behavior highlights the need for making observations in settings like those in which the behaviors evolved. Particularly for studies of ecological correlates of social behavior, this often requires working with wild animals in environments that are as pristine as possible. Captive studies also have an important role, however, since observations can be carried out under optimal visibility (a great advantage in relation to species that live in forest canopies or under water), in greater detail than possible in the field, and with the opportunity for manipulating variables. The two approaches—captive and field—clearly complement each other. We might not want to rely on either one in isolation, but together they form an unbeatable combination.

We hope that the careers of scientists such as Jane Goodall and Ken Norris will inspire those who read this book to pursue new avenues for the perpetually interesting but perplexing study of social complexity and intelligence in animals. In the presence of a second generation of pioneers, now in their fifties and sixties, we set out to break down the taxonomic barriers. There is an ancient tradition of ranking animals from low to high, with humans at the top. This tradition is balanced here by the natural tendency for each researcher to favor their own study animals. To the bird researcher, birds are the most accomplished communicators; to the primatologist, no other animals have such complex social relations; the bat researcher notes that bats can echolocate and fly at the same time; to the marine mammalogists, sperm whales are almost as cultured as humans. Of course, claims about one animal's superiority over another are made partly in jest, but still signify an element of rivalry that we are seeking to overcome. Focus on one's own study animal sometimes leads researchers to limit comparisons to a narrow taxonomic range rather than to perform a broader comparative analysis that includes a variety of species. In an effort to boost the status of their subjects, researchers may even limit these analyses to favorable comparisons with humans. As editors, we have urged contributors to focus on as broad and as balanced a comparative perspective as possible to better understand the evolution of intelligence in complex societies.

There has been renewed interest and controversy about the question of animal cultures. This is an area where the separation of different disciplines may make it difficult to focus on the issue of most interest to us: How and why did societies evolve in which one individual's innovation is adopted by the rest of its group? The use of the term "culture" in some fields is by definition limited to humans. But other disciplines may happily use the term "culture" for the songs that birds learn from one another, leading to geographically distinct dialects. Some fields emphasize the importance of specific mechanisms for social learning, such as imitation and teaching. But it has also been argued that even if a range of learning mechanisms, from simple stimulus enhancement to deliberate teaching, underlies social transmission, the critical point in terms of evolutionary consequences is that behavior is actually being transmitted.

A danger of this definitional morass is the tendency to argue about whether one or another animal species qualifies for culture or imitation. At this early stage of our understanding of these issues, we prefer to focus on the range of ways animals develop traditions and how these traditions affect their survival and reproduction in real life. Although we need to be careful about definitions, and controlled experiments will help verify which learning mechanisms operate in which species and contexts, we agree with contributors to this volume that culture must involve the entire social system, not just particular skills tested in artificial contexts.

No doubt, cultural learning will prove an enduring concept that will be scientifically explored in great depth over the coming decades. Until now, we have only scratched the surface. Such research will change the image of animals as gene machines, programmed to act in particular ways. It will elevate them to a much more flexible status, more like the way we see ourselves. It is ironic that a strengthening belief in cultural learning in animals should parallel amplification of the message that human behavior is a product of evolution, as a stream of popular books exploring human behavior from a biological perspective would suggest.

It is in this context that *Animal Social Complexity* should be viewed. We are presenting data and exploring theoretical ways of looking at individualized societies that apply not just to one taxonomic group—such as the group to which we, humans, belong—but to a vast range of animals. The goal is to find shared principles of sociality and cognition. At a time in which people seem ready to acknowledge their own biology and in which the idea of culture among animals is being debated for the first time, we may be moving toward a truly unified social science, one with a distinctly cross-specific perspective.

Frans B. M. de Waal
Atlanta, Georgia

Peter L. Tyack
Woods Hole, Massachusetts

ANIMAL SOCIAL COMPLEXITY

I

Life History and Brain Evolution

A classic method for studying the evolution of intelligence in different animal species involves comparing the size of their brains based on the assumption that intelligence correlates with brain size. In Chapter 1, Carel van Schaik and Robert Deaner review different hypotheses for the evolution of enlarged brains among mammals, especially among primates. They raise cautions about this approach, which they call the comparative neuroanatomical approach. Larger animals have larger brains, but it is unclear whether or how to correct brain size for variation in body size. Nor is it obvious which brain structures should be used for these studies: the whole brain, the neocortex, and so on.

A variety of studies in the past few decades have used the comparative neuroanatomical approach to test the social intelligence hypothesis, which suggests that the most important selective pressure for increased intelligence involves increasingly sophisticated strategies for competing and cooperating with conspecifics. Although we as editors feel that much of the research reported in this book supports the social intelligence hypothesis, van Schaik and Deaner laudably point out that the comparative neuroanatomical studies do not conclusively demonstrate it to be correct. Along with the problems listed above, indices of social complexity in these studies are often questionable. For example, group size is often used as a proxy for social complexity. Many of the chapters in this book emphasize a particular kind of social complexity in which individuals must keep track of interactions with other individuals with whom they must compete and cooperate. The more partners an animal in such a society must keep track of, the more complicated, but other species may aggregate in large anonymous groups that may involve much simpler interactions than smaller individualized societies.

Van Schaik and Deaner not only argue that it is premature to conclude that the demands of social intelligence led to the evolution of large-brained mammals or primates, but they demonstrate a close relationship between relative brain size and slow life history in primates and all mammalian orders other than bats. This correlation is robust for primates, remaining significant after correcting for body size, phylogeny, and socio-ecological factors such as diet, home range, or group size.

Brain tissue is metabolically expensive, so van Schaik and Deaner emphasize the need to consider the balance between costs and benefits when analyzing the evolution of higher investment in brain tissue for some species. Their life historical perspective suggests several different factors, each of which might lower the cost or increase the benefit of this investment. It also suggests more abstract demographic factors that may be more applicable for broad taxonomic comparisons. For example, primatologists have suggested that an arboreal lifestyle may correlate with the evolution of higher cognitive abilities, but the present authors emphasize that the critical point about arboreality may be a lower risk of predation. This more abstract feature can better be measured for animals that swim or fly or live very different lives from primates. This perspective also helps us understand how cognitive abilities involving many domains may evolve together, even when selection pressures may be domain-specific.

In his case study, Charles Nunn investigates costs of sociality. He analyzed three risk factors for disease in primates: group size, exposure to potentially contaminated substrate, and mating promiscuity. To estimate differences in the chronic risk of disease in different species, Nunn measured the concentration of white blood cells (WBCs), since they are one of the first lines of defense against pathogens. Nunn found a strong correlation of WBC count with mating promiscuity but not with either of the other two hypothesized disease risks.

Marine mammals represent an outlier among big-brained mammals to primatologists and most comparative psychologists. In Chapter 2, Randall Wells highlights lessons from a long-term study of bottlenose dolphins, a study that emphasizes both life history and social complexity. Our conference celebrated 40 years of chimp studies at Gombe National Park. Wells's study of bottlenose dolphins in the waters of Sarasota, Florida is likewise a milestone in cetology. Wells started studying these dolphins in 1970. With each passing year, the depth and breadth of the study have increased, so that now Wells knows the age, sex, genetic relationships, and association patterns of almost all the animals he sees. Such familiarity, after decades of living with a population, is the crucial first step to understanding the social knowledge that our subjects may share.

Wells focuses on life history because he appreciates the same insight demonstrated by van Schaik and Deaner—life history has a profound impact on the opportunities for the kind of individualized societies that are the subject of this book. He thus highlights the longevity, long period of dependency, and long delay for sexual maturity, particularly in males, that characterize bottlenose dolphins. This slow life history is pronounced among many species of dolphin and toothed whales. In Chapter 17, Hal Whitehead emphasizes similar characteristics in sperm whales. These animals invest in parental care for many years, and have relatively low infant mortality. The life history consequences of prolonged parental care are particularly striking among short-finned pilot whales. Female pilot whales go through physiological changes much like menopause in humans. Most stop cycling by 30 years of age, yet they may live to nearly 60 in multigenerational groups. Presumably these reproductively senescent females switch their reproductive effort from having young to taking care of them. The multigenerational groups of some cetaceans provide a setting ripe for cultural transmission of information, and Wells discusses evidence for cultural transmission of feeding behaviors in bottlenose dolphins, following logic similar to that used by primatologists in discussing culture in chimpanzees.

In Chapter 3, Katy Payne discusses another large long-lived and slow-to-mature mammal, the elephant. The most stable social unit in elephants is the family—a group of related females and their young. The oldest female, called the matriarch, is the most dominant member of the family, but her leadership goes well beyond dominance. Payne reports that the matriarch makes many of the decisions about group movements, and she emphasizes the importance of the matriarch's memory for making successful decisions. Memory for ecology may be critical for survival: for example, a matriarch decides where to find water during the worst drought since her youth. Social memory may be vital to deciding which groups to avoid and which to join. Payne cites playback experiments suggesting that it may take decades for an elephant matriarch to build up her model of social networks. The effects of this knowledge on hundreds of daily decisions may be quite important, for families of older matriarchs have more calves per female than families of younger matriarchs.

Payne emphasizes the importance of acoustic communication in mediating social interactions among different groups of elephants, whose home ranges may encompass hundreds to thousands of square kilometers. Elephants produce low frequency calls that can carry for many kilometers. Roving adult males can find receptive females using these calls, and individuals appear to coordinate movements both within and between families with these calls.

Payne observes that different individual elephants often have very different behavioral reactions to the same event. She notes that this behavioral variability may increase the complexity of individualized societies. Some individual elephants can be suddenly dangerous to others nearby when they become aggressive. This suggests another reason why it may be so important for animals in individualized societies to be able to recognize individuals and remember their history of interaction—not just to maintain contact with partners, but also to avoid untrustworthy individuals.

Van Schaik and Deaner caution that each taxon that independently evolved large brains and intelligence may have arrived via a different evolutionary path, each involving an idiosyncratic constellation of selection pressures. In their view, it may be naïve to seek a single evolutionary cause for the evolution of intelligence in all animal groups. This is a useful cautionary tale, but one that runs contrary to the striking commonality of results from the participants of this conference, who have used a diverse array of methods to study such an assortment of mammals and birds. We hope that this section will encourage readers to evaluate the current status of the evolution of intelligence and of complex social behavior and seek new directions for future research.

1

Life History and Cognitive Evolution in Primates

CAREL P. VAN SCHAIK AND ROBERT O. DEANER

The study of animal cognition has become a highly active area (Dukas 1998; Balda et al. 1998; Shettleworth 1998), and primatologists have made some seminal contributions. Perhaps most important, primate researchers have put forth a variety of stimulating hypotheses to account for the apparently exceptional cognitive capacities of primates as a whole and for cognitive variation among primates in particular. These hypotheses focus on selective demands ranging from foraging and arboreal locomotion to sociality, but the common thread is that all are claimed to have selected for some type of fairly generalized ability (see Table 1.1).

In the past decade, several comparative tests of these contending hypotheses have been conducted. These tests were aimed at identifying the hypothesis that best explains variation in the size of the brain or brain structures (Dunbar 1992; Barton & Purvis 1994). Many of these tests have favored the social strategizing or Machiavellian intelligence hypothesis (Byrne & Whiten 1988), which claims that the need for effective competition and cooperation with conspecifics has provided the main selective advantage (i.e., served as the "pacemaker") for evolutionary increases in cognitive abilities. These tests have led workers to reject the other hypotheses or at least suggest that the social strategizing hypothesis holds most generally (Byrne 1995; Barton & Dunbar 1997; Dunbar 1998; Cummins 1998; Whiten 2000).

Table 1.1. Overview of hypotheses for the evolution of cognitive abilities, especially among primates

Hypothesis	Selective demand	Applicability	References
Social strategizing	Predicting and manipulating the behavior of conspecifics	All	Cheney et al. 1986; Byrne & Whiten 1988; Whiten 2000
Spatiotemporal mapping	Monitoring food availability in space and time	All	Clutton-Brock & Harvey 1980; Milton 1988
Extractive foraging	Extracting hidden food items from matrix	All	Parker & Gibson 1977, 1979
Food processing	Manual dexterity and bi-manual coordination to process foods	All? (great apes?)	Byrne 1997
Arboreal clambering	Moving nonquadrupedally in a three-dimensional habitat	Great apes	Povinelli & Cant 1995

The widespread acceptance of one hypothesis may be premature, however, in part because of the nature of the tests used to reach this conclusion. We therefore begin this chapter with a discussion of the key features of the comparative neuroanatomical approach used to test hypotheses for cognitive evolution, and then examine the social strategizing hypothesis in some more detail. This analysis suggests that we should keep an open mind on the pacemakers of cognitive evolution in primates, and mammals in general.

Next we argue that consideration of the costs of cognitive adaptations may be useful. First, we demonstrate that life history and measures of brain size have undergone correlated evolution, indicating that animals with fast life histories are generally unlikely to evolve greater cognitive abilities. Second, we argue that the life history perspective explains why some species show more cognitive adaptations than others despite being subject to the same demands, that domain-general cognitive abilities may arise as a byproduct of slow life history, and that there are many likely, nonexclusive cognitive benefits. Accordingly, a single selective demand (e.g., social strategizing) is unlikely to have universal influence, and natural history characteristics should be expected to produce diverse pressures in different lineages.

Testing Hypotheses of Cognitive Evolution

Although there are several approaches to testing adaptive hypotheses, from a historical perspective (as defined by Coddington 1988) the most rigorous way is to employ the comparative method, asking whether traits have repeatedly evolved in a hypothesized adaptive context (Harvey & Pagel 1991;

Brooks & McLennan 1991; Shettleworth 1998). In the case of cognition, we would like to know whether lineages that faced a particular demand (e.g., spatial mapping, social strategizing) were especially likely to have evolved enhanced abilities. Although there has been some success in applying the comparative approach to the evolution of cognitive abilities (Balda et al. 1996; Lefebvre & Giraldeau 1996), primatologists have generally lacked the relevant data on the socioecological demands and the cognitive abilities. With respect to cognitive abilities, the issue is that the ever-formidable challenge of showing taxonomic differences in ability (compare Macphail 1982; Kamil 1988) becomes yet more difficult when the claim is that the abilities are domain-general (but see Johnson et al. in press). Given these issues, primate workers have used what we call the comparative neuroanatomical approach: rather than assessing cognition through behavioral performance, cognitive ability is assumed to correspond with the size of the whole brain or of brain structures implicated in cognitive tasks; and instead of appraising the socioecological demands empirically, they are assumed to correspond with other, proxy variables (for reviews, see Harvey & Krebs 1990; Barton & Dunbar 1997).

Although there are many difficulties with employing the comparative neuroanatomical approach, they are rarely discussed. With respect to socioecological demands, there is scant information regarding the degree to which the proxy variables used in the tests actually provide valid and reliable estimates of the supposed demands. For instance, tests of the spatio-temporal mapping hypothesis have employed home range size or percentage fruit in the diet on the assumption that the former assays spatial demands and the latter assays temporal demands. It is possible, though, that animals with larger ranges perceive them in a less "fine-grained" manner and so do not face exceptional spatial navigational problems; likewise, despite the impressions of several field-workers (e.g., Eisenberg & Wilson 1981; Clutton-Brock & Harvey 1980; Milton 1988), nonfruit items may also be highly ephemeral (Glander 1981). Similarly, the popular Machiavellian intelligence hypothesis has been tested mainly by using group size to assay social strategizing demands. Although group size is probably related to such things as the need to maintain dominance relationships and establish strategic alliances, the relationships are indirect and there are many exceptions, including gregarious groups that have little apparent complexity or even temporal stability (see Bradbury 1986; Strum et al. 1997).

The validity of brain structures as cognitive assays is also unclear. The un-

derlying assumption—that size matters—is reasonable because the size of brain structure predicts several neurophysiological characteristics, including the number of neurons at the cortical surface and the number of cortical columns (reviewed by Jerison 1991). Likewise, in several contexts, brain structure size corresponds with behavior. For instance, cerebellar size correlates with measures of visual memory and motor dexterity in humans (Paradiso et al. 1997), and avian taxa with superior spatial memory abilities tend to have larger hippocampi (e.g., Basil et al. 1996). Nevertheless, the fact remains that there is little direct indication that the size of brain structures corresponds with domain-general cognition in a comparative context (but see Rumbaugh 1997; Johnson et al. in press).

Furthermore, even if one accepts that the sizes of brain structures provide reasonable proxies for cognition, there are two additional problems. One is that it is unclear which brain structure should be employed. Earlier studies generally used whole brain size (e.g., Clutton-Brock & Harvey 1980; Gittleman 1986), a measure that has several advantages, including easy estimation, even for fossil specimens. More recent studies have generally employed the neocortex or the neocortex minus area V1, chiefly on the grounds that many structures involved in higher-order cognition (e.g., planning, working memory) are located therein (e.g., Sawaguchi & Kudo 1990; Dunbar 1992; Joffe & Dunbar 1997; see Barton 1998). However, many noncortical structures (e.g., cerebellum) are also implicated in higher-order cognitive processes (e.g., Middleton & Strick 1994). Because it is unclear what structure should be used, and because most data are available for whole brain size, we use whole brain size for the analyses in this paper.

A second problem with comparing brain structures across taxa is that species differ dramatically in body size, and it is unclear if and how this fact should be accounted for. Body size and whole brain size, for instance, are highly correlated, and hence most workers advocate scaling techniques (e.g., Jerison's [1973] Encephalization Quotient) to compare brain sizes after all effects of body size have been statistically removed. This approach has several drawbacks, though, including the possibility that cognitive ability and body size truly coevolved, making the control of body size undesirable (Deacon 1997). Several other methods of scaling the brain or brain structures have also been introduced, including the use of ratios (Passingham 1975; Dunbar 1992) that only partially control for body size. Deaner and colleagues (2000) have recently reviewed the various kinds of scaling methods and shown that all are of unproven validity and have potential drawbacks. In the analyses

presented below, we use residuals from the regression of whole brain on body mass. Although this may not turn out to be the best cognitive proxy, it is at least conservative because it strictly controls for body size and thus may prevent spurious correlations.

A final issue in comparative neuroanatomical studies is that once the decision has been made to employ specific socioecological variables and brain structures, species values for these measures cannot be treated as independent data points in a statistical analysis. The reason is that most biological traits are at least partially the product of phylogenetic history, meaning that closely related taxa will often share traits simply due to common descent (Felsenstein 1985). There is ongoing debate regarding how comparative analyses should address this issue, but the most widely adopted approach is to estimate evolutionary changes throughout a phylogenetic reconstruction of a trait's evolution (Harvey & Pagel 1991; Purvis & Webster 1999). In the case of hypotheses about cognitive evolution in primates, the question would be whether evolutionary changes in a socioecological demand consistently co-occur with evolutionary changes in the (scaled) size of a brain structure. Although comparative methods that account for phylogenetic nonindependence can be sensitive to the evolutionary models assumed and the accuracy of phylogenetic information, they are far more likely to provide correct answers than are analyses that do not account for phylogenetic relationships (Harvey & Pagel 1991; Martins & Hansen 1996; Purvis & Webster 1999; Nunn & Barton 2001). In the analyses below, we use the CAIC program (Purvis & Rambaut 1995) to implement the comparative method of independent contrasts (Felsenstein 1985). Nunn and Barton (2001) furnish a clear introduction to the rationale and use of this method.

To summarize, then, although the comparative neuroanatomical approach makes reasonable assumptions, the proxies employed measure the variables of interest with unknown (but possibly sizable) error. Furthermore, the techniques that estimate correlated evolution also include unknown error. For these reasons, the conclusions of all comparative neuroanatomical studies must be viewed cautiously.

Social Strategizing Revisited

As a result of several comparative neuroanatomical studies over the last decade, primatologists seem to be converging on the simple generalization that cognitive evolution, at least in primates, is best explained by responses to

social challenges (Byrne 1995; Barton & Dunbar 1997; Dunbar 1998; Cummins 1998). For several reasons we believe that this conclusion is premature.

The studies on which this conclusion is based (Dunbar 1992, 1995; Barton & Purvis 1994; Barton 1996) have several weaknesses. First, as noted above, group size may not be a good proxy for social demands. Second, the tests that have supposedly differentiated among the hypotheses have each only explored one neuroanatomical scaling method. Recent work has shown, however, that the outcomes of these tests are sensitive to the scaling method employed, so that under several conditions, home range is a better neocortex predictor than is group size (Deaner et al. 2000). Third, group size, home range size, and other socioecological proxies for the hypotheses are intercorrelated (Deaner et al. 2000), making it difficult to differentiate among the hypotheses with these (error-prone) proxies.

Besides these problems with the comparative tests, there are other empirical problems with the social strategizing hypothesis. One is that some major phylogenetic contrasts in primates cannot be accounted for in this way (Byrne 1997). Most obviously, great apes outperform monkeys on cognitive tasks (Byrne 1995; Johnson et al. in press), yet do not show greater social complexity, at least in most species (Tomasello & Call 1994). Similarly, the largest-brained of all the living prosimians—the aye-aye *(Daubentonia madagascariensis)*—is almost entirely solitary. Another problem is that the demands of social strategizing—at least as assayed by demographic variables—appear to have limited explanatory value in nonprimates. Relative brain size in sciurids or birds is not affected by degree of sociality (solitary versus gregarious lifestyle: Meier 1983; Bennett & Harvey 1985). In carnivores, Gittleman (1986) found that relative brain size was larger in multimale than in unimale breeding groups but also noted that other measures of social complexity do not correlate with relative brain size (see also Dunbar & Bever 1998). Marino (1998) reports that brain size and group size are related in odontocetes, but a similar analysis controlling for phylogeny could not confirm this result (R. Deaner, unpublished data).

In conclusion, it is at this stage premature to decide that social strategizing was the only, or even the major, selective pressure on cognitive evolution in primates or other mammals. We would like to stress that we do not claim that social strategizing is not a major force in primate cognitive evolution. It may well be, but we think it is too early to consider the issue settled and to

ascribe all cognitive evolution, at least in primates, to the need for social strategizing.

Life History and Cognition

Would not all animals benefit from being smart? For instance, if the common perception that social primates are smarter than social ungulates is true, then we must ask why ungulates did not undergo selection for similar abilities. Considering this issue underscores a fundamental point of evolutionary biology (and one not incorporated into the hypotheses proposed to date), namely that virtually all benefits entail costs of some kind. Among the many costs of cognitive adaptations, perhaps the most prominent one is that the neural tissue underlying the abilities is energetically expensive to grow and maintain (Aiello & Wheeler 1995; Deaner et al. in press). Thus, the key to understanding when improved cognition will be selected for is gaining insight into the relative costs and benefits of the ability for the lineage in question.

It is our contention that a crucial influence on the relative costs and benefits is life history. Life history summarizes the species-specific statistics of major life events: duration of gestation, age at weaning, age at first reproduction (AFR), interbirth intervals, litter or clutch size, and maximum life span (MLS). It has been well established that life histories, at least among mammals and birds, come in syndromes, i.e., many of the features covary predictably. Most important, we can recognize variation along a slow-fast continuum (Harvey et al. 1987; Read & Harvey 1989; Charnov 1993).

A connection between life history, especially MLS, and brain size has long been suggested (Sacher 1959; Sacher & Staffeldt 1974; see also Allman 1999). However, critics have pointed out that there is no evidence for the direct physiological connections between brain size and life history hypothesized by Sacher (Harvey et al. 1989; Read & Harvey 1989; Harvey & Pagel 1991). Recent life history studies, instead, emphasized the role of demography, especially the rate of ecologically imposed unavoidable mortality, as the main selective pressure on life history (Stearns 2000), and strongly contributed to the acceptance of this theoretical perspective (e.g., Promislow & Harvey 1990).

These recent studies have therefore downplayed the possibility of correlated evolution of life history and brain size. In fact, they suggested that the apparent interspecific correlation between brain size and life history is a statistical byproduct of the correlation between body size and life history (Economos 1980): a stronger correlation between MLS and brain size than with body size could arise because body weight is subject to much more error (= phenotypic variation) than brain weight. With the analyses presented here, we would like to reconsider the old correlation, although we are obviously not arguing in favor of the original proximate constraint hypotheses or against the modern interpretation of life history evolution.

One solution to the problem of correlated errors is to remove the effects of body size on both variables by taking residuals from the regression of log AFR or MLS and of log brain size on log body mass. But then a new problem arises (see Harvey & Krebs 1990; Barton 1999). Because of the large error in body mass estimates, residuals against body size for life history measures and for brain size will share error in the same direction, potentially producing artificially high correlations (see Deaner and colleagues [in press] for a visual illustration). The simplest way to avoid this problem is to calculate the two residuals from independent estimates of body mass.

Recently, Deaner and colleagues (in press) have re-examined the relationship between life history and brain size. We tried to correct for the effects of body size by taking separate body size estimates to calculate residuals of brain and life history variables, and for the effects of phylogenetic nonindependence by using independent contrasts (Harvey & Pagel 1991; Purvis & Rambaut 1995; Nunn & Barton 2001).

We found that the correlation between life history, proxied by log relative MLS, and log relative brain size is real, in primates at least. The correlation between them is highly significant ($r = 0.46$; $n = 56$; $P = 0.0002$; details in Deaner et al. in press), and is maintained if we correct for the body size problem by taking independent sets of body mass estimates ($r = 0.43$; $n = 56$; $P = 0.0005$), and for possible effects of phylogenetic nonindependence and for the body size problem at the same time ($r = 0.31$; $n = 52$; $P = 0.03$; see Figure 1.1). We also found that it was not a byproduct of some obvious socio-ecological factors such as frugivory, home range size, or group size (Deaner et al. in press).

We also looked at the highest taxonomic level in eutherian mammals, i.e., the order. This analysis uses means of all species values within an order, which reduces the impact of error in the estimates of species values. It does not calculate independent contrasts because the phylogenetic relationships among higher taxa are still widely debated. However, since the orders sepa-

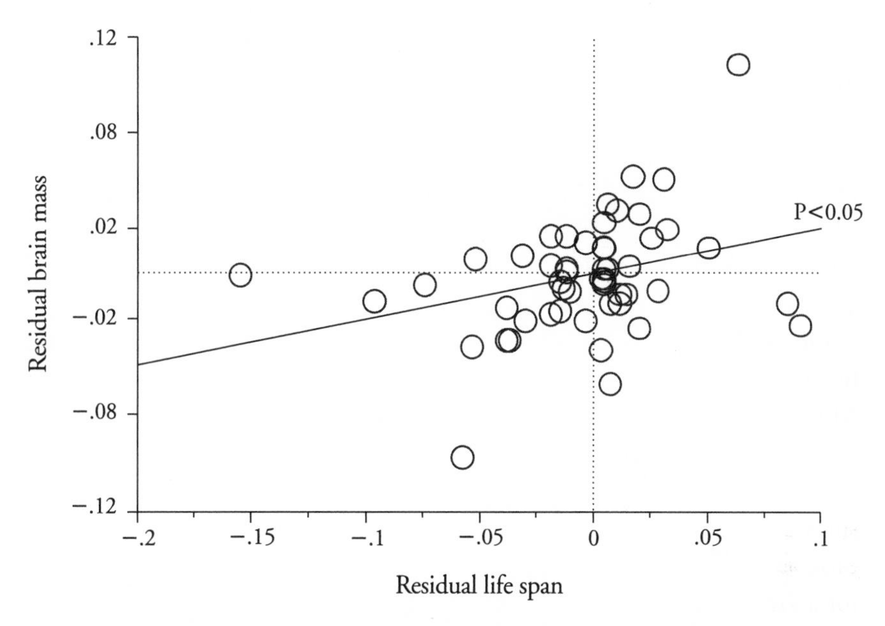

Figure 1.1. The relationship between life history (residual log AFR) and cognitive abilities (residual brain size) in primates, using independent contrasts. Based on Deaner et al. in press.

rated tens of millions of years ago, the problem of nonindependence should be much reduced. At first sight, the relationship between encephalization quotient and longevity quotient (observed maximum life span / expected maximum life span based on the body weight effect; from the database of Austad & Fischer 1992) is weak (see Figure 1.2): $r = +.39$ ($n = 13$, $P = 0.19$). However, there is one strong outlier: the bats (Chiroptera). If the bats are removed, the correlation becomes highly significant ($r = +0.859$, $P < 0.001$) (which is not because of the inclusion of primates: without bats and primates, $r = +0.751$, $P = 0.01$). Thus, after bats are removed, the relationship between life history and cognition is surprisingly strong at this highest taxonomic level.

The conclusion from these comparisons is that there has indeed been correlated evolution between life history and brain, although there is an important outlier, i.e., a clear exception to the rule, which a complete theory needs to explain. Independent support for this conclusion is provided by Promislow's (1991) analysis of senescence in natural populations of 49 different mammal species, which showed that "at least among longer-lived taxa,

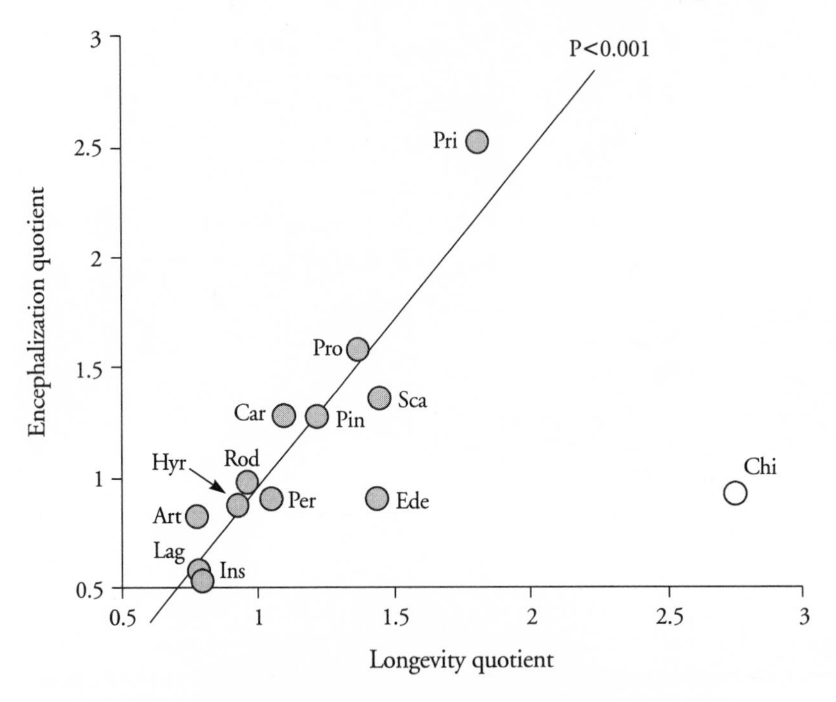

Figure 1.2. Size-controlled life history (longevity quotient) and cognitive abilities (encephalization quotient) at the level of orders of eutherian mammals. Source: Austad & Fischer 1992. *Abbreviations: Art* = Artiodactyla; *Car* = Carnivora (Fissipedia); *Chi* = Chiroptera; *Ede* = Edentata; *Hyr* = Hyracoidea; *Ins* = Insectivora; *Lag* = Lagomorpha; *Per* = Perissodactyla; *Pin* = Pinnipedia; *Pri* = Primates; *Pro* = Proboscoidea; *Rod* = Rodentia; *Sca* = Scandentia.

large brains are correlated with low senescence." Additional support comes from earlier studies that found correlations between relative brain size and other aspects of life history, e.g., litter size, in several mammalian lineages (Eisenberg 1981; Mace & Eisenberg 1982).

This perspective answers at least partly the question of why primates are so smart: they have very slow life history, especially in light of their rather moderate body size (Charnov 1993). It also raises two major new questions: Why do primates have such slow life histories, and more fundamentally, Why did life history and cognition undergo correlated evolution?

Causes of Variation in Mammalian Life Histories

Slow life history can only evolve in animals that have both low unavoidable mortality and distinct advantages to delaying reproduction in terms of more or better offspring (Stearns 1992, 2000). Low mortality is therefore a critical precondition of slow life history.

One of the major sources of reduced mortality is the evolution of larger body size (e.g., Read & Harvey 1989), but another source that does not require increased body size may be an arboreal lifestyle. It can be argued that arboreal animals face fewer predators than a terrestrial animal of the same size, and have more refuges and escape routes. In fact, an effect of arboreality on life history variables such as litter size and gestation length has been suggested before (Eisenberg 1981; Martin 1990; Shea 1987). To investigate this potential relationship systematically, we conducted an ordinal-level analysis of nonaquatic taxa. We considered primates, dermopterans, and bats to be arboreal, although obviously with bats flight adds a confounding effect. We considered the rodents, tree shrews, fissiped carnivores, edentates, and hyraxes to be mixed arboreal-terrestrial, and the other orders terrestrial. At the ordinal level, arboreality has a strong effect on relative longevity ($F_{[2, 13]} = 12.61$, $P < 0.001$), and also a significant effect on relative age at maturity ($F_{[2, 9]} = 8.01$, $P < 0.01$). These patterns are illustrated in Figure 1.3.

Some or all of this effect could be attributed to flight, which is known to reduce mortality rates (Pomeroy 1990). We therefore repeated the analysis with the bats removed, but the effect of arboreality is retained ($F_{[2, 12]} = 8.96$, $P < 0.01$, for relative longevity; $F_{[2, 8]} = 4.84$, $P < 0.05$, for relative age at maturity). Also, we examined the relationship within the order Primates,

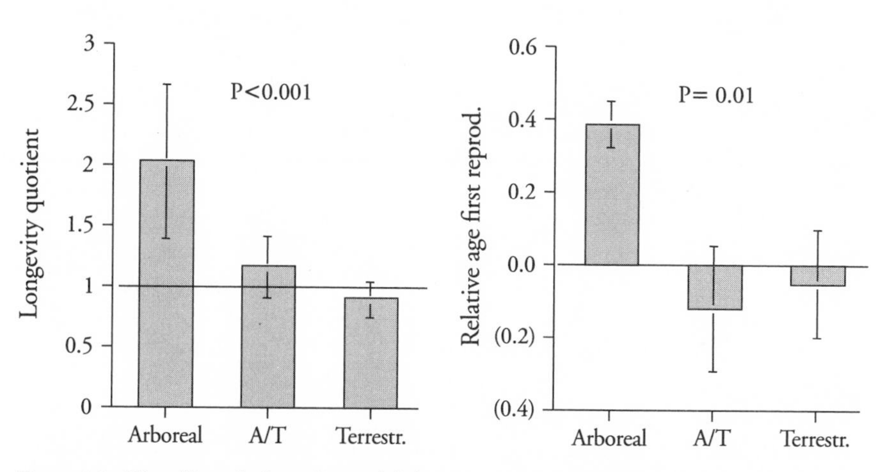

Figure 1.3. The effect of arboreality on life history at the level of orders of nonmarine eutherian mammals. Longevity quotient taken from Austad and Fischer (1992); relative age of first reproduction from Read & Harvey (1989).

where data on substrate use by wild animals is fairly extensive. Among primates, an independent contrast analysis shows the expected relationship between terrestriality and life history (using relative AFR), although it is only marginally significant (Figure 1.4). (The data for this analysis come from Deaner and colleagues [in press]; literature data on percent time spent terrestrially were compiled mainly by Charles Nunn.) Earlier, Ross (1988) had found that size-corrected primate life history speed is correlated with habitat. Species living in savannas, forest edge, and secondary forest were found to have faster life histories. Habitats were classified along these lines to test the predictions of the *r*–*K* model for life history evolution, but this model has now been abandoned because experiments failed to find the density-dependent changes in life history predicted by the model (Stearns 1992). In retrospect, the correlation with terrestrial locomotion is apparent.

These findings indicate that both arboreality and the ability to fly reduce mortality and slow life history. Overall, then, animals that are off the ground

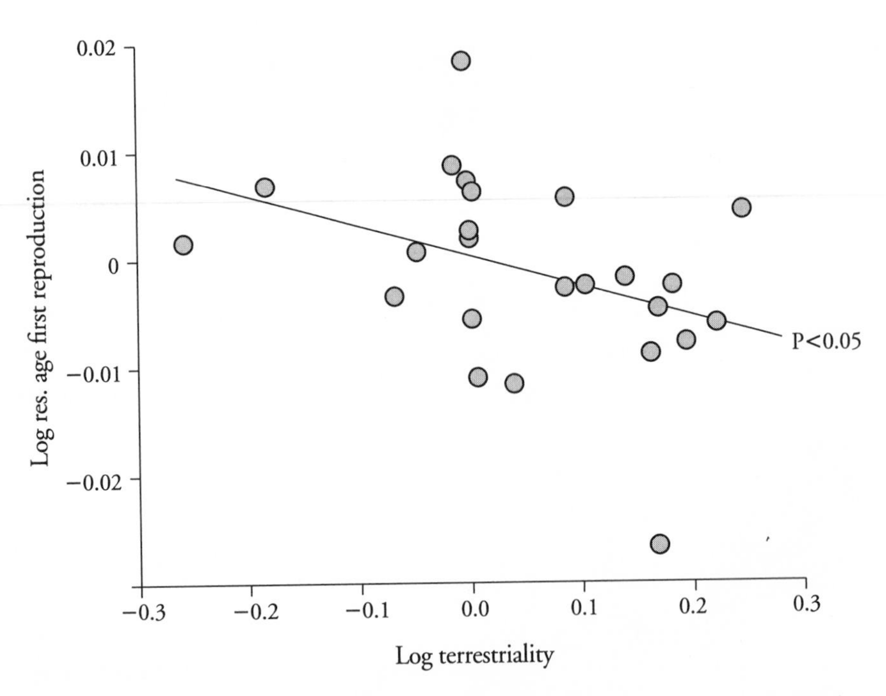

Figure 1.4. The effect of the percentage of time spent as terrestrial on life history (residual age of first reproduction) among primates, using independent contrasts. Life history data are from Deaner et al. (in press); substrate use from C. Nunn (personal communication).

have slower life history (compare Prothero & Jürgens 1987), most likely because they face lower mortality risks. Thus the relative encephalization of the earliest primates can be attributed to the fact that they were one of the few arboreal mammalian radiations (Shea 1987).

One implication of this argument is that there should be a relationship between arboreality and relative brain size. Contrary to this prediction, we did not find any effect of terrestriality on relative brain size in our primate sample. However, at the ordinal level, there is a significant effect after bats are removed ($F_{[2, 9]} = 2.12$, n.s.; after bats are removed, $F_{[2, 8]} = 10.29$, $P < 0.01$). Also, several more limited comparisons did find the expected increase in relative brain size among more arboreal species or even subspecies, compared with the terrestrial counterparts in the same lineage (compiled in Table 1.2).

There are various ways to interpret the absence of a universal effect of arboreality on relative brain size, in spite of the clear-cut effect on life history. A plausible interpretation is that brain size tends to evolve, ratchet-like, in one direction. Throughout the evolutionary history of mammals, there is a

Table 1.2. Relation between arboreality and relative brain size (EQ as defined by Jerison [1973] or Eisenberg [1981]) in mammalian lineages, using (A) comparisons between EQ of terrestrial and arboreal species (*t*-tests); and (B) Pearson correlations with degree of arboreality across species or subspecies (in *Peromyscus*)

A

	Terrestrial	Arboreal	Significance	Source
Neotropical edentates	0.84 ($n = 2$)	1.025 ($n = 2$)	n.s.	Eisenberg 1981
S. African myomorphs	0.33 ($n = 5$)	0.52 ($n = 3$)	$P = 0.01$	Bernard & Nurton 1993
Sciurids	0.96 ($n = 16$)	1.84 ($n = 17$)	$P < 0.0001$	Meier 1983

B

	Correlation with arboreality	Number of species	Significance	Source
Didelphimorpha	+0.72	13	$P < 0.01$	Eisenberg & Wilson 1978
Peromyscus maniculatus	+0.67	18	$P < 0.01$	Lemen 1980
P. leucopus	+0.70	8	$P < 0.05$	Lemen 1980

distinct trend of increasing relative brain size (Jerison 1973), suggesting that once large brains have evolved, they become such an integral and vital part of the organism's functioning that they cannot be reduced by moderate increases in the speed of life history. The testable implication of this suggestion is that transitions toward increased arboreality should be accompanied by increases in relative brain size whereas transitions toward more terrestriality are not necessarily accompanied by reductions in relative brain size. Future work should test this prediction.

Why Do Life History and Cognition Undergo Correlated Evolution?

At this point, we must examine what it is about life history that causes correlated evolution with the brain. All other things being equal, natural selection will favor faster life history because genotypes that reproduce faster will replace others in the population. Thus, there must be positive benefits for slowing down. The advantage most commonly considered is that slower development leads to larger body size, which in turn tends to produce both lower adult mortality rates (thus sustaining a slowing down) and, once reproduction has started, either higher rates of infant production or larger-sized infants. However, as we noted, the fast-slow continuum is retained even if adult body size is statistically removed, so there must also be size-independent benefits to slowing down life history.

We can envisage two possible size-independent benefits. First, if immature mortality is very high relative to adult mortality, selection will favor long life span (but not necessarily delayed maturity) and thus, because of the tradeoff between reproduction and survival (e.g., Stearns 1992), a reduced reproductive effort (Cole 1954). For instance, in sea turtles, infant mortality is orders of magnitude higher than adult mortality. Increased reproductive effort in a single reproductive event (resulting in death) would therefore be unlikely to produce another surviving adult, and hence long life spans and modest reproductive effort would be favored by selection. These conditions are especially likely for species in which entry into the adult niche is extremely difficult because of the absence of parental care, such as in sea turtles.

However, we don't expect these conditions to apply generally to primates or other mammals with extensive parental care. Hence, a second size-independent benefit of slow life history is needed: females that have delayed maturation may produce offspring that are of better quality (Stearns 2000). The nature of this benefit remains largely unexplored (Pagel & Harvey 1993), and tends to be referred to as "experience" (Harvey et al. 1989).

We review various hypotheses that insert some biology into the experience-based, size-independent benefits of slow life history (for complete discussion see Deaner et al. in press). The first two assume that animals evolving larger brains are also expected to evolve slower life histories. The maturational constraints hypothesis claims that immature nervous systems cannot function at the adult level. Thus, large-brained animals whose brain development would not be complete until after commencing reproduction benefit from delaying the onset of reproduction until this is achieved. This hypothesis, then, assumes that slow life history was an evolutionary compromise forced onto animals that were evolving larger brains. The cognitive buffer hypothesis posits that large brains are adaptive because they reduce mortality, especially among adults, thus allowing a slowdown of life history. This hypothesis, then, suggests that slow life history evolved because it was made possible by larger brain size.

The second class of hypotheses assumes that slow life histories relax the costs of brain enlargement, or even favor it. The brain malnutrition risks hypothesis argues that the growing brain's vulnerability to nutrient and energy shortages made it easier to select for larger brain size in organisms with slow life history, because if brains developed rapidly in an energy-poor environment, a considerable proportion of adults would be cognitively impaired and thus unfit. The delayed benefits hypothesis sees slow life history as a critical precondition for the evolution of larger brains. Specifically, it argues that investment in organs that provide their fitness benefits well after they have developed produces greater fitness payoffs in animals with slow life history.

These nonexclusive hypotheses are logically coherent and make assumptions that are reasonably well supported. Their predictions are not all equally supported in all taxa (Deaner et al. in press), perhaps because of the poor quality of life history data (especially longevity), particularly in nonprimates, or because the hypotheses do not apply equally to all taxa (e.g., only to those above a minimum brain size). The hypotheses all suggest that brains and life history evolve in lock-step, but also that the relationship is not quite symmetrical: large-brained organisms must have slow life history to produce demographically viable organisms, but not all animals with slow life histories need necessarily evolve larger brains (as we saw in bats). The delayed benefits hypothesis, after all, assumes there are distinct benefits to the cognitive adaptations that outweigh the obvious developmental costs.

Life history therefore acts as a filter (see Figure 1.5): the same selective pressure will lead to enhanced cognition, and hence enlarged brain size, in a species with slow life history, but not in an otherwise identical species with

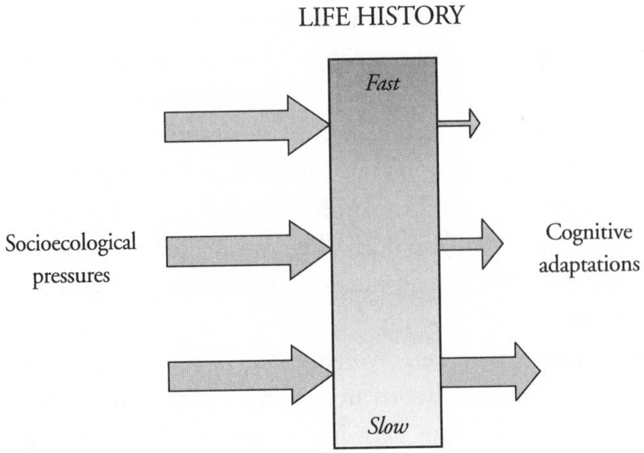

Figure 1.5. Life history as a filter: organisms with slow life history are more likely to respond to social or ecological demands by evolving cognitive adaptations. Various kinds of cognitive adaptations therefore accumulate in organisms with slow life history.

fast life history. The life history perspective thus provides a useful general over-arching framework for cognitive evolution. It does not obviate the need to develop hypotheses that postulate cognitive adaptations to specific external demands, but it does explain which kinds of organisms can attain these adaptations when exposed to the selective pressure, and which cannot. A broader comparative perspective also suggests that there may be numerous possible cognitive benefits, many of them taxon-specific, as explained below.

Reexamining the Selective Advantages of Cognition

Distinguishing among Multiple Hypotheses

It is well known that comparative tests of hypotheses for cognitive evolution face a variety of problems, but our examination of the impact of life history has just made these problems even more daunting. First, statistical models must now explicitly incorporate the interaction effect due to life history. Second, the hypotheses are not mutually exclusive, and statistical models must accordingly be multivariate rather than bivariate (testing hypotheses one by one). We must therefore find ways to improve the resolution of the tests. One way to achieve this is to identify better proxy variables for the socioecological demands and the cognitive variables, as this would reduce the consequences of unequal error variances in multiple regressions.

Another way to improve comparative tests might be to increase the taxonomic scope of the taxa in the test, since this provides more degrees of freedom. However, inclusion of a larger array of taxa also requires explicit recognition of a larger array of hypotheses. All hypotheses listed in Table 1.1 were developed with primates in mind. Some apply only to primates, or even a subset of primates (e.g., the arboreal clambering hypothesis), but others may apply more broadly. On the other hand, in other taxa other benefits may apply. Unfortunately, there have not been many suggested pacemakers for cognitive evolution in nonprimates. Mace and Eisenberg (1982) suggested that intensity of interspecific competition affected relative brain size among rodents, whereas Gittleman (1994) proposed, for carnivores, that unassisted maternal care posed unique cognitive challenges. The important message here is that tests of cognitive evolution require proper identification of all the potentially relevant benefits. So far, we have not reached this stage.

A very different way to sort out the pacemakers of cognitive evolution may therefore be worth considering. One promising avenue is the examination of striking phylogenetic contrasts, in which both cognition and life history show a major divergence between sister species or clades. In principle, the most striking cognitive contrasts should be accompanied by the most easily identified lifestyle differences that have served as the evolutionary facilitators for the cognitive divergence between the two sister taxa. These pacemakers could be divergences in ecology, mating system, social behavior, or any other aspects of their lifestyle that might affect cognition. Having identified the key differences, we can then build a likely scenario that can subsequently be tested for consistency, or add them to a multivariate statistical model.

To illustrate this approach, let us briefly examine one of the most striking phylogenetic contrasts among primates. In primates, the contrast between *Homo* and *Pan* is perhaps the most pronounced, and involves a long list of possible socioecological pacemakers (compare Gibson 1999). Compared to chimpanzees, humans generally rely more on technology for foraging and food processing, show more complex social organization involving more cooperation and exchange, show an amazing capacity for culture, and have an infinitely flexible communication system in language. All of these differences involve cognition. Thus, humans no doubt show more advanced technical intelligence (Byrne 1997), are much better at imitation of complex actions (Tomasello 1999), show more evidence for a theory of mind and multiple levels of intentionality (Dennett 1996), and, Kanzi notwithstanding (Savage-Rumbaugh & Lewin 1994), show much greater linguistic abilities. Thus, there are at least four different major components to the cognitive contrast, the last two almost certainly unique among primates, though not necessarily among mammals more generally (compare cetacea: Tyack 1999).

Both the social strategizing and the food processing hypotheses are implicated in this list of contributing selective benefits, but so are other traits that involve cognition, namely imitative abilities and language. The latter two are not part of the list of general selective advantages for increased cognitive skills among mammals or primates. In other words, it seems as if a unique set of "customized" hypotheses is needed to fully explain this contrast.

If these preliminary analyses are indicative of the general trend, we expect that all better studied contrasts will involve some rare elements. Thus, while it is necessary to make improvements to the comparative tests of the existing hypotheses for cognitive evolution, there is also a need to generate more relevant ones, including perhaps cognitive benefits that may be lineage-specific.

Social Strategizing and Life History

Examination of the interaction between life history and cognition has led us to emphasize the diversity of possible contributing factors, including cognitive benefits that may be fairly idiosyncratic, and thus only relevant to particular lineages. Nonetheless, slow life history may exert a more direct selective influence on cognition that can be placed under the rubric of social strategizing. In general, long-lived animals, unless they are nomadic, will be more prone to form long-term associations with known individuals, leading to many opportunities for social interactions and long-term relationships. Such a setting forms the precondition for social complexity. Moreover, various reproductive parameters are causally linked to life history, but also to the potential for sexual coercion: harassment and infanticide. If this intersexual conflict produces social challenges with possibly cognitive solutions, it may thus affect cognitive evolution.

First, slow life history implies longer interbirth intervals, which in general lead to more male-biased operational sex ratios because fewer females will be ready to conceive at any moment in time. This, in turn, means more potential for intersexual conflict, such as polyandrous mating by females despite attempts at monopolization by dominant males. It may be that behavioral strategies to deal with these problems (by both males and females) have led to solutions that involve social strategizing, including deception.

Second, the benefit of infanticide to the male depends on making the female return to receptivity earlier than without it; and this benefit should strongly depend on the relative duration of lactation and gestation. The relative length of lactation has recently been shown to affect the risk of infanticide by males in mammals (van Schaik 2000). Slow life history leads to slowed-down post-natal development and increased lactation relative to gestation (Figure 1.6). It is conceivable that behavioral strategies to deal with infanticide risk (by both males and females) have led to solutions that involve cognition. If that is so, the threat of infanticide has favored cognitive evolution.

The Evolution of Domain-General Cognition

Most students of cognitive evolution have stressed that cognitive adaptations are domain-specific (e.g., Shettleworth 1998), although cognition in primates may appear to be fairly domain-general (e.g., Johnson et al. in press). We believe that the life history perspective may shed some light on this possi-

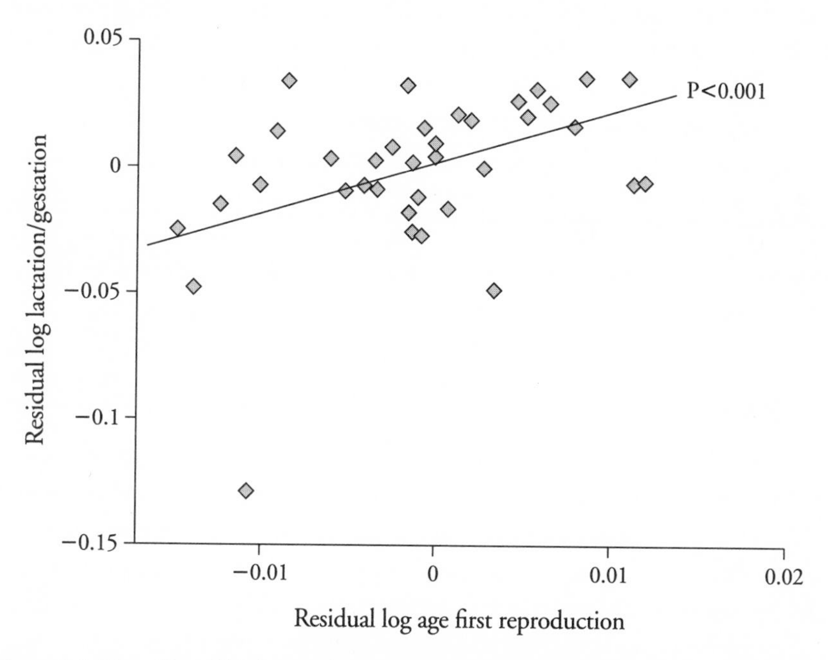

Figure 1.6. The effect of life history (as proxied by residual age of first reproduction) among primates on the length of lactation relative to gestation (also corrected for body weight by taking residuals), using independent contrasts. Life history data are from Deaner et al. (in press); reproductive data from van Schaik (2000).

ble difference. Domain-generality may be real in the sense that different cognitive adaptations share many neural mechanisms (compare Chapter 7), so that selection for one ability also produces improved abilities in another domain. However, domain-generality may also be an illusion. Assuming that approximately the same set of selective pressures operates in species with fast and slow life histories, species with slow life histories will be able to respond to more of them (compare Figure 1.5), leading to an accumulation of various kinds of domain-specific cognitive adaptations in these organisms. As a result, cognitive abilities in various different domains (e.g., ecological, social) will become concentrated in taxa with slow life histories. In taxa with fast life histories, in contrast, most abilities will be absent, and the few that are found will therefore be interpreted, correctly, as domain-specific cognitive adaptations.

True domain-generality can subsequently and gradually arise, or be strengthened, in organisms with slow life histories. Indeed, the presence of multiple specific cognitive adaptations may be a precondition for the evolution of domain-generality, because only in those organisms will there be selection on economizing on the neural substrates for cognition by sharing neural structures and pathways. Domain-general cognitive abilities, i.e., intelligence, may therefore be limited to species with sufficiently slow life histories. It will not be easy to test this speculation because it is hard to distinguish true domain-generality from accumulated domain-specific adaptations. Perhaps the difference can be demonstrated at the neuroanatomical level, or by exploring correlations in cognitive performance on a battery of tests across individuals of the same species.

Conclusion

Recently, the suggestion that cognition evolved in response to social demands has led to great interest in cognitive evolution, especially among primatologists. Although the idea that social strategizing provided the main selective pressure in primate cognitive evolution is now quite popular, we have argued that it is premature to consider the case closed. First, other nonexclusive benefits to cognition exist that may be important in some taxa. Second, all current hypotheses ignore the costs of cognitive adaptations, and therefore overlook the possibility that these adaptations may not evolve in certain taxa in spite of the obvious benefits. We showed that life history and relative brain size have undergone correlated evolution among primates and among mammals in general, implying that only organisms with a slow life history could evolve significant cognitive adaptations.

The life history perspective has already yielded various new insights in cognitive evolution. First, because arboreal ancestry is a major correlate of life history among mammals, the unusual cognitive abilities of primates among mammals can be seen as the outcome of arboreal ancestry. Second, a life history approach suggests that a wide variety of factors may favor cognitive evolution, depending on the details of the taxon's natural history—a suggestion supported by a preliminary examination of the *Homo-Pan* contrast. Third, slow life history also may produce a variety of challenges directly or indirectly linked to social strategizing, e.g., via sexual coercion, strengthening the possibility that social strategizing was a major pacemaker. Fourth, the life history perspective also offers suggestions for the evolution of domain-general cognition.

CASE STUDY 1A

Sociality and Disease Risk: A Comparative Study of Leukocyte Counts in Primates

CHARLES L. NUNN

Most chapters in this book focus on the complexity of interactions among individuals within social groups. By comparison, my research focuses on the emergent properties of these interactions and their implications at macroevolutionary scales, as assessed by comparisons across species and reconstruction of evolutionary change. One of the aims of this comparative research is to understand the factors that best account for differences in sociality across species. In what follows, I consider the role of disease in mammalian social evolution, specifically primates.

Disease is likely to be a potent selective force in human and nonhuman societies (Grenfell & Dobson 1995; Lockhart et al. 1996; Daszak et al. 2000), as indicated, for example, by the existence of costly defenses against disease-causing parasites (Sheldon & Verhulst 1996; Moret & Schmid-Hempel 2000). The immune system is one obvious defense, but certain behaviors also reduce the risk of acquiring infectious disease (Hart 1990; Kiesecker et al. 1999). Among primates, for example, it has been proposed that baboons alter their daily ranging patterns to avoid fecal contaminants in soil (Hausfater & Meade 1982), and howler monkeys expend considerable energy avoiding biting insects that are parasitic and serve as vectors for other parasites (Dudley & Milton 1990). Disease risk also has the potential to influence patterns of sociality and therefore to affect the costs and benefits of group living. For example, the abundance of certain types of parasites increases with group size (Rubenstein & Hohmann 1989; Côté & Poulin 1995), suggesting that animals can reduce parasite loads by living in smaller groups or by interacting in smaller social units within groups.

The potential interaction between parasites and sociality may be important for understanding social interactions, but basic questions about the correlates of disease risk remain unanswered. Thus, a first step toward understanding the effects of parasites on sociality is to examine the correlates of disease risk more generally. Taking stock of current knowledge leads to three hypotheses relevant to primates. Disease risk is expected to increase with group size, use of terrestrial substrates, and mating promiscuity. The effect of group size was discussed above (see also Møller et al. 1993). Consistent use of terrestrial substrates may increase the risk of acquiring infectious disease if such behavior exposes animals to soil-borne parasites, such as those spread through fecal contamination. Finally, many types of parasites are transmitted during mating (sexually transmitted diseases [STDs]; Lockhart et al. 1996), predicting an association between promiscuity and disease risk.

Data on socio-ecological parameters involving group size, substrate use, and mating promiscuity are readily available (van Schaik et al. 1999; Nunn & van Schaik 2001), whereas disease risk is more difficult to quantify. I used information on baseline white blood cell (WBC) counts to test the hypotheses presented above (Nunn et al. 2000; Nunn in revision). The assumption is that increased disease risk will be reflected in higher cell counts. This seems to be a reasonable assumption, given that WBCs are among the first lines of defense against parasites (Roitt et al. 1998), although I discuss other approaches needed to test these hypotheses below. WBC counts were taken from data compiled on zoo animals by the International Species Information System ("ISIS"; Physiological Reference Values CD-ROM 1999, Minnesota Zoological Garden, Apple Valley, MN, USA). ISIS screens data gathered by zoo veterinarians using several criteria to ensure that animals involved are healthy, thus providing robust estimates of "normal" WBC counts for assessing animal health.

I tested the hypotheses after transforming the species data by using phylogenetic comparative methods (Felsenstein 1985; Harvey & Pagel 1991; Nunn & Barton 2001). To calculate independent contrasts, I used Purvis's (1995) composite estimate of primate phylogeny and the computer program

CAIC (Purvis & Rambaut 1995). These tests examined whether evolutionary transitions to larger groups, terrestriality, or promiscuity were associated with evolutionary increases in baseline WBC counts.

The hypotheses and their predictions are not mutually exclusive. Thus, my co-authors and I (Nunn et al. 2000) predicted that WBC counts increase with all three independent variables. To our surprise, however, WBC counts were correlated only with mating promiscuity (Figure 1A.1). In these tests, we used several different measures of mating promiscuity. For example, Figure 1A.1 examines patterns in WBC using relative testes mass (controlling for body mass by taking residuals) as the independent variable. Testes mass is thought to increase in response to sperm competition (Harcourt et al. 1981, 1995), and sperm competition requires that females mate with more than one male. Thus, relative testes mass provides a means to quantify female mating promiscuity. Similar results were obtained using the duration of estrus and a dichotomous classification of female mating promiscuity from van Schaik and colleagues (1999).

These patterns involving WBC counts combine several distinct types of leukocytes. It is also important to consider particular WBC types, as these cells have distinct functions in defending the body from parasites (Roitt et al. 1998). For example, neutrophils and monocytes are part of the innate immune system and involved in phagocytosis of invading parasites, while lymphocytes are involved in adaptive immunity and the recognition of antigens. Neutrophils also are known to be involved in phagocytosis of sperm in the female reproductive tract following mating (Pandya & Cohen 1985; Barratt et al. 1990). Another WBC type, the eosinophil, is thought to fight helminths (Roitt et al. 1998) and therefore may be important in defending against macroparasites acquired from the soil. Using these different WBC types provided no support for the nonmating hypotheses, whereas tests with most WBC types supported the mating promiscuity hypothesis (e.g., relative testes mass and neutrophils: $b = 0.30$, $F_{1,37} = 9.19$, $P < 0.01$).

These results raise several questions for future research. First, if mating promiscuity increases the risk of contracting STDs, then why are some primates highly promiscuous? One reasonable explanation is that the benefits of paternity confusion, especially regarding infanticide risk (van Schaik & Janson 2000), outweigh the costs of promiscuity. This consideration suggests that questions about disease risk should be placed in a cost-benefit framework. It is therefore naïve to ask if disease risk shapes aspects of mating sys-

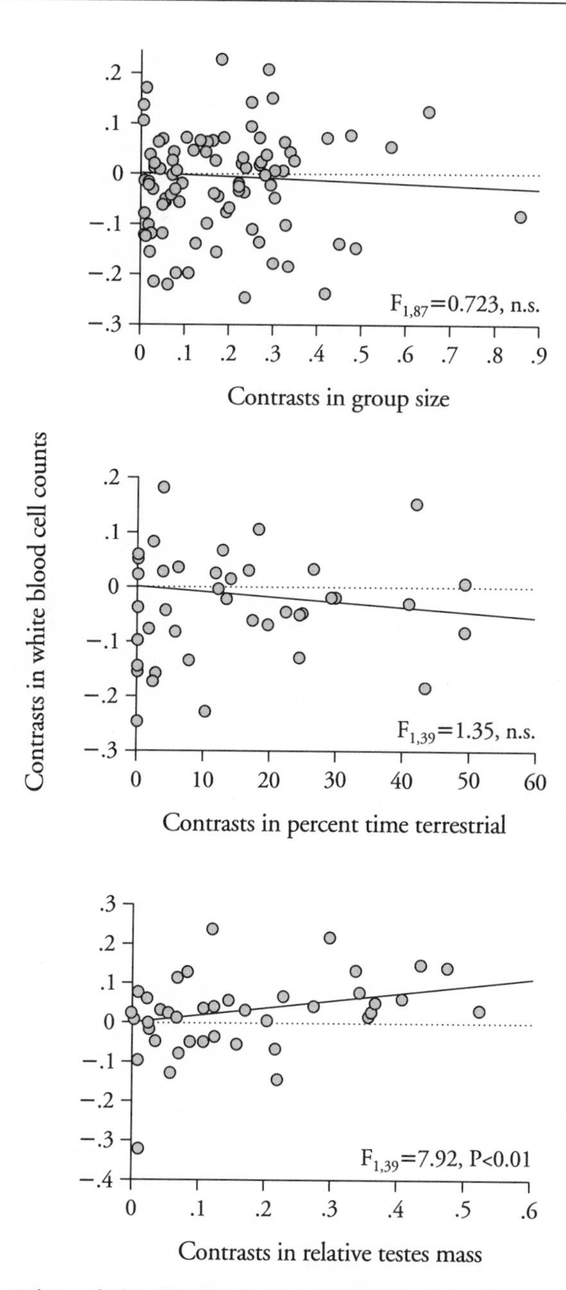

Figure 1A.1. Correlates of white blood cell counts (WBC) in primates. Plots show independent contrasts (Felsenstein 1985) in WBC counts relative to contrasts in socio-ecological variables (group size, percentage of time terrestrial, and relative testes mass). Relative testes mass is a measure of mating promiscuity and is the only significant result. It is also the only result in the predicted direction in these tests. See text for explanation of variables and phylogenetic comparative methods.

tems, versus mating systems shaping disease risk, because both are likely to interact in complex ways and with other socio-ecological parameters.

Second, how might behavioral counterstrategies to parasites modify the risk of acquiring infectious disease? Behavioral counterstrategies operate in conjunction with the immune system to defend the body from parasites, but behavior may be more effective against some parasite transmission modes than others. Parasite avoidance strategies to nonsexual infectious diseases have been discussed extensively in primates, including alteration of ranging patterns (Hausfater & Meade 1982), ingestion of medicinal plants (Huffman & Caton 2001), and avoidance of recent immigrants (Freeland 1976). By comparison, behavioral options against STDs may be more limited, leaving the immune system as the primary defense and possibly explaining the lack of significance for the nonmating hypotheses, in which behavioral counterstrategies may play a greater role. The issue of STD prevention in wild populations is understudied compared to other types of disease, but several possibilities exist. For example, post-copulatory genital grooming reduces STD transmission in male rats (Hart et al. 1987). Similar stereotyped grooming has been documented in prosimian primates (Jolly 1966), but is generally absent in anthropoids. Another possibility is that individuals could inspect potential mating partners to identify those with STDs. Further consideration weakens this as a likely counterstrategy, at least for females, because in the highly competitive social environments of many primate groups, females may be limited in their choice of mating partners (e.g., in single-male groups or under sexual coercion; Smuts & Smuts 1993). Moreover, many STDs in humans (and presumably nonhumans) have no visible symptoms (Holmes et al. 1994), consistent with theoretical models showing that both parasite and host have congruent interests in obscuring cues of infection status (Knell 1999).

Finally, do these results extend to other taxonomic groups? To answer this question, we tested the hypotheses in carnivores and found identical results (C. Nunn, J. Gittleman, and J. Antonovics, unpublished results). These congruent results are obscured by some confounding variables, however, including life history and group size. Another reason for expanding the taxonomic scope is that different hypotheses may apply in other taxonomic groups, thereby shedding light on the multiplicity of factors that influence disease risk. For example, carnivores that eat other vertebrates may experience increased risk of acquiring infectious disease. Our comparative research has revealed limited and inconsistent support for this hypothesis, possibly because carnivores use behavioral and physiological mechanisms to avoid these parasites.

In conclusion, results using baseline WBC counts reveal a striking association with mating promiscuity but fail to support the hypothesis that sociality itself influences disease risk. Testing these hypotheses requires multiple lines of evidence. For example, we need data on individual contact frequencies, compositional turnover, and relative proximity of individuals in social groups to make clear epidemiological predictions. Other means of assessing "disease risk" are needed, including comparable measures of the dynamic immune response, assessment of mucosal immunity in male and female reproductive tracts, and quantitative data on behavioral counterstrategies to parasites. Other data are available for testing these hypotheses, which I am investigating using immune system structures such as the spleen, patterns of parasitism within and across species, and life history correlates of disease. For example, analysis of parasite species richness in primates reveals an association between group size and the number of different parasite species that infect a host taxon (C. Nunn and S. Altizer, unpublished data). These initial results therefore suggest that behavior, disease risk, and immune defense systems operate in complex ways and that multiple approaches are needed to untangle their interactions.

2

Dolphin Social Complexity: Lessons from Long-Term Study and Life History

RANDALL S. WELLS

Much of our understanding of the social complexity and intelligence of terrestrial animals such as chimpanzees has been a direct result of long-term studies. Compiled observations of individuals over time and through sequential life stages have helped scientists see the complex patterns unfold, and have presented opportunities to begin to differentiate among developing patterns of behavior that correspond to life history transitions or ecology, and those that may have a stronger basis in learning. Concurrent comparative studies at multiple field sites have provided opportunities for more clearly identifying cultural transmission of behaviors from generation to generation, where ecological influences can be evaluated (Whiten et al. 1999).

Key to these kinds of studies is access to populations of animals about which detailed observations may be carried out over periods of decades and in which human impacts are minimized or can be evaluated, individuals are readily recognizable, and background information on age, gender, and genetic relationships may be compiled for most of the population members. These conditions are more easily met for studies of terrestrial than marine mammals, but over the last few decades several long-term studies of cetaceans, especially delphinids, have begun to provide some of the requisite information for understanding their complex societies. In particular, three very different species of delphinids—killer whales *(Orcinus orca)* in the northeast

Pacific Ocean, bottlenose dolphins (*Tursiops* spp., Connor et al. 2000a) at several sites, and Atlantic spotted dolphins *(Stenella frontalis)* on the Bahama Banks (Herzing 1997)—have been studied extensively over periods of 15 to over 30 years. As a result of these studies, cetacean societies can be evaluated relative to the better-known terrestrial species, allowing informative comparisons regarding the roles of life history, ecology, and culture in the development of societal complexity.

Studies of delphinid societies have had to overcome many challenges to obtain the kinds of information scientists studying terrestrial mammals might take for granted (see review in Wells et al. 1999). Dolphin ranges are often large, and include habitats that are difficult for humans to negotiate, making it difficult to find and remain with focal animals. Although continuous, often unobstructed views of many terrestrial mammals facilitate behavioral observations, most delphinids are only visible at the water's surface for up to a few seconds during each respiratory cycle. In a few exceptional cases, scientists have been able to work underwater (diving or using vessels with underwater viewing capabilities), where they can observe details of behavior even after the animals submerge if the waters are clear and calm. Elevated vantage points, including cliffs, boat observation towers, aircraft, or video cameras suspended from aerostats, have allowed continuous behavioral observations of some dolphins.

Along with the problems of simply seeing the study animals come difficulties in identifying individuals. Delphinid cetaceans move constantly, and are often obscured from view, precluding the use of fine-scale facial or other features for individual identification. Most, with the notable exceptions of killer whales and spotted dolphins, lack distinctive color patterns. However, many delphinids exhibit individually distinctive patterns of nicks and notches on their dorsal fins, and these fins are visible at the surface each time the animal breathes. Except for a few species that exhibit dramatic sexual dimorphism (e.g., dorsal fins of killer whales), the gender of delphinids must be determined through observations of subtle differences in the configuration of urogenital slits on the ventral surface of the body (or adult females can be tentatively identified in long-term studies through repeated, consistent association with calves). Alternatively, skin obtained through swabbing or biopsy sampling (through darting or capture/release) can provide information on gender, and can help determine genetic relationships within a society.

Delphinid ages can only be determined through monitoring individuals in long-term studies from the time of their birth to mothers who are well known to the researcher, or alternatively from microscopic examination of growth layers in prepared sections of a tooth. The long life spans (up to 50 years or more) of many delphinids mean that long-term study is required to understand behavioral variability relative to life history features. Delphinids do not exhibit the often dramatic, colorful sexual displays of some terrestrial mammals that demonstrate sexual maturity and sexual receptivity, nor can pregnancies be determined readily through observation. Alternatively, retrospective determination of sexual receptivity can be accomplished by subtracting the gestation period from the date of birth, or dolphins can be briefly restrained in shallow water, allowing for safe blood sampling to measure reproductive hormone levels or ultrasonic examination of reproductive organs. Mating system descriptions have been difficult because socio-sexual behavior is common among a variety of age and sex classes, and copulations are not necessarily indicative of paternity. Genetic exclusion testing, through examination of skin or blood samples, can be accomplished for paternity determinations in some cases.

Studies of auditory communication in delphinids have been complicated by the fact that there are few, if any, anatomical indications of sound production. Contact hydrophones and hydrophone arrays have been developed to facilitate localization of the individuals making vocalizations. What follows is an example of how a combination of field techniques described above have been applied to overcome the challenges of studying large mammals in an aquatic environment and to develop an understanding of the social complexity of one delphinid, the bottlenose dolphin *(Tursiops truncatus)*.

The longest-running delphinid study of behavior, social structure, ecology, and life history involves bottlenose dolphins living in and around Sarasota Bay, Florida (Figure 2.1). Individually identifiable dolphins residing year-round in this shallow region of sheltered bays, channels, passes, and the open waters of the Gulf of Mexico (Figure 2.2) have been studied since 1970 (Irvine & Wells 1972; Wells et al. 1980; Irvine et al. 1981; Scott et al. 1990a; Wells 1991). As of this writing (2001), about 120 individuals use this approximately 125 square kilometer region regularly, including half of the dolphins originally identified through tagging during 1970 and 1971, along with three subsequent generations.

Information on the dolphins of Sarasota Bay derives from five primary sources:

Figure 2.1. Bottlenose dolphin (adult male).

1) photographic identification of distinctive natural markings during surveys to record ranging and social association patterns (Figure 2.3; Scott et al. 1990b; Würsig & Jefferson 1990);

2) occasional capture, sample, mark, and release efforts to obtain data on age, sex, genetic relationships, reproductive condition, health, body condition, acoustic patterns and responses to whistle playbacks (Sayigh et al. 1999), and concentrations of environmental contaminants in blood, milk, and blubber (Figure 2.4; Wells 1991);

3) tagging and tracking via small short-range or satellite-linked radio transmitters (Figure 2.5; Scott et al. in press);

4) focal animal behavioral observations (Altmann 1974); and

5) acoustic recordings of free-ranging animals via single hydrophones or localizing hydrophone arrays, or digital data logger tags deployed for periods of hours on dolphin dorsal fins following release (Figure 2.5; Nowacek et al. 1998).

Acoustic recordings are typically made in combination with focal animal behavioral observations from small vessels (Sayigh et al. 1993) or, for con-

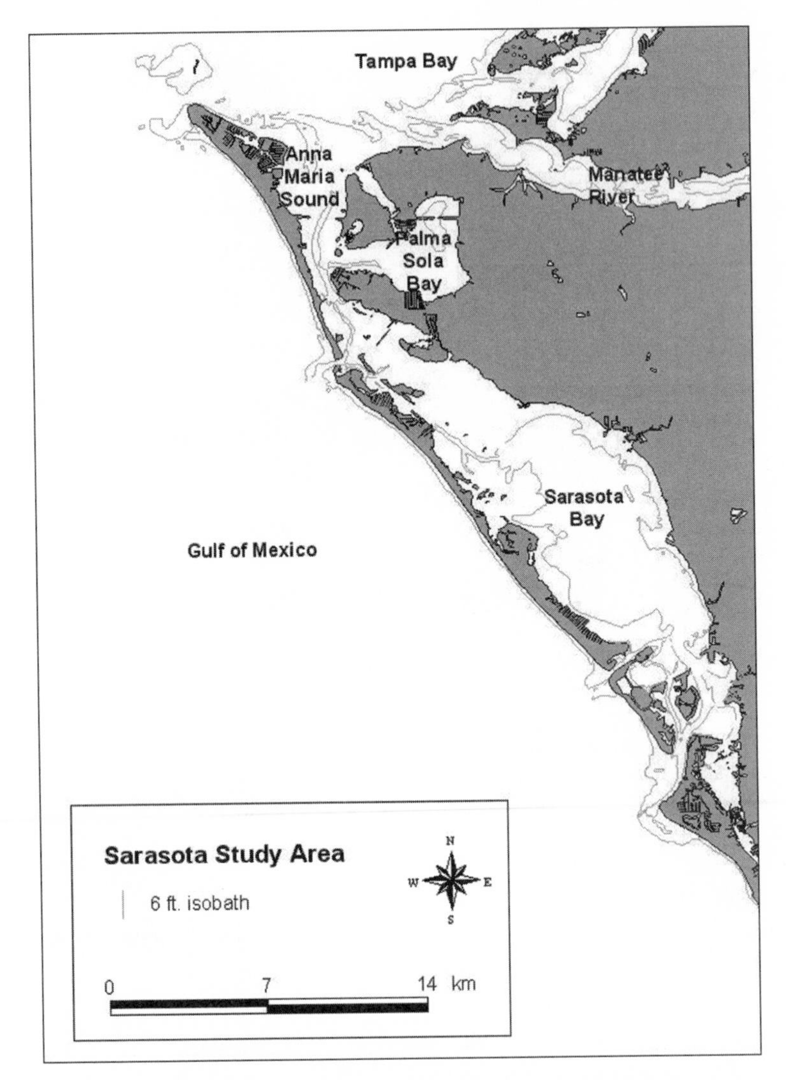

Figure 2.2. Sarasota, Florida long-term study area, including six-foot isobath.

tinuous observations even while the animals are submerged, via a remote-controlled digital video recording system suspended from a small tethered aerostat (Figure 2.6; Nowacek et al. 2001a). These efforts have resulted in identifications of more than 2,500 dolphins from the central west coast of Florida in the waters to the north, west, and south of Sarasota Bay. Some of the resident Sarasota Bay dolphins have been re-sighted more than 750 times over periods of decades. Based on observations and brief handling events, the

Figure 2.3. Distinctive natural dorsal fin markings on bottlenose dolphins.

Figure 2.4. Bottlenose dolphin capture, sample, mark, and release efforts. Photograph courtesy of Flip Nicklin.

Figure 2.5. Digital data logger tag attached via suction cups to a dolphin dorsal fin. Also note the small VHF radio transmitter attached to the top of the fin.

ages, sex, and genetic relationships have been determined for more than 90 percent of the resident dolphins of Sarasota Bay.

The natural laboratory situation of Sarasota Bay provides the right mix of ingredients for beginning to examine the roles of life history, ecology, and culture in the development of social complexity. How has the bottlenose dolphin society been shaped by genetic or ecological factors associated with age, growth, and reproduction? To what extent has social complexity been increased by the cultural transmission of knowledge vertically from one generation of these long-lived mammals to the next, and spread horizontally through the society? The answers to these questions lie in our ability to find and study multigenerational lineages of dolphins of known backgrounds on a relatively predictable basis, to follow individuals as they reach and pass life history milestones, and then to compare the findings to other populations. The following description provides a framework to begin to address some of the questions about the role of culture in the complex societies of dolphins.

Figure 2.6. Tethered aerostat video recording system being used to record behavioral responses to vessel approaches.

Reproduction and Life History

Bottlenose dolphin life history and reproductive patterns have been summarized recently by Wells and Scott (1999). Reproduction tends to be seasonal in Sarasota Bay (Wells et al. 1987; Urian et al. 1996), with most calves produced during late spring through summer. Females are spontaneous ovulators and seasonally polyestrous. Males exhibit seasonal changes in testosterone concentrations and testis size (Figure 2.7). Bottlenose dolphins are

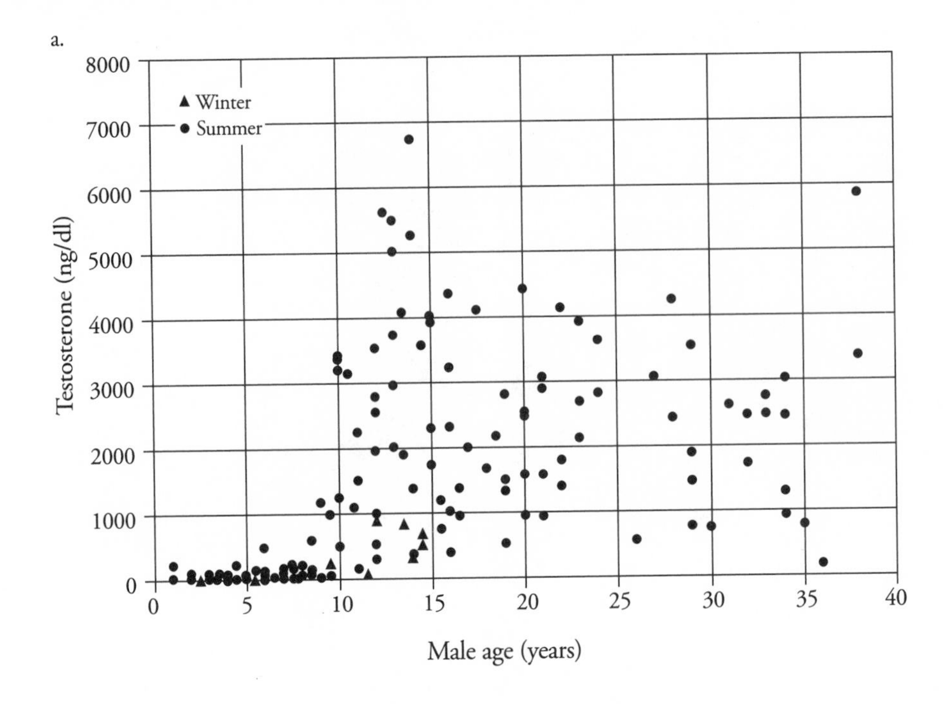

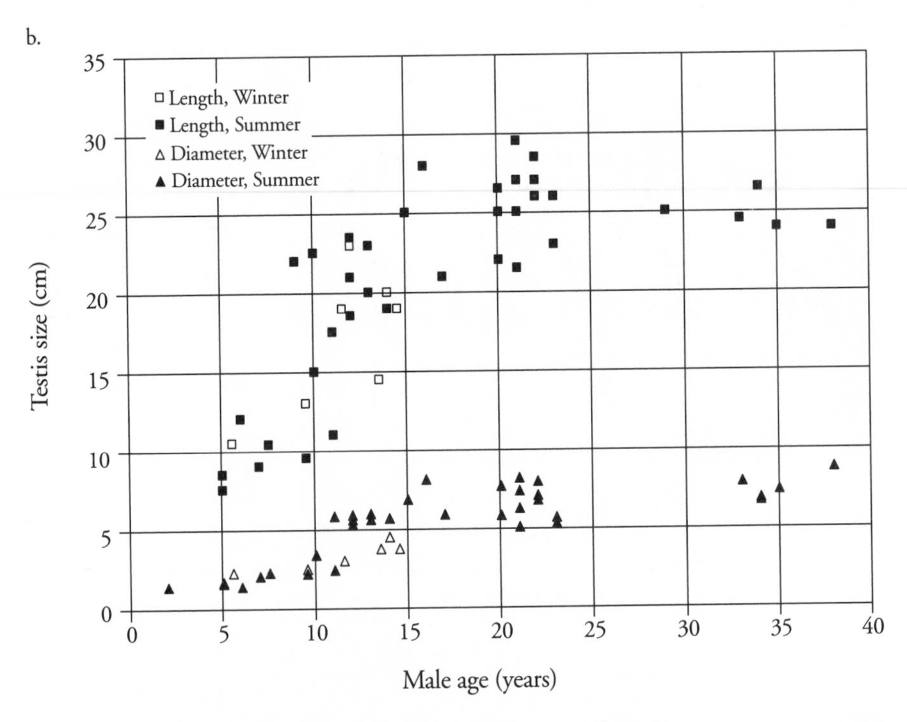

Figure 2.7. Sexual maturity of male bottlenose dolphins as indicated by testosterone concentrations *(a)* and ultrasonic measurements of testis dimensions *(b)*.

typically born singly following a 12-month gestation period. Though precocial swimmers at birth, they require nutritional support from their mothers for at least the first year of life (and may continue to nurse at a reduced level for up to seven years or more), and remain with their mothers for 3–6 years on average (Wells et al. 1987). Primiparous calves rarely survive to independence, but calf survivorship improves with the age of the mother.

Females reach sexual maturity as early as five years of age, and begin to produce calves soon thereafter. Maximum body size for females is achieved typically by 10 years of age (Read et al. 1993). Females continue to reproduce through their entire adult lives, with some females as old as 48 years giving birth and successfully rearing calves for several more years.

Males reach sexual maturity at about 10 years of age, as indicated by testosterone concentrations and ultrasonic measurements of testis dimensions (Figure 2.7). Males achieve their maximum body size between 10 and 20 years of age (Read et al. 1993), with adult lengths and weights significantly exceeding those of females, and allometric increases in some features related to propulsion, defense, and agonistic interactions (Tolley et al. 1995). Male mortality tends to occur earlier in life than for females. Few males live beyond 40 years of age, and adult sex ratios favor females by approximately two to one (Wells et al. 1987).

Mortality at an average annual rate of about 4 percent results from a variety of natural and anthropogenic sources (Wells & Scott 1990). Natural causes of death include shark predation, disease, stingray barbs, and likely also include occasional mortality from red tide toxin. Entanglement in fishing gear (Wells & Scott 1994; Wells et al. 1998) and collisions with boats (Wells & Scott 1997) also lead to serious injuries or deaths. The long-term implications for survival and reproduction from repeated short-term disturbance from sources such as vessels remain to be seen (Nowacek et al. 2001b). The importance of environmental contaminants as factors influencing dolphin health and reproduction in Sarasota Bay has been suggested through studies of immune system function relative to contaminant concentrations (Lahvis et al. 1995) and concentrations relative to age, sex, and reproductive histories (Vedder 1996; Küss 1998).

Social System

Long-term site fidelity is one of the strongest features helping to shape our understanding of bottlenose dolphin social structure along the central west coast of Florida. Dolphins residing in these bays, sounds, and estuaries live in a mosaic of slightly overlapping "communities" (Wells et al. 1987). A dolphin community is defined as "a regional society of animals sharing ranges and social associates, but exhibiting genetic exchange with other similar units" (Wells et al. 1999), and it is considered to be a subset of a population. The 120 members of the Sarasota dolphin community inhabit the waters between southern Tampa Bay and southern Sarasota Bay, including Gulf of Mexico waters within several kilometers of shore (see Figure 2.2). Although most of the dolphins are present year-round, seasonal shifts in habitat use occur, involving more extensive use of shallow bays and seagrass meadows during summer months, likely as a result of prey fish distribution and abundance (Irvine et al. 1981; Barros & Wells 1998). The 125 square kilometer community range has remained relatively stable over the three decades of our research, though the northern extent of the range was not well documented during the first few years of study because of logistical constraints (Wells et al. 1980, 1987). Other bottlenose dolphin communities inhabit the adjacent waters of Tampa Bay and the Gulf of Mexico, sharing borders and mixing with Sarasota community members in about 17 percent of sightings involving Sarasota dolphins (Wells et al. 1987).

Community membership also appears to be relatively stable over time, maintained across multiple generations. Half of the dolphins first identified 30 years ago are still present. Several resident matrilines spanning four generations have been observed, as have numerous three-generation lineages. Genetic studies suggest long-term differentiation between communities, based on blood protein electrophoresis profiles and mitochondrial DNA haplotype frequencies (Duffield & Wells 1991; Duffield & Wells in press; Sellas 2002).

Geography plays a strong role in patterns of association among coastal dolphins. Community ranges in the bays, sounds, and estuaries tend to be defined at least in part by physiographic features that influence the movements of the animals, such as water depth, inlets, and sandbars. Within the larger community range, residents interact with one another to a much greater extent than they do with inhabitants of adjacent waters. Individuals emphasize core areas within the community range, but some, especially maturing and adult males, may move outside the community range for months at a time. Infants start life with a core area defined by their mothers' movements during their period of association, but core areas may change over time, with males often demonstrating more range expansion with age than females. Immigration and emigration are low, with annual rates of less than 3 percent (Wells & Scott 1990). Permanent range shifts from one community to another are rare, but some seasonal shifts have been noted in recent years. Females from the next community to the north have begun to use the northern bays of the Sarasota community each summer, for example. At the same time, some Sarasota females previously using the northern waters have begun to use Sarasota Bay to a greater extent. Whether the range shift of one set of females is directly related to a shift by the others, or whether the shifts are related to environmental changes has not yet been determined. In recent years, prey fish in Sarasota Bay appear to have become more abundant as a result of a commercial fishing ban in 1995, large predator (shark) abundance in Sarasota Bay and surrounding waters has declined dramatically from directed and incidental fishing takes, and the numbers of vessels (personal watercraft, flats boats, and air boats) using the shallow northern waters has increased. These factors may be driving or encouraging the Sarasota dolphins to use the deeper (up to four meters), more open waters of Sarasota Bay proper to a greater extent than in the previous 25 years.

Bottlenose dolphin group size throughout the species' range is highly variable (as are definitions of "group"), but typical groups range in size from one to 20 individuals (Shane et al. 1986; Connor et al. 2000a). In Sarasota Bay, typical group size is five to seven dolphins (Wells et al. 1980, 1987). Though some social associations may last for years or even decades, in general the dolphins of Sarasota Bay live in a fission/fusion society, with group composition changing over periods of minutes or hours. Sexual segregation provides a basic framework through which the fluid groupings are formed, at least for adults. Nursery groups of females with their most recent calves are a prominent feature of the society, as are strongly bonded pairs of adult males. Juvenile groups of mixed sex are the third common component. Interactions among these units occur from time to time, but as the mixed group splits, it tends to be reduced into these basic units.

Female Social System

Reproductive status and shared core areas appear to be two of the primary factors governing female-female associations in Sarasota Bay (Wells et al. 1987). Females sharing largely overlapping core areas form a pool of potential associates, referred to as a "band" (Wells et al. 1987). Practically speaking, all of the members of a band are rarely found together on any given day, but individual members are much more likely to associate with band members than with other females that may enter the area. Because use of core areas is often continued across generations, and because the dolphins' long reproductive life span means several generations within a matriline may be reproducing concurrently, a number of related as well as unrelated individuals may comprise the pool of potential associates. However, the actual frequency of close associations is dynamic and based more on the presence and ages of calves than on kinship. Females with calves of similar age may be found together, consistently or repeatedly, for as long as their reproductive cycles are synchronous. Similar patterns hold for pregnant, receptive, or resting females. At any given time, adult females in different reproductive states may be found in large groups, but close associations within subgroups and repeated associations over time tend to be related to reproductive status.

Although maternally based genetic features such as mt-DNA help to distinguish among dolphin communities, further genetic structuring that would differentiate female bands has not yet been found within the community (Duffield & Wells in press). Band membership appears to reflect finer-scale long-term social relationships, as well as familial relationships. This idea is supported by the findings from signature whistle playback studies, in which recorded whistles from related or familiar individuals elicited stronger behavioral responses than did less familiar whistles (Sayigh et al. 1999).

Band structure is changeable over periods of decades, due in large part to demographic changes. Changes within two of these bands over the last 20 years illustrate the pattern. During the 1970s and 1980s, membership in two bands accounted for about 78 percent of the well-known noncalf females of Sarasota Bay (Wells et al. 1987). The Anna Maria band (named for its Anna Maria Sound core area) included seven females; the Palma Sola band (named for its Palma Sola Bay core area) was composed of 14 females, including several of the community's oldest individuals. Maturing females are often recruited back into their natal bands. During the 1990s, mortality and lack of recruitment through reproduction reduced the Anna Maria band to only three individuals, whose core areas have become increasingly restricted to the northern margin of the community range (Figures 2.8a, 2.8b). The Palma Sola band flourished during the 1980s because of successful recruitment of female offspring (Figure 2.8a). With the increase in numbers and the loss of the two oldest matriarchs within a six-month period in 1993–94, the band began to divide into a number of smaller units, whose core areas expanded into the waters previously occupied extensively by the Anna Maria females and into Sarasota Bay proper (Figure 2.8b). Whether these units develop the kind of band structure evident during previous decades remains to be seen, but the situation may provide a tremendous opportunity to learn about the factors involved in this process.

One of the hallmarks of dolphin society is a high degree of variability. This variability both enriches the social fabric and complicates the researcher's desire for simple, absolute descriptions. Not all females can be clearly identified as belonging to a band on the basis of our current criteria. Females employ a variety of approaches to calf rearing, ranging from those who attempt to rear them alone to those who rear them with band members in nursery groups of various and changeable sizes. Overall, females rearing calves in larger, more stable groups enjoy significantly greater reproductive success than do others (Wells 2000). Thus, female reproductive success seems to be enhanced by group living, likely through improved protection of calves from predation and other threats and calf exposure to other individuals for socialization, learning, and possibly allomaternal care. Likewise, reproductive success increases with mother's age and maternal experience (Figure 2.9). In an examination of the behavioral differences between primiparous and experienced mothers, Owen (2001) showed that experienced mothers tended to maintain greater synchrony with and closer distances to their calves, thereby providing increased control over their calves' environment. Experienced mothers also tended to include other mothers with calves as close associates.

Learning appears to play an important role in successful calf rearing. Female calves have the opportunity to observe rearing within nursery groups, and in several cases female offspring have remained with their mothers and younger siblings, keeping in close contact with the siblings at times when mother is engaged in other activities. It is difficult to clearly define this specific activity as "babysitting," but the overall exposure to an experienced mother's calf rearing presumably accrues benefits to the mothers-to-be. Clear benefits to calves from associations with nonparturient females may come about from induced lactation, as described from captive situations (Ridgway et al. 1995). This phenomenon has yet to be described from the wild. In fact, in two cases of orphans in Sarasota Bay, neither was adopted by related adult females. One of these two, orphaned at 16 months of age, survived without the consistent aid of others, however, and is now rearing her own calf.

On average, calves remain with their mothers for three to six years, until the birth of the next calf (Wells et al. 1987). Nursing typically provides primary nutrition for the first few months of life, then is supplemented by fish,

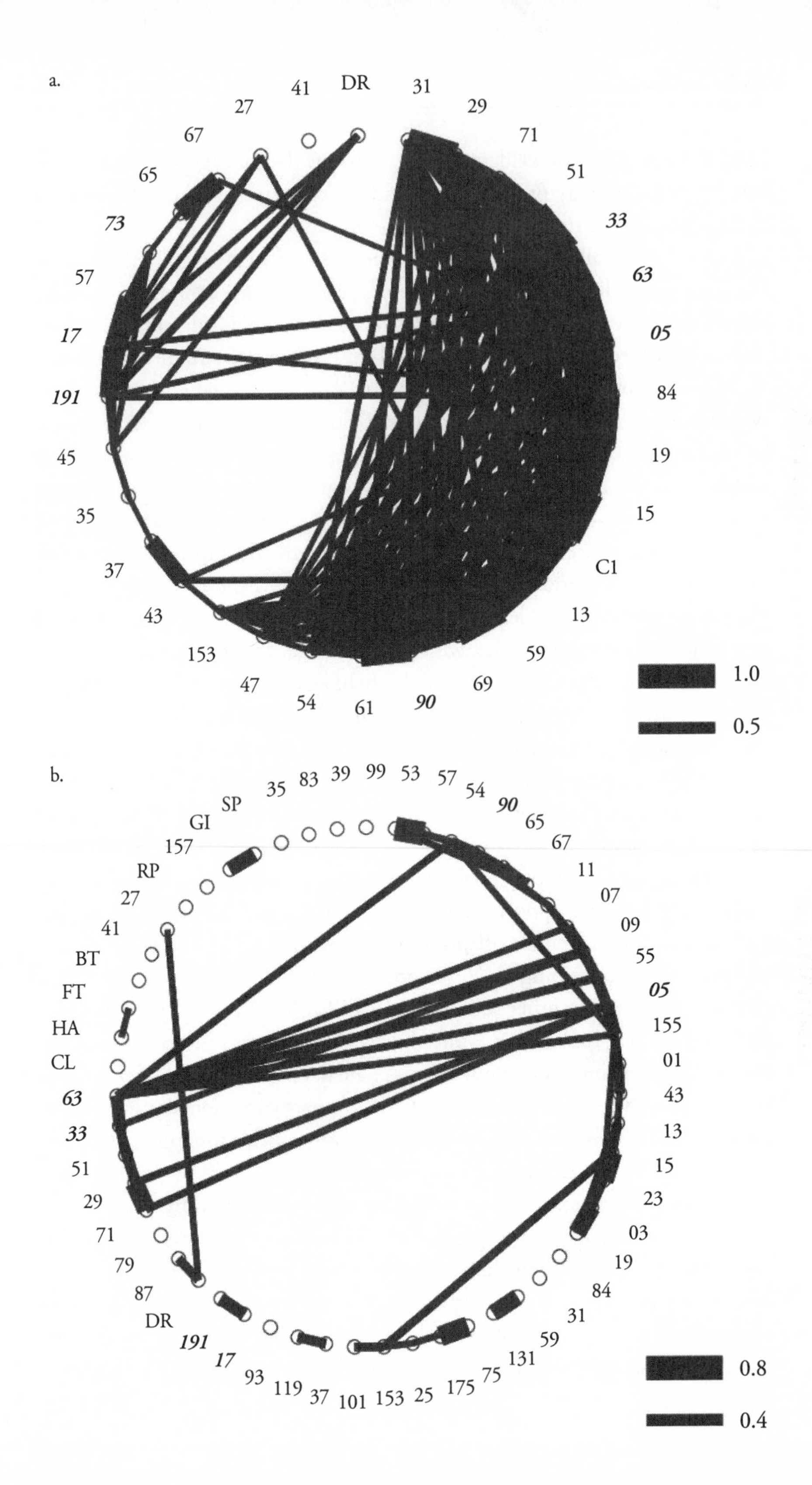
a.
41
DR
31
29
27
71
67
51
65
33
73
63
57
05
17
84
191
19
45
15
35
C1
37
13
43
59
153
69
47
1.0
54
61
90
0.5
b.
35
83
39
99
53
57
54
90
SP
65
GI
67
157
11
RP
07
27
09
41
55
BT
05
FT
155
HA
01
CL
43
63
13
33
15
51
23
29
03
71
19
79
84
87
31
DR
59
191
131
17
75
175
93
119
37
101
153
25
0.8
0.4

Figure 2.8. Sociograms of resident Sarasota Bay female bottlenose dolphins, based on half-weight coefficients of association (Cairns & Schwager 1987). Examples of Palma Sola female band members include most of the dolphins in the figure, with the following serving as illustrative examples: 05, 90, 63, and 33. Long-term examples of Anna Maria female band members are limited to 17, 191, and 73. *a)* 1980–1985; *b)* 1990–1995. Note the high degree of structuring in the early sociogram, and reduction 10 years later, especially for the Anna Maria female band. Ordering of dolphins around the sociogram and indicated levels of association are automated functions of software producing the sociograms.

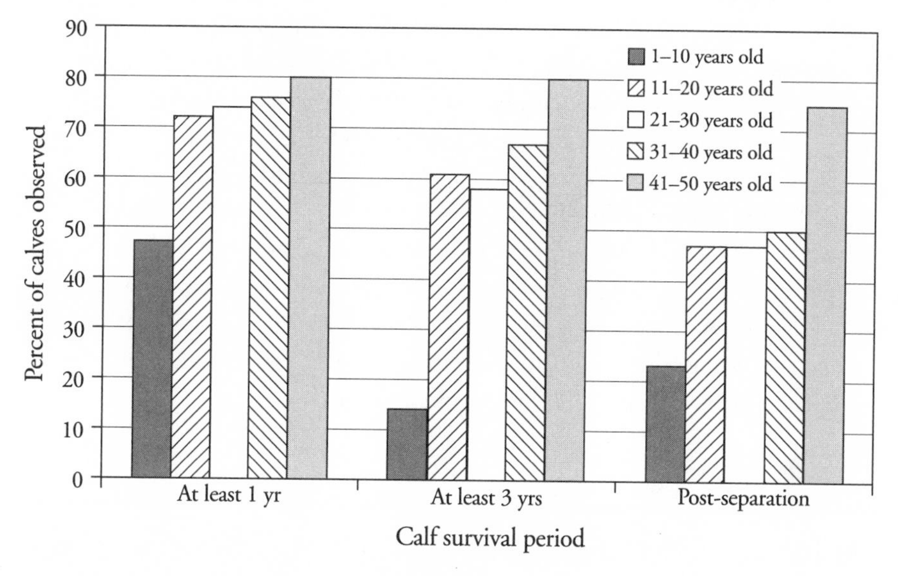

Figure 2.9. Calf survivorship relative to mother's age (n = 112).

and ultimately replaced by a diet of fish when the dolphins reach one or two years of age (Wells & Scott 1999). The successful orphan mentioned above apparently survived strictly on fish she caught herself, without milk, from 16 months of age. As evidenced by these observations, the period of maternal investment extends well beyond nursing, suggesting the importance of benefits other than milk, benefits such as social learning (Brodie 1969).

Juvenile Social System

Once calves separate from their mothers, they typically spend a period of months on their own within their mother's core area or nearby before joining groups of other juveniles. These mixed-sex groups include young ranging from three years of age through early or mid-teens. Juvenile females may stay with or return to their mothers for periods of time, especially if their mother is accompanied by another calf. Though compositions of juvenile groups may be fluid on a day-to-day basis, associations during this period of development are recurring. Typically, the composition of juvenile groups is skewed toward males. Females mature at a younger age, and as they become pregnant and give birth to their first calf they are likely to shift their interactions to other reproducing females, especially members of their natal bands.

Males interact extensively within their groups, and most begin to develop close associations with another male of similar age (within one to three years, typically) as they reach sexual maturity. Candidates for developing partnerships often originate from the same female band, suggesting the importance of long-term relationships and development of cohorts during calf rearing. In all of the cases examined to date, males in pairs have not been closely related: they have different mothers and genetic tests suggest they also have different fathers. Maturing males spend decreasing amounts of time with juveniles as they approach both sexual and physical maturity. In the early stages of pair-bonding, they begin to move more widely through the community range, and may occasionally shadow older adult males.

Male Social System

The bonds of longest duration among Sarasota Bay dolphins involve pairs of adult males. Males are defined as paired if they share half-weight coefficients of association (half-weight, where CoA = 0.00 indicates animals never seen together and CoA = 1.00 indicates animals always seen together; Cairns & Schwager 1987) of at least 0.80 maintained over multiple years (Wells et al. 1987; Wells 1991). Most males that will pair with another have done so by 20 years of age, such that 58 percent of adult males are paired at any given time (Figure 2.10). Most males are involved in a pairing at some time during their lives. Pairs developed at sexual maturity often are maintained through life, until one member dies (Figure 2.11). Some pairs in Sarasota Bay have been observed for more than two decades. In some cases, when one member is lost, the remaining male develops a pair bond with another unpaired male. Unpaired adult males tend to be those that have lost and not replaced a partner or those born during years of few births, resulting in small cohorts of potential partners. Being of similar age seems to be a less important criterion in

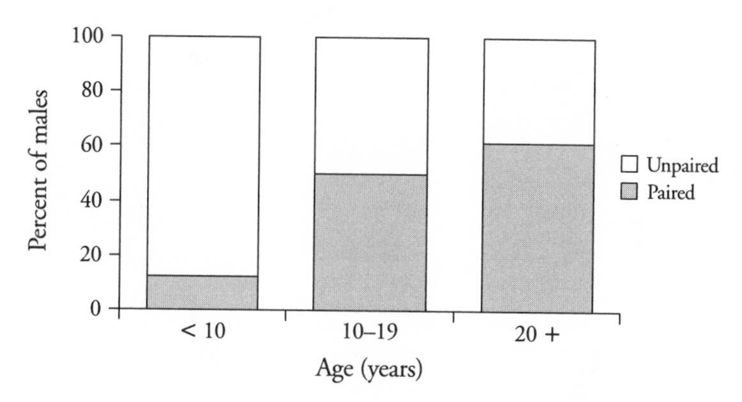

Figure 2.10. Proportion of paired males relative to age.

the development of subsequent pair bonds for individuals that have lost a partner.

Adult males tend to range either as individual pairs or singletons through the core areas of a variety of females, interacting briefly as they encounter females. Encounters with receptive females may lead to extended periods of association (days to weeks). In some cases males leave the community home range for periods of time and interact with females in adjacent communities. This pattern, if reciprocated by males from adjacent communities, may account for the approximately 50 percent maximum rate of genetic exchange reported from Sarasota Bay, based on the proportion of calves for which paternity tests have excluded all Sarasota males as potential sires (Wells & Duffield in prep.; it should be noted that this is a maximum rate because some resident adult males may have sired calves included in the sample but died before they themselves could be sampled). To date, the most distant documented movements by adult Sarasota males outside of the community home range have involved pairs.

The functions of male pair bonds in Sarasota Bay remain to be defined conclusively. However, observations suggest several functions. For example, paired males have been observed engaging in highly synchronous cooperative feeding behaviors, suggesting improved feeding effectiveness (Nowacek 1999). Mutual protection from predatory sharks or aggressive conspecifics is another likely function. Some attacks by sharks are survivable, as documented by Wells and colleagues (1987): 22 percent of noncalf dolphins bore shark bite scars. The possible role of a partner in reducing the occurrence or seriousness of these attacks remains to be examined. On the rare occasions

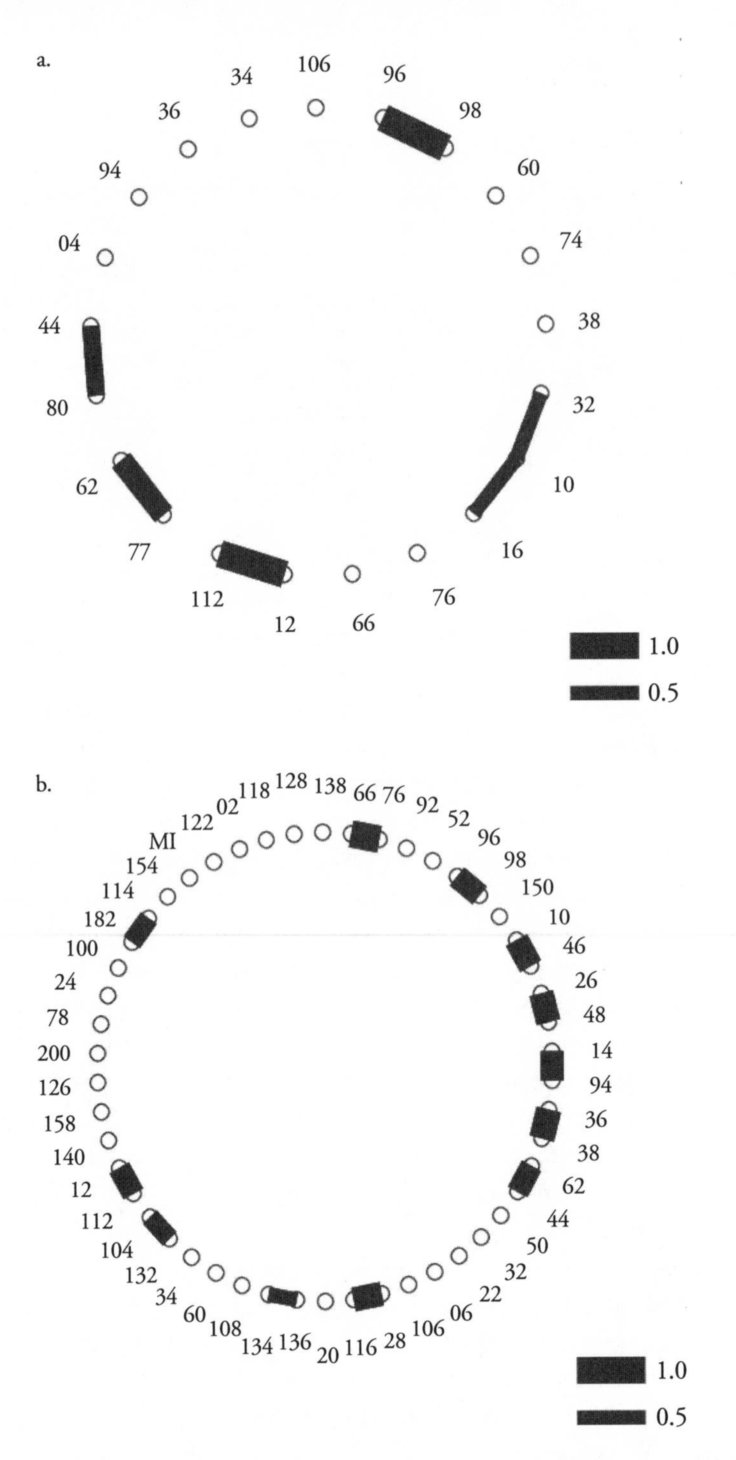

Figure 2.11. Sociograms of resident Sarasota Bay male bottlenose dolphins: *a)* 1980–1985; *b)* 1990–1995.

when battles between males from adjacent communities have been observed along shared community borders, these have tended to involve pairs of Sarasota males; it is likely that the incidence of agonistic encounters between males increases as nonresident males penetrate community ranges. Pairing may also improve mating opportunities, such that males may be able to better control the movements of the female, or other males' access to receptive females, by working together. Extreme examples of complex interactions between alliances of males in gaining and controlling access to female dolphins have been reported from studies in Shark Bay, Australia, as reported elsewhere in this volume (Connor et al. 1992a, 1992b, 1999, 2000a).

Mating System

The behavior of adult Sarasota males observed during the breeding season involves intense mate guarding (Moors 1997). Receptive females are flanked by adult males during their presumed period of receptivity (as defined from hormone measurements or retrospective determinations of conception dates). With variations, this positioning typically puts each male within several meters of each side of the female, and slightly behind her. Such flanking allows the males to detect and take advantage of mating opportunities. The close proximity also aids males in preventing males other than the guarding pair from mating. Compared to Shark Bay, the level of aggression between males and receptive females in Sarasota Bay appears minimal, suggesting that female choice may play an important role in determining mating opportunities (Connor et al. 2000a).

Findings from genetic paternity testing nearing completion (Wells & Duffield in prep.) provide further details about the mating system. For these analyses, 62 mother-calf combinations including calves born during 1973–1996 are being tested against 47 males of at least ten years of age, which includes all current resident adult males. It is clear that monogamy is not a feature of the social system. More than one calf of a particular female may be sired by a single male, but in no cases were all of a female's calves sired by a single male. Early findings with a more limited sample (Duffield & Wells 1991) indicated that only males older than 20 years of age sired calves, suggesting that some behavioral mechanism or social factor prevented physiologically capable males from breeding for ten or more years following sexual maturity. This delay in achieving social maturity was thought possibly to be related to the later onset of physical maturity. The preliminary results of the current synoptic genetic analyses suggest a refinement of this conclusion. Tests conducted to date indicate that males as young as 13 years of age have sired calves, but only in cases in which the male is unusually large for his age. Thus, it appears that body size may be a more important criterion in male breeding success than age, at least for younger males. Potential relationships between male body size and attractiveness to females or physical competitive ability relative to other males remain to be determined.

Social complexity may also be evidenced through male reproductive success. Though the analyses are not yet complete, preliminary findings indicate that both paired and unpaired males sire calves, but paired males sire proportionately more calves than solitary males. Within pairs, a variety of siring situations occur. In some cases only one member has been documented as a sire, but in others, both members have been sires. When only one member has been the sire, this male has been the larger or the smaller, the older or the younger. Genetic analyses, currently underway, may provide data for a more detailed interpretation. The ability to directly measure reproductive success of unpaired males and both members of a pair will facilitate future evaluations of the costs and benefits of living in male pairs.

Social Complexity and Culture

How much of the social complexity described results from cultural transmission of information in bottlenose dolphins? Whiten and colleagues (1999, p. 682) identified a cultural behavior as one "that is transmitted repeatedly through social or observational learning to become a population-level characteristic." Such behaviors are expected to be "customary or habitual in some communities, but are absent in others where ecological explanations have been discounted." Variations in age-related patterns of social association occur among the sites where bottlenose dolphins have been studied in greatest detail: Sarasota Bay; Moray Firth, Scotland; and Shark Bay, Australia (Wells 1991; Wilson 1995; Connor et al. 2000a). Evaluation of these variations as cultural phenomena is complicated by the fact that ecological differences occur from site to site, and our current understanding of the influence of ecology on group structure and social behavior is limited. In Sarasota Bay, longitudinal study across multiple generations has clearly established that basic population-level characteristics of age-related social association patterns are maintained generation after generation. Ontogenetic changes in these patterns occur consistently across individuals within a sex, and are strongly correlated with life history milestones such as separation from mother, sexual maturity, physical maturity, and reproductive status. These basic biological correlations suggest a genetic basis to social association patterns. The role of social learning or observational learning in the expression of these traits remains to be determined.

It may be more fruitful to examine behaviors that are thought to be largely independent of life history changes as possible products of cultural transmission of knowledge. Feeding behaviors, for example, are used throughout life post-weaning and are strongly influenced by ecological factors. Based on observations of colonies in oceanarium settings, feeding behaviors have been used to exemplify the high degree of behavioral flexibility of dolphins (Caldwell & Caldwell 1972). Table 2.1 is modeled after Table 1 of Whiten and colleagues (1999, p. 683). It shows variation in occurrence of nine feeding behavior patterns across 12 study sites for bottlenose dolphins. The selection and scoring of this sample of behaviors was accomplished in a less rigorous manner than that of Whiten and colleagues (1999) because of the incomplete knowledge of behaviors and individuals at each site, different research objectives at different sites, and different methodology and duration of study at each site. In addition, Table 2.1 has not been subject to the level of communication among principal investigators involved in the derivation of the table by Whiten and colleagues.

In spite of the relative weaknesses of the evidence from dolphins, the table may have some utility in identifying possible examples of cultural behaviors. Not surprisingly, some behaviors are common across all sites. At the other extreme, "sponge use" (Smolker et al. 1997), in which a sponge is carried on the dolphin's rostrum, has been identified at only a single site, where its use is habitual. This behavior requires sponges, which may not be present at many of the other sites. Even within Shark Bay, it is only practiced by a subset of the population, and appears to be passed from mother to young (Smolker et al. 1997). The feeding function of this behavior remains to be confirmed, but alternatives seem unlikely. Other behaviors, such as "fish-whacking" (Wells et al. 1987) and "kerplunking" (Nowacek 1999; Connor et al. 2000b), have been reported from only a few sites, and are absent from others where no ecological explanation is apparent. These behaviors involve the use of the fluke to strike fish or to drive them from cover. Both behaviors can be used by single individuals, but often they involve multiple individuals using the behaviors concurrently. It appears that one animal may come up with a new foraging behavior and others may pick it up through obser-

Table 2.1. Bottlenose dolphin feeding pattern summary

Behavior pattern	Sarasota Bay Florida USA	Tampa Bay Florida USA	Sanibel Island Florida USA	Shark Bay Australia	Moreton Bay Australia	Northern Bahamas	Texas
Accelerate/lunge	C	U	C	C	U	P	P
Root/drift/crater feed	C	U	C	C	U	C	C
Pinwheel/sideswim	C	U	C	C	U	U	P
Kerplunk	H	H	P	H	U	U	A
Fishwhack	H	U	H	H	U	P	A
Cooperative herding	P	U	P	A	U	H	P
Drive against barrier/ strand feed	P	U	U	H	U	C	E
Fishery interactions	P	U	U	P	C	U	H
Sponge use	U	U	U	H	U	U	U
References:	Nowacek 1999	Nowacek 1999	Shane 1990a	Smolker et al. 1997	Corkeron et al. 1990	Herzing 1996	Shane 1990b
	Wells et al. 1987			Connor et al. 2000b		Rossbach & Herzing 1997	
	Nowacek pers. comm.			Connor pers. comm.		Herzing pers. comm.	

Note: Behavior category codes (after Whiten et al. 1999)
Customary (C): behavior occurs in all or most able-bodied members of at least one age/sex class
Habitual (H): behavior is not customary but has occurred repeatedly in several individuals
Present (P): behavior neither customary nor habitual, but clearly identified
Absent (A): behavior not recorded and no ecological explanation is apparent
Ecological explanation (E): absence of behavior is explicable because of a local ecological feature
Unknown (U): behavior has not been recorded, but may be due to inadequacy of relevant observational opportunities

vational learning. Along the central west coast of Florida, kerplunking was first observed in Tampa Bay. Within the next few years, one adult female who frequently engaged in this behavior began to include Sarasota waters in her summer range, and she continued this behavior in comparable habitat. Rough approximations of this behavior have now been exhibited by two of her offspring in Sarasota waters. It will be interesting to see whether the behavior spreads to others within the Sarasota dolphin community.

Other feeding patterns listed in the table have appeared so recently as to raise the question about their rapid development and transmission through populations. In particular, dolphin interactions with a large variety of human fisheries have become common in many parts of the world. In some cases, these interactions have occurred with artisanal fisheries over several (human and dolphin) generations (e.g., Busnel 1973; Pryor et al. 1990). More commonly, dolphins have learned to supplement their diets from larger fisheries that have only existed for several decades. Could widespread use of these fisheries have resulted from cultural transmission of information developed originally by a few individuals?

One can speculate about the possible mechanism by which cultural transmission of knowledge about feeding behaviors might have occurred. The phenomenon of observational learning by dolphins is well known to those who work with them in oceanarium settings, where dolphins that have not been trained to perform certain behaviors nonetheless demonstrate proficiency in the behaviors for which poolmates have been trained. The opportunity for observational learning to occur in the wild exists in the prolonged period of mother-calf association, up to several years beyond weaning, and it provides a possible explanation for this maternal investment.

Although we know little about the transmission mechanisms of dolphin foraging behaviors, there is strong evidence for social learning of vocalizations used in acoustic communication among wild dolphins (see Chapter 13). There is evidence for both vertical transmission of vocalizations from the mother and horizontal or diagonal transmission from unrelated dolphins. In large colony pools, where a captive dolphin mother has little control over the acoustic environment of her calf, calves develop whistle vocalizations that differ from their parents, but match acoustic models in the natal pool (Tyack 1997; Tyack & Sayigh 1997). In Sarasota, most daughters and about half of sons develop whistles different from their mother and the other half of the sons learn to produce a whistle similar to their mother (Sayigh et al. 1995). Some of the calves that develop whistles different from the mother develop whistles very similar to other members of the community, suggesting diagonal transmission of vocalizations from unrelated adults. S. Watwood (personal communication) has shown that as dolphin males form a pair bond, their whistles converge, indicating horizontal transmission of vocalizations. Similarity of whistles correlates with a strong social bond, a pattern of vocal accommodation similar to that seen in humans, chimpanzees, and some birds. Data on dolphin vocalizations show cultural transmission of communication patterns, and data on dolphin foraging suggest cultural transmission of foraging behaviors.

As observations of dolphins at long-term study sites continue, as more sites are developed, and as researchers communicate their findings to one another following the model provided by primate researchers, the possibility of identifying cultural behaviors for dolphins should improve. It seems likely that the level of social complexity demonstrated by these animals provides sufficient opportunities for social and observational learning to occur and spread through long-term social units of dolphins.

3

Sources of Social Complexity in the Three Elephant Species

KATY PAYNE

Of the hundreds of proboscidean species that once populated all land masses except Antarctica, Australia, and New Zealand, three descendants survive today (Roca et al. 2001). These live only in Asia and sub-Saharan Africa, but in a wide array of ecological habitats. The African savanna elephant, *Loxodonta africana,* is found in East and southern African savannas, with one population in the extreme desert of western Namibia. The African forest elephant, *Loxodonta cyclotis,* lives in equatorial rainforests in West and Central Africa. The Asian elephant, *Elephas maximus,* lives in forest and forest edges in the Indian subcontinent, and in continental Southeast Asia and Island Asia.

Elephants are long-lived, large-brained mammals organized in female-bonded kin groups. In at least one population of each species, the life histories and behaviors of hundreds of identified individuals have been documented throughout more than a decade. Cynthia Moss and her associates have been studying the demography and behavior of African savanna elephants in Kenya's Amboseli National Park since 1973 (Moss 1988). Andrea Turkalo has continuously observed forest elephants in the Nouabale-Ndoki National Park, Central African Republic, since 1991 (Turkalo 1996). R. Sukumar has been documenting the lives of the Asian elephants in the Mudumalai Wildlife Sanctuary in Tamil Nadu, India since 1983 (Sukumar 1989). These long-term research projects, together with several dozen shorter ones, confirm that most aspects of elephants' life history and behavior are fundamentally similar across the three species.

The largest of all terrestrial mammals, elephants have a natural life span of 60–70 years. They mature slowly, reaching puberty in their early teens. Few offspring are produced per female lifetime—up to 12 for a female African savanna elephant (Moss 1992). Males leave their natal families in early adolescence, after which their social activities consist mainly of competing for rank in male dominance hierarchies and roving among family groups in search of estrous females. Adult bulls have no involvement in the care of young. Females, on the other hand, spend their lives immersed in a social network that encompasses many families and several levels of competitive and collaborative relationships. Communal care of young is a conspicuous feature of the female society.

In this chapter, I will provide a comparative description of elephant society in the three species, while highlighting two sources of social complexity that seem to characterize them all. Complexity results from elephants' ability to recognize and track individuals over long periods of time through changes of age, status, and condition. This promotes the development of multiple and many-layered social relationships. Interacting with this source of complexity is a high level of variability in the behaviors of individual elephants. This variability, which may have a variety of origins, multiplies the possible complexity of all interactions. I will close this chapter with a detailed example illustrating behavioral diversity among individual forest elephants.

The Female Social System

Families

In all three species, the family is the most stable social unit in an elephant population; however, the distances between family members are in constant flux. Nursing calves generally remain within a few body-lengths of their mothers or the females who assist in their care (allomothers), but other family members may drift as much as half a kilometer away for separate foraging several times a day. Less predictably, male visitors arrive and depart in the course of a few hours or days. Thus the social organization of elephants is dynamic, and any definition, even of a "core family group," is somewhat arbitrary.

Nonetheless, estimates of average family size have been made for each species. In the Amboseli savanna elephant population, an average dry season family size of 9.3 individuals was reported (Moss & Poole 1983). In India, family groups of Asian elephants averaged 5.8 and 8.8 individuals in consecutive wet seasons, and 8.2 in a dry season (Sukumar 1989). In the Central African Republic, an average family size of 2.8 individuals was estimated for African forest elephants (Turkalo 1996). I suspect, however, that this estimate is not strictly comparable to the estimates for African savanna and Asian elephants, for it was made in a mineral lick area where competition for digging spots frequently separates family members for short periods of time. Stable forest elephant families with as many as seven members are not uncommon, and their structure and internal interactions closely resemble those observed in African savanna and Asian elephants (A. Turkalo and K. Payne, personal communication 2000).

Allomothering—caring for the young of others—is a prominent aspect of female life in elephants of all three species, and reveals evidence of the advantages of large group size. Juvenile and adolescent females guard, instruct (by manipulating with a trunk or a foot), and play with newborn and infant calves. The result is that young calves are kept from wandering, from being left behind, and from other stressful or dangerous situations, and their mothers are often released to forage at a distance. Closely related adult females (aunts, great-aunts, and grandmothers) occasionally nurse the infant of a close relative (Lee 1987 for savanna elephants; A. Turkalo, personal communication 2000 for forest elephants); comfort-suckling is often performed by females too young to lactate, and adoption of the infants of lost relatives occasionally occurs (Moss 1988). During the first two years of an African savanna elephant calf's life, when it is most vulnerable (Lee & Moss 1986), it spends roughly as much of its nonnursing time with allomothers as with its mother, and much less time with other elephants (Lee 1989). Calf mortality declines as the number of potential allomothers in the family unit increases (see Figure 3.1), indicating the adaptive value of these behaviors.

The frequent reinforcement of the bonds within families relies on individual recognition involving all sensory modalities—smell, taste, hearing, vision, and touch. All are apparent, for instance, in the emotional greetings among family members when they join after separations of several hours or more. The gathering animals run together rumbling and screaming, circling, clashing tusks if they have them, entwining trunks, sniffing, stroking, shov-

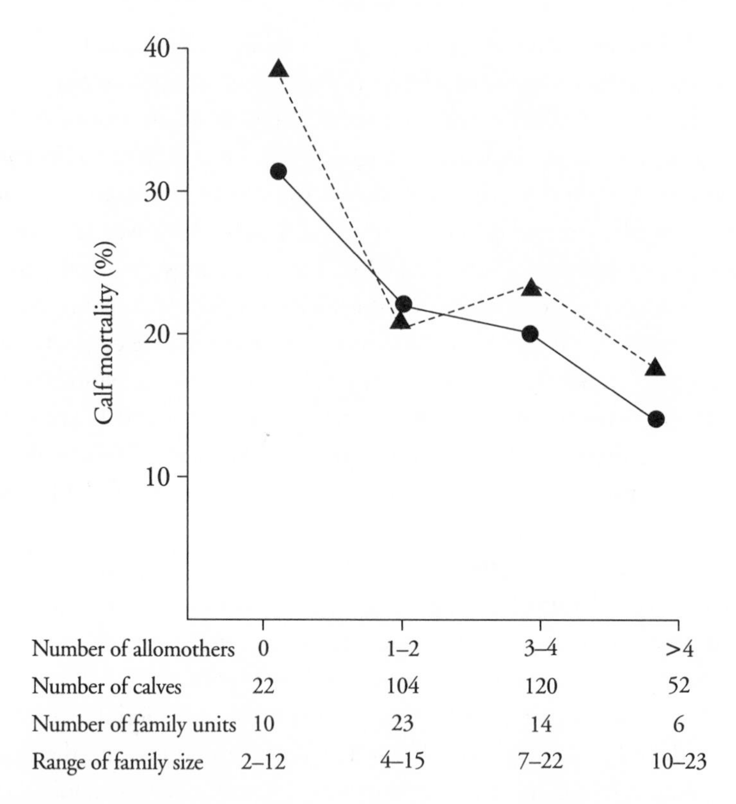

Figure 3.1. The percentage of mortality in the first 24 months of life for elephant calves born to families with different numbers of juvenile and adolescent females (allomothers) *(circles).* The mean mortality per family unit averaged over 12 years for the median availability of allomothers is also presented *(triangles).* The range of family sizes for each category is given. Data from C. J. Moss, long-term records (from Lee 1989).

ing, bumping rumps—expressing excitement in social behaviors that range from affectionate to aggressive—while individually secreting from their temporal glands, and urinating and defecating.

A varying number of individuals can be involved in greetings: Figure 3.2 illustrates a greeting between two frequently associated adult forest elephant females following a separation. Standing side by side with eyes wide, they wave their raised ears rapidly (as often happens during excited vocalization), secrete from their temporal glands (releasing pheromones), and produce a series of alternating, slightly overlapping broadband calls through which each may be identified to the other and their reunion possibly announced to distant listeners (A. Turkalo and K. Payne, personal communication 2000).

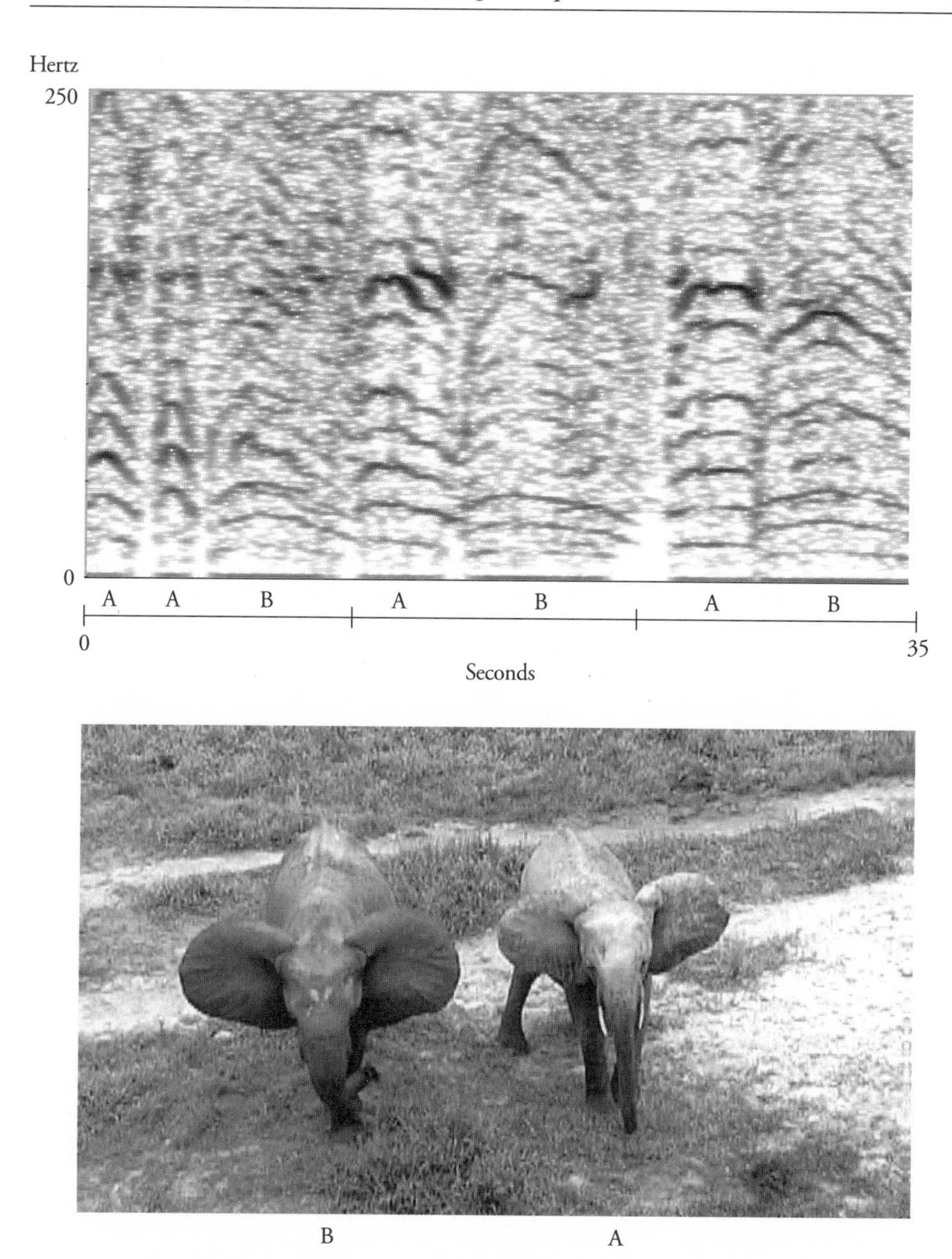

Figure 3.2. A pair of closely associated adult female forest elephants in the Dzanga clearing in the Central African Republic exchange greeting rumbles after a period of separation. Spread, rapidly waving ears and abutting or overlapping low rumbles are characteristic of this behavior, which often involves many individuals. Unpublished data from K. Payne and A. Turkalo, 2000.

Another indicator of the association among elephant family members is an acoustically triggered behavior that elephant researchers translate familiarly as "Let's go." Often after an elephant family has been drinking at a waterhole, an adult female will draw away from the group, face outward, and repeatedly utter a series of long, unmodulated, low-pitched rumbles. One by one, all members of the family will cluster around her facing in the direction she has indicated. This process may take as long as 20 minutes. When, but not until, all members are gathered, a final rumble is heard and the group moves away in unison (Payne 1998).

Greetings, synchronized movements, and a number of other intra-family behaviors are coordinated acoustically. Other aspects of family synchronization are coordinated hormonally. Adult females in a family have been found, for instance, to synchronize their periods of estrus (Moss 1988). This may result in a higher chance for each to be fertilized, because estrous females attract large, noisy groups, which in turn attract mate-searching males.

Matriarchs

The oldest female in an elephant family, the matriarch, tends to act as its leader. Her dominance is expressed not only in cooperative but also in competitive interactions. At a limited resource (e.g., an elephant-dug well or mineral lick), the matriarch may bully (kick, shove, threaten) other members of her family, including her own offspring, in order to secure the first or best drink for herself. She may then defend the resource to assure subsequent drinks for her closest family members. When the family is threatened by any outside factor, the matriarch becomes an altruistic leader, taking risks on their behalf. In such situations the rest of her group follows her lead and amplifies the consequences of her decisions. Eltringham (1982) describes a number of intense encounters with matriarchs during darting and culling operations that point vividly to the authority of the matriarch's decisions in stressful situations.

> [S]he will defend the group and the rest look to her for guidance. If she runs away, they will follow but, if she turns around and attacks, they will charge with her . . . If one darts a young female in one group, it is extremely difficult to chase off the matriarch. She will stand guard over the stricken animal, trumpeting loudly and attempting to lift it to its feet. The others follow her example and one is then faced with a milling

> mass of furious beasts . . . An alternative is to dart the matriarch herself. The difference in the reaction of the elephants is striking. In this situation, they seem to have no idea of what to do and rarely make any show of resistance. (p. 52)

In ordinary times, the matriarch makes most of her group's decisions about where and when to move in search of food and water and how long to stay. Her memories of seldom-visited water sources may increase her family's chances of survival in times or places of drought, as seems particularly likely in the Namibian desert elephant population. Their enormous home ranges (expanding over two-year periods to encompass up to 3,000 km^2) reflect the distribution of water sources, some of which are visited less than once a year (Viljoen 1989). The survival of the desert elephants depends on their knowledge of these drinking sites, as became clear in a 1981 drought. More than 85 percent of the desert-adapted herbivores (e.g., gemsbok and springbok) in the elephants' home range died from lack of available food in the vicinity of the remaining water-holes (Viljoen 1982). Meanwhile the radio-collared elephants expanded their daily treks to include water-holes far removed from their food resources, and lost none of their members (Viljoen 1990).

The desert elephant example emphasizes the importance to a family group of its collective *ecological* memory. In a slightly different vein, a study by McComb and colleagues (2001) illustrates the probable impact of a matriarch's *social* memory on her family's welfare. They found, over a seven-year period in Amboseli, that the age of a matriarch was "a significant predictor of the number of calves produced by the family per female reproductive year." This correlation seems significant in the light of Lee's discovery that group size affects a calf's chances of survival (Lee 1989, as illustrated in Figure 3.1). McComb and colleagues related their finding to the results of a playback experiment during the same period, in which old (e.g., 55-year-old) matriarchs appeared better able to distinguish between the contact calls of familiar and unfamiliar members of the social network than did young (e.g., 35-year-old) matriarchs. Families led by young matriarchs tended to bunch—a fearful response—when they heard calls of any other families, whereas old matriarchs initiated bunching only in response to calls of unfamiliar families. A matriarch's social memory, the authors concluded, progressively improves her family's inclusive fitness. Dominance, like social memory, contributes to a family's fitness, and both social memory and dominance tend to increase with age.

Following the death of a matriarch, several females may temporarily share group leadership or the group may break into fragments (Moss 1988). In these as well as in unstressed times, it becomes apparent that other individuals also have the capacity for autonomy and leadership, and that more than one animal is influential. After a large family has satisfied its thirst and hunger, it is not unusual, for instance, to see periods of hesitation followed by what appears to be a group decision about where to go. Out of a calm clustered group, a subset of several elephants will emerge heading in one direction; these soon return to the group as another subset heads in a second direction. Each subset is led by an adult or subadult female. The second subset may be joined by members of the first, or both may reintegrate themselves as a third subset starts off in a third direction. This process may continue for half an hour or more, until all at once the entire group moves off in a unified procession (Payne 1998). Events of this sort, illustrating a temporary fluidity in the social hierarchy, call to mind the process by which swarming bees arrive at apparent consensus about a new hive site (Seeley & Buhrman 1999), and by which hamadryas baboons reach travel decisions (Kummer 1968).

Bond Groups

A level of close association known as the "bond group" (Moss & Poole 1983) or "kinship group" (Douglas-Hamilton & Douglas-Hamilton 1975) has been identified in African savanna elephants. A bond group consists of two or more families led by closely associated matriarchs who spend 35–70 percent of their time in close proximity (Lee 1991). Members of bond groups sometimes greet excitedly, indicating family-like recognition, when they join after a separation of a few hours or more. When the emotion subsides, the enlarged group settles in to grazing or drinking as if a single family, but when they move off they do so as separate units, each accompanied by its matriarch and containing only its own members.

In at least two cases, the origin of savanna elephant bond groups has been documented as the gradual splitting, or fission, of a single rapidly growing family group (Douglas-Hamilton & Douglas-Hamilton 1975; Moss 1988). It seems likely that this is the usual mechanism by which bond groups form; however, Moss (1988) notes the occasional occurrence of bond groups containing maternally unrelated elephants. These are probably lasting relationships that are reinforced and perpetuated through reciprocal altruism. The propensity of female elephants who lack close genetic relatives to form mutually advantageous relationships involving allomothering and other family-like behaviors has also been documented in free-ranging forest elephants (A. Turkalo, personal communication 2000) and in semi-domestic Asian elephants in Tamil Nadu, India (K. Payne, personal communication 1997). Many examples of the same phenomenon have been documented in zoos.

In a radio-tracking study, Charif and colleagues (in review) located radio-collared savanna elephants from 13 family groups every three hours for two and a half months. Both the spatial and the temporal aspects of the data were analyzed quantitatively, revealing dynamic aspects of extended family relationships that would not have emerged in a coarser-grained study. Some elephant families moved in such a way that they remained within infrasonic hearing range of one another more of the time than would be expected by chance. Others shared home ranges but moved about inside them independently of one another. The authors speculated that coordinated movements are an aspect of bond group behavior.

The benefits of bond group relationships are most apparent in savanna elephants, who often need help protecting their calves from predators. Three savanna predators—lions, hyenas, and wild dogs—hunt in groups whose members simultaneously attack from several directions. Elephants are slow compared to these predators, and a lone elephant calf is extremely vulnerable. The adult elephants' protective strategy involves tight bunching in an outward-facing formation, with their calves sequestered in the middle. The number of adults determines the effectiveness of such a barrier in protecting the enclosed calves. Emboldened by greater numbers, the adults become more aggressive, reducing the chance that predators will isolate and kill a calf.

Along with this distinct benefit comes the potential cost of increased crowdedness that is inevitably associated with any kind of alliance, be it at the level of the family or the bond group. In fact, given that a single adult elephant may consume 300 pounds of vegetation and drink 30–50 gallons of water every day (Moss 1982), it is remarkable that a social life has evolved for elephants at all. What appears to make the difference is a capacity for long-distance communication, which enables elephants to remain in acoustic range of one another while foraging separately. A discussion of how this communication system works will be presented later in this chapter.

The advantages and disadvantages of social alliances are probably not the same in savannas and forests. Because forest predators are few and tend to be solitary, it seems likely that bond groups might not convey advantages great enough to offset the costs of coordinated (thus, limited) movements in forests. Fernando and Lande (2000) made a similar point when they found no evidence of bond groups in the four families they sampled in a radio-tracking and observational study of Asian elephants in Sri Lanka. In African forest elephants, on the other hand, repeated greetings among certain families provide evidence of some inter-family affiliations (A. Turkalo, personal communication 2001). Suffice it to say for now that the fission-fusion nature of elephant societies is apparent in all three species: affiliations form, dissolve, and re-form opportunistically, providing evidence of mutual recognition in a large social network.

Clans

A number of authors (e.g., Douglas-Hamilton 1972; Moss & Poole 1983; Sukumar 1989) refer to families that share portions of their dry season home ranges as members of a "clan." Charif and colleagues (in review) identified three geographically coherent clans in their study of elephants' spatial behaviors, but during their two-and-a-half-month study found neither genetic nor behavioral homogeneity at the level of the clan (see Figures 3.3 and 3.4). Given the fluidity of elephant societies, however, one can imagine that members of a clan may sometimes unite to act as a social unit when the need arises.

Family Group Home Ranges

Figure 3.3 shows the home ranges of 11 savanna elephant families and the probable matrilineal kinship of the elephant families occupying them (Charif et al. in review). The locations and shapes of home ranges were directly related to the locations of water sources, as was the segmentation of these home ranges into three clans. Home ranges varied sixfold in size, with the smallest ones in the Sengwa clan centered on an important resource—the only extensive open pools within several miles (indicated by an asterisk in Figure 3.3). It was interesting that the radio-collared families occupying these home ranges all shared a single mitochondrial haplotype, suggesting that the locations of female home ranges may reflect matrilineal dominance. The intriguing appearance that one matriline has a monopoly on the best water sources could not be confirmed, however, as not all elephant families were collared.

Figure 3.4 shows a dendrogram derived from the same home range corre-

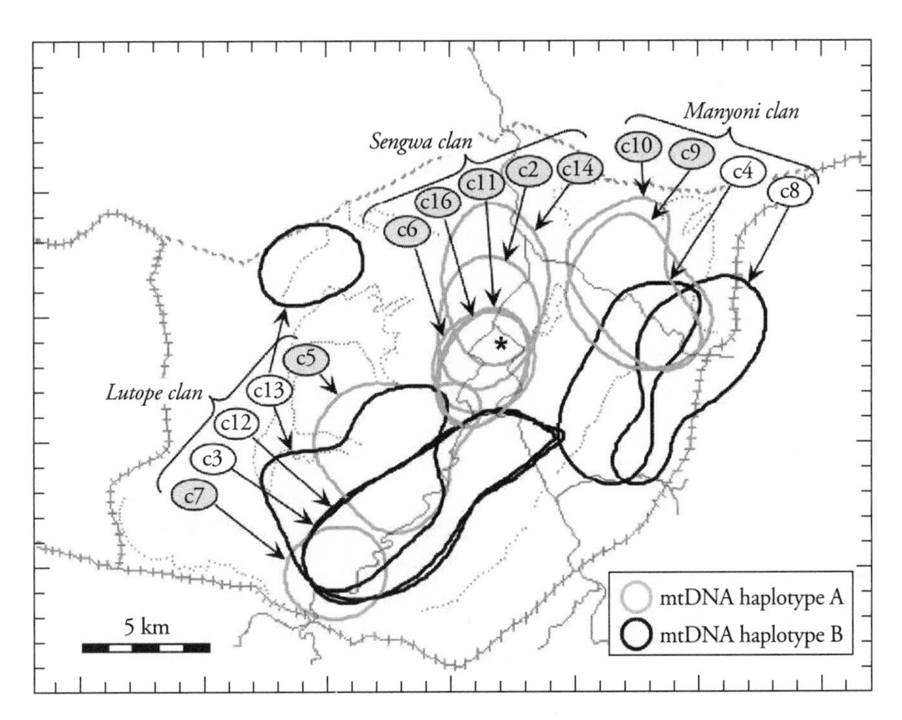

Figure 3.3. Simplified representation of the 50 percent home ranges of 14 radio-collared elephants in three clans, based on utilization distribution data gathered in the Sengwa Wildlife Research Area, Zimbabwe, August–October, 1990. The asterisk near the intersection of two rivers shows the location of the most abundant water in the area. Modified after Charif et al. in review.

lations and relates these to genetic relationship and movements within home ranges. Temporal analyses of the data showed that while some elephants were sharing space by coordinating their movements with one another, others were sharing space but using it independently. Most pairs of elephants who coordinated their movements (indicated by arrows on the left side of Figure 3.4) had the same mitochondrial DNA haplotype and were thus probably related to one another, but one out of the five coordinating pairs came from different matrilines (C5-Munya and C12-Chis). These distinctions reinforce our impression that female elephants assess, remember, and participate in a large number of relationships that may have different origins and different bearings on their behaviors.

Only in the Zimbabwean savanna elephant population has family home range behavior been studied with adequate resolution to reveal the temporal and spatial relationships related above (Charif et al. in review). The study

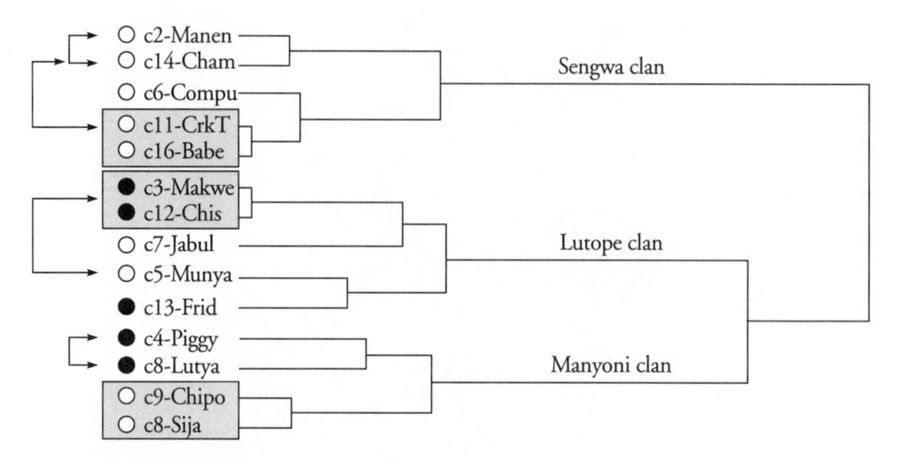

Figure 3.4. A dendrogram generated from the above data by clustering elephants according to home range correlation. Elephants are identified by collar number and abbreviated name. Open symbols indicate mtDNA haplotype A; filled symbols indicate haplotype B. Boxes indicate pairs of elephants in the same families. Arrows at left indicate pairs of elephants in different families that coordinated their use of shared space. From Charif et al. in review.

lasted only one three-month dry season in a forested savanna, however. In another study, by Viljoen (1989), the movements of elephant families in a Namibian desert elephant population were documented over a two-year period, but the data were collected on a coarser scale. It is possible that the general pattern of spatial use is fundamentally similar to that of elephant families in savanna environments. In Viljoen's words:

> Important components of the short-term spatial patterns are the apparent ability of elephants to sense localized rain storms or river floods over considerable distances, and their readiness to move to the areas affected. Range use by elephants is therefore based on fixed immutable dry season ranges from where they make irregular use of more distant areas where rain has fallen or floods have occurred. These movements are largely unpredictable in terms of timing and direction, being governed by short-term rainfall patterns. They are, however, predictable in terms of overall location according to individual home ranges. Return movements are predictable with considerable reliability if the dry season ranges of individual elephants are known. This flexible system of movements and the ability to move long distances to and from a fixed home base enable the elephants to make use of resources that are available

> only temporarily while at the same time reducing their impact on the area on which they depend for dry season survival. (p. 17)

Viljoen's radio-tracking records revealed identical home ranges for three pairs of families, but it is not known whether these families were coordinating their movements and remaining in acoustic range of one another while making long treks, or using the shared space independently.

Dominance Hierarchy

Female elephants' multilevel social organization reflects on the one hand varying degrees and types of collaboration, and on the other a dominance hierarchy resulting from competition within and between families. On the level of the individual, age and body size affect the relative ranks of females, who compete for local resources, for access to mates, and for the privilege of allomothering a newborn calf—a privilege whose benefits appear to be defined in terms of access to the resources available to the calf's mother (Dublin 1983). On the level of the group, family size and the age and rank of its matriarch affect the competitive positions of families in relation to resources (Dublin 1983, A. Turkalo, personal communication 2000).

Although minor shovings and exclusions are common in any group of elephants, competition is on the whole less overt among females than among males. The mechanisms through which females enforce a dominance hierarchy on larger temporal and spatial scales than those apparent in physical altercations are not yet fully understood. An interesting hypothesis emerged from a two-year study revealing a relationship between the dominance ranking of females and the survival of their young (Dublin 1983). A correlation was found among the social status of mothers, the timing of when they gave birth, and the survival of their calves. The calves of subordinate females were generally born after the peak of the long rains and suffered higher mortality than did the calves of dominant individuals, who gave birth earlier. This relationship suggests the presence of social and/or hormonal interactions that suppress, delay, or disrupt the reproductive efforts of subordinate females.

Male Social System

In adulthood the social relationships of a bull elephant are very different from those experienced during his first decade as a member of a stable family group. Having reached sexual maturity at around 17 (Laws 1969), the young bull roves independently from family to family in search of estrous females. When he encounters one he mounts her, unless rejected in favor of older males. When he meets another male he often flees or threatens or fights, depending on his position in a dominance hierarchy. Off and on, he may spend time in a loosely organized, impermanent all-bull association, or "bull group." In his late twenties he has his first period of musth, a hormonal condition that, like rut in deer, occurs annually and is characterized by aggressive and sexual competition (Poole 1987). From then on he spends an increasing portion of every year in musth, engaging with other musth bulls in serious fights, and guarding and impregnating as many females as possible. Ultimately, his breeding success is determined by his position in the dominance hierarchy and by female choice. Such social structure as we find among adult male elephants thus takes two forms: the bull group and the dominance hierarchy.

Bull Groups

African savanna bull groups range in size from a pair of bulls to several dozen, although in one extreme case a group with 144 individuals was noted (Estes 1991). These associations last from a matter of hours to several days. The main occupations in bull groups are feeding, resting, and harmless jousting, and moving in a somewhat coordinated fashion from one place to the next. The ranging patterns of the males in such groups reveal bull areas that often overlap (Poole 1982).

Members of African savanna bull groups sometimes cooperate in aggressive coalitions. On one occasion, two such coalitions were observed engaging in a confrontation with each other. Some of the individuals were radio-collared, so that their home ranges and some evidence of previous association had been established. In a suddenly violent encounter, one group threatened and then chased the other away from the traditional border between their home ranges (Martin 1978). Group crop raiding is a more common kind of coalition behavior, reported in African savanna elephants (Osborn 1998), Asian elephants (Sukumar 1994), and African forest elephants (R. Barnes, personal communication 2001). It will be interesting to learn about the stability of these groups, the roles of individuals in them, and about the costs and benefits of membership in such groups.

Dominance Hierarchy and Musth

Throughout his life, a bull elephant in any of the three species expends much time and energy assessing his strength relative to that of other bulls. In the first 25 years this is a rather straightforward process, involving jousting and shoving matches whose outcomes are directly related to size and weight. Serious injuries rarely occur during these contests. But as the bull approaches the age of 30, a new factor begins to assert its influence on his dominance relationships. This is musth, an annually recurring hormonal condition analogous to rut, characterized by aggressive competition for females (Poole 1987). Testosterone levels in the urine of musth males are significantly higher than in sexually active nonmusth males. These in turn have higher testosterone levels than do sexually inactive nonmusth males. Musth bulls typically dominate bulls who are not in musth, and engage in dangerous contests with others of roughly equal status.

Each year a bull's period of musth increases in duration. In the first decade it spans only a few days or weeks, but by the time a bull is 40 musth lasts two to four months. The brevity of early musth periods is due in part to hormonal inhibition in the presence of older bulls (Poole 1987). Slotow and colleagues (2000) noted a startling example of this inhibition in a group of 85 male elephants who were confined together after having been orphaned as juveniles in Kruger Park culls. Growing up without older males around, these young bulls started coming into musth at the age of 17, and were breeding successfully by the age of 18—a decade earlier than would be typical in an undisturbed population. The introduction of six older males into the population significantly reduced and delayed the timing of the young males' periods of musth.

Male elephants use both olfactory and acoustic signals to advertise their sexual status: both modes seem particularly well-adapted for mate-finding over long distances. Secreting from their temporal glands and dribbling urine continuously, musth males lay pheromone-rich odor trails (Rasmussen et al. 1996), and both their listening and their vocalization rate increase (Poole 1989). Very low-frequency calls announce the musth condition (Figure 3.5). These probably also announce position in the dominance hierarchy, for playback experiments show that musth rumbles attract male rivals of equal status but discourage males of lesser status (Poole 1999), reducing the occurrence of fights between unequals.

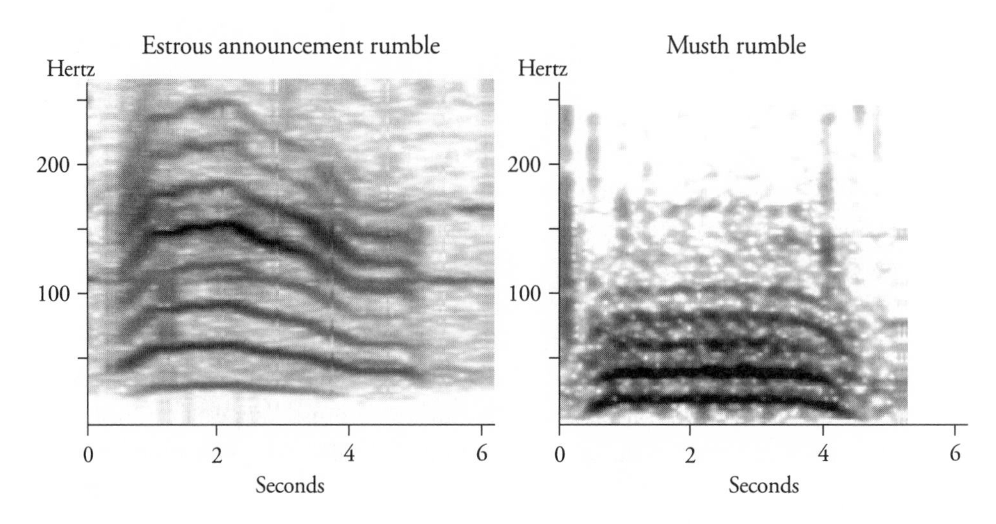

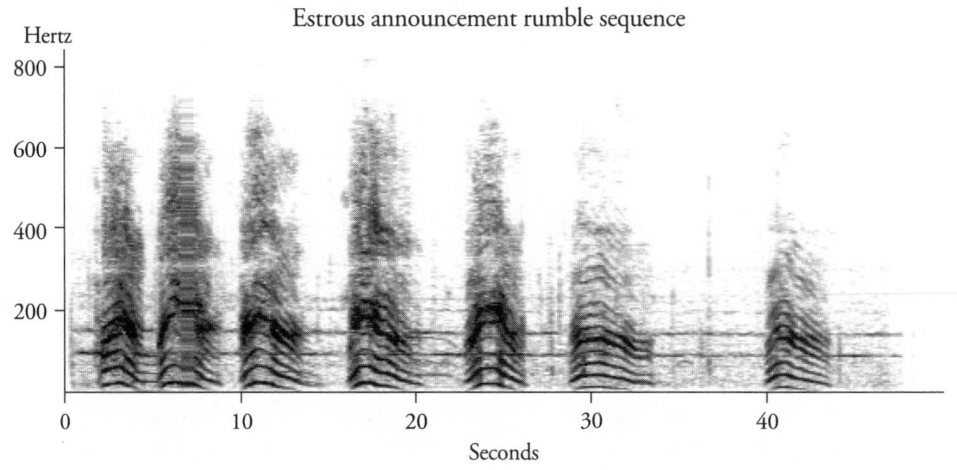

Figure 3.5. *Top left:* Single call in the estrous announcement sequence of a female savanna elephant when she is ready to mate but not yet guarded by a male in musth. *Top right:* A bull's musth announcement call. *Bottom:* A seven-call estrous announcement sequence. Such a sequence can continue for up to 45 minutes. Recorded by K. Payne in Amboseli Park in 1985.

Toward the end of a long period in musth, a male's energy reserves and testosterone level decline, and his status declines accordingly. Younger and/or smaller males who yielded to him when he was first in musth may challenge him successfully one or two months later even if he is still in musth. This is typical of the complexities affecting dominance among bulls. Several factors affecting status interact with one another: relative ages and weights, relative status when out of musth, relative status when one and not the other is freshly in musth, the same in reverse, the same when both are freshly in musth, the effect of time on the rate at which strength and aggressiveness wane when either or both are in musth, and so forth (Poole & Moss 1989). As a consequence, dominance relationships among males change much more frequently than those among females.

A male elephant's competitive agenda probably does not require him to remember other individuals to the extent that females do, although Poole (1982) reports that "several pairs of males [in Amboseli] were observed in association greater than 30 percent of the time that each were observed in the company of other males." As presently understood, the bull dominance hierarchy reveals very little evidence of social structure above the level of one-on-one contests. Yet bull groups form, foster companionship, and serve as shifting reservoirs of potential allies who may join in spontaneous ventures that equally benefit all participants. This is cooperation on a different level from many of the affiliations found among female elephants, who enter into costly collaborations with benefits that are reaped by kin.

The differences between the types of affiliation found among males and females in the three elephant species are summarized in Table 3.1. Fitness costs and benefits are suggested for each kind of affiliation, leading to a guess about the mechanism through which it has evolved.

Communication

Elephants in all three species make calls with fundamental frequencies below the lower limit of human hearing (20 Hz), in the range called infrasound (Payne et al. 1986). Some of these calls are quite powerful (90 to 117 dB SPL). Because low-frequency sound attenuates very little with distance, elephants' powerful infrasonic calls enable them to stay in contact as they move separately over large areas of savanna or forest. A preliminary study of hearing in elephants indicates unusual sensitivity to frequencies as low as 17 Hz; lower frequencies were not explored (Heffner & Heffner 1980). A playback experiment indicates that free-ranging savanna elephants respond to one another's calls over at least two and probably four kilometers during daylight hours (Langbauer et al. 1991).

As we consider the size of elephants' sensory world, the timing as well as the frequency and power of their vocalizations turns out to be important.

Table 3.1. Adaptive perspectives on affiliative behaviors in the three species of elephants

Level of affiliation	Indicator of affiliation	Potential fitness benefits to performers	Potential fitness costs to performers	Potential fitness benefits to receivers	Potential fitness costs to receivers	Mechanism for selection	Found in:
Family	Matriarchal leadership	Kin survival	Energetic cost of attentiveness and decision-making Vulnerability during group defense	*Followers of matriarch* Best access to resources Instruction re: resources and social networks Defense of home range Defense against predators	*Followers of matriarch* Risk of following unfit matriarch	Kin selection	*L.a.*[a] *L.c.* *E.m.*
	Allomothering	Access to resources of mother Practice in maternal care Reciprocity Kin survival	Energetic cost of attending infant Loss of foraging time Delayed reproduction	*Mother and calf* Infant protection Infant instruction Increase in foraging range for mother	*Mother and calf* Risk of unfit allomother	Kin selection	*L.a.* *L.c.* *E.m.*
	Greeting	Recognition of kin Reinforcement of kinship bond	Announcement of presence to predators	*Distant listeners* Recognition of associates vs. strangers Location of associates vs. strangers	*Distant listeners* None	Kin selection	*L.a.* *L.c.* *E.m.*
	Mating pandemonium	Attraction of musth males Reinforcement of kinship bond	Announcement of presence to predators	*Distant listeners* Location of fertile females	*Distant listeners* None	Kin selection	*L.a.* *L.c.* ?

	Synchronized estrus	Enlargement of group size Greater chance of attracting musth males Synchronization of births: benefit from group care	Synchronization of births: synchronization of environmental risks	*Visiting males* Concentrations of fertile females	*Visiting males* None	Kin selection	*L.a.* ? ?
Bond group	Greeting	Recognition of associates Reinforcement of family and group bonds	Announcement of presence to predators	*Distant listeners* Recognition of associates vs. strangers Location of associates vs. strangers	*Distant listeners* None	Kin selection Reciprocal altruism	*L.a.* ? ?
	Coordinated movements over long distances	Proximity of associates maintained with less resource competition Defense of home range	Resource sharing Self-limited range	NA[b]	NA	Kin selection Reciprocal altruism	*L.a.* ? *E.m.*
Bull group	Resource-defense coalition	Improved access to resources Communal risk affords dilution effect	Resource sharing	*Competitors for resource* None	*Competitors for resource* Decreased access to resource	Reciprocal altruism	*L.a.* ? ?
	Crop-raiding coalition	Improved access to resources Communal risk affords dilution effect	Resource sharing Increased risk of human detection	NA	NA	Reciprocal altruism	*L.a.* ? *E.m.*

a. *Loxodonta africana, Loxodonta cyclotis, Elephas maximus.*
b. Not applicable

The propagation of very low-frequency sound varies with atmospheric conditions, which change on a diurnal schedule. On a typical dry season evening in the savanna a temperature inversion forms, potentially increasing the listening area of elephants as much as tenfold—from 30 square kilometers at midday to 300 square kilometers in the same evening (Larom et al. 1997). In light of this fact it is interesting that savanna elephants make most of their loud low-frequency calls during the hours of best sound propagation (Larom et al. 1997). We do not know whether this is an innate or opportunistic response to fluctuations in the size of their communication area, but in either case it is clear that as the area shrinks and expands, so does the network of potential associates and mates.

The repertoire of elephant calls has been best studied in savanna elephants (Poole et al. 1988; Poole 1999) and appears to be very similar in the other two species (K. Payne, personal observation 2000). Many dozen call types can be defined by their behavioral context: of these, the majority are made by females and function in group coordination or reproduction. Powerful, low-frequency contact calls enable females to identify one another acoustically, as demonstrated for savanna elephants in playback experiments (McComb et al. 2000): these enable families and bond groups to forage separately—sometimes miles apart—yet remain accessible to one another for group defense.

Long-distance communication also enables male and female elephants to find one another for mating even as they move about over large areas independently. That such a system has evolved is particularly striking in light of the fact that a female elephant typically spends only one period of two to five days every four or five years in estrus. The reason for the delay is that (at least in Amboseli, where these data were gathered) a female is almost invariably fertilized during her estrous period. Almost two years of gestation follow, and then two years of lactation: both conditions inhibit ovulation. The result is that there is a very narrow window of opportunity in which males must find and mate with estrous females. Yet available females are almost always inseminated (Moss 1982).

The details of the communicative events surrounding reproduction are as follows: a female in estrus announces her condition by uttering a long (sometimes 45 minutes), predictable, and sometimes repeated sequence of powerful low-frequency calls (Poole et al. 1988) (Figure 3.5). These calls, apparently unique to estrus, are the most powerful calls we have recorded from individual elephants. They attract males from several kilometers away, as demonstrated in playback experiments (Langbauer et al. 1991). All males who approach the site compete for access to the calling female. She selects a mate—typically the most powerful musth bull in the vicinity—who then guards her for the duration of her estrus (Figure 3.6), sharing resources such as water from a well with her, resources from which he systematically excludes others, and periodically mounting her. As she is mounted, numerous members of the female's family may make loud, excited, overlapping calls ("the mating pandemonium"), which in turn may attract other adult males to the site.

Considering the importance of long-distance communication to elephants' reproduction system, the tendency of males to listen a great deal and vocalize very little is interesting, for it adds to our impression of male elephants as fundamentally solitary. Whether in bull groups or alone, adult bulls vocalize far less frequently than do females (Payne et al. in review).

Figure 3.6. A dominant savanna elephant bull in musth *(left)* guards a young female in estrus *(right forward)* in an aggregation of families. Urine-dribbling, temporal gland secretions, and a swaggering gait are all indicators of the male's condition; the female's raised stiffened ears show her attentiveness. Photograph by K. Payne, Amboseli Park, 1985.

Nearly all bull calls are solo: they include musth rumbles, which are the lowest-frequency calls so far recorded from elephants—fundamental frequencies at 5 Hz have been recorded from African forest elephants, and at 12 Hz from African savanna elephants. These function in spacing among males, as shown in playback experiments by Joyce Poole (1999).

Rates of calling have been measured both in African savanna and in African forest elephants (Payne et al. in review; A. Turkalo and K. Payne, personal communication 2000). In both species, calling rates increase progressively with numbers of individuals present. Females are responsible for this effect. In females but not in males, calls frequently overlap with or abut the calls of others, forming clusters. In savanna elephants the rate at which call events occur (where a call event is defined as either a solo call or a call cluster) mirrors the absolute number of individuals present (Payne et al. in review) (Figure 3.7).

The robustness of the relationship between group size and rates of calling

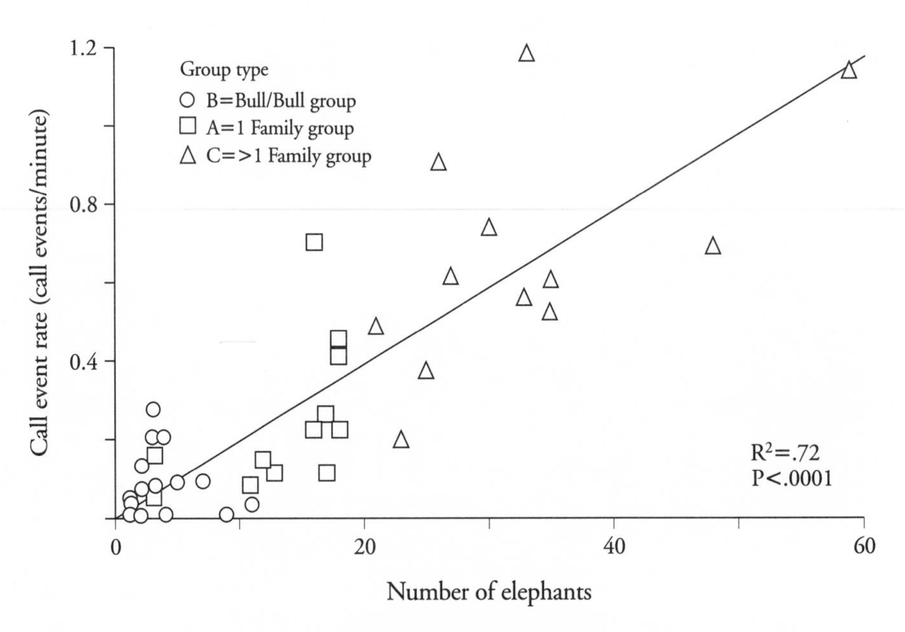

Figure 3.7. Linear regression analysis of data gathered in Etosha National Park, Namibia. Each point represents the call rate during a continuous period (Average 51 minutes, range 14–202 minutes) when a certain combination of elephants was visible and (as indicated by locations from an acoustic array) responsible for all the counted calls. From Payne et al. in review.

suggests a speculation about long-distance communication: in choosing their destinations do mate-searching bulls make use of the relationship between calling rates and group size? Poole and Moss (1989) found that when in musth, males spend a significantly larger portion of their time listening, rumbling, and joining female groups than when they are not in musth. They also found that females spend their estrous periods in larger-than-average groups. Our finding (Payne et al. in review) suggests that bulls in search of mates could locate estrous females by assessing rates of calling even when conditions would not permit detailed call recognition.

Other communication channels are undoubtedly involved in mediating elephants' social interactions, olfaction probably being the most important. Using their trunks to selectively focus olfactory attention, elephants continually check the air, the ground, and one another's bodies for information encoded in pheromones (Rasmussen et al. 1996; Rasmussen & Krishnamurthy 2000). It is only because more is known about the details of acoustic signals that I emphasize elephants' calls over their smells: evidence exists that both contain cues enabling elephants to recognize individuals and that both function over long distances. It seems likely that the infrasonic components of family greetings, bond group greetings, musth calls, estrous calls, and mating pandemonia are all involved in supporting the expansion of elephant societies, in all their complexity, into home ranges commensurate with their ecological needs. In this respect it is reassuring to note a fairly close correspondence between the area in which elephants potentially hear one another and the sizes of the female home ranges (Charif et al. in review), suggesting that inter-elephant communication has played a limiting role.

Behavioral Variability among Individuals

All field observers have noticed variation in the responses and responsiveness of different individual elephants. Insofar as this variation increases the unpredictability of social behavior, it contributes to complexity. There is a challenge, then, to quantify variability in behaviors. No one has attempted to do such a thing, as far as I know, and the attempt is beyond the scope of this chapter. What I offer instead is a unique example that permits the comparison of the serial responses of a large number of elephants to the same natural event. The point of including this example is to illustrate my notion that a substantial portion of the complexity observed in elephant societies may ultimately be traced back to individual diversity.

Table 3.2. Individual responses to the death of an infant: Four males and four females, Dzanga-Ndoki National Park, 26 June 2000

Identity of elephant	Strong response Number of times behavior occurs				Weak response Number of times behavior occurs			
	Orial, adult female	Unnamed adult female	Sappho II, subadult male	Waldo, adult male	Unnamed adult female	Phyllis 2, adult female	Unnamed subadult male	Unnamed subadult male
Sniffs air	1	1		1				
Raises tail	1							
Rapidly approaches body			1					
Tentatively approaches body		1						1
Sniffs body	1	1	5	2	1	1	1	
Touches body with trunk		1	1					
Touches or hovers foot over body		9	1					
Tastes trunk after body touch		2						
Screams, rumbles, trumpets	2T, 2R							
Threatens calf's mother away		1	4	1				
Threatens others from nearing, appears to guard body		1	1	1				
Lifts body with foot		3	25					
Lifts body with trunk	6		1					
Lifts body with tusks			31					
Time attending to body (mins:secs)	3:00	4:00	10:22	3:15	0:15	0:06	0:40	1:35

129 Responses to the Death of an Elephant Calf

The following events took place in the Dzanga Sangha forest clearing in the Central African Republic, where Andrea Turkalo has been documenting the demography and behaviors of a population of African forest elephants for a decade. Several colleagues and I joined Turkalo for three months in the spring of 2000. On June 26, we observed and videotaped the death of a yearling elephant calf on a spot about 100 meters from our observation platform, and (on that and the next day) 129 visits to the body by elephants of both sexes and all ages.

The cause of death was probably starvation. We had observed the calf with her mother (named "Morna I" in Turkalo's demographic study) and a well-fed older sister several times in the weeks preceding death, noting that the mother was rejecting the infant's attempts to nurse while favoring the sister's attempts. As her calf died, within two meters of a much-traveled elephant trail, the mother and sister stood over or near her. Both tended to draw aside when other elephants approached the trio on the path. The visiting elephants tended to arrive one at a time on a hard sandy path flanked by soft mud; thus their perceptions and responses were to an unusual extent independent of one another.

Table 3.2 summarizes the behaviors of all elephants ("visitors") who walked up the path during a 5.25-hour recording period on the day of death (June 25) and a 6.5-hour period on the next day. The last visible movements of the calf occurred about two hours before the end of the first observation period. Not knowing exactly how to determine the moment of death, we have divided the data into the "day of death" and "day after death."

On the day of death we documented 56 visits to the body by 38 individuals, including the calf's mother and sister, who made six of the visits. On the next day we documented 73 visits by 54 individuals. The calf's mother and sister had presumably left the area: we did not see them again during our remaining 22 days of observation.

Of the 129 visitors, 128 changed their behaviors as they approached the body. The behavioral changes included:

exploratory behaviors in 80 percent of the encounters (obvious air and body sniffing, hastened approaches to body, touching or hovering foot over body, touching body with trunk, tasting trunk after body touch);
fear/alarm behaviors in 50 percent of the encounters (sidling or faltering

approaches, backing away from body altogether, avoidance in detours off all paths, ears lifted, tails lifted, trumpeting and rumbling, body jolts, hasty departures after exploration);

efforts to lift the dying calf in 18 percent of the encounters, using foot, trunk, or tusks;

body-guarding reactions in 15 percent of the encounters (threatening others—including the calf's mother—away from the body and standing over or next to it for a period of time);

aggression toward the body (tusk-stabbing and ripping the body apart) in a few prolonged visits by one individual, a subadult female who had in the previous year mauled a dying calf, and whose behaviors toward human observers were also aberrant.

The nature, extent, and coupling of these responses differed dramatically from individual to individual, as did the time spent attending the body and the number of returns to the body. The most extreme responses were given by a subadult male, Sappho II, and a young nulliparous adult female, Miss Lonelyheart. Sappho II visited the body five times (a total of 13 minutes), specializing in exploration and potential succorance, including 57 attempts to raise the calf to her feet. Miss Lonelyheart (so named because some of her other behaviors had been noted as bizarre since 1998 when she was first identified) spent 250 minutes at the body, tearing it apart and occasionally putting bits of it into her mouth. This individual was the only one showing aggression toward the calf. Other visitors expressed a broad range of fearful, affiliative, and exploratory behaviors, in various combinations. Morna I, the calf's mother, was among the individuals who spent the most time with the body and performed the most affiliative behaviors, exceeded in the first capacity only by Miss Lonelyheart and in the second only by Sappho II.

Long-term records indicate dates on which identified individuals have been seen in the Dzanga clearing (A. Turkalo, personal communication 2001). An association matrix was made from these, and used to evaluate the relative probability of genetic or social relationship between the dead calf's mother and each of the elephants identified in the clearing on June 25–26. A comparison of association levels of those who did and did not visit the body yielded a *P*-value of .544 (two sample *t*-tests), indicating that visiting elephants were not more closely associated with Morna I than the elephants in the clearing who did not visit the body. This confirms our impression that first encounters with the body were, except in the case of the mother and sister, the consequence of the body's location next to a well-used trail. Returns to the body were, of course, another matter, suggesting interest and intention.

Of all visitors, only Morna's two calves, who were always with her, had coincided with her in more than 27 percent of their appearances at the clearing throughout the period since Morna's first identification in 1991. In spite of these low levels of association, one might expect some correlation between the nature and intensity of visitors' responses and their history of association with Morna I. One might also expect males to be less responsive than females. Figure 3.8 shows, however, an absence of relationship between Morna I–visitor association levels and each of three factors: the time spent at the body, the number of returns to the body, and the number of affiliative behaviors performed. Gender did not offer a predictor of behavior either. This series of encounters seems, then, to offer evidence of individual variability beyond what one would expect as the consequence of differences in age, sex, and degrees of relatedness. Probably it is evidence of variability in several interactive domains, including cognition, emotion, and the physiological factors that account for levels of reactiveness.

The value of this example lies in its documentation in elephants of what we would broadly refer to, in human subjects, as personality differences. Undoubtedly, such differences are also reflected in elephants' more ordinary behaviors, but rarely does one get the chance to eliminate the contribution of such factors as age, sex, and relatedness as cleanly as in this example. We cannot, however, assess the extent to which these idiosyncrasies may have been shaped by prior experience. Given their long lives, their social memory, and their capacity for learning, it would be remarkable if elephants did not develop conspicuous personality differences in the ordinary course of their lifetimes.

The responses illustrated in this example take at least one form that seems maladaptive, namely Miss Lonelyheart's persistent attacks on an unrelated female's dying calf. This unexpected dimension sheds new light on the importance of individual recognition and remembered interactions—adaptive traits for members of a society in which sociopathic behaviors can develop. Here we see another way in which the increasing social memory of a matriarch probably contributes to the inclusive fitness of her family.

Slow Life History, Large Brain, Social Complexity, and Individuality

A number of authors (e.g., Dunbar 1992) have linked social complexity to large brain size. Van Schaik and Deaner in Chapter 1 correlate social com-

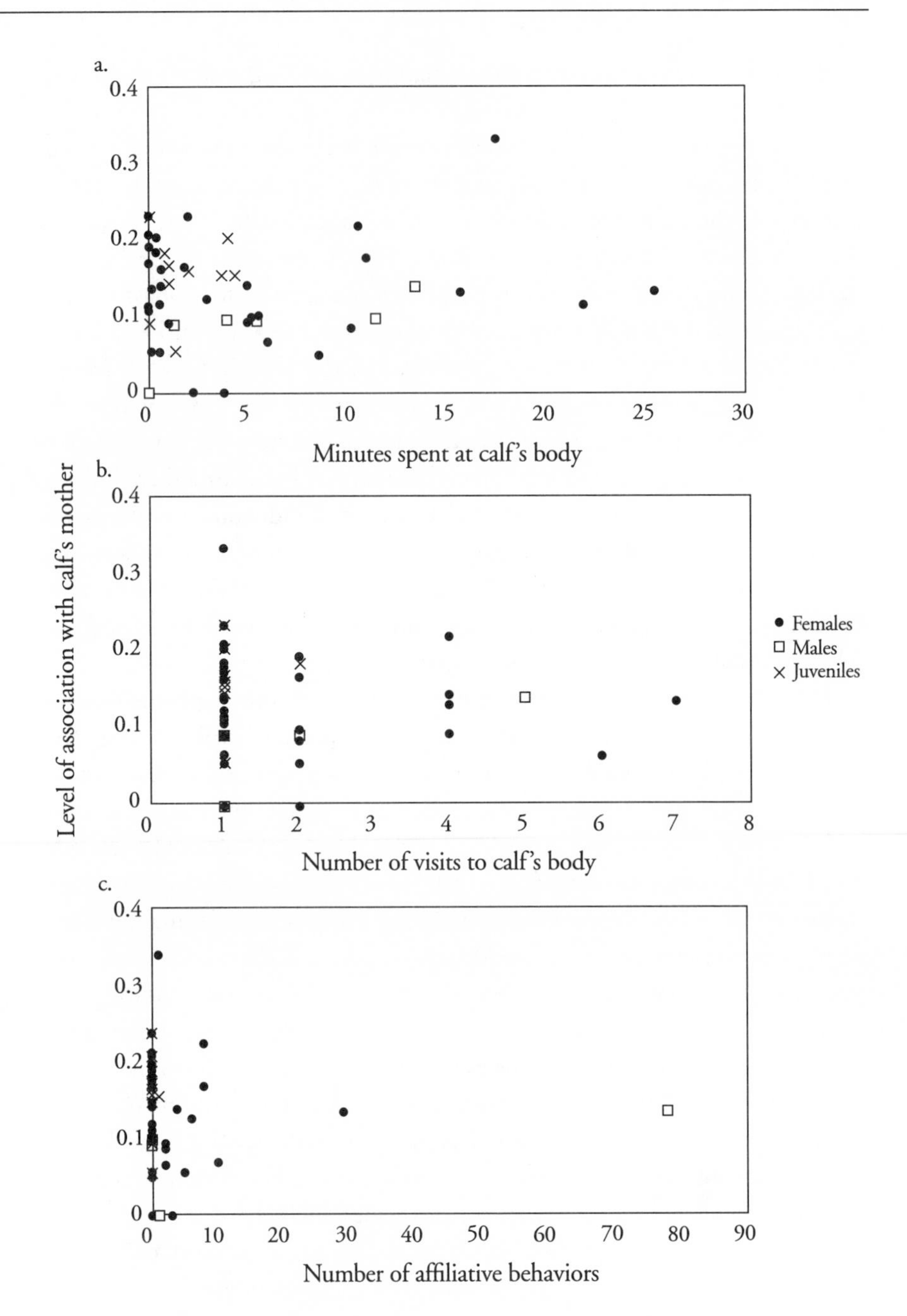
a.
0.4
0.3
0.2
0.1
0
0 5 10 15 20 25 30
Minutes spent at calf's body
b.
0.4
0.3
0.2
0.1
0
0 1 2 3 4 5 6 7 8
Number of visits to calf's body
c.
0.4
0.3
0.2
0.1
0
0 10 20 30 40 50 60 70 80 90
Number of affiliative behaviors
Level of association with calf's mother
Females
Males
Juveniles

Figure 3.8. Intensity of behavioral responses to the body of a dying/dead elephant calf, in relation to the level of association between the responder and the calf's mother. (Level of association is defined as the number of days when visitor and the mother were both in the clearing divided by the total number of days when the visitor was in the clearing.) No correlation is apparent between that index of association and *a)* the time a visitor spent attending to the body; *b)* the number of times the visitor returned; and *c)* the number of affiliative behaviors (body-lifting, body-guarding, and defense of the body). The calf's mother, not included in these charts, spent 198 minutes at the body, made six visits, and performed 56 affiliative behaviors. One other individual was also off the top of the chart, a young adult female with an association index of .064 who spent 250 minutes at the body tearing it limb from limb.

plexity and large brains with a slow life history, a necessary precondition for the development of both. With the brain primed for social learning, sources of social complexity multiply during the long lifetimes of individual elephants. New status, new alliances, and new rivalries develop, adding to and altering existing ones as the result of hormonal changes and increasing experience. Old females become repositories of ecological as well as social information, and take on different social roles from those of younger relatives. The learning that supports these changes occurs in the context of overlapping generations whose members spend much of their time in close proximity. A large capacity for memory supports cumulative learning, and this, together with a system for communicating over long distances, enables elephants to exploit large, variable home ranges. At the same time it results in a high level of behavioral variability among individuals. Thus social complexity compounds itself in a network of temporal, spatial, and cognitive dimensions, each of which is, by virtue of its interactions with the others, flexible.

II

Evolution of Cooperative Strategies

Cooperation has long been thought to play a role in the evolution of complex social behavior. Many early papers recognized the complexity inherent in societies where individuals have to balance the need to cooperate with partners with whom they also compete. In Chapter 4, Christophe Boesch focuses on the details of cooperative behavior, starting with a taxonomy of cooperative behaviors from relatively simple to more complex. The most complex form of cooperative foraging considered by Boesch involves hunters performing different complementary actions toward the same prey. This represents an important mode of foraging for human hunters and a few chimpanzee populations, but Boesch argues that other social carnivores use simpler forms of group action. If cooperation in which different partners play different roles is so beneficial, then why is it so rare? Boesch reviews cognitive tests with primates and concludes that it is particularly difficult for primates to take into account the perspectives of others, which may be critical for understanding the partner's role when that role differs from one's own.

Boesch has studied chimpanzees in the Taï forest of the Ivory Coast for the past two decades. He suggests that chimpanzees can develop complex cooperative networks when the ecological setting favors it: when hunting is difficult, adult male chimpanzees cooperate to hunt for meat. In the Taï forest, they consistently catch more meat than they eat, and the excess is distributed to females and younger males, who depend upon this important source of food. Chimpanzees are at a disadvantage hunting red colobus monkeys because red colobus are lighter and better able to move and jump between trees than chimpanzees. This means that a red colobus can usually escape from the pursuit of a single chimpanzee. Chimpanzees can overcome this problem by coordinating in a team to surround a group of monkeys. The chimpanzees play different roles in the hunt: a driver may move the monkeys in a specific direction, chasers may pursue a monkey rapidly, and ambushers may anticipate the direction and speed of escape. It is difficult for chimpanzees to learn to hunt cooperatively and they may continue to improve over 20 years or more. The most difficult aspect of this hunting involves the need to predict how the prey will respond, and how the other chimpanzees will drive the monkeys. Boesch suggests that this kind of hunting is particularly difficult to learn because it requires many correct steps to lead to capture, and no reinforcement follows unless all of the steps are performed correctly by all the members of the team.

When two communities of chimpanzees interact, adult males defend their territory in groups. Boesch's data suggests that chimpanzees are attuned to the relative balance between the strength of their own group and that of their opponents—they change their strategy depending upon the size of their party. Larger groups have the advantage in territorial defense, and females at times will join in. Chimpanzees will also often forage in larger groups if there is a heightened risk of confrontation with another group.

Taï chimpanzees use rock tools to open energy-rich nuts. Adult females open a surplus of nuts, which they share with their young. This form of parental care extends well after weaning—as long as 12–13 years. In summary, Boesch describes a fascinating pattern where females use technology to produce and share rich food, whereas males use complex social interaction to catch meat and to defend their territories. Sex roles are flexible, however, for males use the same tools to open nuts, and females can join the hunt or battle.

In their chapter on spotted hyenas, Christine Drea and Laurence Frank echo the main message of this book: even if hypotheses relating the evolution of intelligence to life history, feeding ecology, and social complexity may have originated among primatologists, they are best tested among as broad a taxonomic base as possible. Drea and Frank approach this issue by championing the view that their study animal, the spotted hyena, compares favorably with primates. Similar to the chimpanzees studied by Boesch, spotted hyenas must cooperate to kill their prey. Drea and Frank emphasize that hyenas are more dependent upon this cooperative hunting than primates, and have impressive abilities to solve cooperation problems in the lab. Hyenas travel long distances in search of prey, and must travel through the territories of other clans. Drea and Frank suggest that the problems of orienting and responding correctly to predators, prey, and hostile groups of conspecifics during long movements in times of scarcity may have selected for increased intelligence in hyenas as has been suggested for some primates.

Drea and Frank agree with van Schaik and Deaner's argument in Chapter 1 that animals with "primate-like" intelligence share slow life histories—they develop slowly, grow to large size, have few offspring, and show prolonged parental care. One of the benefits of broadening primatological studies to new species is the way in which different constellations of traits help us sharpen our view of the essential elements of social concepts. For example, in species with prolonged parental care, the young might be expected to develop a strong bond with the caregiver. Most work on the "attachment" between mother and young involves old world primates, in which the infant physically attaches to the mother. By contrast, hyenas are born in a natal den; the den rather than the mother represents safe haven to the infant. Yet in spite of this, hyena cubs are distressed by separation from the mother, even if a sib is present. Cetaceans present yet another perspective on parental care: even though the young are precocial in their sensory and locomotor abilities, they require prolonged parental care. The young can neither cling to the mother, nor rely on a refuge in the open ocean. As Hal Whitehead describes in Chapter 17, caring for the young is particularly problematic for sperm whales: young calves cannot follow the mother when she dives to feed for an hour many hundreds of meters below the sea. Adult females stagger their foraging dives when calves are in the group, apparently to provide more continuous care and defense for the calves, who stay near the surface. We can only begin to guess how this broadening of parental care may affect attachment in a species in which parental care may extend for longer than a decade.

Animals that live long enough to have societies with several overlapping generations add an extra dimension to social complexity. Social life can be particularly complicated for animals living in multigenerational groups that include a broad array of kinship relations. Sarah Mesnick and colleagues present genetic evidence in Case Study 6A that sperm whale groups are made up of individuals with a complex mixture of degrees of relatedness. The ecology of sperm whales selects for groups large enough to include several matrilines. Whales in these groups are dependent upon one another, especially for defense of the young, with cooperation likely maintained by a mixture of kin selection, reciprocity, and mutualism.

Anne Engh and Kay Holekamp caution in Case Study 5A that even though different animal species may be faced with social situations of similar complexity, they may deal with these problems with different levels of cognitive complexity. For example, playback experiments with vervet monkeys reveal that vervets understand the kinship relations of other members of their group. Similar experiments show that spotted hyenas recognize their own kin, but fail to show that hyenas recognize kinship relations in others. It is possible that the vervets may have a fuller understanding of the kinship relations of each animal in their group, whereas the hyenas may solve complex social problems by simpler rules of thumb than those employed by the vervets.

Spotted hyenas form clans with multiple adult males and multiple adult females. Females have a fascinating set of features typically viewed as masculine. They grow larger than males and are dominant to them. Hyenas in a clan may join forces to bring down a kill, but they then compete fiercely over the carcass, with the highest ranking female and her offspring having priority of access. Hyenas are communal breeders. Dominant females do not suppress reproduction in subordinates, but higher-ranking females breed earlier and have more offspring. As in cercopithecine primates (e.g., baboons and macaques), a spotted hyena female acquires her rank from her mother. In their case study, Engh and Holekamp review the process by which spotted hyenas acquire their rank, and they conclude that the process is similar in many aspects to that found in cercopithecine primates.

Rank is so central to access to food and reproductive success that many hyena social interactions involve fights to maintain rank. At times, two or more individuals (often littermates) will gang up to threaten a third party. Drea and Frank identify these as coalitions similar to those seen in many primate species. Although most work on coalitions has involved studies of Old World monkeys and apes, in Case Study 4A, Susan Perry discusses coalitionary aggression in the New World white-faced capuchin monkey. In Case Study 4B, Richard Connor and Michael Krützen describe detailed observations of coalitions in wild bottlenose dolphins, supporting the view that these behaviors are not limited to primates. As the scope of coalitionary behavior expands, Engh and Holekamp's question of whether different animals may employ behavioral strategies using different levels of cognitive complexity to engage in behaviors that look similar becomes especially pertinent.

Most of the animal species discussed in this book form dominance hierarchies through repeated agonistic interactions. Dominant animals have been viewed as winners in a competition for power and resources, with subordinates being unambiguous losers in social competition. The benefits of dominance are often obvious, but Scott Creel and Jennifer Sands take a nuanced view in Chapter 6, balancing the cost/benefit relations of dominance versus subordinate social status. They focus on species of mammals and birds that breed cooperatively, a situation in which dominant animals often monopolize reproduction.

It has often been thought that being subordinate is inherently more stressful than being dominant. When one houses unfamiliar rodents or primates together, fighting increases particularly among males. This increase is accompanied by increasing levels of circulating adrenal glucocorticoid (GC) hormones that stimulate physiological responses associated with stress. GC levels are higher in losers than winners in this situation. Early work on baboons also showed higher GC levels in animals of low or decreasing rank.

Creel and Sands report that dominance is associated with high endocrine indicators of stress in three species of cooperatively breeding mammal. They initially studied dwarf mongooses: dominants of either sex fight nearly three times as often as subordinates. This led them to predict that the higher rates of aggression among dominants led to the higher GC levels. This prediction is weakened by the observation that the association between rank and GC levels only holds for females in this species. Studies in African wild dogs and wolves also show a clear correlation between rank and GC levels, but there is not a stable association between aggression and stress in these species.

Creel and Sands conclude from this result that the stress response of an animal depends upon "knowing what an animal is thinking, rather than what it is doing." Some individuals may have personalities that make the same social situation more or less stressful. Creel and Sands also point out that animals with poorer abilities to understand a social situation—"is the rival nearby threatening or not?"—had higher chronic levels of stress than animals that could distinguish how threatening a social situation really was. Since stress has such a clear impact on survival and reproduction, this provides yet another selection pressure for social intelligence.

These three chapters move in scope from a comparison between chimpanzees and hominids to hyenas and other social carnivores. This section highlights the importance of testing precisely what different species understand about social contexts that appear to us to be similar. As Creel and Sands argue, it may not be enough to describe the situation objectively. One must know what the animal thinks about it. Engh and Holekamp point out that even though different animal species may be faced with social situations of similar complexity, they may deal with these problems with different levels of cognitive complexity. These broad comparisons should improve our understanding of general principles underlying the evolution of social behavior and also our understanding of how and why different animal groups have evolved different cognitive abilities to deal with similar problems of social life.

4

Complex Cooperation among Taï Chimpanzees

CHRISTOPHE BOESCH

The quest to understand the factors that led to the evolution of modern humans has placed a new emphasis on the importance of social factors (Wilson 1975; Trivers 1985; Foley 1995). According to this approach, the social demands of living in large and complex groups elicit the evolution of elaborate faculties unneeded in interactions with inanimate objects. This role of the social environment favors the appearance of special cognitive abilities of the type we see in great apes and humans (Humphrey 1976; Whiten & Byrne 1988; Byrne 1995; Mithen 1996). In primates, the size of the cerebral cortex correlates with the number of available social partners, and primate species with a larger cortex also use deception more frequently (Dunbar 1992; Byrne 1995). The coalition, in which one individual joins forces with another one, is a social behavior that we expect to occur only in species with a certain level of cognitive abilities that allow them to take into account the contributions of others. Hence, coalitions are more frequently observed in species with a relatively larger brain and the complexity of coalitions seems to be greater in primates than in other species (Harcourt & de Waal 1992; but see Connor and Krützen, Case Study 4B).

Cooperation occurs when at least two individuals act together in order to achieve a common goal. In a mutualistic interaction, each individual does better by acting in concert than alone. However, these definitions do not fully take into account how individuals cooperate. In fact, cooperative actions can take different forms. Boesch and Boesch (1989) distinguished four levels of cooperation: in "similarity" all hunters concentrate similar actions on the same prey, but with neither spatial nor time coordination among them. However, at least two individuals always act simultaneously. In "synchrony" hunters concentrate similar actions on the same prey and try to relate in time to the others' actions. In "coordination" hunters concentrate similar actions on the same prey and try to relate in time and space to others' actions. Lastly, in "collaboration" hunters perform different complementary actions all directed toward the same prey. Some studies illustrate the difficulty that may exist when two individuals coordinate their actions: Capuchins coordinate similar actions when acting in parallel, but seem unable to do so in a task where both participants face each other (Mendres & de Waal 2000; Chalmeau 1994; Chalmeau et al. 1997). A comparison of the hunting behavior in social carnivores and primates has revealed that collaboration is common only in a few of the known chimpanzee populations, whereas social carnivores rely on simpler forms of group actions (Stander 1992; Creel & Creel 1995; Boesch & Boesch-Achermann 2000). Because of this relative paucity of observation in nonhuman species, cooperation has been proposed to be a typical human ability with a central role in human evolution (Isaac 1978; Foley 1995; Mithen 1996). New analysis of hunter-gatherer societies has shown that the product of cooperative foraging, which often involves active participation of many individuals, results in major food intake for all individuals in the society (Kaplan et al. 2000).

Why is collaboration so rarely or never performed in many species, when it is so clear that the product of cooperation can be beneficial? Perhaps understanding the partner's role is difficult. One has to take into account what others do so that the combined actions can lead to success. In a comparative captive study, it was shown that whereas chimpanzees could collaborate by adopting different complementary roles, orangutans could only perform similar roles in a synchronised way, and capuchin monkeys had difficulties in coordinating their actions with those of others (Chalmeau 1994; Chalmeau et al. 1997; but see Mendres & de Waal 2000). Chimpanzees have been observed to collaborate in a variety of experiments as well as under natural conditions (Boesch & Boesch 1989; Chalmeau 1994; Chalmeau et al. 1997; Menzel 1974; Savage-Rumbaugh et al. 1978). The ability to take the perspectives of others into account has been proposed to be limited in most animal species, but pivotal in humans, which may explain the common occurrence of collaboration in humans (Carruthers & Smith 1996; Heyes 1998; Tomasello & Call 1997).

In this chapter, I will concentrate on both the payoffs of cooperative acts and the methods of collaborative hunting. We studied the behavior of the Taï chimpanzees, a population known to cooperate regularly in hunting, for two decades (see Boesch & Boesch-Achermann 2000). I will review how much group members gain from cooperative action and elucidate some of their underlying cognitive capacities. I shall limit my discussion about cooperation in wild chimpanzees to two contexts: the hunting of small mammals for meat and territorial defense against other groups.

Dependence upon Cooperation in Taï Chimpanzees

In a recent review, Kaplan and colleagues (2000) explored the contributions of group members to the diet of others in human hunter-gatherer societies. They showed that in those societies for which data are available, mature adult men tend to be important providers of rich food sources, mainly meat, to other group members, males and females alike. Women do also produce an abundance of food that they share with others, but their contribution seems to be energetically less important since the type of food is less rich (Kaplan et al. 2000). At the same time, based on crude assumptions, the authors compared chimpanzee estimations with the human data and proposed that in the former species all individuals eat only what they can acquire on their own and that cooperative actions play no role, nor does any individual significantly profit from the effort of others (Kaplan et al. 2000). This simplistic view of the chimpanzee does not improve our comprehension of species differences with regard to cooperation, but it raises the possibility that cooperation plays a different role in the food acquisition of the two species.

To elucidate this issue, I analyzed the food acquisition by Taï chimpanzees of two rich food sources: meat and nuts. The results are presented in Figures 4.1a and 4.1b and Figures 4.2a and 4.2b. The picture that emerges is that in chimpanzees, as in human hunter-gatherers, individual group members are very dependent on the food acquisition skills of others to gain access to rich foods that are either hard to acquire, such as meat, or hard to process, such as nuts. For meat, male chimpanzees over eight years of age consistently produce, or secure through their actions, more food than they consume. This meat surplus is distributed within the group: younger males and females of all ages receive a share (Figure 4.1a). We have shown that males eat

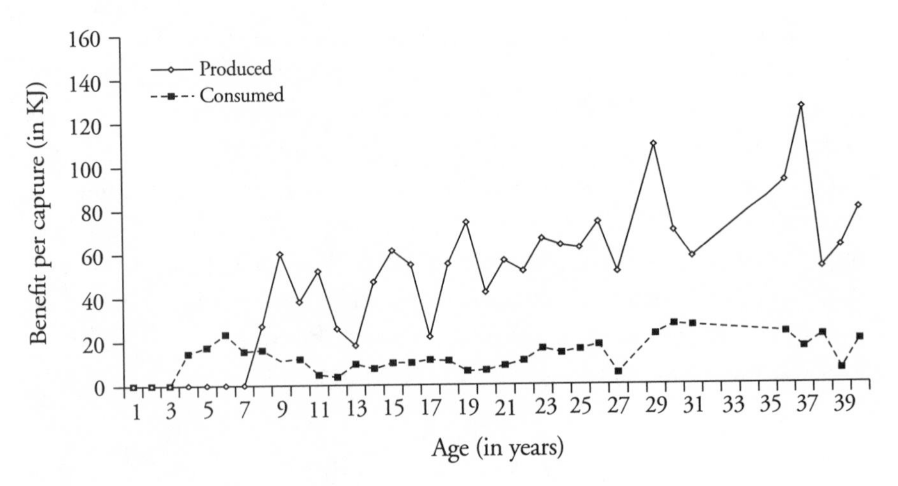

Figure 4.1a. Meat eating: amount produced and consumed by male Taï chimpanzees.

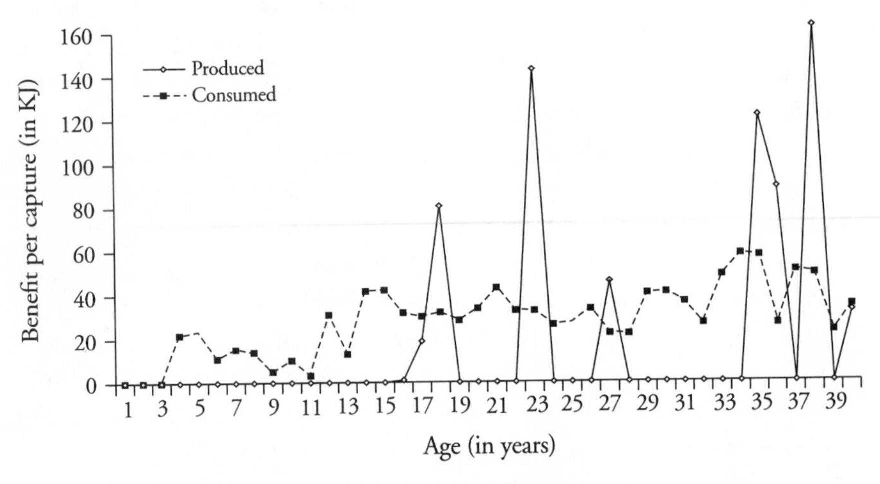

Figure 4.1b. Meat eating: amount produced and consumed by female Taï chimpanzees.

on average 186 grams of meat per day throughout the year, which is possible only through meat sharing within the group (Boesch & Boesch-Achermann 2000). Females rarely capture prey themselves, but are granted about 25 grams of meat per day throughout the year (Figure 4.1b). The distribution of meat is uneven among females; some eat four times the average. This result confirms what is generally known for chimpanzees: males are the main hunters and usually the owners of the largest pieces of meat (Goodall 1986; Teleki 1973; Stanford 1998; Mitani & Watts 1999; Uehara et al. 1992).

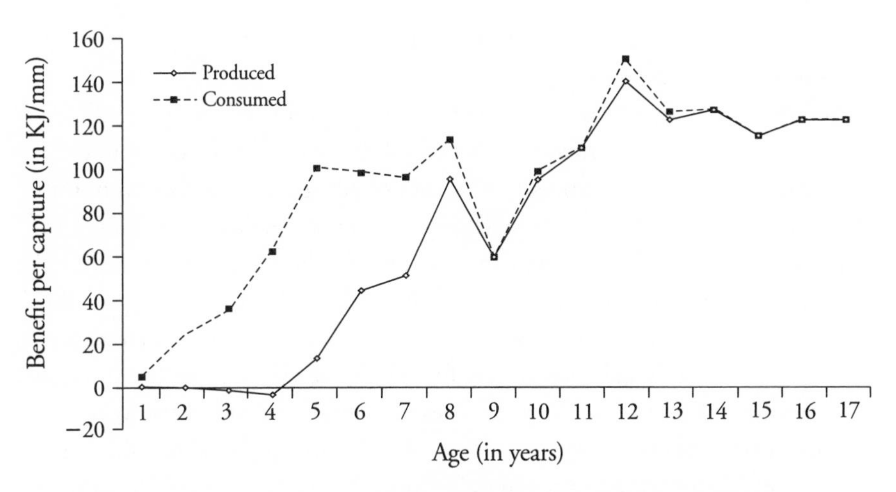

Figure 4.2a. Nut eating: amount produced and consumed by male Taï chimpanzees.

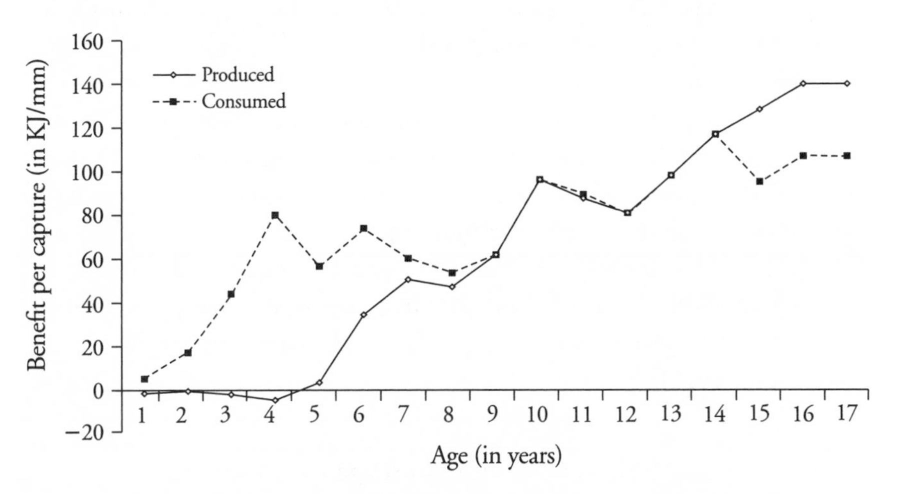

Figure 4.2b. Nut eating: amount produced and consumed by female Taï chimpanzees.

However, Figure 4.1a shows that adult males of all age classes consistently produce a surplus: older males are as successful as prime males. This is different from the human situation, in which a reduction of food production appears with older age (Kaplan et al. 2000; Hawkes et al. 1997). Meat is distributed throughout the whole social group—it is not restricted to kin (see also Boesch & Boesch-Achermann 2000). In addition, although adult female chimpanzees hunt rarely, they do sometimes contribute to the surplus production of meat within the group. Even if this is irregular, it is strikingly different from human hunter-gatherers, where the division of labor between the sexes seems more absolute.

For nuts, the situation is reversed. Females produce a surplus of nuts within the social group, whereas males eat all they produce once able to crack nuts (Figures 4.2a and 4.2b). Adult females of all ages use the surplus of nuts cracked to provide for their young (Boesch & Boesch-Achermann 2000). Taï chimpanzees are the only chimpanzees known to share food that is acquired through the use of tools (Goodall 1986; Nishida & Turner 1996): probably similar to what is known of human societies. Females support the development of male and female youngsters for the first eight years of life through nut production and help them to a lesser degree for additional years. As weaning occurs at an average age of five years in chimpanzees, this illustrates how maternal investment can extend beyond the typical nursing investment found in other mammals. Human mothers in hunter-gatherer societies like the Hazda and the Achè show prolonged offspring provisioning, but in the light of our data on Taï chimpanzees, this might not be unique to humans, as proposed by some anthropologists (Hawkes et al. 1997; Kaplan et al. 2000).

Despite the restriction of the present analysis to two food sources, it is clear that among chimpanzees food acquisition decisively depends on the contributions of other group members. Dependence on contributed food indicates that cooperation for meat acquisition was important in hominoid evolution—perhaps for a very long time. Such cooperation could only occur once the social relationships were tolerant enough to make food sharing possible. The despotic social relations seen in baboons and macaques seem to have prevented the appearance of regular food-sharing exchanges (Strum 1981; Teleki 1975).

Coordination of Actions in Group Hunt against Monkeys

During a hunt, often more than one chimpanzee takes an active part, and the participants may change as the hunt develops. In Serengeti lions, individuals have been observed to refrain from participating depending upon the presence of other hunters (Scheel & Packer 1991), whereas individual lions in Botswana have been shown to specialize in performing specific roles (Stander 1992). In chimpanzees, it can sometimes be difficult to follow what each individual does, either because of the limitation in the visibility of the habitat, or because of the high number of potential hunters (Stanford et al. 1994; Mitani & Watts 1999). However, in the Taï forest it is possible to follow all males present during a hunt, which has revealed that males vary their participation according to the situation, hunt for different amounts of time, and perform different roles (Boesch 1994; Boesch & Boesch-Achermann 2000). I will offer a detailed description of the actions performed by chimpanzees in the Taï forest, Ivory Coast, in two different contexts: a group hunt for monkeys, and macro-coalitions defending the territory. For both scenarios I address the following questions: How do the different individuals of the community perform a cooperative act? How do the young chimpanzees learn it?

Chimpanzees' main prey is the red colobus monkey, who spends most of its time up in the canopy of the highest trees of the forest. This species reaches 8 to 13 kg when adult. In the Taï forest, hunted red colobus did their best to escape from the hunters, using their great abilities to move and jump between trees. Chimpanzee hunters are four to five times heavier than a colobus, and in principle a colobus monkey could just sit and wait safely on a branch that is too thin to support the chimpanzee's weight. Chimpanzees need to overcome this disadvantage when hunting arboreal prey, and have been observed to do so in two ways. The first is by hunting for red colobus monkeys predominantly when the forest canopy is interrupted or irregular, so that by chasing the monkeys in the appropriate direction they have a fair chance of cornering them, at which point the likelihood of a capture increases. This seems to be the main solution adopted by chimpanzees when hunting in Gombe National Park and Mahale National Park in Tanzania, as well as in the Kibale forest in Uganda (Goodall 1986; Nishida et al. 1983; Takahata et al. 1984; Uehara et al. 1992; Mitani & Watts 1999).

The second possibility is to hunt in groups as a coordinated team so that the hunters can surround the prey even within a forest with a continuous canopy. This tactic is regularly observed in the chimpanzees of the Taï forest, Ivory Coast (Boesch 1994; Boesch & Boesch-Achermann 2000).

In the Taï forest, chimpanzees surprise their prey by approaching soundlessly, remaining on the ground, scrutinizing the vegetation for the colobus, and concentrating their attention on monkeys that are most numerous and lowest in the trees. When the colobus are really low in the trees, some of the chimpanzees may rush up in an attempt to catch one by surprise, which is rarely successful. Otherwise, one of them starts to climb slowly about five meters high, usually unnoticed by the colobus (a second chimpanzee may climb another tree in coordination with the first one, but this is rare). Other chimpanzees move on the ground in apparent *anticipation* of the possible escape routes of the colobus, ready to join the pursuit. When the monkeys spot the climber, he rushes up. The climber's contribution then consists mainly in keeping the monkeys moving in a given direction, while the other chimpanzees on the ground follow and undertake different blocking moves, checking regularly where the monkey moves. The *driver* usually follows the monkeys through the branches without trying to capture one on his own.

At this stage the colobus are usually still in a large group. The chimpanzees try to keep them moving in one direction. If they try to escape in two or more directions, a *blocker* may climb a tree to block an escape route by his mere presence—this chimpanzee does not move once he has adopted his position. As the hunt progresses, some chimpanzees take turns in performing driver actions as they climb up under the escaping colobus, while other chimpanzees assume a *chaser* role, attempting to catch a monkey by a rapid pursuit. The chimpanzees usually select and try to isolate a prey, often a mother with her baby, or a small group of individuals. Once they have separated these from the main group, the hunt accelerates, with chasers coming up from different directions. But the most difficult task remains: a chimpanzee must anticipate the location of the tree to which the colobus are going to flee, and be there to block them before the first monkey shows up. The chimpanzees that have chased the monkeys up to now have little chance to achieve a capture themselves. The *ambusher* is the hunter that anticipates the escape route of the quarry far enough in advance either to force it to turn backward toward its pursuers or to move downward into the lower canopy, where the chimpanzees have a good chance of catching it: in the continuous tree cover at this lower level chimpanzees run faster than colobus monkeys.

Anticipation is further complicated by the fact that the hunter not only has to anticipate the direction in which the prey will flee (recorded as a *half anticipation*), but also the speed of the prey, so as to synchronize his movements to reach the correct height in the tree before the prey enters it (recorded as a *full anticipation*). This is more complex than it sounds—the hunt happens in three-dimensional space, and the hunter has to convert the speed of the colobus into the distance he has to run ahead of the colobus, mostly out of sight on the ground, and to take in account the time he will need to climb a tree high enough to be able to lay an ambush, unnoticed by the monkey. For the human observer, to distinguish between half and full anticipation is relatively easy: a chimpanzee that makes a half anticipation typically waits for the colobus to confirm his prediction by staying at the base of the tree for the monkey to reach it and then rushes up as quickly as possible. Those making a full anticipation climb into the tree before any colobus enters it and do so slowly and carefully so as not to alert the monkeys to their presence by moving branches. We have also once recorded a *double anticipation* when a hunter not only anticipated the actions of the prey, but also the effect the action of other chimpanzees would have on future movements of the colobus; i.e., the male not only anticipated on the basis of what he saw (the escaping colobus) but also on the basis of how the movements of another chimpanzee hunter would influence the escaping monkeys.

Learning hunting behavior is a long process: most aspects of the hunt are progressively acquired over a 20-year period. Young males are keen to hunt, and there is a tendency for young males to perform more movements per hunt than the older ones. The proportion of ambushes and blocks performed by males correlates strongly with age. Older males performed such demanding roles three to four times more often than young males.

The correlation with age is equally strong for anticipations. Males performed half anticipations quite early in life. The frequency increased in old adolescents, and 18-year-old males performed half anticipations most frequently. The two oldest males used fewer of them, because when they anticipated, they did so fully. The frequency of full anticipations increased steadily with age and only the oldest males (over 30 years) used them frequently. Twenty-year-old males occasionally showed full anticipations, but also made errors either by selecting trees to which no colobus were coming or by changing their minds and climbing more than one tree during the same move. Long-distance anticipations, in which the hunter chooses a tree far ahead of the colobus, were seen only in older males so confident of their choice that, unlike younger ones, they did not need to keep an eye on the colobus. We saw only eight moves of double anticipation, and the two oldest hunters made all of them. However, we probably underestimated the frequency of double anticipations, for the observation conditions prevailing during the hunts usually made it difficult to determine that the conditions for double anticipation were fulfilled.

Thus, learning hunting behavior is an exceptionally slow process. It starts when the young males are eight to ten years old and then lasts for about 20 years. Cooperative tactics are learned very slowly because elaborate understanding of the prey's perspective and that of the other hunters is required to coordinate actions properly. The 20-year learning process seen in Taï chimpanzees is mirrored by an equally long process in human hunting as seen in Achè and Hiwi hunters who need a similar amount of practice before being fully proficient (Kaplan et al. 2000).

Coordination of Actions in Territorial Defense among Taï Chimpanzees

Taï chimpanzees have been followed during 129 territorial activities, of which 91 included at least auditory contact between the two communities over a period of 14 years (Boesch & Boesch-Achermann 2000). I shall here discuss only the reaction of the chimpanzees as they defend their territory to the presence of strangers. Two general points need to be stressed: First, territorial defense in chimpanzees is a group activity performed by adult males (Table 4.1). I never saw a lone male patrol or approach strangers.

Second, groups of males react as macro-coalitions adapting their behavior to the fighting power they represent: A small coalition of three or fewer males contents itself to carefully listen for strangers (i.e., "Check" in Table 4.1). A medium coalition of four to six males is more active and controls the border more systematically (i.e., "Patrol" in Table 4.1). Large coalitions are the ones that seek physical contact with strangers and do so in about two-thirds of all observations. If we do the same analysis but consider both adult males and females present in the party, no significant difference is observed. This suggests that Taï chimpanzees take into account the number of adult males and not the total number of chimpanzees that are present when selecting a strategy. At various times, we saw parties of four males waiting, we presumed, for more males to join them, but as none came, they limited themselves to checking or patrolling the area. Such a flexible response of macro-coalitions illustrates a decision process that takes into account both the likelihood of making contact with an unknown number of strangers and the absolute power of the coalition. In the course of our study, the community size declined, as did the number of adult males (Boesch & Boesch-Achermann 2000). As the males became less numerous, they changed their strategy:

Table 4.1. Territorial strategies used depending upon the number of adult males present

Strategy	1–3 males	4–6 males	7–9 males
Check/Avoidance	67%	35%	17%
Patrol	17%	37%	20%
Attack	17%	28%	63%
Total	18	76	30

Note: Statistics: $\chi^2 = 21.07$, $df = 4$, $p < 0.001$. Residuals: check/avoidance for 1–3 males = 2.22, attack for 7–9 males = 2.67.

when the number of adult males fell below eight in 1988, they started to be more careful, investing more time in patrols and less in direct confrontations. Strikingly, from 1992 onward avoidance behavior, in which the chimpanzees retreated in silence from the strangers, became the dominant strategy, and avoidance replaced the checks used before. This suggests that the chimpanzees were aware of the imbalance of power and avoided risks.

The macro-coalition of males within the community adapts its behavior to play the best game with the cards they hold. They seem to do so without precise knowledge of the power of their opponents. When the overall number of males at Taï decreased, and they became more careful when facing strangers, the frequency of encounters with neighbours nevertheless remained the same, and the territory size decreased only moderately. It seems that by patrolling to surprise strangers and by having direct confrontations with lateral attacks (see below), the males developed a confrontation strategy to hold their territory despite the reduction in intrinsic power. This was only possible to a certain limit. When the group declined further to four or even two adult males, males began to avoid confrontations. The higher frequency of avoidance tactics by small groups illustrates how the territory size decreases, and how neighbors progressively extend their territory without encountering resistance.

Once a coalition has established auditory contact with strangers, it can adopt different strategies to fight the intruders. In one-third of the auditory contacts, no visual contact was made between the two communities, but they exchanged aggressive drumming for up to one hour. When they heard strangers, males usually gathered to reassure each other, sometimes uttering little screams, and then moved silently toward the strangers. This distance to the strangers varied from 80 meters to 2 kilometers, depending both on the distance between the two communities when they were heard, and on the motivation of the males approaching. They then usually started to drum very loudly and maintained this auditory contact for up to about 30 minutes. Subsequently, they progressively moved apart, with the males joining the females that had stayed behind. Drumming during territorial encounters is especially loud with markedly more hits on buttresses than usual and is accompanied by typical climax screams following pant-hoots that contain many loud high-pitched waa-barks, which are absent in the pant-hoots done in ordinary social contexts.

In 48 percent of the auditory contacts, the chimpanzees of different communities came into visual contact with one another. The decision to seek visual contact with the strangers was taken very quickly, possibly within minutes of noticing them. The number of males present plays a decisive role (see Table 4.1). In all instances, the strangers that first saw the approaching chimpanzees uttered long loud screams and ran away. These screams have a special intensity. Within seconds the other community members know what is going on, and they immediately rush in support. In our sample, we never observed an attack without a counter-attack, and support was regularly provided by nearby community members. When the study community attacked, we distinguished five strategies (Table 4.2).

The Frontal Attack

The most straightforward strategy is the frontal attack. The coalition simply heads directly toward the strangers and when they see them, start to charge. These frontal attacks were mainly used when the study community contained more than eight adult males, which suggests that this is a strategy for strong communities. Except for the first emotional screams, the march toward an attack is always totally silent. The males walk closely one behind the other, regularly turning toward the following male with a fearful open-grin face, seeking reassurance. The same male regularly leads the procession throughout a whole attack. When more calls are heard from the strangers, they carefully listen and adapt their course so as to head directly toward them. When almost within sight of the strangers, the attackers slow down even further and fan out so as to charge in a line. Because we are following behind we cannot judge what exactly the chimpanzees see. They seem to judge the number of chimpanzees and, if the estimate is high, start the charge from far away and make aggressive waa-barks early. When they estimate a small number of opponents, they go much closer, start the attack silently, and try to catch them.

Table 4.2. Strategies used by Taï chimpanzees when attacking neighboring communities

Strategy	Number observed
Frontal	16
Back and forth	9
Rearguard	9
Lateral	6
Commando	5

During a frontal attack, all males rush toward the strangers, giving loud attack calls, powerful high waa-barks made with the mouth wide open but the teeth fully covered by the lips. We heard these attack calls only in attack and counter-attack situations. The surprise effect is impressive and attacked individuals always retreat first. These initial attacks are also the ones in which physical contact is most frequent: within a minute chimpanzees can be badly bitten. The most effective scenario is one in which the attackers hide and surprise strangers coming down from a tree where they were feeding. For example, two males from the southern community once caught Macho alone in a tree unexpectedly. An adult female and three males who arrived to support him within the next minute almost rescued him. However, Macho had already been bitten 19 times and he was bleeding from all the wounds. None looked particularly bad, but one bite had missed his eye by about one centimeter.

While retreating, the attacked individuals probably judge how close and how numerous the aggressors are. Sometimes they stop their flight after about 50 meters, at other times they flee for up to 400 meters. If the retreat is long, the situation is clear and counter-attacks are brief. The communities rapidly part. If, however, the strangers stop, the pursuers do so too and wait for reactions. The strangers might be more powerful and may chase the attackers for a long distance. In this case the situation is also clear and the communities part rapidly. If, however, the powers seem equal, probably judged by the mutual abundance of screams and the number of individuals that are seen (it is impossible in the forest to see all of them), then the frontal attack changes into a back-and-forth attack (see below). Females with and without infants may join a frontal attack or counter-attack, but they tend to avoid direct physical contact with members of the other community. We have seen mothers of the study community taking part in an attack. Stranger mothers may also join with stranger males to attack the study community. Because of the poor visibility in the forest, the chimpanzees do not always have time to check on the number of adult male pursuers and may run away, incorrect in their estimate of potential attackers.

Back-and-Forth Attack

In the case of a back-and-forth attack, both coalitions present are about equal in power and motivation and they attack mutually. In extreme situations, one can observe two lines with all the adult males and some females facing one another, the attacks alternating from one side to the other (Table 4.2). In other situations, the chimpanzees are more spread out in the forest, and we have seen parties of two to three males attacking the other side. These attacks seem to be coordinated vocally through attack calls. In two of these back-and-forth attacks, lasting over 20 minutes, the opponents calmed down, only facing and threatening each other. Five young estrous females quietly crossed the lines to join the other males, mated with one or two of them and returned calmly back to their community. These were the only cases in which we witnessed sexual activities during inter-community aggression. None of the females transferred for good, nor did these visits to our knowledge result in conceptions.

Rearguard Supported Attack

During encounters in which many females are present, the coalition fighting in front may receive the support of the females, who start moving toward the battle line making very loud waa-barks and drumming frequently (Table 4.2). The females listen to the calls of the males and each time their males attack they amplify their calls and rush forward. Often the adolescent males or one of the older males join in with the females and they produce a very impressive rearguard that progresses toward the front. If battle screams increase, the rearguard accelerates, and the young males may even rush to support the fighting males in the front. This appears to generate significant support for the males, and there is no way for the strangers to know for sure how many of the drummers in the rearguard are actually males that might provide physical support for the fighting males.

This strategy seems deceptive when some males purposely remain in the back, drumming and repeatedly calling loudly, while other males move silently toward the opponents. There the front males wait silently for the strangers advancing to surprise the noisy rearguard, unaware of the close presence of the silent males, who may then ambush unexpectedly.

Lateral Attack

Once the study community had fewer than nine adult males, Brutus, at the time alpha male and leader in most inter-community encounters, started to lead lateral attacks instead of frontal ones (Table 4.2). A lateral attack occurs when the advancing and strictly silent coalition aims its progression not straight toward the audible opponents but to one side. In this way, it avoids the noisiest and possibly also largest party and looks for silent individuals in smaller parties that they might defeat. In three such lateral attacks, they found a small silent party of strangers and chased them for many hundreds of meters. The main party, probably containing most of the males, knew by the screams that some of their community members were being attacked and had to head backward to support them.

The lateral attacks appear to be deliberate. While going sideways the males constantly look toward the direction of the noisy stranger calls, as if making sure not to encounter them. This strategy allows a community in decline to win fights that might have been more difficult to win in a frontal attack. Brutus, the best hunter of the community, initiated this strategy. He also made most of the full anticipations during the hunts and the only double anticipations we saw. Brutus might have understood that there were no longer enough adult males in his community for frontal attacks and therefore favored the lateral strategy.

Commando

The last strategy and the most dangerous one is the commando (Table 4.2). In a commando, a coalition of adult males makes a deep incursion into a neighboring territory looking for strangers. Such incursions may last up to six hours. When they find some strangers, the goal seems to be to scare them as much as possible, not to attempt to physically defeat them. Therefore, they pursue strangers as long as they run, but when they are counter-attacked, they simply run away until they stop being pursued and return into their territory. Such a commando is very silent and invests much time in searching for strangers. Twice we saw such parties find isolated mothers that they managed to keep prisoner, preventing them from moving away, biting and hitting them. The bites were concentrated on the head, shoulders, feet, and genital swelling, and did not appear serious.

Discussion

The analysis presented here shows clearly that cooperation could play a very important role in chimpanzees. Meat is eaten by the vast majority of group members only thanks to the collaborative effort of others and of generous meat-sharing rules. Nuts eaten by youngsters up to nine years old are usually processed by others, normally the mother. Both food sources are rich but difficult to acquire. For both, chimpanzees are strongly dependent on cooperation by group members. For most other food sources chimpanzees tend to provide for themselves. Thus, we see that, when needed, chimpanzees can develop an important cooperative network. But when do chimpanzees rely on cooperation for their food? First, mothers share hard-to-process food more readily with their offspring (Silk 1978; Nishida & Turner 1996). Such food-sharing by the mother may extend beyond the weaning period if the food is valuable enough for the offspring. For example, panda nuts are shared regularly with 12- and 13-year-old offspring (Boesch & Boesch-Achermann 2000).

Second, for food sources that are not easily produced by a lone adult, such as meat, chimpanzees develop a complex cooperation network that allows both the cooperation to be stable and the product of the cooperative actions to be widely distributed within the social unit (Boesch 1994; Boesch & Boesch-Achermann 2000). This network rests on mutualism for the hunters and on social trade-offs between males and females. It could in principle be used to produce other resources, if the circumstances were similar to those of meat acquisition. A comparison with Gombe chimpanzees shows that if the hunting conditions become easier, the cooperative network disappears, since it is not essential anymore (Boesch 1994). A similar trend has been observed in social carnivores—lions, wild dogs, and hyenas—in which cooperative hunting frequency diminished drastically when the conditions made hunting easier, e.g., in habitat with high prey densities and good visibility (Packer et al. 1990; Cooper 1990; Creel 1997; Boesch & Boesch-Achermann 2000). Human hunter-gatherer societies also rely on cooperative networks for food acquisition, although disagreement prevails among scholars about how this works. Some claim that sharing of the resources is forced on the producers, because access to large prey cannot be controlled (Hawkes 1991; Hawkes et al. 1997), whereas others argue that processes similar to reciprocal altruism are at work, under which good hunters gain more offspring or sexual partners and receive more support when ill, compared to bad or stingy hunters (Gurven et al. 2000; Kaplan & Hill 1985). We know that strategies in humans can be flexible: for example, men may trade parenting effort for mating effort when they have greater mating opportunities (Marlowe 2000). Thus, it seems that in social carnivores and humans the reliance on cooperative actions follows economic rules, by which the level of cooperation increases when it pays. In humans, a strict division of labor has been observed in most societies, which relies on cooperation, whereas in chimpanzees the solution seems to remain more flexible and is based on sexual specialization. One possible explanation for this difference is ecological. Humans live in habitats where larger home ranges are necessary to support the social groups, and where women with their dependent offspring cannot range as widely as men, leading to a strong sexual division of labor. Chimpanzees live in smaller home ranges, and female ranges in the rich tropical rainforest are similar to those of males, giving them the ability to produce most of their food if needed.

What cognitive abilities are needed to perform a cooperative act? Studies with captive capuchin monkeys illustrate this point nicely. When asked to pull two strings on an apparatus that displayed no hint of success until the task was completed, capuchins were not able to collaborate (Chalmeau et al. 1997). When asked to pull from the same side of a single tray they could see coming toward them (which indicated progress in the task), they were able to collaborate and shared more readily with their partner (de Waal & Berger 2000). So cooperation is more difficult if no clues are available indicating that progress is being made. In hunting, success is only achieved with a capture and no simple clue can predict it. In addition, experiments with human children have shown that imitation tasks which require children to copy not only the goal (touching the ear) but also previous sub-goals (such as using the left or the right hand) are very complicated. Children perform them accurately four years later than they are able to perform a task that involves just a single goal (Bekkering et al. 2000). This may explain why collaboration, the performance of different complementary roles involving different successive sub-goals, is more difficult to perform than synchrony, the performance of the same role by different individuals at the same time. Furthermore, collaboration in the Taï forest is often performed amid very low visibility, so that the contribution of others cannot be directly observed and must be inferred.

The ability to cooperate is limited by the understanding one has not only of the task to perform but of others' intention to perform complementary actions. The study of how chimpanzees learn hunting tactics demonstrated how difficult it might be for members of this species to collaborate. The anticipation of hunters must take into account the reactions of the prey, a different species. Hunters must also consider the influence other hunters may have on the prey's movements, and all while the prey is running as fast as it can to escape through the forest canopy, a low-visibility environment

(Boesch & Boesch-Achermann 2000). Full anticipation is a challenge for young chimpanzees that requires up to 20 years to master. Human hunter-gatherers encounter similar difficulties when hunting, especially in the Achè, for whom arboreal monkeys are a staple. Achè hunters also need 20 years of practice to reach peak efficiency (Kaplan et al. 2000).

Territorial defense is the other context where cooperation is of utmost importance, since it has been shown in many species that larger coalitions are more likely to win an encounter (Grinnell et al. 1995; Manson & Wrangham 1991; van der Dennen 1995). Lions hearing roars of strangers within their territory will approach them at a rate that depends upon the number of intruders and the size of their own group (Grinnell et al. 1995). Males approach more slowly when they are outnumbered. Animals tend to forage in groups larger than the expected optimal size if they face the possibility of an inter-group confrontation (Zemel & Lubin 1995). Despite such evidence, the presence of macro-coalitions in territorial defense has rarely been reported and, when it was observed, it was restricted to very simple strategies like frontal attacks with loud vocal exchanges (van der Dennen 1995). It has been suggested that the occurrence of lethal violence in this context is another common characteristic of humans and chimpanzees (Wrangham & Peterson 1996). The absence of any lethal violence in 20 years of observations in the Taï forest, despite very regular encounters between neighboring groups, indicates that chimpanzee populations may differ and that lethal violence is not always part of the chimpanzee repertoire (Boesch & Boesch-Achermann 2000).

To conclude, the frequency of use of cooperation in animals seems to be influenced by two factors. First, social cognitive abilities are required that allow each animal to take into account the actions of others. Second, cooperative actions will happen only in situations in which a benefit could be achieved. Group actions are fairly commonly observed in animals, but the performance of different complementary actions is less frequent. Chimpanzees use cooperative actions systematically in territorial defense, whereas they use it in hunting only when conditions are such that group hunting is necessary for success. In the case of hunting for meat, cooperative actions can be very successful and provide other group members with a significant amount of food.

CASE STUDY 4A

Coalitionary Aggression in White-Faced Capuchins

SUSAN PERRY

Until quite recently, little was known about the dynamics of social relationships in New World monkeys, and the vast majority of research on social complexity in primates was conducted on Old World monkeys and apes. Because of their extraordinarily large encephalization quotients (which can be calculated from data in Harvey et al. 1987), proponents of the social intelligence hypothesis would certainly expect to find social complexity and cooperation in the gregarious capuchin monkey, genus *Cebus.* Indeed, capuchins, unlike most other New World monkeys, form frequent coalitions in a wide variety of contexts, and it seems reasonable to assume that skill in forming and maintaining alliances is critical for enhancing reproductive success. Here, I present data from a wild population of white-faced capuchins, *Cebus capucinus,* studied at Lomas Barbudal Biological Reserve, Costa Rica, from 1990–2001.

Capuchins have an elaborate repertoire of gestures for coalitionary aggression (see Figure 4A.1). They form coalitions in a broad array of social contexts. About 25 percent of female-female coalitions versus conspecifics occur in the context of feeding competition (Perry 1995). Even when the monkeys are foraging in trees large enough to hold the entire group plus a few more individuals, the lowest-ranking male and female are typically barred from entering the tree by coalitions.

Figure 4A.1. The "overlord," the most frequent coalitionary posture in capuchins.

Redirected aggression toward subordinates is common in capuchins, and many coalitions occur in the wake of other fights or in other socially tense situations. Rank reversals always entail extensive coalitionary involvement from multiple individuals. Although the female dominance hierarchy is relatively stable and linear, changes occur when adolescent females ascend the hierarchy, and when key members of social networks die, leaving their former allies vulnerable to challenge (Perry 1996a; Manson et al. 1999). Unlike most Old World monkeys with despotic ranking systems, capuchin females do not appear to support their own daughters preferentially in the rank acquisition process (Manson et al. 1999). All adult males are dominant to adult females in most capuchin groups (Fedigan 1993; Perry 1997). The alpha male has a very distinct role from subordinate males, but the relative rankings of subordinate males are sometimes difficult to discern (Perry 1998a). Both of the two overthrows of the alpha male that were witnessed at Lomas Barbudal were bloody and involved large coalitions of both males and females. The alpha male was killed in one of these reversals (J. Gros-Louis, unpublished data), and in the other the alpha male was temporarily evicted but returned as a subordinate male later (Perry 1998b).

One coalitionary context is particularly noteworthy because of its parallel with chimpanzee coalitionary strategies (de Waal 1978; Nishida & Hosaka

1996): separating interventions. Separating interventions are quite common in capuchins, and although they are employed by both males and females, they are a particularly common strategy for alpha males (Perry 1998a,b). When the alpha male sees two subordinate males affiliating or engaging in a sexual interaction, he threatens one of the two and solicits aid from the other one, thus ending the interaction. The alpha male is remarkably inconsistent with regard to which males he supports, even over short time intervals, and this coalitionary fickleness, combined with the fact that the alpha male rarely permits subordinates to associate, may account for the lack of clarity in subordinate males' dominance relationships (Perry 1998a,b).

According to the standard definition of coalitions (de Waal & Harcourt 1992, p. 3), a coalition involves aggressive cooperation between group-mates against conspecifics. However, capuchins engage in the same motor patterns as found in true coalitions when they cooperate in menacing heterospecifics, and these pseudocoalitions are far more common than true coalitions. For example, 25 percent of all male-male coalitions are directed at monkeys, and the rest are directed at heterospecifics, including both predators and harmless creatures. Although all group members mob predators to some extent, adult males are the most frequent participants (Perry 1995; Rose 1998). When adult females form coalitions with adult males against potential predators, they do so preferentially with the alpha male (Perry 1997).

Capuchins form pseudocoalitions against harmless animals such as frogs, deer, anteaters, howlers, and primatologists, and sometimes even inanimate objects. For example, capuchins sometimes form vehement and prolonged coalitions directed at a pile of monkey feces, some innocuous-looking insect, a dead or sleeping animal, an egg, or a bare patch of dirt. In fact, these pseudocoalitions are far more common than coalitions against conspecifics. Pseudocoalitions most likely serve to communicate something about the relationship between the two participants. It is possible that they provide practice in coordinating aggression for future coalitions against other monkeys and predators. Because pseudocoalitions are often prolonged and can consume a fairly significant portion of the activity budget, the opportunity costs entailed by engaging in them may also make them useful signals of commitment to alliances. One of the most common contexts of pseudocoalitions is when a subordinate monkey is foraging peacefully and a dominant comes into view, heading his way. Then the subordinate will start threatening the harmless target and solicit aid from the dominant, who almost invariably responds by joining in the threats and perhaps forming an overlord, an aggressive coalitionary posture. Pseudocoalitions may be an effective tactic for subordinates to avoid displacements by dominants at feeding sites, because the dominant often leaves the subordinate to feed peacefully following a pseudo-coalition.

The last main context of cooperation is male-male coalitions versus extra-group males (Perry 1996b). During intergroup encounters, the females and juveniles flee while the resident males of each group form coalitions with one another against the males of the other group. Intergroup fighting sometimes results in severe wounding and is probably very important in preventing foreign males from immigrating. Immigration of a group of males generally results in the eviction of some or all of the resident males, and there are sometimes infanticides associated with turnovers in male membership (Fedigan et al. 1996, unpublished data; Rose 1998). Two infanticides have been witnessed at Lomas Barbudal (J. Gros-Louis, personal communication), and several infanticides have also been reported at nearby Santa Rosa (Rose 1994; K. MacKinnon and K. Jack, personal communication).

Partner choice in coalition formation follows a set of fairly rigid rules in capuchins. When a monkey witnesses a fight between a female and a male and opts to take sides, she or he sides with the female 85 percent of the time, despite the fact that males are dominant to females. In the fights that did not conform to this pattern, the victim was a low-ranking female and the male's supporter was a female who ranked higher than the victim. When fights break out between same-sexed partners, however, capuchins almost always support the dominant opponent. This is true in about 89 percent of male-male fights, and 98 percent of female-female fights, and most exceptions to this rule consist of an adult supporting a juvenile against an adult. Capuchins also tend to preferentially support monkeys with whom they have more affiliative relationships (as measured by "proportion of interactions that are affiliative"); in 85 percent of those fights in which a witness chose to join in the fight, she supported the monkey with whom she had the more affiliative relationship. All three of these "rules" (preference for females, dominants, and affiliative partners) exert independent effects on partner choice.

Capuchins clearly have complex social relationships and social organizations, and their coalitionary strategies parallel those of chimpanzees in many ways. The extent to which cognitive complexity is involved in creating the observed patterns of behavior is less clear, and further research in captive settings probably will be necessary to resolve this issue.

CASE STUDY 4B

Levels and Patterns in Dolphin Alliance Formation

RICHARD C. CONNOR AND MICHAEL KRÜTZEN

Real Notch is a male bottlenose dolphin (*Tursiops aduncus*) in Shark Bay, Western Australia. Since 1986 his nearly constant companion has been another male named Hi. What is the function of their bond? At least part of the answer lies in their mating relationships with females. Real Notch and Hi form consortships with individual females that may last for hours to over a month (Connor et al. 1992a,b, 1996). These consortships are often initiated and maintained with aggression, and the males cooperate to capture and subsequently herd the female. Like other stable pairs and trios of males in Shark Bay, Real Notch and Hi share association coefficients of 80–100 (Connor et al. 1992a,b, using the "half-weight" coefficient, calculated for two individuals, A and B, as $100 \times 2N_T/(N_A + N_B)$ where N_T is the total number of groups in which A and B are found together and N_A and N_B are the total number of group sightings for each individual, respectively; Cairns & Schwager 1987). Males in stable alliances are more closely related to each other than males from the population paired randomly (Real Notch and Hi are estimated to be half-sibs), but it is equally important to note that stable alliance partners are sometimes not related.

Each stable pair or trio of males spends considerable time with another alliance. In the main breeding season it is especially common to see two alliances (four to six individuals), one or both with female consorts, either together or shadowing each other up to a few hundred meters apart. This pattern is both offensive and defensive, because teams of alliances will attack other alliances to take their female consorts. Connor and colleagues (1992a) referred to the pairs and trios that consort females as "first-order" alliances and the teams of alliances that compete for females as "second-order" alliances. Intriguingly, although alliances herd only one female at a time, an alliance that already has a female may help an alliance without a female steal one. Rewards for such assistance could take the form of inclusive fitness benefits, mating "rights," or reciprocity in later thefts (Connor et al. 1992b). As with first-order alliances, males in second-order alliances are more related than expected by chance, but are sometimes unrelated.

A somewhat different pattern is exhibited by one observed 14-member super-alliance (Connor et al. 1999). A three-year study (1995–1997) showed that super-alliance males preserve both the first-order and second-order levels needed for stable alliances. Trios and occasionally pairs of super-alliance males consort females, and teams of these trios cooperate to attack other alliances. At each level super-alliance trios are labile, often changing composition between consortships, and the sheer size of the super-alliance (up to 14 males) is enough to overwhelm other groups (Table 4B.1). For example, an attack by one trio of super-alliance males put to flight the second-order alliance of Real Notch, Hi, and another male pair, likely because at the moment of attack at least eight other members began leaping in the direction of the conflict. The super-alliance males have been al-

Table 4B.1. Males in stable alliances compared with males in the super-alliance

	Stable alliances	Super-alliance
Number of consortships	16–24	10–30
Number of alliances	1–2	5–11
Number of alliance partners	1–3	5–11
Percent consortships with primary alliance	84–100	17–57
Mean male group size	3.6	6.1

Note: Males in stable alliances: 8 males, 62 consortships. Males in the super-alliance: 14 males, 100 consortships. The mean male group size was larger for members of the super-alliance ($N = 57$) compared to five stable alliances ($N = 115$; Mann-Whitney U Test, $p < .001$). Stable alliances were never documented in male groups of 10 or more individuals but 25 percent of groups with at least one member of the super-alliance contained 10 or more of the 14 males. (Reprinted with permission from *Nature.*)

lied for at least 14 years. Ten super-alliance males were photographed together in 1987, and at least 10 are still together in 2001. In contrast to stable-alliance males, super-alliance males are not more related than expected by chance.

Both patterns of alliance formation in Shark Bay, the stable first-order alliances that team up with other stable alliances, and the labile first-order alliances of the large super-alliance, may turn out to be extremes of a continuum. Differences in relatedness among males in stable versus labile alliances are intriguing but demand a larger sample size of groups of labile alliances. Simple ecological explanations for differences in second-order alliance size (14 for the single group of labile alliances versus four to six for stable alliances) are difficult to invoke because the two alliance types are found in the same habitat. The range of the super-alliance overlaps extensively with that of three pairs of stable alliances (Connor et al. 1999).

The potential significance of multilevel alliances for social intelligence is obvious: individuals might have to simultaneously consider the consequences of actions at each level of alliance they participate in. However, the cognitive burden of an additional alliance level depends on the complexity of interactions within each level, and this may vary considerably. Humans have a remarkable capacity to form nested levels of alliance with complex triadic interactions at different levels (see Connor et al. 1992b). For example, Yanomamo villages contain two levels of male alliance (male lineages and factions within lineages), but villages may also form alliances against each other on an opportunistic basis. By comparison, male chimpanzees exhibit two levels of alliance formation, one within and another between communities. Triadic interactions, although complex (de Waal 1982; Nishida 1983), are observed only within communities. Evidence for triadic interactions between chimpanzee communities would require observations of males from two communities forming an alliance against a third, the equivalent of an alliance between two Yanomamo villages.

The significance of the two levels of male alliance in Shark Bay (pairs and trios that herd females and teams of alliances that steal them) hinges on the fact that both levels occur within a social network (although probably not a closed one). This allows for triadic interactions not only within but also between alliances, as is found in humans (Connor et al. 1992b).

Triadic interactions, however, do not automatically yield complex strategies. For example, two levels of within-group alliances (each with triadic interactions) have been described in female-bonded Old World monkeys (Chapais 1992; see review in Connor et al. 1992b). First-order alliances consist of females with their matrilineal relatives; alliances of matrilines form the second level. Most of the alliance behavior in this two-level structure can be explained if the monkeys are following two simple rules: ally with members of your matriline against other females, and ally with higher-ranking nonrelatives against lower-ranking nonrelatives (Chapais 1992). Experimentally induced violations of the second rule indicate that females will opportunistically ally with lower-ranking individuals against those of higher rank, allowing the occasional change in what are otherwise stable relations among matrilines (Chapais 1992). These overthrows are described as "mob attacks," usually directed against one individual at a time. Thus, the complexity of interactions among female monkeys appears to differ between levels, being greater in second-order compared to first-order interactions.

Kinship, although likely important, is variable in dolphin alliances. Other "simple" rules that might contribute to our understanding of dolphin partner choice include size, fighting ability, and dominance rank (see Keller & Reeve 1994), but these factors remain unexplored. Even these simple factors may yield complex strategies when, for example, individuals compete for the same large or high-ranking alliance partner and the high-ranking individuals can take advantage of their popularity (e.g., Noe 1990). Such complex social strategies may be revealed in alliance shifts (e.g., Nishida 1983), especially when they occur independently of changes in size and so on.

Although the frequently shifting alliance partnerships in the super-alliance might suggest complex social strategies, a more parsimonious hypothesis is that members are essentially exchangeable for the purposes of consorting females. From a cognitive perspective, this might be accomplished with a simple equivalence rule, in which individuals are classified as members or not, and all members are treated the same (Chapter 7; Schusterman et al. 2000). This was not the case, however. Within the super-alliance males clearly preferred certain males as alliance partners and avoided others (Connor et al. 2001). Males also differed in alliance stability and males that formed more stable alliances spent more time consorting females. There was no correlation between association in first-order alliances and relatedness in the super-alliance. Together, these observations suggest a complex social structure within the super-alliance. Relationships with males outside of the super-alliance were neither simple nor consistent. We observed, albeit rarely, super-alliance males in resting or traveling groups with nonmember males they had attacked on other occasions.

What of Real Notch and Hi and other stable alliances? If shifts reflect complexities in alliance relationships (such as competition for preferred allies, see Harcourt 1992), then perhaps long-term stability reflects simpler alliance strategies. However, a long-term record of the alliance relations of Real Notch and Hi and their male associates reveals important shifts, especially between alliances. Between 1984 and 1985, Real Notch and Hi were members of two different but closely associated pairs. Their partners disappeared at different times in 1986. One left after it was attacked by Real Notch, Hi, and another trio the two had begun associating with. That trio had been closely allied with another trio, but from 1987–1989 the bond between the trios was fractured. Real Notch and Hi began splitting their time with both trios—on one occasion initiating aggression with one trio against the other.

The triangle between Real Notch and Hi and the two trios came to an end in 1989 with the disappearance of two members of one trio. The third male from that trio, Bottomhook, then joined Real Notch and Hi to form a new trio, which traveled with the other trio. However, in 1990 the other trio disappeared, leaving Real Notch, Hi, and Bottomhook alone. This void was filled quickly by another pair, who traveled with the three until 1994. Then Bottomhook and one of the males from the pair, Pointer, formed a pair, apparently at the other male's expense. In the first weeks following the shift, this male could often be seen trailing the others, as though trying to maintain his alliance membership. Following the disappearance of Bottomhook in 2000, Pointer rejoined his former partner briefly before again leaving him and joining Real Notch and Hi to form a trio. In sum, since 1985 Real Notch and Hi lost their original alliance parters, formed a pair, formed a trio with another male on two separate occasions, and had a total of 10 different first- and second-order alliance partners.

It seems unlikely that the overall complexity of social strategies among male dolphins would vary between stable and labile alliances in the same population. What probably matters most is not the absolute frequency of shifts but the cognitive challenges associated with maintaining and forming bonds that are useful, but risky. In the case of female Old World monkeys, relations within the first-order alliances, involving matrilineal relatives, are relatively risk-free. Greater risk and opportunity are found in alliance relationships between females of different matrilines. It follows that, outside of learning to recognize relatives and rank, the greatest cognitive challenges for the monkeys involve recognizing opportunities to elevate the rank of their matriline or the need to counter such efforts. We may yet find some combination of simple rules for alliance formation among male dolphins, but the shifts in membership at both levels, the presence of two patterns in the same population, and variable patterns of relatedness suggest that their strategies might be quite complex.

5

The Social Complexity of Spotted Hyenas

CHRISTINE M. DREA AND LAURENCE G. FRANK

This chapter focuses on the social intelligence of spotted hyenas *(Crocuta crocuta),* as inferred by the manner in which they solve daily problems arising from behavioral interactions. For comparative purposes, we frame our review in the context of evolutionary models of primate intelligence. We begin with a presentation of spotted hyena natural history, underscoring some of their unusual attributes, followed by a discussion of this species' life history variables, highlighting certain features shared with primates. The ensuing commentary on social organization and behavioral ecology centers on the balance between aggression and affiliation, and provides an account of various mechanisms that contribute toward maintaining group cohesion. Our final discussion of cooperative hunting and commuting addresses the cognitive implications of elaborate foraging strategies. Throughout, we consider aspects of the spotted hyena's behavioral repertoire that reflect the complexity of social interaction and the capacity for individual storage and retrieval of information about a changing environment. We propose that current hypotheses relating life history variables, feeding ecology, and social complexity to the evolution of primate intelligence should be tested against other taxa in which species display similar attributes.

The Evolution of Intelligence: Are Primates Unique?

In considering the evolution of intelligence, primates historically have been set apart from other mammals, with cognitive supremacy most often attributed to their large brains (Jerison 1973; Passingham 1978). Brain size, however, correlates with certain life history variables, such as an extended period of infant dependency, delayed sexual maturity, and longevity (Chapter 1; Parker 1990; Allman 1999; Kaplan et al. 2000). Moreover, various ecological demands and feeding strategies have been linked to cognitive development (Clutton-Brock & Harvey 1980), such as the increased need for spatial and temporal mapping in frugivores (Milton 1988) or reliance on manual food processing and extractive foraging in tool-using primates (Parker & Gibson 1977; Gibson 1986). Whereas some theorists hypothesize that the exploitation of environmental resources is the main selective force driving intelligence, others cite cognitive benefits arising from the Machiavellian exploitation of conspecifics (Byrne & Whiten 1988). Thus, the complexity of primate social systems, behavioral interactions, social strategizing, and scorekeeping might also be spurring intelligence (Jolly 1966b; Humphrey 1976; Cheney & Seyfarth 1990).

Whereas all of these attributes no doubt contribute to shaping the cognitive skills of primates, many are not unique to primates. Detailed comparisons of mammalian ethograms had dismissed early preconceptions about the exceptional complexity of primate behavioral repertoires (Eisenberg 1973) and more recently, as reflected by the chapters in this volume, perspectives on primate cognitive uniqueness are also changing as we learn more about the social systems of other animals. If intelligence evolved in primates to solve specific social and ecological problems, it follows that other mammals facing similar pressures might share comparable cognitive abilities. Thus, through a comparative approach, species similarities and differences can be profitably applied to test the generality of existing theories on the evolution of intelligence. Toward that end, we present some of the unusual traits and shared primate-like characteristics of a social carnivore, the spotted hyena. We discuss the relations among life history variables, complexity of social organization, and essential aspects of their behavioral repertoire (including cohesiveness, communication, and foraging), and explore the potential cognitive implications of these attributes.

The Natural History of Spotted Hyenas: Some Singular Attributes

The spotted hyena belongs to a small family, Hyaenidae, comprising only four extant species (Jenks & Werdelin 1998): the striped hyena (*Hyaena hyaena*), brown hyena (*H. brunnea*), spotted hyena *(Crocuta crocuta)*, and aardwolf (*Proteles cristatus*). Within Hyaenidae, mating and social systems are varied (Mills 1978): Aardwolves (Kruuk & Sands 1972; Richardson 1987) and striped hyenas (Kruuk 1976) are pair bonded, brown hyenas breed cooperatively within promiscuous (Mills 1990) or uni-male polygynous (Owens & Owens 1984, 1996) systems, and spotted hyenas are polygynous (Frank 1986b). Unlike their relatives, but in common with many cercopithecine primates, spotted hyenas show no pair bonding or paternal care.

The stable social groupings characteristic of primates are also found among social carnivores (Gittleman 1989). Thus canids and hyenids display long-term bonds, maintaining year-round integration of the sexes and adult age classes. Whereas most social carnivores live in small groups, however, spotted hyenas are extremely gregarious, forming multi-male, multi-female "clans" (Kruuk 1972). These clans are territorial and defend core space against intruders (Kruuk 1972; Henschel & Skinner 1991; Hofer & East, 1993a,b; Boydston et al. 2001). As in cercopithecine primates (Pusey & Packer 1987), clan females are philopatric, but natal males disperse (Mills 1985; Frank 1986a; Henschel & Skinner 1987; Holekamp & Smale 1998b).

Moreover, whereas other hyenids obtain their diet primarily through solitary foraging and scavenging (Mills 1978), spotted hyenas gain the majority of their food through active predation, including cooperative hunting of large mammals (Kruuk 1972; Bearder 1977; Tilson et al. 1980; Henschel & Tilson 1988; Cooper 1990; Henschel & Skinner 1990b; Mills 1990; Gasaway et al. 1991; Hofer & East 1993a; Holekamp et al. 1997; Cooper et al. 1999). Exceptionally strong jaws and a specialized digestive tract allow them to eat and assimilate entire prey, including the skeleton (Sutcliffe 1970; Skinner et al. 1986). Due to their hunting success and dietary efficiency, spotted hyenas can be extremely common in fertile areas where game is abundant. Clans in the sparsely populated Namib Desert might number as few as three animals (Tilson & Henschel 1986), but those in the prey-rich Ngorongoro Crater may number 100, with hyena densities running as high as 1.7 per square kilometer (Kruuk 1972).

Most remarkable, however, is the suite of masculine characteristics, including unique reproductive anatomy, of the female spotted hyena. Since the time of Aristotle, this species has stirred controversy for its reputed hermaphroditism (Glickman 1995). Females sport dramatically "masculinized" genitalia (Watson 1877; Matthews 1939; Neaves et al. 1980; Racey & Skinner 1979; Frank et al. 1990). An external vagina is absent and the clitoris is hypertrophied, fully erectile, similar in appearance to the male penis, and traversed by a central canal through which the female urinates, copulates, and gives birth. The developmental mechanisms of this extraordinary syndrome have been extensively studied (Glickman et al. 1987, 1992a,b, 1993, 1998; Licht et al. 1992, 1998; Yalcinkaya et al. 1993; Drea et al. 1998) and have significant implications for social behavior.

Similarly, whereas other hyenids are size monomorphic (Drea et al. 1999), female spotted hyenas weigh approximately 12 percent more than males, both in the wild (Kruuk 1972) and in captivity, where the sexes have equal access to food (Glickman et al. 1992b). The magnitude of this sexual dimorphism is unusual among species displaying similar reversals (Ralls 1976), especially because it is not a byproduct of differential nutrition, as was previously suggested (van Jaarsveld et al. 1988). Besides being morphologically masculinized, female spotted hyenas are behaviorally masculinized, showing the opposite of typical mammalian patterns of sexually dimorphic behavior (Glickman et al. 1992a, 1993). Females are more aggressive than males, dominate adult males in virtually all social interactions (Kruuk 1972; Tilson & Hamilton 1984; Frank 1986b; Frank et al. 1989; Mills 1990; East & Hofer 1991b; Holekamp & Smale 1993; Smale et al. 1993), engage in more rough-and-tumble play than males (Pedersen et al. 1990), and take the lead in territorial defense (Henschel & Skinner 1991; Holekamp et al. 2000; Boydston et al. 2001).

Masculinization of morphology and behavior gives females leverage in reproductive decisions (Drea & Wallen in press). The male spotted hyena shows extreme approach-avoidance conflict when courting a potentially estrous female (Frank 1986b; Mills 1990). His obsequiousness could complicate reproduction were it not for behavioral and physical accommodation by the female. Her seemingly passive body posture signals sexual receptivity (Glickman 2000; E. M. Coscia, unpublished data) and, to allow penetration by the male, she retracts her clitoris so that it is flush with the abdomen (Neaves et al. 1980). Because copulation through the small and inconveniently located opening of a peniform clitoris is challenging, mating requires female consent and collaboration (East et al. 1993; Drea et al. 1999).

Perhaps more important, social dominance allows females to gain the upper hand in resource competition. Although clan members may join forces to bring down large ungulates, they are fiercely competitive over the spoils. Many hyenas can quickly gather at a kill, attracted by the sounds of the feeding melee. Thirty gorging hyenas can eliminate an entire zebra *(Equus burchelli)* in less than 30 minutes, each animal attempting to feed as rapidly as possible (Kruuk 1972; Mills 1990). As the carcass diminishes, competition escalates. Lower-ranking individuals are forced off until the remnants are in the sole possession of the highest-ranking female and her offspring (Frank 1986b). Competition in smaller parties is similarly reflected by rank-related priority of access to the carcass (Tilson & Hamilton 1984). We are unaware of any mammal in which scramble competition for food is as intense as in the spotted hyena (Kruuk 1972; Holekamp & Smale 1990; Mills 1990). This competition has been hypothesized to underlie many aspects of their morphological and social evolution (Frank 1996, 1997).

Life History Variables Shared with Primates

Like many other mammals that have a slow life history, spotted hyenas additionally express a suite of traits theorized in the primate literature to drive cognitive evolution (Chapter 1). These involve key life history variables, such as body mass, longevity, litter size, gestation length or fetal development, maternal investment or infant dependency, age at sexual maturity, fecundity, and patterns of lifetime reproductive success. By comparing spotted hyena life history variables to those of other carnivores (particularly hyenids) and cercopithecine primates we aim to point out similarities and differences that may prove relevant for testing such theories.

Certain hallmarks of anthropoid primates include a relatively large body, long life (Harvey et al. 1987), small litters, and precocial infants (Harvey et al. 1987). Likewise, spotted hyenas are the largest (weighing 40–75 kilograms) and longest-lived members of Hyaenidae (van Jaarsveld 1993; Drea et al. 1999), with life spans of up to 19 years in the wild (L. Frank, unpublished data) and 41 years in captivity (Jones 1982). Longevity is not unusual among carnivores, but coupled with social stability it translates to overlapping generations. Moreover, spotted hyenas are considered to be the only precocial fissiped carnivore (Kruuk 1972; van Jaarsveld et al. 1988). Like other carnivores (Ewer 1973), female aardwolves, brown hyenas, and striped hyenas have multiple sets of functional teats. Their litters can number up to six cubs and they gestate for about 90 days (Drea et al. 1999). By contrast, female spotted hyenas have only one set of functional teats. They produce litters that usually comprise twins or singletons, and have a gestation of approximately 110 days (Kruuk 1972).

As a consequence of prolonged fetal development, spotted hyena cubs are also relatively large at birth (Figure 5.1) and unusually precocial, with eyes open and teeth erupted (Matthews 1939; Pournelle 1965). Moreover, neonates exhibit more physical coordination and mobility than normal, altricial carnivores. Most surprisingly, they fight violently at birth (Frank et al. 1991; Smale et al. 1995; Drea et al. 1996a). The "bite-shake" attack of neonates is

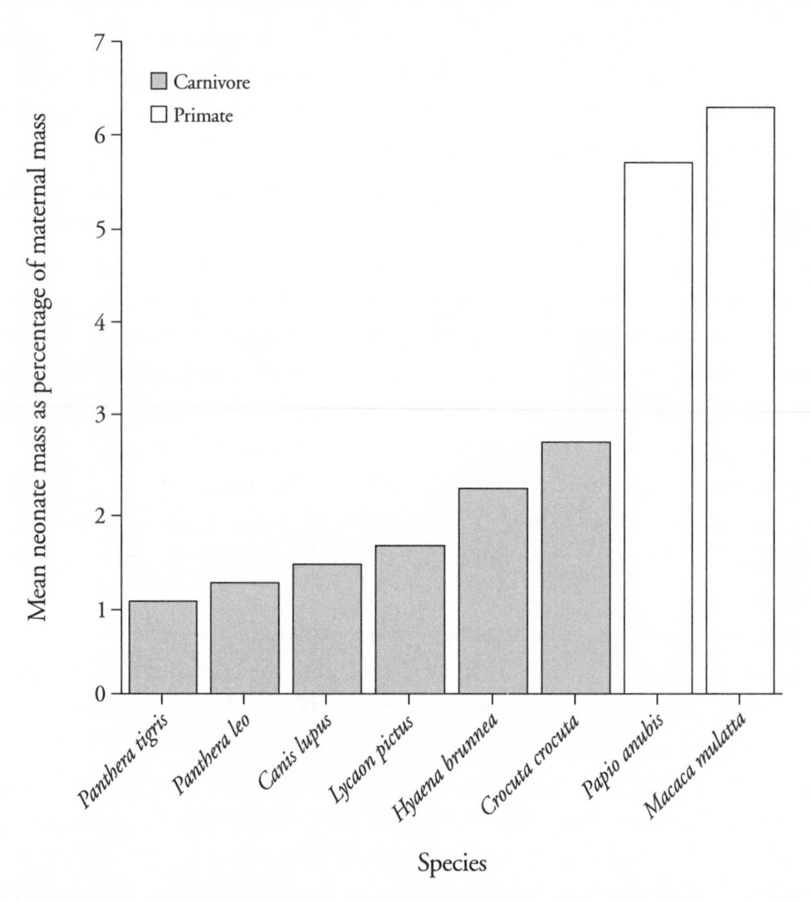

Figure 5.1. Mean birth mass of neonate carnivores and primates as a percentage of mean maternal mass. Carnivore data are from Oftedal and Gittleman (1989) and primate data from Hendrickx (1971), Bourne (1975), and Altmann and Alberts (1987).

strikingly similar to that of adults, so motor patterns of intra-specific aggression do not require extensive practice as in other species (Drea et al. 1996a). Neonatal aggression establishes intra-litter dominance within days or even hours of birth and may, in some cases, lead to infant death (Frank et al. 1991; Frank 1996; Golla et al. 1999; Smale et al. 1999).

Despite their comparative head start in life, hyena cubs have an unusually long period of dependency (van Jaarsveld 1993) involving various discrete developmental stages (Hofer & East 1995; Holekamp & Smale 1998a). Rather than being at their mother's bosom, as in anthropoid primates (Hinde & White 1974), much of their infancy is spent in a crèche or protective den (Kruuk 1972; East et al. 1989; Henschel & Skinner 1990a). Parturition occurs in private, usually at the mouth of a cave or abandoned burrow, which adults can enter only part way (Skinner et al. 1986), and is followed by extended periods of mother-infant contact and nursing above ground. During the mother's absence, young vulnerable cubs seek refuge from larger animals and shelter from the elements by crawling into (and further excavating) the small subterranean tunnels of their lair.

Cubs are initially sequestered at the natal den for two to four weeks. Thereafter, their mother transfers them to the communal den, the focal point of clan activity, where they and the nursing cubs of related females remain stationed for many months (Kruuk 1972; East et al. 1989; Henschel & Skinner 1990a; Holekamp & Smale 1998a). In striped (Kruuk 1976) and brown (Owens & Owens 1979, 1984; Mills 1990) hyenas, other clan members participate in feeding dependent cubs, either through allonursing or provisioning, and in mated pairs of aardwolves, the male guards the young (Richardson 1987). By contrast, in spotted hyenas there is little or no cooperative rearing (Kruuk 1972; Mills 1985; Holekamp & Smale 1990). Rare exceptions of allonursing illustrate unusual behavioral flexibility in this species (Knight et al. 1992). Although cooperative breeding is common in carnivores, its near absence in spotted hyenas is consistent with litter reduction and larger group size (Packer et al. 1992).

Time spent at the communal den involves an intense period of social learning; thereafter, young hyenas begin the weaning process and start exploring the physical environment at large (Hofer & East 1995; Holekamp & Smale 1998a). Juveniles will accompany adults on hunting forays at 6–12 months of age; nevertheless, cubs of lower-ranking females rely on maternal milk for at least 12–15 months (East et al. 1989; Mills 1990; Holekamp & Smale 1993). Other carnivores typically wean their young much earlier

(Ewer 1973). As spotted hyena milk has the highest protein content of any fissiped carnivore (Hofer & East 1995), it requires a correspondingly high output of maternal energy (Oftedal & Gittleman 1989). Thus, as in primates, the extended period of dependency in spotted hyenas constitutes an enormous drain on lactating females, reflecting both parental investment and the profound influence of maternal care on offspring survival (van Jaarsveld et al. 1988). An interesting paradox, however, is that female spotted hyenas can lactate and gestate concurrently (Kruuk 1972; van Jaarsveld et al. 1992).

Unusually for mammals, female spotted hyenas attain sexual maturity later than males (at three versus two years, respectively: Matthews 1939; Frank 1986a)—the delay in female puberty possibly owing to their unusual hormone profiles (Glickman et al. 1992b). Female age at first breeding ranges from 2.5 to 4.3 years (Frank et al. 1995a), which is comparable to patterns observed in cercopithecines (Pusey & Packer 1987). Nevertheless, the onset of successful reproduction is often postponed past sexual maturity, owing to frequent stillbirths in nulliparous females: squeezing an exceptionally large fetus through a ridiculously small opening during initial clitoral parturition can have lethal complications (Frank & Glickman 1994; Frank et al. 1995b).

In many social carnivores only the dominant female breeds, suppressing reproduction in subordinates (Rood 1980; Packard et al. 1985; Rasa 1987; Creel et al. 1991, 1997; Asa & Valdespino 1998). Among spotted hyenas, however, all adult females breed successfully, but the cubs of high-ranking females are more likely to survive to adulthood (Frank et al. 1995a; Frank 1996). Compared to subordinates, higher-ranking females begin breeding at younger ages, are more apt to support pregnancy and lactation concurrently, experience shorter inter-birth intervals, and are less vulnerable to fluctuations in the food supply (Holekamp et al. 1996). Similar rank-related differences in reproductive success are evident in cercopithecine primates (Altmann et al. 1996). In addition, as female rank in spotted hyenas is maternally acquired and maintained for life (Frank 1986b; Holekamp & Smale 1993), relative ranks of the clan's matrilines are remarkably stable over generations. Therefore the dominant matriline continually gains members while lower-ranking ones dwindle and disappear because of poor recruitment (Frank et al. 1995a). As female rank is central to reproductive success, much of the spotted hyena's social system revolves around aggression-mediated rank maintenance.

Because males lose their maternal rank upon dispersal at two to five years of age, the formula for male reproductive success is quite different. Joining a new clan at the bottom of the male hierarchy, young immigrant males are at first chased by residents, but if they persist, they gradually rise in rank as senior males disappear (Frank 1986a). Male mating opportunities increase with social rank, which in turn is a function of residence time in the clan, suggesting that lifetime male reproductive success correlates with length of tenure (Frank 1986b; East & Hofer 2001). That male reproductive choice reflects a preference for high-ranking females (Szykman et al. 2001) places an additional premium on dominant status in females.

Prolonged infant dependency, longevity, social stability, overlapping generations, and male presence are cited as key features contributing to the complexity that characterizes primate societies and enhances the learning environment (Humphrey 1976). For instance, lengthening the period of postnatal maturation provides the time necessary for infants to learn appropriate social and survival skills. Moreover, only seasoned individuals possess relevant information about environmental events that occur rarely enough to skip generations. As repositories of information, elders of both sexes play crucial roles in the welfare of younger group members (Kummer 1971). Through various modes of social learning, knowledge gained in the course of a lifetime's experience can be passed on to future generations, sometimes taking the form of cultural traditions. If the same logic relating life history variables and social complexity to enhanced learning environments applies across taxonomic groups then spotted hyena society should provide an interesting test case of this relationship.

Dominance Hierarchies and Social Organization

Dominance hierarchies describe patterns of enduring relationships between animals engaged in repeated social encounters in which status is not predictable from physical characteristics alone (Ralls 1976). Increasing predictability in the outcome of dyadic interactions reduces the need for harmful conflict. As animals must learn their place in society through gradual associative processes, hierarchies connote a level of complexity, involving individual recognition and long-term memory at the very least. Our comparative analysis of social complexity therefore continues with an examination of hyena social organization.

Although the trophic ecology of a predator on large mammals is entirely different from that of primates, spotted hyena social organization is strikingly similar to that of cercopithecine monkeys (reviewed in Holekamp & Smale 1991). Highly structured hyena clans comprise several matrilines and resident or transitory males, each sex showing a strong and consistent linear dominance hierarchy (Frank 1986a,b). As in strepsirrhine primates (Jolly 1966a; Richard 1987; Kappeler 1990), female spotted hyenas rule the roost; otherwise, the mechanisms of dominance are virtually identical to those operating in, e.g., baboons, vervets, and macaques.

Dominance among female hyenas is maintained by aggression against lower-ranking animals, in the form of displacements, threats, and biting attacks. Because appropriately timed submissive signals might prevent physical retribution, subordinates are frequently appeasing, even in the absence of overt aggression, suggesting an ability to predict behavior based on prior experience. Subordination notwithstanding, two or more individuals might gang up to threaten a third hyena, buttressing each other in an intimidating "parallel walk" that can escalate into violent physical attacks (Kruuk 1972; Zabel et al. 1992). Thus, like many primates (Harcourt & de Waal 1992), spotted hyenas form coalitions.

As in macaques, where sibling alliances are most common (Bernstein & Ehardt 1985), coalition partners in hyenas are frequently littermates, suggesting preferential kin support (Smale et al. 1995). Moreover, maternal presence during their offspring's aggressive interactions strongly influences the outcome. Thus, as in primates, rank is acquired through defensive maternal interventions and coalitionary support (see Case Study 5A; also Engh et al. 2000). Juveniles therefore gradually attain positions in the female dominance hierarchy adjacent to those of their mothers (Frank 1986b), with rank among peers and adult females established by about six months of age (Holekamp & Smale 1993). Although maternal influence suggests an important role for maternal experience, experimental studies have shown that a matrilineal social organization can emerge de novo among captive, peer-reared hyenas (Jenks et al. 1995). Thus, matrilineal rank acquisition may be an inherent outcome of relationships generated between mother-offspring units and other group members, independent of maternal experience. Likewise, a species-typical social system can develop in captive, peer-reared rhesus monkeys *(Macaca mulatta)* united after many years of social deprivation (Suomi 1991), further illustrating possible equivalence of social processes across taxa.

Coalition formation follows different patterns depending on the target of aggression. Within a clan, the rule of thumb for an onlooker is to join an attack as long as the targeted animal ranks lower (Zabel et al. 1992). The best strategy for the target, however, is redirected aggression—diverting aggressive attention onto an even lower-ranking dupe. The original victim thus solicits support from its own attackers and quickly reorients aggression by forming transitional coalitions. The orderly, trickle-down pathway of aggression reduces the risk of retaliation and reinforces the existing hierarchy. This typically unidirectional pattern of aggression may differ from that displayed by chimpanzees *(Pan troglodytes),* for example, for whom dominance relations vary over time. Alliances are therefore more fickle, and aid is often given to the victim of an attack (de Waal 1982; Hemelrijk & Ek 1991). Variance in coalition support may explain why rank reversals in primates can be more common. Hyenas might support an offense against a higher-ranking individual, possibly because "aiding-at-risk" provides a safer way to challenge the hierarchy (Zabel et al. 1992; Engh et al. 2000). Nevertheless, in spite of the obvious advantages gained by dominant status, hyenas rarely attempt overthrows, and rank reversals within clans are extremely rare.

Coalitions against intruders take the same physical form as they do within the clan: When potential allies are nearby a hyena will solicit aid by pressing against its neighbor's body; vocalizations serve to recruit distant supporters (East & Hofer 1991b; Henschel & Skinner 1991). Nevertheless, the rules differ insofar as all clan members rally in cooperative defense (Kruuk 1972; Boydston et al. 2001). Fierce border clashes or territorial wars between neighboring clans may involve scores of hyenas chasing each other across clan boundaries or involve costly, physical combat.

Dominance-subordination interactions, maternal rank acquisition, coalition formation, redirected aggression, preferential kin support, and defense of central living space are features of an aggressive behavioral repertoire that spotted hyenas share with certain primates (e.g., Bernstein & Ehardt 1985; Lee 1987). The question remains whether simple rules of associative learning (Heyes 1994), including the formation of equivalence classes (Chapter 7; Schusterman & Kastak 1998), can explain complex hierarchical classifications. Or are higher-order cognitive skills, such as associative transitivity, understanding of ordinal position (D'Amato & Colombo 1989) and relational concepts (Cheney & Seyfarth 1990) required to handle the sheer volume of social interactions? In primate troops and hyena clans of 100 or more individuals, the number of potential dyadic and triadic relationships is staggering. Cognitive mechanisms mediating the recognition of status or kinship relationships may be unclear, but similarities in social organization and behavior highlight the need for continued comparative studies.

Affiliation, Group Cohesion, and Other Social Behavior

Overt conflict erupting from the blatantly bellicose nature of spotted hyenas might be considered divisive, but behavioral mechanisms mediating social cohesion exist, although in more subtle form. This section examines aspects of the affiliative repertoire of spotted hyenas and explores how friendly interactions might cement social bonds, facilitate integration, maintain solidarity, and resolve social conflict, all of which permit continuance of social relationships throughout various stages of life.

Attachment

Because primate infants seek proximity to and comfort, shelter, and security from their primary caretaker, mothers and their infants form enduring attachment relationships (Bowlby 1969). Attachment is a special type of affectional bond (distinct from affiliatory bonds) that exists when the tie between two individuals is long lasting and noninterchangeable, such that separation causes emotional and physiological distress. Infant primates seek comfort from other relationships, but stress is primarily reduced upon reunion with the mother—the secure base—independent of normal care-giving activities such as nursing. These attachment relationships traditionally have been viewed as unique to Old World primates (Suomi 1995, 1999).

As is typical of species that occupy burrows, infant hyenas maintain proximity to the den rather than to their mother and, consequently, the former seemingly provides the safe haven (East et al. 1989); nevertheless, spotted hyena cubs also form attachments to their mothers (Rifkin 2000). For instance, temporary separation from the mother distresses young cubs, regardless of whether they are twins or singletons. The presence of a sibling is insufficient to mollify the effect of separation from the mother, and infant vocalizations that promote proximity to the mother increase during the period following reunion. Moreover, as cubs eventually physically outgrow the benefits of small dens, being imprinted on the lair could, in time, prove to be maladaptive (Holekamp & Smale 1998a).

Despite profound differences in mother-infant contact time between Old

World primates and carnivores, the mother-infant attachment bond in spotted hyenas may be qualitatively similar to that displayed by cercopithecines. In spotted hyenas, mother-infant attachment may facilitate social cohesion (e.g., ensuring infant transfer during frequent den moves: Holekamp et al. 2000) and may be pivotal in promoting costly, long-term maternal care and behavioral support. If so, rather than representing a recent evolutionary adaptation in primates (Suomi 1995), attachment may be characteristic of animals living in stable societies distinguished by life-long familial ties.

Social Play

Young spotted hyenas experience a dramatic transition and behavioral reorganization early in their social development that coincides with their transfer from the natal to the communal den (Drea et al. 1996a; Holekamp & Smale 1998a). At the cloistered natal den the cub's social environment is limited to contact with the mother and possibly a sibling, and initial sibling interactions are surprisingly hostile. By contrast, the communal den is a hub of social activity where interactions involve all clan members and the cub's behavior is generally appeasing. As the cub begins the process of socialization into the clan (Cooper 1993), its behavioral repertoire is likely to require more than fight and flight responses.

Indeed, once dominance is established between neonates, fighting grades into play (Drea et al. 1996a). In captivity, low-intensity prosocial interactions emerge as early as the second week of life and then gradually become more vigorous. By one month of age, cubs engage in more frequent bouts of rough-and-tumble play and therefore have an affiliative system in place by the time they are introduced to peers and older clan members. The emergence of social play in spotted hyenas occurs early by comparison with other carnivores (Bekoff 1974; Barrett & Bateson 1978; Biben 1983; Loeven 1994), including other hyenids (Owens & Owens 1978; Rieger 1981; Mills 1990). The timing suggests that, as in primates (Lee 1983; Fagen 1994), play in spotted hyenas serves both an immediate socialization function, facilitating integration into the clan, establishing friendships, and strengthening social bonds (Drea et al. 1996a), as well as a delayed physical function, enhancing motor skills (Holekamp & Smale 1998a).

Play may partially fulfill these roles by relaxing the rules of social conduct. In young captive hyenas, the dominant sibling initiates over 90 percent of aggressive interactions, whereas vigorous play is reciprocal, initiated equally by either participant (Drea et al. 1996a). A dominant hyena that might otherwise be intolerant of physical contact with a subordinate, in play will accept being plowed into, knocked over, wrestled to the ground, and gently mauled by that same underling. Likewise, typically deferential yearling males in captivity become more assertive during play and are invigorated by the playful behavior of females (Pedersen et al. 1990). It might seem paradoxical that play would be important to a species in which behavior is often constrained by aggressive enforcement of individual dominance status, but not if one considers the need for balance. To the extent that the rules of play reflect reciprocity and trust, and may provide insights into the evolution of primate social morality, the implications of fair play should be examined from a comparative perspective (Bekoff 2001).

Greeting Ceremonies and Reconciliation

Unlike cercopithecine societies, in which troop members are never far from one another, spotted hyenas have a fission-fusion society, such that clan members spend significant periods of time alone and isolated (Kruuk 1972; Holekamp et al. 2000). Such fragmentation allows individual hyenas to forage for smaller food items that cannot be shared, such as carrion or small antelope. Foraging alone also increases the likelihood of encountering prey, which then allows the lone forager to recruit help if needed. Thus, whereas cercopithecine troops are continuously cohesive, spotted hyena clans repeatedly fragment and coalesce and have developed means of negotiating the transition from solitary to social existence (Kruuk 1972; Glickman et al. 1997).

Spotted hyenas that spend time alone or in small groups eventually reconvene at the communal den, where they engage in ritualized greeting behavior (Kruuk 1972; Mills 1990; East et al. 1993). These ceremonies involve animals of either sex standing head to tail, with legs lifted in reciprocal presentation of the external genitalia and anal scent glands. To fully appreciate the significance of this act one must consider the details of spotted hyena reproductive anatomy, as virilization of female genitalia allows both sexes to use the erect phallus in social displays. These organs are mutually inspected according to strict rules of social etiquette (East et al. 1993). Subordinates must lift their legs first, exposing their entire reproductive future to the bone-crushing teeth and powerful jaws of a higher-ranking animal.

Despite the destructive potential of this intimidating act, elaborate greeting ceremonies serve a socially cohesive function, much like the grooming behavior of primates. Like grooming in some monkeys (Seyfarth & Cheney 1984), hyena greetings involve kin more often than nonkin (East et al. 1993). Moreover, hyenas reunited after even a brief experimental separation of a few minutes will readily initiate a greeting (Krusko et al. 1988; Drea et al. in press a). As an overt and mutual expression of the participants' relative status, the greeting ceremony offers a means of repeatedly advertising and reaffirming social relationships (East et al. 1993).

The greeting ritual, like various other behavioral displays, may also serve a conciliatory function in conflict management. Signaling the end of a conflict (Silk 1997) or mending social bonds and reducing tension following a dispute are crucial processes for the social well-being of primates and other animals (Aureli & de Waal 2000). Primate antagonists might extend an arm, kiss, mount, rub genitals, or groom one another to resolve their differences. Likewise, spotted hyenas that engage in aggressive interactions are subsequently more likely to initiate greeting ceremonies with their former opponents (Hofer & East 2000; see also Case Study 9A). This correlation suggests that greetings might also provide a means of repairing social relationships and achieving balance between aggression and affiliation.

Integrating Three Types of Social Communication

Primate societies derive complexity from the repetitive nature of social interactions with known individuals. It follows that species with similarly complex societies must have means of recognizing each other (Chapter 13). Whereas cercopithecine primates are diurnal and rely heavily upon visual cues, spotted hyenas are crepuscular (largely to avoid hunting in the heat of the day) and rely both on visual and olfactory cues. Accordingly, spotted hyenas have an elaborate system of communication that combines visual, vocal/auditory, and olfactory modalities.

Visual Communication

Spotted hyenas apparently recognize each other by sight at a distance of several hundred meters (Kruuk 1972). They may use individual shapes, postures, and gaits or may rely on unique coat patterns. If human observers can use spot distribution to identify individual hyenas (Frank 1986a), it is likely that their subjects can too. Studies of spotted hyena retinal morphology show that spot recognition is well within the visual capabilities of this species (Calderone et al. 1995).

As in primates and other carnivores, hyenas use various body postures, tail positions, and facial expressions to signal dominance or submission (Figure 5.2), and aggression or affiliation (Kruuk 1972; East et al. 1993). Specific actions, postures, gaits, or grimaces also precede play bouts (Drea et al. 1996a), functioning as signals that communicate the playful intention of a potentially harmful act (Bekoff 1972, 2001). Human observers perceive redundancy in functionally similar displays that may instead suggest the potential for subtle contextual differences in visual cues. This potential merits further attention in primates (Maestripieri & Wallen 1997) and other animals.

An issue that has long intrigued primatologists is whether nonhuman primates communicate about their environment and if visual signals can be modified by experience. If so, can visual signals be concealed or used to give false information? In a replication of Menzel's (1974) classic study of chimpanzee spatial knowledge and nonvocal communication, Sonja Yoerg (unpublished data) found that spotted hyenas were seemingly deceptive about

Figure 5.2. The profiles of an aggressive interaction between two captive, adult spotted hyenas at the Field Station for Behavioral Research, University of California at Berkeley. The aggressor *(right)* is standing tall, with head lifted, ears pointing forward, mane piloerected, and tail up. By contrast, the recipient *(left)* is crouching, with head lowered, ears down, hair flat, and tail tucked under. The recipient's mouth is open in a characteristic appeasement grimace. Photograph by C. M. Drea.

their knowledge of the environment, depending on the social circumstances. When a dominant hyena was informed about the location of food hidden among various potential caches, she approached the baited cache directly, whether alone or accompanied by naïve group members. By contrast, a subordinate hyena tested under identical conditions initially led naïve group members astray, and later surreptitiously returned to the baited site to claim the prize. Mangabeys *(Cercocebus torquatus torquatus)* tested on the same paradigm displayed similar behavior after several trials, making it difficult to distinguish between learning and an understanding of deception (Coussi-Korbel 1994). Although the cognitive mechanisms of deception remain poorly understood, at the very least, spotted hyena visual communication, as in primates, can be subtle and socially modulated.

Spotted hyena behavior can also follow simple rules, such as "do what the other hyena is doing" (Glickman et al. 1997). Joining in the activity of another individual is a particularly salient feature of hyena behavior (Kruuk 1972). Contagion, synchrony, and social or response facilitation are apparent during coalition-formation (Zabel et al. 1992), eating (Yoerg 1991), greeting, scent marking (Woodmansee et al. 1991), play (Pedersen et al. 1990), cooperation (Drea et al. 1996b), rolling, drinking, urinating, and even regurgitation (Glickman et al. 1997). The same rule may also explain the "social rallies" preceding group travel and the coordination of den moves (Holekamp et al. 2000). Such coordination is typical of social carnivores and may serve to promote interdependencies, strengthen bonds, and maintain cohesion (Kuhme 1965; Zabel et al. 1992; Glickman et al. 1997). Given the ubiquity of social facilitation in spotted hyena behavior, perhaps the most revealing cues of social manipulation are to be found among the subtleties of individual variation—when animals do not follow the rule.

Vocal Communication

Spotted hyenas have a rich vocal repertoire, containing a dozen calls, all of which inter-grade (Kruuk 1972). The long-distance "whoop" is individually recognized by other clan members and serves to display identity, request support, or advertise caller location (East & Hofer 1991a,b). Mothers respond to the distress calls of their own young, ignoring those of unrelated cubs (Frank 1986b), which demonstrates that they can sense individual variation in infant vocalizations (Rifkin 2000). The latter observations have been confirmed through playback studies of prerecorded cub whoops played to mothers and other breeding females (Holekamp et al. 1999; see also Case Study 5A). Although cercopithecine primates recognize third-party social relationships among group members based on vocalizations (Cheney & Seyfarth 1990), to date comparable evidence in hyenas is lacking (Holekamp et al. 1999). Nevertheless, the richness of the hyena vocal repertoire suggests that further research in contextual modulation of vocal communication may be fruitful.

Olfactory Communication

In contrast to cercopithecine primates, odors play a prominent role in the lives of carnivores (Ewer 1973). Spotted hyenas have a keen olfactory sense and rely on odors to locate food, assess the environment, navigate, and warn of danger (Kruuk 1972; Mills 1990; Drea et al. in press a). For instance, differential responses to categories of environmental odors suggest that captive hyenas discriminate or classify odors (Drea et al. in press a). In addition, spotted hyenas use olfactory cues for intraspecific communication and to synchronize or facilitate group behavior, define territorial boundaries, advertise ownership, and maintain contact or solidarity (Kruuk 1972; Mills & Gorman 1987; Mills 1990; Henschel & Skinner 1991; Woodmansee et al. 1991; East et al. 1993; Holekamp et al. 2000).

Intraspecific olfactory communication among animals of both sexes is accomplished via an elaborate scent-marking repertoire involving long-lasting, composite signals (Kruuk 1972; Mills 1990). Hyenas urinate and defecate in latrines, and as their calcium-rich droppings are resistant to degradation, scat remains visible for up to 14 months (Bearder & Randall 1978). Hyenas also deposit odor from interdigital glands by scratching the soil vigorously enough to leave scented furrows (Kruuk 1972). Finally, hyenas possess a pair of large supra-anal sacs that produce a thick, creamy, glandular secretion or paste (Matthews 1939; Buglass et al. 1990). These sacs open into an anal pouch that can be everted to deposit its contents. In a behavior termed "pasting," hyenas use a stereotyped pattern of walking in a semi-squatting position, with the tail curved up, and the anal pouch bulging over the item being marked (Kruuk 1972). Pasted scent marks are visible from a distance of several meters and can remain potent for a month or longer (Mills et al. 1980; Apps et al. 1989).

Olfactory cues are often deposited during border patrols, when hyenas cruise the boundaries of their territory pasting and adding feces to communal latrines along the way (Kruuk 1972; Boydston et al. 2001). Patrolling animals are goal-directed: Scent marking during these outings is not haphazard and all the while clan members scan the horizon for trespassers. Patrols involve both preventative and exploratory behavior, serving to minimize the risk of costly wars by fortifying existing boundaries and to assess neighboring territories. Silent incursions (Henschel & Skinner 1991; Boydston et al. 2001) possibly explain how newly vacated space quickly becomes occupied. Patrolling, which likely requires cognitive mapping skills, is common in social carnivores (e.g., Sillero-Zubiri & MacDonald 1998) and also has been described for chimpanzees (Wrangham & Peterson 1996).

Gosling (1982) argues that for scent marking to function in territorial advertisement, marks should be individually identifiable. Indeed unique odor cues are contained in hyena paste (Mills et al. 1980; Apps et al. 1989; Richardson 1990). Moreover, these scent signatures convey detectable information on an individual's familiarity, sex, and identity, as reflected by differential responding to various classes of conspecific odor in aardwolves (Sliwa & Richardson 1998), brown hyenas (Mills et al. 1980), and spotted hyenas (Figure 5.3; Drea et al. in press b). Variation in scent marking behavior suggests a variety of functions. Whereas scent marks deposited along territorial boundaries likely advertise ownership between social groups, when deposited within the territory they may convey information to clan members on indi-

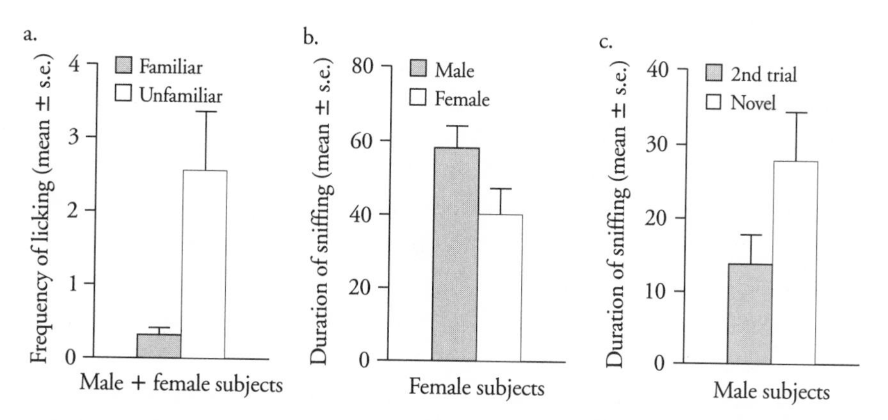

Figure 5.3. Responses of captive spotted hyenas to conspecific paste derived from the anal scent glands of various classes of individuals. *a)* Male and female subjects licked the paste of unfamiliar animals more than that of familiar animals, $t_7 = 2.54, p < .05$, regardless of sex. *b)* Female subjects sniffed the paste of males longer than the paste of other females, $t_3 = 10.69, p < .005$, regardless of familiarity. *c)* When presented with the paste of various unknown females, male subjects sniffed samples encountered for the first time (novel) longer than samples encountered only once before (2nd trial), $t_3 = 4.55, p = .02$.

vidual reproductive status (Drea et al. in press b) or recent foraging activity (Gorman & Mills 1984).

Unlike scent marking that transfers the animal's odor to its surroundings, rolling transfers odors from the environment to the animal's body (Rieger 1979). Rolling in stinky matter is common in carnivores (Brown & Macdonald 1985) and spotted hyenas find this activity particularly delightful (Kruuk 1972). Rolling may allow hunters to camouflage their own odor, to better sneak up on their prey (Rieger 1979), but spotted hyenas neither stalk nor lie in wait like cats. Moreover, they show a preference for potent animal-based odors (Drea et al. in press a), including their own regurgitated hair masses—choices that would function as a beacon rather than a blind. Instead, rolling appears to confer a social advantage (Drea et al. in press a). Regardless of social status, hyenas that roll in potent odors become more attractive to conspecifics, thereafter enjoying the benefits of affiliative interactions and allogrooming—privileges typically reserved for dominant animals.

Humans are endowed with relatively limited olfactory abilities and a narrowly defined auditory range, so we are naturally inclined to recognize our own attributes in other animals, particularly in Anthropoid primates that evince similar subtleties in visual communication. This bias should not prevent us, however, from appreciating the complexities involved in other modalities.

Cooperation, Commuting, and Conflict in Feeding Strategies

We have seen that, like primates (Walters & Seyfarth 1987), hyenas display a delicate balance between aggressive and affiliative components of their social behavior (Smale et al. 1995; Glickman et al. 1997). Here we further address the seeming paradox of this highly competitive yet cooperative species by investigating foraging behavior. Like many primates, social carnivores live and forage in groups. Unlike most primates, however, the food supply of carnivores is capricious, mobile, and retaliatory. Moreover, because migratory prey are ephemeral, hyenas must often travel great distances through hostile territory in search of food. Here we highlight the complexities of cooperative hunting, foraging for transitory resources, and contest competition.

Cooperative Hunting

As obligate carnivores spotted hyenas consume approximately four kilograms of food per day (Henschel & Tilson 1988). The proportion of carrion in the total biomass consumed can be as little as 5 percent (Cooper et al. 1999); the rest is obtained by hunting. Like wolves (*Canis lupus:* Mech 1970), spotted hyenas are cursorial hunters, depending on speed and endurance to run down prey, bringing their victim to the ground by tearing at the flanks and belly until it falters. Whereas a single hyena can successfully hunt wildebeest *(Connochaetes taurinus)*, a hunting party can bring down more formidable prey, such as the massive African buffalo *(Syncerus caffer)* or the cantankerous zebra. Although solitary hunters are capable of amazing feats, increasing party size significantly improves hunting success (Holekamp et al. 1997).

Long life spans and enduring relationships are crucial to cognitive development, and are reportedly essential for successful group hunting forays in primates. For instance, collaborative hunting behavior of chimpanzees in the Taï forest requires many years to perfect (Boesch & Boesch 1989; see also Chapter 4) and human hunter-gatherers only master food-acquiring skills after years of practice (Kaplan et al. 2000). Likewise, a learning component is involved in perfecting the hunting skills of spotted hyenas, with young animals only achieving adult competency after five to six years (Holekamp et al. 1997). The importance of learning is further evidenced by significant plasticity in hunting behavior. Notably, spotted hyenas tailor the number of hunters in their parties to match the type of prey pursued (Kruuk 1972; Cooper 1990). Gazelle (*Gazella thomsoni*) are most often hunted alone, parties of 4–20 animals will attack zebras, and a rhinoceros *(Diceros bicornis)* is always assailed by many hyenas simultaneously (Kruuk 1975). Most striking about their group hunting is that the target appears predetermined—the hunting party forms long before the quarry is selected and while other prey species are present (Kruuk 1972). Cognitive skills involved in such preparatory behavior likely include prey representation, predicting profitability, and intra-specific communication.

The demographics of the hunting party also influence behavior. For instance, Tilson and Hamilton (1984) report an interaction between hunter status and prey size, and suggest that subordinates should participate in hunts of prey large enough to satisfy the needs of all hunters because they are completely excluded from small carcasses. In addition, Holekamp and colleagues (1997) found that kin tend to hunt together and low-ranking females hunt more frequently than higher-ranking females. These observations suggest that differential access to the kill may influence party composition and hunting rates. Various prey species also adopt different escape and anti-predator behavior, and require different hunting or killing tactics, including variation in the degree of cooperation among hunters. Individual learning is clearly involved in adapting to unique features or local distribution of resources. However, one needs to consider social learning to explain why certain clans are more staunch hunters of one type of prey than are neighboring clans if preferences are irrespective of prey density or location (Kruuk 1972). Similar "cultural traditions," so well studied in primates (Whiten et al. 1999), have been reported for wild dogs (*Lycaon pictus:* Malcolm & van Lawick 1975) and wolves (Haber 1996).

Cooperative Problem-Solving for Food

Behavioral ecologists studying carnivore group hunting and psychologists or anthropologists studying primate cooperative problem-solving typically address the same phenomenon—animals working together to obtain food—from two very different perspectives. Whereas the former approach typically frames questions in a game theoretical context (Packer & Ruttan 1988), the latter focuses on the cognitive implications of cooperative behavior (Chalmeau & Gallo 1996). This section presents research that attempts to bridge the taxonomic divide by modeling the group-hunting behavior of carnivores as a cooperative problem-solving task and analyzing performance from a cognitive perspective. Accordingly, for hyenas to be cooperating, partners must understand each other's purpose and exchange information to adjust to each other's behavior. Their actions must meet the three criteria required of primates: common goal, reciprocity, and communication (Chalmeau & Gallo 1995). While pursuing the same goal, the level of behavioral organization among the participants can vary in temporal and spatial complexity, increasing in complexity from simple similarity of action, to synchrony, then coordination, and, finally, collaboration—the pinnacle of cooperation (Chapter 4; Boesch & Boesch 1989).

Cognitive studies of primate cooperative problem-solving frequently turn to the laboratory, with the paradigm of choice requiring that two animals perform similar or complementary actions in unison to secure a food reward. Experimental studies of cooperation in monkeys have produced conflicting results, with many reporting no evidence of cooperation (Mason & Hollis 1962; Fady 1972; Burton 1977; Petit et al. 1992; Chalmeau et al. 1997). Notable exceptions of cooperation in monkeys come from reports of collaboration in baboons (*Papio hamadryas:* Beck 1973) and synchrony in capuchins (*Cebus apella:* Mendres & de Waal 2000). Moreover, when monkeys successfully work together, some researchers suggest the result is accidental because individuals fail to understand their partner's purpose (Mason & Hollis 1962; Petit et al. 1992; Chalmeau et al. 1997). By contrast, other studies provide evidence that partners take each other's behavior into account (Mendres & de Waal 2000). More consistently, captive chimpanzees solve various problems, such as simultaneous pulling or lever pressing to obtain food, that show evidence of understanding each other's purpose (Crawford 1937; Chalmeau & Gallo 1993; Chalmeau 1994). Nevertheless, chimpanzees typically have to be trained extensively to work together and when given the choice, prefer to work alone (Crawford 1937; Chalmeau 1994).

Captive spotted hyenas readily cooperate to solve comparable tasks for food, and do so on short order, without modeling, shaping, or specific training (Drea et al. 1996b; Drea unpublished data). They successfully solve synchrony tasks (requiring temporally synchronous production of similar behavior—see Figure 5.4) as well as more complex coordination tasks (requiring temporally *and spatially* synchronous production of similar behavior). The latter reveals a sophisticated level of organizational complexity. Hyenas also display behavioral flexibility while cooperating, modifying their actions to accommodate various partners (including naïve individuals), adopting different leadership roles, switching positions when necessary, and visually monitoring their partner's actions. Coordination, reciprocity, and visual communication displayed by captive hyenas are all features of cooperative hunting described in the wild for lions (*Panthera leo:* Stander 1992).

Consistent with hyena hunting behavior in nature (Tilson & Hamilton 1984; Holekamp et al. 1997), cooperative problem-solving in captivity also varies with group size and composition (Drea et al. 1996b). Thus, the simple addition of animals (from two to four) increases success and efficiency, and social dominance relations among cooperators modulate performance (Drea et al. 1996b). Because high-ranking hyenas can gain access to a resource by displacing subordinates with targeted aggression, greater disparity in partners' ranks correlates with poorer performance (e.g., greater latency to solve the task). In a study of the effects of social context on learning and performance in group-tested rhesus monkeys *(Macaca mulatta),* deference to higher-ranking animals similarly explained the poor performance of subordinates (Drea 1998; Drea & Wallen 1999). Thus, rank-related differences in performance can reflect interference through overt aggression by dominant animals or passive deference by subordinates.

If one were merely to interpret performance as cognitive ability, then the facility and efficiency with which spotted hyenas solve complex cooperation

Figure 5.4. Two captive, adult spotted hyenas cooperating to solve a synchronous rope-pulling task to open a trap door that drops food to the ground at the Field Station for Behavioral Research, University of California at Berkeley. Photograph by C. M. Drea.

tasks might seem remarkable compared to some primate studies. However, we believe that performance differences between taxonomic groups reflect the animals' divergent lifestyles in nature rather than differences in cognitive ability per se. For instance, cooperation is unlikely to have a selective advantage as an adaptation to the handling of food in herbivores (Kruuk 1975). Therefore, many primate species may show little or no cooperation because, as plant foragers, they are less likely to benefit from teamwork. According to this line of reasoning, it should come as no surprise that the few primate species showing evidence of cooperation are those that naturally engage in predatory behavior.

Differences among species in the facility with which cooperation is expressed in the laboratory may reflect the biological significance of cooperative hunting in the wild. The consumption of flesh by primate hunters, including chimpanzees, baboons (*Papio* spp.), and capuchins (*Cebus* spp.), although far from negligible, represents a supplement to a basic vegetarian diet (Teleki 1973), with meat comprising a smaller percentage of the diet than in most modern hunter-gatherers (Hill 1982). Moreover, the targeted prey species of primates are smaller and weaker than are the hunters (King 1980). Among primates only humans hunt prey larger than themselves or even large enough to require team effort (Butynski 1982). By contrast, obligate carnivores, such as spotted hyenas, *must* obtain meat for survival, so they have hunting returns in the range of human hunters (Hill 1982). More to the point, they are unable to kill certain prey without assistance. If meat-eaters are better prepared by their evolutionary history to hunt socially, then studies of cooperation could greatly benefit from adopting a more comparative approach (King 1976, 1980; Kruuk 1975; Schaller & Lowther 1969; Thompson 1975).

Our point in making these comparisons is to raise the following issue for consideration: Either the cognitive implications attributed to primates evincing cooperation should be extended to other animals, so that species solving similar problems are recognized as possessing at least comparable skills, or we should consider the possibility that solution of such tasks reveals little about higher-order cognitive function. We recognize that these interpretations are not necessarily mutually exclusive and raise another question: Should intelligence be measured by how well animals solve everyday problems or should it be assessed by the ingenuity with which behavior is modified to solve novel problems?

Commuting and Hostile Encounters

Because in many ecosystems the most important prey species are migratory, spotted hyenas may travel over long distances in search of prey. Thus, when food supply is disrupted, feeding ranges become decoupled from territoriality. Eloff (1964) found hyenas travelling as far as 80 kilometers nightly in the Kalahari. Likewise, in the Serengeti, where wildebeest and zebra undertake long seasonal movements, females with young cubs commute between their dens and distant herds (Hofer & East 1993a,b,c, 1995). These travels take hyenas across the territories of clans whose members are hostile to the commuters.

Lions, humans, and other hyenas are a spotted hyena's primary threats (Frank et al. 1995a). Although hyenas normally flee humans (at least by day), the tactics used against their enemies may vary widely depending on complex circumstances. Both lions and intruding hyenas may steal prey from clan residents. The reaction of the original possessor to either lions or foreign hyenas may depend on such variables as presence of prey, number and group composition of the enemy, and number and group composition of allies. For instance, spotted hyenas normally flee from a male lion, but mob and chase lionesses if the hyenas are sufficiently numerous (Cooper 1991). Similarly, the reaction to a group of foreign hyenas may be fearful or aggressive, again depending on relative numbers and identity, location in the clan territory, or presence of prey.

To forage successfully, spotted hyenas likely maintain cognitive maps of very large areas, over which they must move, find prey, hunt, and avoid hostile resident hyenas as well as lions. In general, carnivores have large home ranges that exceed those of primates (McNab 1963). Finding distant trees that bear fruit at different times of the season and processing these food items has been suggested as one explanation of primate cognitive evolution. Very similar processes clearly operate in the feeding strategies of carnivores and enable individuals to survive during periods of scarcity.

Testing the Theories

In this chapter, we have described a carnivore social system that shares certain complexities found in many primate societies. The spotted hyena occurs in large groups, in which each member has its own personality (Gosling

1998) and unique relationship with every other member, relationships that can last for nearly two decades. In spite of the differences in trophic ecology, most aspects of social organization are identical in cercopithecine primates and spotted hyenas: male dispersal, female philopatry, matrilineal dominance structure with maternal rank acquisition, social stability, familial continuity, polygyny, and rank-related differences in reproductive success. Although sociality in spotted hyenas may be linked to hunting (Tilson & Hamilton 1984), these organizational similarities suggest that this complex system does not result from aspects of feeding ecology, but from social group size, life history variables, and the lack of female reproductive suppression.

Dominance status, mediated by aggression and learned through early experience and maternal intervention, determines everything from an individual's access to highly contested food to lifetime reproductive success. Monkeys forage as individuals (albeit in groups), spending several hours per day finding small dispersed food items that are consumed individually and with little competition. Spotted hyenas, by contrast, forage socially, cooperating to kill large mammals in a brief and violent hunt. Hyena food is thus extremely concentrated in space and time, and the resulting competition is extremely fierce. Interestingly, the only occasion when feeding competition (and food sharing) is intense among primates is precisely when they behave like hyenas: after male baboons or chimpanzees have killed a monkey or small antelope (Butynski 1982; Boesch & Boesch 1989).

As a social hunter, the spotted hyena must learn to cooperate to bring down a variety of dangerous prey species. In turn, as the victim of lions and humans, it must learn to avoid becoming prey itself. As a member of a social group that lives in severe hostility with surrounding neighbors, it must learn the identities of its comrades and enemies, as well as the boundaries of its own territory, using at least three sensory modalities. As a predator of migratory animals, it must also learn the vagaries of various foes through whose territory it must pass during times of prey shortage.

The physical environment of most primates may be more complex insofar as they inhabit three-dimensional space and manually manipulate food or other objects; however, it would be difficult to argue that the social environment of a baboon is significantly more complex than that of a spotted hyena. Undoubtedly, social intelligence is indispensable if an animal is to cooperate, compete, and reproduce successfully in a large group of long-lived conspecifics. We should emphasize that, although we have focused on the spotted hyena to represent an extreme in carnivore social complexity, group size, and longevity, other carnivore species (e.g., the dwarf mongoose, *Helogale parvula;* banded mongoose, *Mungos mungo;* meerkat, *Suricata suricatta;* coati, *Nasua* spp.) also share many characteristics of cercopithecine primates (Bekoff et al. 1984; Holekamp et al. 2000). Research on intelligence and social complexity in these and other species should be highly rewarding.

CASE STUDY 5A

Maternal Rank "Inheritance" in the Spotted Hyena

ANNE ENGH AND KAY E. HOLEKAMP

Hyena biologists often think of spotted hyenas *(Crocuta crocuta)* as baboons with big teeth and relatively small brains. Like baboons and many other cercopithecine primates, hyenas live in large, complex social groups consisting of multiple matrilines of adult female kin and their offspring, as well as multiple adult males (Frank 1986; Melnick & Pearl 1987). Both taxa are characterized by linear dominance hierarchies, in which social rank determines individuals' priority of access to resources (Wrangham & Waterman 1981; Tilson & Hamilton 1984; Frank 1986). In both taxa, high-ranking individuals are preferred over low-ranking individuals as social companions (Seyfarth 1980; Holekamp et al. 1997), and coalitionary interactions are commonplace (Chapais 1992; Zabel et al. 1992). Additionally, patterns of greeting behavior observed in hyenas are similar to patterns of social grooming observed in vervet monkeys, in which individuals prefer to direct affiliative behavior toward high-ranking nonkin (Seyfarth & Cheney 1984; East et al. 1993). Finally, in both hyenas and cercopithecine primates, youngsters "inherit" their mothers' ranks. That is, they eventually attain social ranks adjacent to those of their mothers (Horrocks & Hunte 1983; Holekamp & Smale 1991, 1993; Smale et al. 1993; Jenks et al. 1995).

Because the social lives of hyenas and cercopithecine primates are so similar, we recently inquired whether young hyenas "inherit" their ranks via the same processes that promote rank inheritance in monkeys (Engh et al. 2000). In some primates, adult females direct high rates of unprovoked aggression toward the offspring of lower-ranking females (Horrocks & Hunte 1983), but this is not observed in hyenas. Adult female hyenas rarely direct unprovoked aggression at cubs, regardless of the cubs' maternal ranks. In all other respects, however, the mediation of maternal rank inheritance is strikingly similar in both hyenas and primates. In both taxa, maternal interventions in disputes and coalitionary interactions appear to play important roles. Female hyenas frequently intervene in disputes between their offspring and conspecifics (Figure 5A.1). High-ranking females intervene more successfully and more frequently than low-ranking females, particularly when the dispute involves food. Maternal interventions appear to be particularly important to cubs less than 18 months old, and the threat of maternal intervention may influence the behavior of clan members even in the absence of the mother. In addition to maternal interventions, agonistic support also plays an important role in rank acquisition. Coalition formation usually follows the "rules" of

Figure 5A.1. An adult female hyena *(left)* intervenes in a dispute between her cub and a lower-ranking clan member. Maternal intervention and coalition formation are the primary mechanisms by which young hyenas "inherit" their mothers' social ranks. Photograph by K. E. Holekamp.

maternal rank inheritance. That is, cubs are often supported by both kin and nonkin when they behave aggressively toward individuals ranking lower than their mothers. Coalitions against cubs form most often when cubs engage in disputes with individuals of higher maternal rank. Finally, cubs tend to join coalitions directed at animals lower-ranking than their mothers. Cubs of high-ranking mothers have both larger numbers of allies and higher-ranking allies than do cubs of low-ranking females.

As occurs in many primates, rank acquisition by young hyenas takes place in a stepwise process of associative learning. When a month-old hyena cub first arrives at the clan's communal den, it typically directs appeasement behavior to all of the individuals it encounters. As the cub matures, much of its submissive behavior is extinguished, until appeasement is directed only toward those individuals higher-ranking than the cub's mother. By the time that the cub is three months old, it begins to attack unrelated hyenas. At first, these attacks are directed with equal frequency against peers of higher and lower maternal rank (Holekamp & Smale 1993). However, aggressive behavior directed toward individuals of higher maternal rank is quickly extinguished through punishing counterattacks or ineffective maternal interventions. In contrast, cubs' attacks on individuals of lower maternal rank are reinforced through effective maternal protection and coalitionary support from other members of the clan. Ontogenetic trends suggest that the process of maternal rank inheritance begins with defensive interventions in which the cub does not participate. Later in development, cubs start to join in attacks when their mothers intervene on their behalf. Eventually, cubs begin to initiate attacks on conspecifics, and they often receive support from their mothers or other allies when they attack individuals of lower maternal rank. This sequence of events leads to rank reversals, first with peers, and then with larger and older individuals (Smale et al. 1993). Once these reversals have occurred, cubs maintain their ranks by winning in dyadic encounters with low-born individuals, as well as by gaining third-party support and by joining attacks against low-born animals. The processes by which maternal rank is acquired in hyenas are strikingly similar to the mechanisms of maternal rank inheritance documented in Old World monkeys (Engh et al. 2000).

Given that hyenas resemble cercopithecine primates not only with respect to the complexity of their social system, but also the way in which their social behaviors develop, it seems reasonable to expect that social interactions in hyenas and primates might be mediated by similar cognitive abilities. Certainly, if the impressive cognitive abilities of primates evolved as a result of selection pressures associated with the labile social behavior of conspecific group members (reviewed by Byrne 1994; Tomasello & Call 1997), then hyenas should be as "smart" as monkeys, since they have been exposed to the same selection pressures. To date, the cognitive abilities of hyenas and other social carnivores are poorly understood. Although coordinated hunting behaviors in hyenas and other carnivores appear to require complex mental processes, these behaviors can be explained most easily with a few simple rules of thumb. For example, the coordinated movements of lions hunting in groups may be the result of following the rule "move wherever you need to in order to keep the selected prey animal between you and another lion" (Holekamp et al. 2000). Female vervet monkeys responded to playbacks of distress calls of unrelated infants by looking toward the infant's mother, indicating that they recognized kinship relations between other group members (Cheney & Seyfarth 1980). Whereas similar field playback experiments clearly showed that hyenas can recognize their own kin, they failed to indicate that hyenas can recognize more complex, third-party relationships (Holekamp et al. 1999). Studies of the mental abilities of hyenas and other nonprimates are critical to our understanding of the evolution of intelligence. Although the social behavior of hyenas and monkeys appears nearly identical in many respects, hyena behavior may result from following simple rules, rather than from understanding the complexity of their social milieu. If hyenas can accomplish the same social feats as those performed by primates without invoking complicated mental algorithms, then we might need to revise our thinking about the evolution of social intelligence. The evolution of social complexity need not necessarily require the concomitant evolution of intelligence, and many species without large, complex brains may engage in complicated social interactions.

6

Is Social Stress a Consequence of Subordination or a Cost of Dominance?

SCOTT CREEL AND JENNIFER L. SANDS

Among social carnivores, groups are typically stratified by a dominance hierarchy. For many species, the aggression and stylized displays that mediate social status are apparent in even the most casual observations (Figure 6.1). Because dominant animals are more likely to win contests for resources such as food or mates, they should generally have greater reproductive success or survival than subordinates. Dominance has particularly strong effects on fitness among cooperative breeders, where dominants can monopolize reproduction almost completely.

From an evolutionary perspective, it seems obvious that being dominant is better than being subordinate. However, this does not mean that dominance carries no costs. The fitness benefits of dominance have been heavily studied, whereas the costs (if any) have received little attention. The conclusion that dominance has benefits is secure on the basis of many studies, but it is logically plausible that struggling for dominance yields high benefits at high cost, whereas accepting subordination yields lower benefits at lower cost. If so, this might explain the perplexing willingness of social subordinates to accept their status without obvious resistance, in cases where the inclusive fitness costs of subordination are apparently large.

For example, dominance and age are highly correlated in dwarf mongooses *(Helogale parvula),* with age explaining 68 percent of the variance in

Figure 6.1. Displays of dominance in social carnivores.

rank (Rood 1990; Creel et al. 1992). Within each group, only the socially dominant individual of each sex is guaranteed of breeding, and genetic data show that dominant individuals produce 75 percent to 85 percent of all offspring raised (Keane et al. 1994), although they comprise only 20 percent of the population. Subordinates gain indirect fitness benefits by helping to raise the dominants' offspring, but the "offspring equivalents" accrued by helpers are substantially lower than the reproductive success attained by breeders (Creel & Waser 1994). Logically, one might expect a strong relationship between age and rank when comparing subadults or young adults to older animals, because size and fighting ability are still increasing, but it is difficult to see why age and fighting ability would remain closely related among older adults. Nonetheless, in 13 years of study we observed no cases in which a dominant mongoose was deposed by a younger pack mate through an internal coup (Rood 1990). In the most extreme case we observed, a 13-year-old female who was visibly senescent (for example, she had difficulty moving with the pack as it foraged) remained at the top of a hierarchy that included seven prime-aged females. It remains a mystery why prime-aged subordinate mongooses respect the age convention. Perhaps social status becomes self-reinforcing as a pair of individuals with a long history of interaction develop social inertia. If so, this could be considered a simple form of deceit.

In discussing the correlates of rank in female baboons, Packer and colleagues (1995; also see Wasser 1995) suggested that elevated androgens may place limits on the aggressive behavior needed to establish and maintain rank. Their argument was that high androgen levels and aggression might yield high rank, but also might interfere with reproduction or maternal behavior. If rank and androgen levels are indeed related, this general argument can be extended to other costs; for example, elevated androgens might compromise immune function (Wingfield & Ramenofsky 1999) and thus might reduce survival. Some empirical studies show an association between rank and androgen levels (e.g., Creel et al. 1997), but this relationship is notoriously variable (Bercovitch 1993; Creel et al. 1993; Arnold & Dittami 1997; Ginther et al. 2001).

Dominance might also carry hidden physiological costs mediated by the hypothalamic-pituitary-adrenal axis. Animals respond to a stressor with a series of endocrine responses that increase the immediate availability of energy, in part by inhibiting physiological processes that are not required for immediate survival (Munck et al. 1984; Sapolsky 1992; Wingfield 1994). One of the primary responses to stress is an increase in the activity of the hypothalamic-pituitary-adrenocortical axis, producing an increase in the concentration of circulating adrenal glucocorticoids (GC). In the short term (hours to days), GC elevations redirect resources to mobilize energy that can be used to resolve the stressful situation. If the stressor is not eliminated and GC levels remain high chronically (weeks or longer), a broad range of harmful consequences ensue, including immune suppression, loss of muscle mass, and reproductive suppression (Sapolsky 1992; Chrousos & Gold 1992; Pottinger 1999).

It is widely believed that social subordination is stressful; indeed the term social stress is often used to mean the stress of subordination. The idea that social stress might fall more heavily on dominants than on subordinates has received less attention, but we argue that among cooperative breeders, social stress is often a cost of dominance, rather than a consequence of subordination. Although relatively few species have been studied (Table 6.1), recent field data strongly challenge the traditional view of social stress. On the basis of logic alone, it is difficult to predict whether domination or subordination should be more stressful. For example, domination might be stressful if it requires frequent fighting. On the other hand, losing a fight might be more stressful than winning two fights: the algebra of winning and losing is not transparent, and requires empirical study.

Before presenting data, it is interesting to examine the origins of the conventional view that subordination is stressful. For more than 30 years, we have known that aggressive or agonistic interactions can provoke large and persistent increases in GC secretion. Influential early work on this issue was conducted with captive rodents and primates, often by grouping unfamiliar individuals, observing the fights that ensued, and comparing the GC levels of winners and losers (Bronson & Eleftheriou 1964; Louch & Higginbotham 1967; Manogue 1975). In this situation, both winners and losers show a strong stress response, but the response is larger among losers. In these early studies, losers were generally called subordinates, and winners were called dominants. These were important studies of behavioral interactions and stress responses, but they do not necessarily reveal the consequences of living as a subordinate in a social group with a settled dominance hierarchy. The rate and severity of fighting are high immediately after strangers are grouped, particularly among males, which were the focus of most winner/loser studies. For example, in a recent study by Blanchard and colleagues (1995), grouped rats fought approximately 30 times per day. After 13 days, they had an average of 17 body wounds and had lost more than 20 percent of their initial body mass, despite being removed from the colony on four days to feed for eight hours. Aggression this severe would be unusual for a social species in the wild, where agonistic encounters rarely escalate to the point of wounding, and rates of aggression are much lower (Creel et al. 1992, 1997). Captive subordinates cannot avoid dominant individuals as effectively as they can in the wild. In the wild, moving away is a common means of terminating an attack, if behavioral appeasement does not work. Of course, the ultimate form of moving away from dominant individuals is dispersal, an option not open to captive animals. Despite these complexities, captive winner/loser studies are the original basis of the common argument that the stress of subordination or psychological castration might underlie reproductive suppression among subordinates in cooperatively breeding species.

In his groundbreaking study of the endocrine correlates of social status in free-living olive baboons *(Papio anubis),* Robert Sapolsky (1982, 1983, 1992) found elevated basal GC levels in individuals of low rank and in those that were losing rank (regardless of their position in the hierarchy). Interestingly, the correlation between status and stress disappeared during periods of

Table 6.1. Relationships between basal glucocorticoid levels and social status within cooperatively breeding groups

Species	Sex	Basal GC pattern	Reproductive suppression of subordinates	Captive or wild study	Method of sampling GCs	Notes
Rodents						
naked mole-rat[1] *Heterocephalus glaber*	F and M	sub > dom	high skew	captive	urine	long-term studies
Alpine marmot[2] *Marmota marmota*	M	dom > 3 types of sub dom < 1 type of sub	low skew	wild	blood, with long and variable lag from trapping to sample	categorized subs by yearling/adult and son/nonson
Primates						
common marmoset[3] *Callithrix jacchus*	F	dom > sub	high skew	captive	blood	
black tufted-ear marmoset[4] *Callithrix kuhli*	F	dom = sub	high skew	captive	urine	
	M	dom > sub	high skew	captive	urine	additional data from J. French
ring-tailed lemur[5] *Lemur catta*	F	dom > sub	low skew	wild	feces	
cotton-top tamarin[6] *Saguinus oedipus*	F	paired > sub with male	high skew	captive	urine	
Birds						
white-browed Sparrow weaver[7] *Plocepasser mahali*	F and M	dom = sub	low skew	wild	blood	very low GC levels for all ranks

Florida scrub jay[8] *Aphelocoma c. coerulescens*	M	dom = sub *(see notes)*	low skew	wild	blood	dom > sub at mating stage
Harris's hawk[9] *Parabuteo unicinctus*	F and M	dom = sub *(see notes)*	high skew	wild	blood	NS ANOVA for GC in breeders, auxiliaries, and juveniles
Carnivores						
wolf[10] *Canis lupus*	F and M	dom = sub *(see notes)*	high skew	captive	blood	small sample; in wild packs, dom > sub
wolf (this study)	F and M	dom > sub	moderate skew	wild	feces	in captivity, dom = sub
dwarf mongoose[11,12] *Helogale parvula*	F	dom > sub	high skew	wild	urine	acute GC: dom > sub
	M	dom = sub	high skew	wild	urine	acute GC: dom < sub
African wild dog[12,13] *Lycaon pictus*	F and M	dom > sub	high skew	wild	feces	

Note: This table excludes studies in which individuals in different groups or on different territories are compared, for example comparisons of unmated, monogamously mated, and polygynously mated, male birds on neighboring breeding territories. GC = glucocorticoid; NS = not significant at $P = 0.05$. High skew refers to species in which reproductive success is highly skewed within groups (subordinates rarely breed). Low skew refers to species in which subordinates often breed.

Sources: 1 Faulkes & Abbott 1997. 2 Arnold & Dittami 1997. 3 Saltzmann et al. 1998. 4 Smith & French 1997, and personal communication J. French. 5 Cavigelli 1999. 6 Zeigler et al. 1995. 7 Wingfield et al. 1991. 8 Schoech et al. 1991. 9 Mays et al. 1991. 10 McLeod et al. 1996. 11 Creel et al. 1992. 12 Creel et al. 1996. 13 Creel et al. 1997.

social instability. From a historical perspective it is interesting to note that many subsequent field studies have found substantially different relationships between rank and stress, but Sapolsky's study was the first to address this question in the field. Because the baboon research was well-executed and convincing, the conclusion that subordination is stressful became quite strongly entrenched.

Given this background, it is interesting to ask what recent field studies reveal about the endocrine correlates of rank in social species, particularly cooperative breeders (Table 6.1). In this chapter, we summarize three field studies of the behavioral, endocrine, and demographic correlates of rank in cooperatively breeding carnivores (dwarf mongooses, *Helogale parvula;* African wild dogs, *Lycaon pictus;* and wolves, *Canis lupus*). These three species have similar social systems, with stable packs that typically hold several adults of both sexes (although it is not unusual for a wolf pack to have only one adult of a given sex). There is a clear dominance hierarchy within each sex, and only the dominant individual of each sex is assured of reproducing, though subordinates of both sexes occasionally breed. Dispersal is common for both sexes, though some individuals remain in their natal pack for their entire lives, and nondispersers can accrue inclusive fitness benefits by helping relatives to raise nondescendant kin (e.g., see Creel & Waser 1994 for calculations).

Dwarf Mongooses

On our study area in Serengeti National Park, dwarf mongoose packs held an average of 9.0 ± 0.3 adults and yearlings, with an even sex ratio (Rood 1990; Creel & Waser 1994). Dwarf mongooses are obligately cooperative breeders: unaided breeding pairs are rare (12 cases in 202 pack-years of observation) and almost never succeed in raising offspring to independence (mean annual reproductive success = 0.07 ± 0.07 offspring). Of 302 pregnancies, 219 (72.5 percent) were by alpha females, even though subordinates outnumbered alphas 3.5-fold (Creel et al. 1992). In 11 cases in which subordinates gave birth out of synchrony with the dominant female, no offspring survived to the age of emergence from the den at four to six weeks. In 51 cases, at least one subordinate gave birth in synchrony with the alpha female. The mean size of joint litters (3.2 offspring at emergence) was significantly larger than alpha-only litters (2.4 offspring; $t = 2.92$, $P < 0.05$). Thus, on the occasions that subordinates become pregnant, there is some evidence that the off-

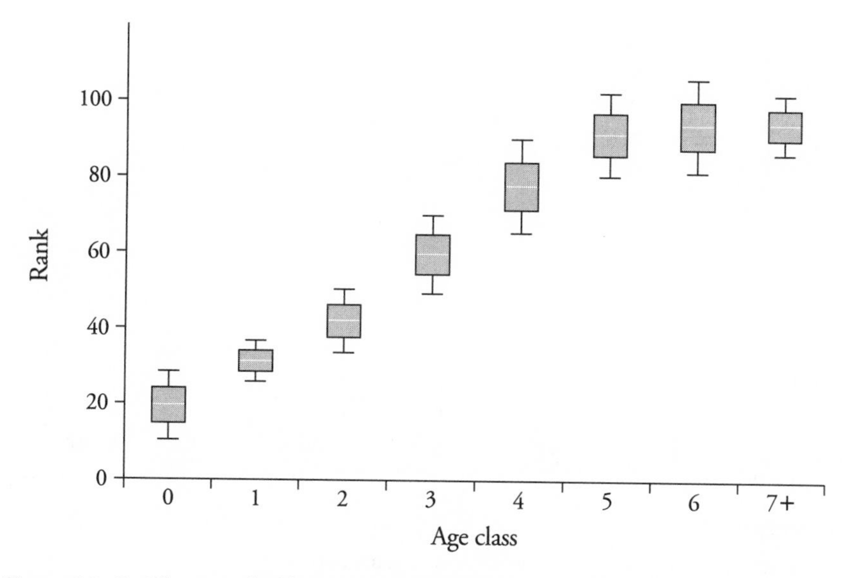

Figure 6.2. Social status is highly correlated to age in dwarf mongooses. The alpha in any pack had a rank of 100, with quantitative gaps between all lower-ranking individuals determined by the Batchelder-Bershad-Simpson method (Jameson et al. 1999) from data on wins, losses, and rank of the opponent. Boxes show mean (central line), one SE (box) and two SE (whisker).

spring of subordinate females occasionally survive, but only when litters are creched. Moreover, joint litters are substantially smaller than would be expected if subordinates produced a number of young equal to alpha females. Genetic data showed that 15 percent of all offspring are produced by subordinate females, while 25 percent are fathered by subordinate males (Keane et al. 1994).

The demographic and morphological correlates of rank are similar for males and females, which is not surprising for a cooperative breeder with little sexual dimorphism. By itself, age explained 69 percent of the variance in rank across the population at large (Figure 6.2). Within single packs, the relationship was even stronger: we detected no exceptions to the rule that the oldest mongoose within a specific pack was dominant. After controlling for the effects of age, body mass explained a significant portion (14 percent) of the variance in dominance: within an age-class, heavier mongooses tended to be dominant.

The behavioral correlates of dominance were broadly similar for males and females, but differed in some ways that might affect the endocrine correlates

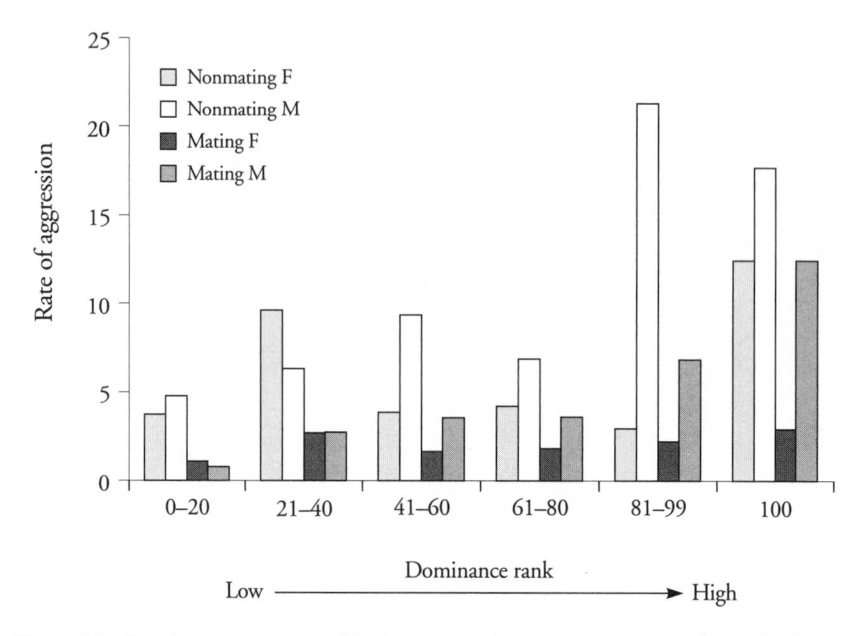

Figure 6.3. Dominant mongooses of both sexes engage in more aggression than subordinates do in mating periods and nonmating periods. See caption for Figure 6.2 for an explanation of how rank was determined.

of rank. Dominant mongooses of both sexes engaged in aggression at significantly higher rates than subordinates (Figure 6.3) during periods of mating and nonmating. On average, dominant males fought 2.7 times more often than subordinates, whereas dominant females fought 2.0 times more often than subordinates. If fighting affects basal GCs for winners as well as losers, then we might expect to see elevated GCs in dominant dwarf mongooses of both sexes.

This expectation is met for females, but not for males. High-ranking females have higher basal urinary GC levels than subordinates, and the difference is particularly pronounced for alpha females (Figure 6.4a: regression, $F_{1,99} = 40.6$, $P < 0.001$: Creel et al. 1996). Despite elevated basal GCs, dominant female mongooses produced higher cortisol levels than subordinates in response to the short-term stress of trapping ($F_{1,74} = 9.36$, $P < 0.005$), a result that runs contrary to the general pattern that chronically elevated basal GCs weaken the acute GC response (Creel 2001). In contrast, there was no detectable relationship between rank and basal urinary GCs for male dwarf mongooses (despite statistical power almost identical to the test

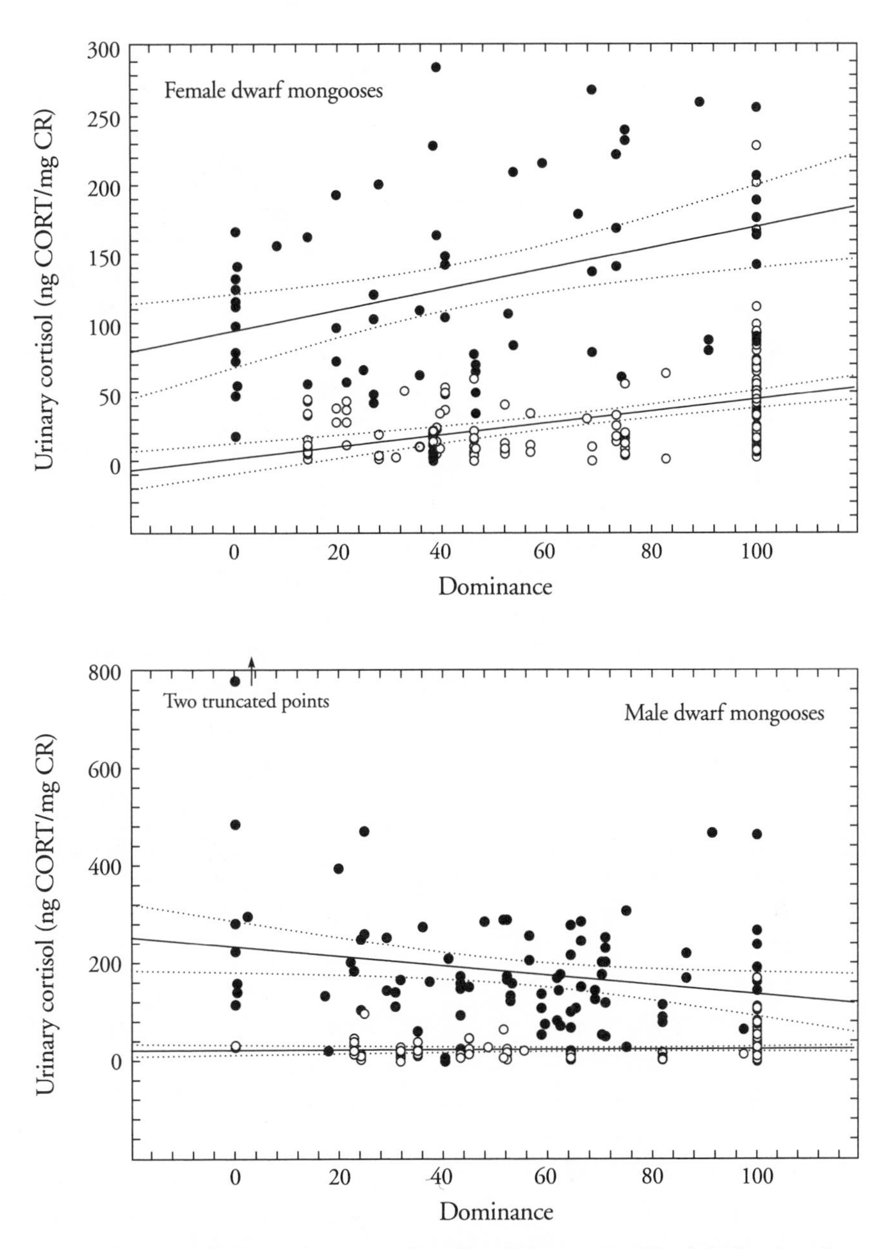

Figure 6.4a. The relationship between rank and basal glucocorticoid levels in female and male dwarf mongooses. The alpha in any pack had a rank of 100, with quantitative gaps between all lower-ranking individuals determined by an iterative procedure (Friend et al. 1977) using data on wins, losses, and rank of the opponent. GC levels are plotted independently for baseline samples (open circles = collected noninvasively from untrapped mongooses) and for acute stress responses (filled circles = collected from trapped mongooses).

for females), but dominant males showed low peak cortisol levels in response to trap stress (Figure 6.4a: $F_{1,91} = 9.06$, $P < 0.005$), suggesting that their acute stress response was somehow compromised. It is not clear to us why male and female mongooses differ so clearly in the *endocrine* correlates of rank, despite being monomorphic and having strong similarities in the *behavioral* correlates of rank. Nonetheless, this result fits a pattern that effects of rank on endocrine function are more common in females, whereas purely behavioral mechanisms are more common in males (Creel et al. 1992).

This general pattern may arise because selection for reproductive restraint is driven by reproductive constraints: without constraints, there is no selection in favor of restraint. Reproductive constraints may be greater for subordinate females (i.e., it is probably easier for dominants to enforce reproductive suppression among females) simply because it is easier to identify maternity than paternity. This allows alpha females to kill offspring other than their own. In contrast, alpha males are often poorly positioned to use infanticide to enforce reproductive suppression. Paternity can be mixed within litters for essentially all carnivores studied to date. If the alpha male mates more often than other males, but cannot directly determine the paternity of individual offspring, then infanticide is not a viable method of enforcing reproductive suppression (Creel & Waser 1996). It is possible that a father can recognize his own offspring even within litters of mixed paternity, but the evidence for such capabilities is limited, and it seems very likely that confidence of paternity will be lower than confidence of maternity.

African Wild Dogs

Wild dog packs averaged nine adults and yearlings on our study site in the Selous Game Reserve, with a range of 2 to 27. Across several populations, mean pack size was 6.8 ± 0.8 (Creel & Creel in press). Alpha females produced 81 percent of 85 litters in Kruger National Park (M. G. L. Mills, personal communication) and 75 percent of 57 litters in Serengeti National Park (Malcolm & Marten 1982; Burrows 1995). In the Selous Game Reserve, alpha females had an annual probability of breeding of 81.5 percent ± 7.0 percent, far higher than subordinates (6.4 percent ± 2.4 percent; $z = 7.36$, $P < 0.001$). We observed five cases in which no litter was produced, all in newly formed packs that waited until the next rainy season to breed, rather than mating out of season. In stable packs, the alpha female invariably gave birth once a year, whereas seven litters were born to subordinate females

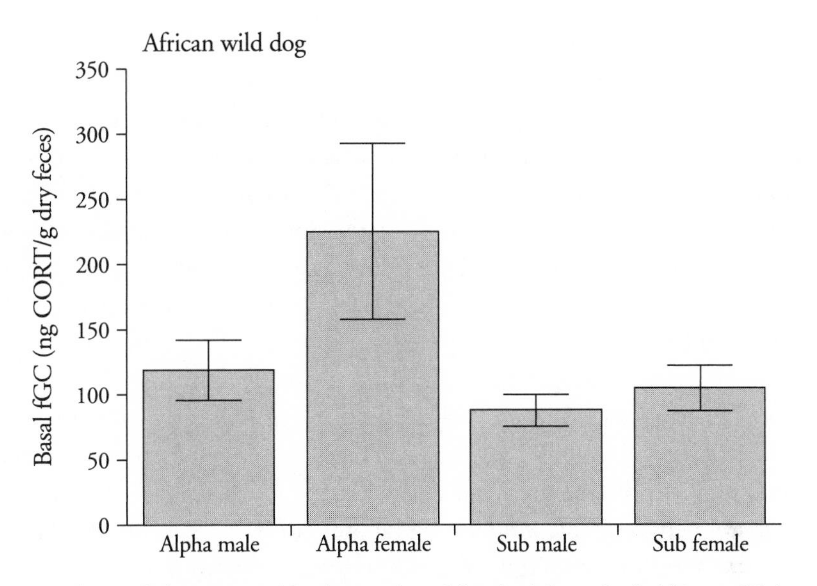

Figure 6.4b. Basal glucocorticoid levels in male and female alpha and sub African wild dogs.

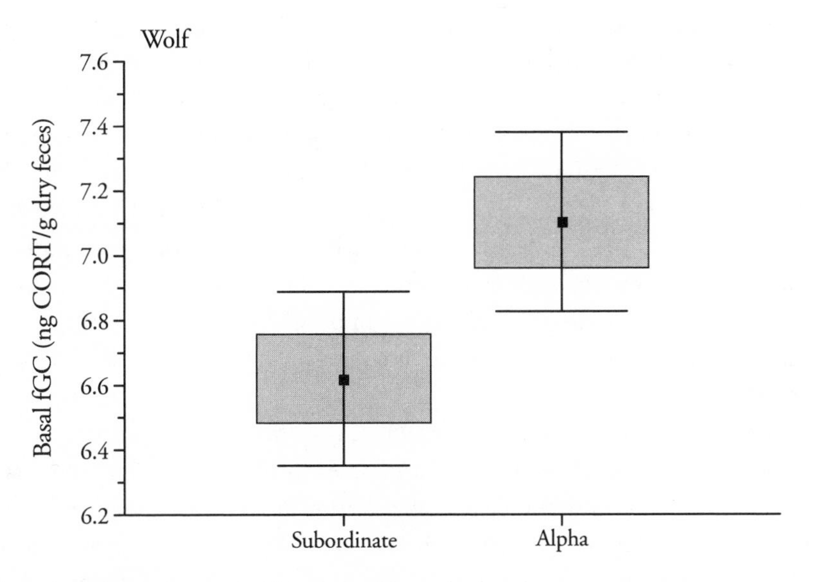

Figure 6.4c. Basal glucocorticoid levels in alpha and sub wolves.

in 110 individual years of observation. When a subordinate female gave birth, she usually did so several days after the alpha female. In Selous Game Reserve as elsewhere, subordinates' pups were sometimes killed, but were sometimes creched with the dominant female's pups and raised (van Lawick 1973; Frame et al. 1979; Fuller et al. 1992). Behavioral and genetic data show that most offspring are fathered by the dominant male (Girman et al. 1997; Creel & Creel in press). Like dwarf mongooses, wild dogs are obligately cooperative breeders: no unaided pair has been observed to raise offspring to independence in any population.

Among females, the oldest individuals were usually dominant. In contrast, males of intermediate age were highest ranking, with older alpha males often being deposed by younger packmates (Creel & Creel in press). During nonmating periods (the great majority of the year) there was no detectable association between rank and fighting rate for either sex. During mating periods alphas of both sexes fought significantly more often than subordinates (Creel et al. 1997). Although rank and aggression were related only over a narrow time window, fecal GC levels were higher in dominant dogs throughout the year (Figure 6.4b: factorial analysis of variance (ANOVA) including rank and sex: rank effect $F = 6.76$, $P < 0.01$, Creel et al. 1996).

Wolves

In Yellowstone National Park, the average size of our three focal packs was 8.7 adults and yearlings (range 6–12). Over two years, we collected 345 fecal samples and 375 hours of behavioral observations in these three packs (Druid, Leopold, and Rose). Of these 345 samples, 122 came from 20 known individuals, whose social status was determined from aggressive and agonistic interactions using the Batchelder-Bershad-Simpson method of assigning ranks (Jameson et al. 1999). For cortisol radioimmunoassay methods and validation, see Creel and colleagues (in press).

Like wild dogs and female dwarf mongooses, dominant wolves of both sexes had higher basal glucocorticoid levels than subordinates (Figure 6.4c). We used analysis of covariance to examine the effect of rank on GCs, controlling for the properties of the fecal samples (proportion indigestible matter, percent water, and coefficient of variation for repeated within-sample measurements), and for variation among packs and years. GC values were log transformed prior to analysis to obtain normality. In the analysis of covariance (ANCOVA), social status had a significant effect on GCs (Figure

6.4c: $F_{(1,96)} = 4.60$, $P = 0.038$), with dominant animals having higher levels (1,876 ± 286.7 ng cortisol/g dry feces) than subordinates (1,381.8 ± 212.0 ng cortisol/g dry feces). This pattern was consistent between years (interaction: $F_{(1,96)} = 0.17$, $P = 0.68$) and across packs (interaction: $F_{(2,96)} =$ 0.58, $P = 0.56$).

Though the endocrine correlates of rank are similar for these three carnivores, the behavioral correlates are more variable. For wolves, our behavioral data come from one pack that we observed for a total of 270 hours. Using multiple regression, we tested whether the combined rate of all agonistic and aggressive behaviors was associated with GC levels, again controlling for properties of fecal samples (as above). GC levels were not detectably related to the rates of these behaviors (Figure 6.5: partial correlation, $F_{(3,78)} = 2.06$, $P < 0.12$, $R^2 = 0.04$). To test whether specific agonistic or aggressive behaviors showed a stronger relationship, we examined partial correlations between GC levels and dominance, submission, stylized aggression, and attacks. Again, we detected no relationships.

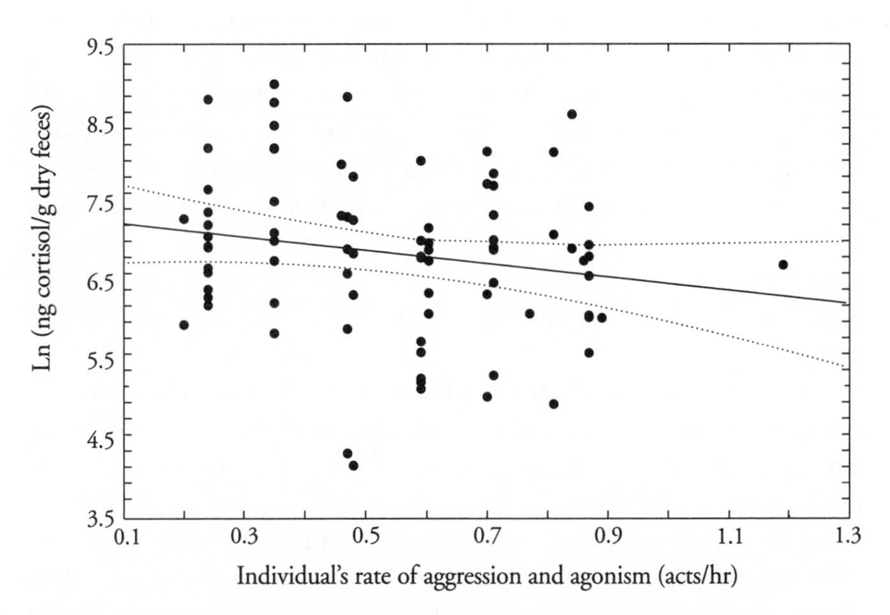

Figure 6.5. The relationship between basal glucocorticoid levels and rates of aggressive and agonistic behavior in wolves: simple regression with 95 percent confidence limits.

Conclusions

Noninvasive field studies of social carnivores and other cooperative breeders clearly show that elevated GC levels are often a cost of dominance. GC levels are higher in dominant wolves and wild dogs of both sexes, and in female dwarf mongooses. When we first detected this pattern in dwarf mongooses, we were quite surprised, as it opposed the prevailing view that subordination provokes a chronic elevation of GCs. However, subsequent field studies of cooperatively breeding birds, carnivores, and primates have shown that elevated GC levels in dominants are not unusual (Table 6.1). Seeking to explain this pattern, and based largely on data from dwarf mongooses, we suggested that higher rates of aggression among alphas might be the cause. With data from wild dogs and wolves, this explanation now seems incorrect, or at least not general.

In wild dogs, the GC levels of alphas are higher year-round, but they fight more frequently only during the mating period, which lasts just a few weeks. For wolves, rates of agonistic and aggressive behavior were not related to GC levels. Though they win more often, dominant wolves do not engage in agonistic or aggressive behavior more often than subordinates. This leads us to an interesting logical dead end. If we applied the same (apparently reasonable) logic as the original studies, concluding that losing or subordination is stressful, we would be forced to conclude that winning fights is more stressful than losing at the same rate. This seems quite unlikely. It is more likely that some aspects of maintaining social status are not easily captured in simple measures of the frequency and outcome of agonistic interactions (as argued by Sapolsky 1983; Virgin & Sapolsky 1997). Rank is proving to be a good predictor of GC levels in cooperatively breeding carnivores, but the underlying behavioral mechanisms that affect rank and GC levels are not as easily pigeonholed.

We believe that quantifying the pattern of wins and losses is effective in evaluating social dominance, which in turn is a good predictor of GC levels. Unfortunately, this same pattern of wins and losses is not clearly and consistently related to GC levels. Although social stress is often a cost of dominance for cooperative breeders, the behavioral aspects of social status that drive this relationship are not yet clear. We suspect that it will be difficult to identify the behavioral driving forces, because the answers probably depend on knowing what an animal is thinking, rather than what it is doing.

For example, Sapolsky (1990) suggests that certain behavioral traits or "personalities" are associated with elevated GCs in baboons, regardless of the rank of the individual that has these traits. This conclusion is well supported by data from baboons (Sapolsky 1990; Sapolsky & Ray 1989). We have not tested it directly for our three species of social carnivores, but it seems unlikely to hold, simply because relationships among age, rank, and GC levels are strong. Consequently, we would expect to find a relationship between personality and GC levels only if personality changes with age in a predictable manner. Under this scenario, it would be difficult to distinguish between the effect of personality and the effect of age itself.

Finally, we think there are difficult problems of interpretation in the measurement of some of the aspects of personality that appear to affect GC levels. To illustrate, GC levels were higher in baboons with a poor "ability to behaviorally differentiate between winning and losing a fight," and in baboons with a poor "ability to behaviorally differentiate between a threatening interaction [versus] the neutral presence of a rival male" (Sapolsky 1990, p. 873). To evaluate these aspects of personality requires a belief that we understand the social interactions of another species well enough that we can determine when an individual is making a mistake in its interpretation of the behavior of a conspecific. This is a high standard, particularly if the species is significantly different from humans in its sensory capabilities (for example, a canid might use its well-developed sense of smell to determine that an individual is likely to be aggressive, even when its overt behavior is apparently neutral to human eyes). If basal GC levels do indeed relate to personality more closely than to rank, it will be intriguing to see if the behavioral traits associated with stress responses are consistent among species. Do the same things worry an elephant and a mongoose? Maybe, maybe not.

CASE STUDY 6A

Sperm Whale Social Structure: Why It Takes a Village to Raise a Child

SARAH L. MESNICK, KAREN EVANS, BARBARA L. TAYLOR, JOHN HYDE, SERGIO ESCORZA-TREVIÑO, AND ANDREW E. DIZON

At sea, female and immature sperm whales are typically found in cohesive groups of about 10–40 individuals that move and act together in a coordinated manner (Best 1979; Whitehead et al. 1991). These groups are the social entities in which calves are born and raised, individuals forage, and mating takes place. Members are known to exhibit a wide variety of aid-giving behaviors, including allomaternal care (e.g., "babysitting" and perhaps allonursing; Whitehead 1996) and behaviors that result in communal protection of the group, sometimes to the point of self-sacrifice (e.g., issuing alarm calls, forming defensive formations, and aggressive or defense attack; Scammon 1874; Caldwell & Caldwell 1966; Caldwell et al. 1966; Berzin 1972; Pitman et al. 2001). Traditionally, these behaviors were easy to interpret as kin helping kin in stable matrilineal groups (Ohsumi 1971; Best 1979). Results from recent photo-identification and genetic analyses, however, show that these groups are neither particularly stable nor matrilineal (Richard et al. 1996; Christal et al. 1998; Mesnick 2001). If kin selection is not a viable hypothesis, then other explanations are required.

Adult female and immature sperm whales of both sexes appear to live in a fission-fusion society with the observed "group" being comprised of temporary associations among more stable social "units" (Whitehead et al. 1991; Whitehead & Kahn 1992). We used both mitochondrial (mtDNA) and nuclear markers (dinucleotide repeat microsatellite loci) to investigate relatedness in "groups" and "units" of female and immature sperm whales sampled during mass strandings (all beached animals sampled) and from live animals at sea (partially sampled units and groups). Results are consistent within and between the two data sets. Here, we focus on the data obtained from one of the mass strandings (from Stanley, Tasmania) and from one "unit" known from long-term photo-identification records (Galápagos Island samples; see also Chapter 17) to illustrate the main findings.

The cluster diagram of pairwise relatedness values for 10 adult females from the Stanley stranding is shown in Figure 6A.1. This analysis indicates that there were some close relatives among the stranded females but that not all individuals were closely related. Females whose pairwise relatedness values fall within the range expected for first-order relatives include a mother-fetus pair (9669 and 9673) and a cluster of four closely related adult females (9664, 9666, 9665, 9667). Individuals whose closest relative falls within the range expected for a second-order relation include an older female (9671) with 61 growth layer groups (GLG; the number of growth layer groups identified in tooth sections) who could potentially be the grandmother or aunt of several other females in the stranding (9663, 9665, 9668, and 9669). There was one female, with 46 GLG, not closely related to any of the others (9670). All individuals in the stranding shared a single mitochondrial haplotype. This latter finding is unusual among the groups we have looked at; all other units and groups from which we have collected five or more samples, except this one, contained two to four mtDNA haplotypes ($n = 3$ stranded groups and 10 groups and 1 unit sampled at sea; Mesnick 2001).

The cluster diagram of pairwise relatedness values for five of nine adult females in one Galápagos "unit" is shown in Figure 17.2 of Chapter 17. Two mtDNA haplotypes were present in the unit. No pairwise relatedness values fell within the range expected for first-order relations although all the individuals were known from photo-identification records to be long-term associates.

Our analysis of sperm whale samples and four other recent studies, covering three different ocean basins, show no genetic evidence of a strictly or largely matrilineal unit or group of sperm whales, albeit no living unit or group has been completely sampled (Dillon 1996; Richard et al. 1996; Christal 1998; Bond 1999; Mesnick 2001). Although we do not know

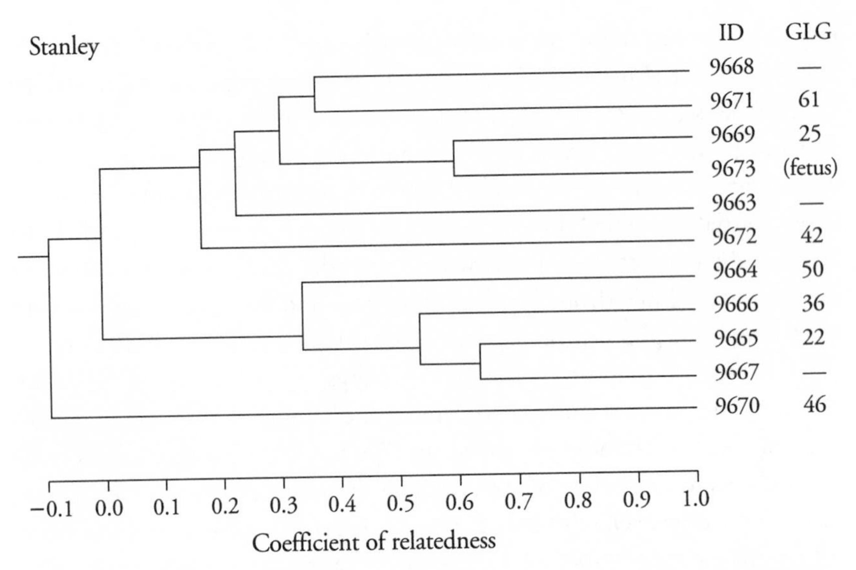

Figure 6A.1. Relatedness among the 10 adult females (and one fetus) that stranded at Stanley, Tasmania, in 1998. The strength of genetic relationships was calculated by estimating the probability of a microsatellite allele co-occurring in two different animals; higher values represented greater than average relatedness (Relatedness 5.0; Queller and Goodnight 1989). On this scale, first-order relations (parents and offspring) would theoretically have a relatedness value of 0.5 (full sibs would have a mean value of 0.5); second-order relations (grandparents and grandchildren, aunts or uncles and nieces or nephews or half sibs) would have a mean relatedness value of 0.25, and so on; observed values are distributed around these means. Individuals with high degrees of relatedness appear close to one another on the cluster diagram. Listed at the right are estimates of age based on the number of growth layer groups (GLG) identified in tooth sections. UPGMA clustering was done using MEGA 1.01 (Kumar et al. 1993).

whether these groups are representative of historical (pre-whaling) sperm whale social structure, contemporary groups of female and immature sperm whales generally contain more than one matriline, as indicated by the presence of multiple mtDNA haplotypes. Both units and groups contain clusters of closely related animals, but some individuals have no close relations. The view of sperm whale social structure that is emerging is one of kith (close, but not genetically related companions) and kin.

These results present us with an intriguing question, rare among nonhumans: why do individuals, who may be unrelated, come to the aid of others, and sometimes even risk their lives to do so? Aid-giving behavior can arise along three major pathways: kin selection, reciprocity, and mutualism maintained as a by-product of selfish acts (Dugatkin 1997). Individuals may derive inclusive fitness benefits by helping relatives (kin selection; Hamilton 1964). They may have an opportunity for repeated interactions and reciprocation (reciprocal altruism; Trivers 1971). Finally, it may be mutually beneficial, either incidentally or when the well-being of one's companions has an important effect on one's own fitness (Trivers 1971; Connor 1986; Lima 1989). Although we do not know whether some of the most spectacular "altruistic" behavior, such as risking one's life for another and allonursing, happens outside of matrilines, our data raise this possibility.

Hamilton (1964) and Trivers (1971) predicted that aid-giving behavior will be selected for in species that are long-lived, reside in relatively stable, small social groups, have the ability to recognize individuals, and have long periods of parental care; all characteristics attributable to groups of female sperm whales (Connor & Norris 1982). Aid-giving behavior is also favored when individuals cooperate in defense against a common enemy. Hamilton (1971), explaining the evolution of gregariousness in animals, suggested that the selfish avoidance of a predator can lead to aggregation and that this tendency is intensified when the individuals become mutually dependent on the presence of one another for a successful defense. He coined the term "selfish herd" to emphasize that there is nothing in the least altruistic about the behavior of individuals in these aggregations. Rather, it becomes worth keeping others alive to protect yourself (Connor 1986; Lima 1989). In short, "selflessness is a form of selfishness" (Robert Pitman, personal communication).

This behavior is particularly evident in open habitats. Among terrestrial mammals, the most gregarious are those that inhabit open grassy plains rather than forests. Among fish, schooling is particularly evident in open water (Hamilton 1971). In cetaceans, group size is significantly larger farther from shore; the most gregarious species are those that live in the pelagic realm, the open ocean far away from any cover (Norris & Dohl 1980). Protection from predation by sharks and killer whales is hypothesized to be the causal factor for the evolution of cetacean schools, which function as protective systems for their members (Connor & Norris 1982).

For sperm whales, Whitehead & Weilgart (1991) suggested that calf protection is likely to be the most important function of sociality among females. Whitehead and colleagues (1991) and Whitehead (1996) argue that this may well be adaptive in a species in which adults make long feeding dives to depths where their calves cannot follow. At the surface, the calves are vulnerable to killer whales and other potential predators. Sociality is further enforced by evidence of allomaternal care, the individuals becoming mutually dependent on one another for communal care of the young as well as defense. It is interesting to note that in primates, young, nulliparous females most frequently display allomaternal or "aunting" behavior (Nicolson 1987) whereas our data suggest that for cetaceans older females may exhibit such care. Conceivably, female sperm whales and their dependent young reside in a social, ecological, and physical environment in which all three major pathways for the evolution of aid-giving behavior (kin selection, reciprocity, and mutualism maintained as a by-product of selfish acts) may operate, reinforced by defense against common enemies.

We suggest that for the sperm whale and other cetacean species that live, raise young, and contend with predators in the pelagic realm, it is the bonds developed through association, cooperation, and lactation that hold these sometimes unrelated females together. Indeed, in the case of sperm whales, it may take a group to raise a calf.

III

Social Cognition

As far back as the 1960s, it was proposed that a complex social milieu promotes brain evolution. For animals to come up with adjustments to the shifting friendships and hierarchies in which they live, they need intimate knowledge of the social environment. One of the original proponents of this view—a Swiss primatologist named Hans Kummer—joined our meeting in Chicago to remind us of the early work that inspired this now popular theory. Among the observable items often mentioned as signs of complexity are alliance formation (i.e., cooperative aggression), reconciliation after fights, and general knowledge of social relationships in the group. Nonhuman primates demonstrate all three: they build intricate societies so rife with jockeying for position that the word "politics" comes to mind. Based on current knowledge of other taxons (e.g., Case Study 4B), there seems no reason to reserve such terminology for primates.

In the following three chapters, the authors attempt to break down into simpler components the cognition underlying social life in individualized societies, or at least organize the requisite cognition such that fundamental principles become recognizable. Chapters 7 and 8 look at the acquisition of knowledge: how do animals sort through the social information that they gather in order to detect regularities and categorize relationships? Chapter 9 looks at the other side of the cognition coin, that is, the application of knowledge toward solutions to problems posed by the social environment.

Based on many years of creative and careful experiments with sea lions, Ronald Schusterman, Colleen Reichmuth Kastak, and David Kastak conclude in Chapter 7 that these animals generalize from known to unknown associations between different items, meaning that larger classes emerge from a smaller set of trained relationships. In addition, associations may vary by context, making for flexible equivalence classes. The authors then take us on an excursion into the social complexity of primates viewed from this particular perspective, in which individuals categorize the group members around them based on perceived equivalence relationships that may shift with context. The authors also explore applying this idea to predator-specific alarm calls. In a striking illustration, they highlight the tendency of wild chimpanzees to attack empty nests left by rival groups. Perhaps because chimpanzees divide others simply into friends or foes, these nests get lumped with the latter.

After this extrapolation from learning experiments to the cognitive demands of life in complex societies, Robert Seyfarth and Dorothy Cheney travel in the opposite direction in Chapter 8, linking their fieldwork on the social knowledge of vervet monkeys in Kenya to recent insights into how rats in the laboratory exploit radial mazes. As perhaps no other research team in the world, Seyfarth & Cheney have mapped the cognitive demands posed by life in a matrilineal society, that is, a matrifocal, multigenerational society in which females connect with daughters, granddaughters, sisters, and mothers. This makes for a society in which it is critically important not only to belong to a matriline, and to protect its status among other matrilines, but also to know to which matriline each other individual in the group belongs, and where each matriline ranks relative to one's own matriline and all the others. Such categorization is a complex task, and the ingenious experiments of the authors, who use playbacks of species-specific vocalizations, have illuminated many subtle aspects of it, opening our eyes to the truly impressive knowledge possessed by these unremarkable-looking monkeys.

Referring to maze experiments in which rats reduce the load on memory by clustering visits to different parts of the maze based on the type of food placed there, Seyfarth and Cheney propose that the spontaneous organization that the rats impose on this task provides a good model for how primates classify group members. Similar to the rats' activity, vervet monkeys may seek to reduce the amount of specific information needed by imposing a hierarchical classification. Seyfarth and Cheney note that with increasing group size it is virtually impossible to keep track of all dyadic (let alone triadic) relationships, so that some reduction of information processing is required.

Suggesting that the "chunking" demonstrated in rats provides the answer, they argue against Schusterman and colleagues' alternative proposal for the role of equivalence relationships. It remains to be seen to what degree these two views are mutually exclusive—some researchers see "chunking" as part of the equivalence paradigm. If the two mechanisms are true alternatives, we can only hope that placing both chapters side by side will inspire experiments to distinguish between "chunking" and equivalence learning as ways in which animals organize information on group mates.

In Chapter 9, Frans de Waal investigates the flexible applications of social knowledge. Whereas Seyfarth and Cheney suggest a hierarchical organization of incoming information, de Waal looks at the hierarchical arrangement of bits of information for the purpose of successful action. It has long been noted that the great apes in particular are extremely adaptive in their application of knowledge, rearranging it in their heads until they hit upon the solution. Wolfgang Köhler, who first discovered this process in the 1920s in relation to tool use, described it as creative and insightful. We now have enough knowledge of the social domain to apply similar ideas to the solutions observed there, such as mediated conflict resolution or the formation of political alliances. The apes' solutions can be so innovative that de Waal has called upon the writings of Niccolò Machiavelli to make sense of the observed complexities.

De Waal compares the "if-then" conditionality of social problem-solving with the tree-like structure of "syntax" or "action grammar" (see also Chapter 14), suggesting that this terminology, adopted from the linguistic domain, may be more than a metaphor. Perhaps the syntax of human language arose out of the organization of action sequences. Even though de Waal focuses on some of the more complex social mechanisms found in nonhuman primates, he at the same time argues that the lives of other long-lived animals pose very similar challenges. The findings of Sofia Wahaj and Kay Holekamp (Case Study 9A) on reconciliation behavior in spotted hyenas drive this point home. Similarly, for other sophisticated social mechanisms, such as reciprocal exchange of benefits, there is a growing body of evidence covering a variety of nonprimates.

In sum, Chapters 7–9 address the active structuring of social data by animals faced with a multitude of organizational levels (e.g., the dominance hierarchy, the kinship network, alliances) about which they need to make quick, well-informed decisions in order to succeed. Whereas none of the contributors offer proof that animals handle social information exactly the way they propose, their ideas provide an excellent starting point for further research that may take us beyond vague statements such as that observed strategies are "smart" or "complex." The proposals formulated here are far more specific, and may inspire empirical testing in both the field and laboratory.

7

Equivalence Classification as an Approach to Social Knowledge: From Sea Lions to Simians

RONALD J. SCHUSTERMAN, COLLEEN REICHMUTH KASTAK, AND DAVID KASTAK

Despite the fact that no other living species has communication skills that rival human language, the Cartesian position that nonhuman animals lack cognitive abilities can no longer be defended. On the contrary, many species show varying degrees of proficiency at forming abstract concepts and manipulating symbols. Indeed, the past quarter century has seen a wealth of research on learning, memory, and concept formation, demonstrating that animals as diverse as sea lions and simian primates are able to solve abstract problems. Strong evidence of complex cognitive skills can be found in the laboratory and in the field when circumstances require an animal to form representations of its earlier experiences.

Recently there have been several findings indicating that animals not only classify objects along physical dimensions, but can also classify them according to their function. Thus, dissimilar stimuli can be perceived as constituting an equivalence class. An equivalence class is defined by equivalence relations that emerge between dissimilar stimuli that share common spatiotemporal or functional contingencies. These stimuli can include signals, individuals, objects, behavioral events, and reinforcers. Relations among the stimuli comprising an equivalence class enable individuals to apply similar meaning to structurally disparate class members. Thus, stimulus equivalence, or more simply, equivalence, allows animals to rapidly form natural categories.

The behavioral theory of equivalence relations developed from research dealing with reading comprehension in human subjects. Thirty years ago, Sidman (1971) reported the emergence of reading comprehension in a severely retarded teenage boy. Through reinforcement training with a procedure called matching-to-sample (MTS), the boy was taught to select pictures of objects and printed names of objects upon hearing spoken object names. Following this training he spontaneously related the written words to the corresponding spoken English words as well as the objects they represented. The critical node or link in this paradigm was the spoken word; it allowed the boy to spontaneously connect the printed names and pictures of objects with one another even though he had not been able to relate them before.

Equivalence as a conceptual process has some of its historical roots in the notion of mediated generalization, in which different stimuli become related to one another through their connection with a common element. For example, various food items may be rapidly generalized, not on the basis on their physical similarity, but rather, on the basis of a common response, "salivation," that mediates the relationships among the different foods (see for example Pearce 1994, p. 128). Similarly, with respect to language, humans tested via galvanic skin response show greater generalization between the printed words "urn" and "vase" than the printed words "urn" and "earn." In this case, common linguistic meaning, rather than structural similarity, serves to mediate the relationship between the different words (Riess 1940).

A revised model of equivalence, as developed by Sidman (1994), is particularly useful for describing the symbolic behavior of verbally competent subjects. However, the formation of equivalence relations is far from being the sole province of humans. There is now accumulating evidence that at least some nonhuman primates, marine and terrestrial mammals, and birds form equivalence classes in natural as well as laboratory studies.

In this chapter we will explain how equivalence relations are established in the laboratory as well as in the field. We will describe how stimulus equivalence phenomena show that associative learning mechanisms, far from being simplistic, can account for novel and complex behaviors. This approach provides a useful framework for understanding how a variety of animals organize information about social groups as well as social relationships involving themselves and others.

Equivalence in the Animal Laboratory

Sidman's original formulation of equivalence classes relied on the integration of experimental results and mathematical theory. This led to a behavioral definition of equivalence that was based on the demonstration of three emergent pairwise stimulus relations (reflexivity, symmetry, and transitivity) from a subset of directly trained relations. Procedurally, this meant that a formal demonstration of equivalence class formation could take place only within the context of MTS. Such an MTS procedure involves the training of conditional discriminations, or "if-then" rules, between stimuli by requiring the subject to select one of two or more comparison stimuli as the correct match to a sample stimulus. A given comparison stimulus is thus selected conditionally upon the presentation of a particular sample (Carter & Werner 1978). Based on the formulation by Sidman and Tailby (1982), equivalence relations can be demonstrated following MTS training of the two relations A→B and B→C (that is, if stimulus A is the sample, select stimulus B as the comparison; if B is the sample, select C as the comparison). The reflexive relationships A→A, B→B, and C→C, symmetrical relationships B→A and C→B, transitive relationship A→C, and symmetry/transitivity combination C→A should all emerge without further training. These emergent relationships demonstrate that the three stimuli have become equivalent, or substitutable, for one another. This procedure is especially suited to studies of linguistic behavior because verbal labels, pictures of objects, and printed words can easily be designated as sample and comparison stimuli.

Potential members of an equivalence class in an experimental setting can be arbitrary shapes, sounds, or any one of a variety of different stimuli; however, some psychologists view the ability of human subjects to form equivalence classes a result of linguistic competence (see, e.g., Horne & Lowe 1996, 1997 and associated commentaries). This view is not supported by recent research showing that verbally deficient humans, and even humans lacking basic linguistic skills, can form equivalence relationships between items in their environment (e.g., Carr et al. 2000). Thus, equivalence classification, rather than being a cognitive by-product of linguistic competence, is more likely a basic ability that underlies or structures complex human behaviors such as language. Given this perspective, the issue of how nonhuman animals organize perceptual information into meaningful categories takes on new comparative significance.

An example of an MTS procedure adapted for use with nonhuman subjects is shown in Figure 7.1, which illustrates a sea lion matching visual stimuli. Animals have had some success in demonstrating the properties of equivalence within the context of MTS procedures. Reflexivity, or generalized identity matching, has been convincingly demonstrated experimentally by sea lions (Kastak & Schusterman 1994), bottlenose dolphins (Herman et al. 1989), and common chimpanzees (Oden et al. 1988). Symmetry and transitivity have been more difficult to demonstrate in MTS procedures (but see results of D'Amato et al. 1985; review by Zentall 1998). Thus far, only a single sea lion has shown all the emergent properties of equivalence relations (Schusterman & Kastak 1993). Further, the same sea lion subsequently transferred her knowledge of the equivalence relations from MTS to a two-choice simple discrimination task (Schusterman & Kastak 1998). The success of the sea lion in the original MTS study (Schusterman & Kastak 1993) can be attributed to several factors, including the large sample size of stimulus sets, exemplar training with a subset of those stimulus sets, and the subject's experience performing identity matching prior to symmetry and transitivity testing. Rather than being an isolated case of success, we expect that, given the same extensive and appropriate training and testing methodology, many other taxa will also show evidence of equivalence classification, or what could also be termed classes or categories of object relations.

This view is bolstered by a variety of studies with animals showing categorical behavior in experimental procedures other than MTS. For example, Schusterman and Gisiner (1997) have suggested that the grammatical sequences of gestural signs or lexigrams used in animal language research may lead to the formation of functionally equivalent classes (see also Herman et al. 1984). In studies with language-trained sea lions, referential signs of a given type (e.g., objects, actions, or modifiers) could be interchanged without disrupting the resulting performance of the animal. For example, any sign representing an object, whether a "ball," "cone," or "cube," or even a novel object, generated an object-oriented response when placed in the correct position in an instructional sequence. However, if the signs for an object and an action were transposed in a sequence, the sea lion often balked. Schusterman and Gisiner cited the apparent substitutability of signs sharing a common sequence position as evidence of functional class formation, with

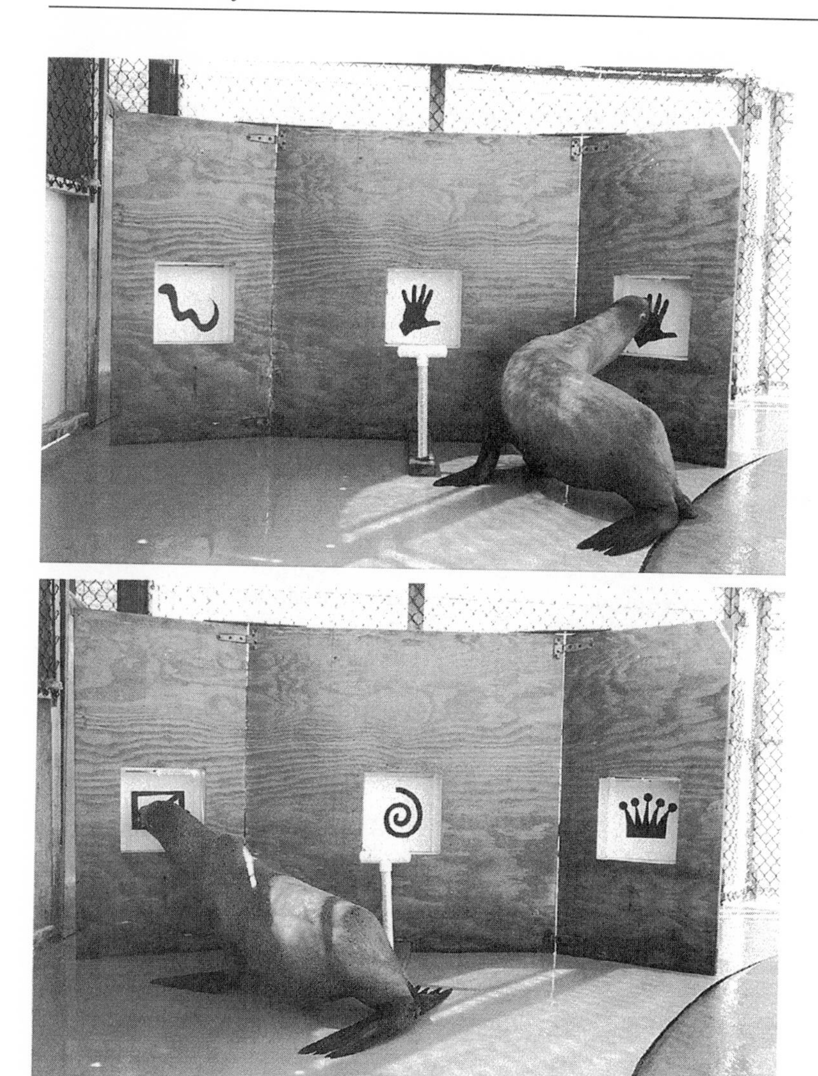

Figure 7.1. A California sea lion relating stimuli in the context of a matching-to-sample procedure. The subject positions her head at the station in front of the center panel, views the sample stimulus *(center)* and two comparison stimuli *(sides)*, and then selects one of the comparison stimuli by touching it with her nose. The upper photograph shows an identity match (A→A, or in this case, hand→hand). The lower photograph shows an arbitrary match (A→B, or in this case, spiral→rectangle). The procedure can be extended to allow for testing of symmetrical and transitive relations.

the different stimuli in a class sharing general response topographies that do not extend to stimuli occupying other sequential positions. Further support for the idea that functional classes can arise from sequential procedures in nonhuman animals comes from studies with rhesus monkeys showing a transfer of function between stimuli sharing the same ordinal positions in different stimulus sequences (Chen et al. 1997). After learning to organize multiple sets of stimuli into desired sequences, a rhesus monkey can immediately organize a recombined set of stimuli (drawn from different stimulus sets) into the correct order on the basis of their prior ordinal positions. Figure 7.2 illustrates this finding with a hypothetical example showing how sequence training can generate emergent equivalence classes based on common ordinal positions. Both sea lions and simians understand that disparate stimuli appearing in a given sequential position are related to one another.

Vaughan (1988) was the first to argue that the distinction between functional classes and equivalence classes was artificial, and based primarily on differences in procedure. He trained pigeons on a discrimination reversal procedure in which the subjects were presented with a sequence of 40 different slides of trees that were divided into two arbitrary sets of 20 slides each. The pigeons were conditioned to peck at any of one set of slides, designated as positive, and to withhold pecking when presented with any of the other set of slides, designated as negative. Following learning of the positive set, the reinforcement contingencies were reversed, and responses to members of the

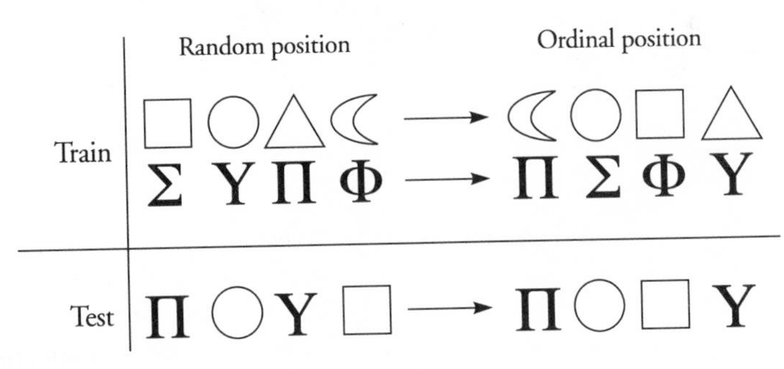

Figure 7.2. An illustration demonstrating how, hypothetically, sequence training can generate emergent equivalence classes based on shared ordinal positions. In the *training* phase, the animal learns to place four randomly assorted geometric shapes or Greek letters into appropriate ordinal positions. Afterward, in a *test* phase, the animal demonstrates that it can spontaneously substitute a geometric shape for a Greek letter and vice versa by placing the stimuli presented into the four appropriate ordinal positions.

formerly negative set were reinforced as positive. After repeatedly shifting the reinforcement contingencies between the two sets of stimuli, the pigeons began changing their responses to all members of a set after being exposed to the reversed contingency with just a few. Thus, the reversed contingency for slides presented at the beginning of a session predicted reversed contingencies for slides presented in the remainder of the session. Although the stimuli in each set were not physically similar or directly related, they became interrelated through their common behavioral functions. Vaughan concluded that the functional classes formed by his pigeons in the context of a simple discrimination/reversal procedure were the same as equivalence classes demonstrated in MTS procedures. However, his interpretation was not widely accepted because he failed to determine if the two functional classes formed by his pigeons would transfer to equivalence classes as demonstrated by MTS.

Sidman and his colleagues (1989) replicated and expanded Vaughan's procedure to determine if human subjects would show transfer to MTS. In this experiment, the functional classes formed by human subjects in a simple discrimination/reversal procedure transferred immediately to conditional discriminations in a MTS procedure. Further, the functional classes were expanded through traditionally defined equivalence relations. Thus, for human subjects, functional classes and equivalence relations do appear to comprise the same behavioral processes (Sidman 1994).

Recently, we adapted Sidman and colleagues' (1989) procedure to test whether functional classes formed by another nonhuman species, the California sea lion, in a simple discrimination/reversal procedure would also generate traditionally defined equivalence relations (Reichmuth Kastak et al. 2001). Using a repeated reversal procedure with stimuli designated "letters" and "numbers," we first documented the formation of functional classes in two California sea lions. This was accomplished through training on two-choice discrimination trials where a number was always presented with a letter. Once each subject consistently responded to the letter class, the reinforcement contingencies were reversed, and subsequently, only selections of number stimuli were rewarded. Following many reversals in reinforcement contingencies from one positive class to the other, we found that the sea lions began to treat all of the stimuli in a class alike; that is, the responses established for one class member immediately transferred to all other members of that class (see Figure 7.3). We next tested our sea lions to determine if class membership would transfer across procedures to an MTS context. We found

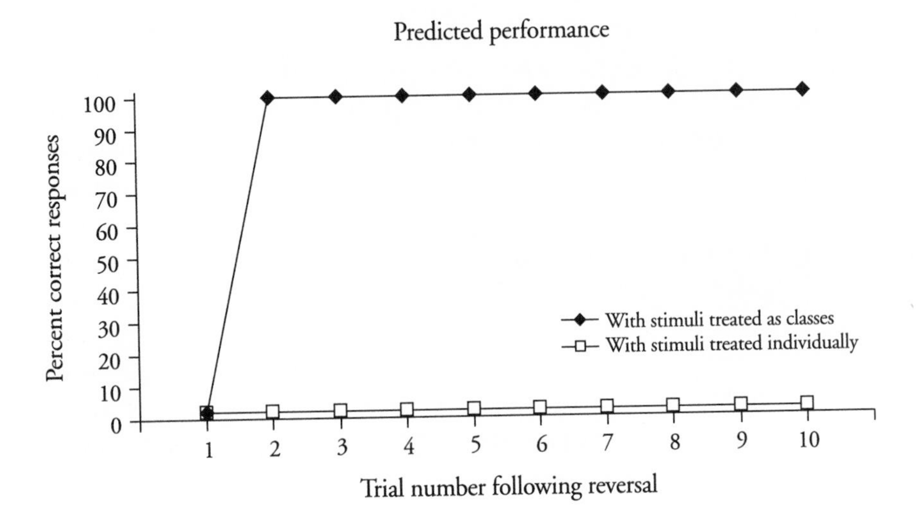

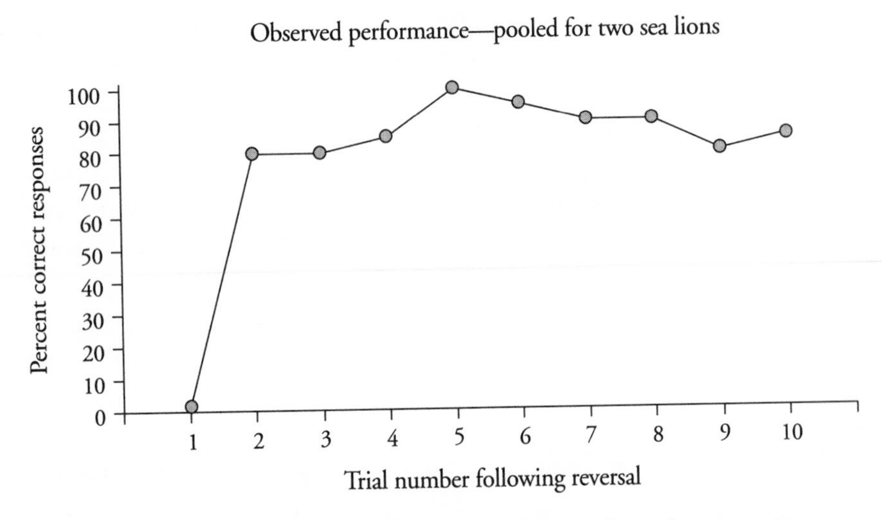

Figure 7.3. Predicted and observed performance showing equivalence classes emerging within a repeated reversal experiment. Two sea lions were trained with 20 dissimilar stimuli divided into two sets of 10. Two-choice simple discrimination trials were presented with one stimulus from each set, with stimuli from the same set always reinforced as correct choices. Following acquisition of responses to one set, reinforcement contingencies were repeatedly alternated between sets. The upper panel shows predicted performance on the 10 unique trials following a set reversal with subjects treating the stimuli either as individual problems or as members of a class. The lower panel shows the observed performance pooled for two sea lions, which is strikingly consistent with the model based on equivalence classification.

that the sea lions immediately related the letters and numbers into class-consistent conditional discriminations; that is, given three stimuli, the sea lions matched the sample stimulus (either a letter or a number) to the comparison stimulus belonging to the same class. Further, we found that these classes could be expanded through equivalence relations. Once a new stimulus was related to an existing class member, new relationships emerged between the new stimulus and all the other members of the class without further training. In all of these experiments, appropriate within-class responding produced class-specific food reinforcers, that is, correct responses to the letter class were followed by one type of fish reinforcement (herring), whereas correct responses to the number class were followed by an alternate type of fish reinforcement (capelin). These results show that equivalence classes can be readily formed when stimuli are related to the same biological reinforcer. Further, these findings show that the functional classes formed by our sea lions were the same as equivalence classes. Whether this result would be obtained with pigeons remains to be determined.

Although the majority of laboratory experiments investigating abstract classification in animals have not been conducted with an equivalence model in mind, the results of many of these studies also describe new relationships emerging from directly trained relationships (see Zentall 1998 for a partial review). These studies include operant procedures and classical conditioning paradigms, as well as complex categorical problem-solving tasks (see, e.g., Savage-Rumbaugh et al. 1980; Wasserman et al. 1992). The sum of what we presently know about relational learning in a range of animals leads us to believe that relational learning abilities are not unique to any single animal taxa studied in experimental contexts. Whether this ability extends to more natural and complex social situations is the central theme explored in this paper.

Equivalence and Contextual Control

Even the expanded model of equivalence, which includes relations based on function, responses, and reinforcers, may not, at first glance, appear to account for the flexibility with which some animals classify stimuli in their environment. To facilitate an understanding of how equivalence applies to complex behaviors seen in the field, in this section we introduce the ideas of contextual control and behavioral contingencies, which are important in determining equivalence class membership under a variety of conditions.

According to the mathematical definition, an equivalence relation satisfies the properties of reflexivity, transitivity, and symmetry. As described earlier, this definition is restrictive in both a theoretical and an experimental sense. However, Sidman more succinctly described an equivalence relation as "defined by the emergence of new—and predictable—analytic units of behavior from previously demonstrated units" (1994, pp. 387–388). Thus, after training A→B and B→C, the relation C→A emerges without training. This may also occur when members of a potential class share a common reinforcement history, temporal or spatial contiguity, or a common response. The colloquial definition of the word "equivalence" is similarly restrictive. In lay terminology, equivalence implies that members of a class are the same, or at least treated as such. Such misunderstandings likely stem from the idea of mutual substitutability or interchangeability and can lead to confusion of equivalence with failure to discriminate. This type of confusion has resulted in only limited acceptance of equivalence as a useful model for describing social relations (Connor et al. 2001; Seyfarth & Cheney in press).

Ideas regarding applications of equivalence in a social setting would be nearly useless if not for contextual control, which allows the content of an equivalence class to change based on context, or the environmental conditions that share some spatial, temporal, or situational contingency with the class members themselves. For example, in one context, a hammer and a screwdriver might be members of the class "tools," defined by a common general function. These tools might also be related by spatial contiguity (e.g., they occupy the same drawer in a toolbox), or shared association with a third stimulus (e.g., they were manufactured by the same company). However, in a new context, based on specific function, these two tools occupy different classes (i.e., one is used to drive nails, the other is used to turn screws). Thus, the idea of contextual control allows equivalence classes to be both flexible (class membership can change based on context) and hierarchical (stimuli can belong to more than one class at the same time).

In operant terminology, the equivalence relation and contextual control can be exemplified by a four-term behavioral contingency, conditional stimulus—discriminative stimulus—response—reinforcer (see Sidman 2000 for a review of the relation between four-term contingencies and equivalence relations). The three-term part of the contingency (discriminative stimulus—response—reinforcer) describes generally understood operant behavior in which the discriminative stimulus sets the occasion for a particular response, which is then reinforced. The response is strengthened, but only in the presence of the appropriate discriminative stimulus. The four-term contingency adds contextual control to the three-term contingency in the form of the conditional stimulus. In the presence of one conditional stimulus, a given discriminative stimulus occasions the appropriate response. In the presence of a second conditional stimulus, a different discriminative stimulus becomes relevant, and thus a different response is reinforced. In every case, the conditional and discriminative stimuli, appropriate response, and contingent reinforcer become members of a class, and the defining equivalence relation includes each stimulus pair as well as each pairing of stimulus and reinforcer.

Connor and colleagues (1999; Case Study 4B) have described the formation of super-alliances and shifting alliance partners in bottlenose dolphins. A hypothetical example of a four-term contingency inspired by this observation of alliance formation is shown in Figure 7.4. In this example, a male dolphin chooses alliance partners based on their usefulness in helping to obtain access to two types of resources: food and ovulating females. The presence of either food or ovulating females determines the salience of one group of male dolphins or another. If ovulating females are present (conditional

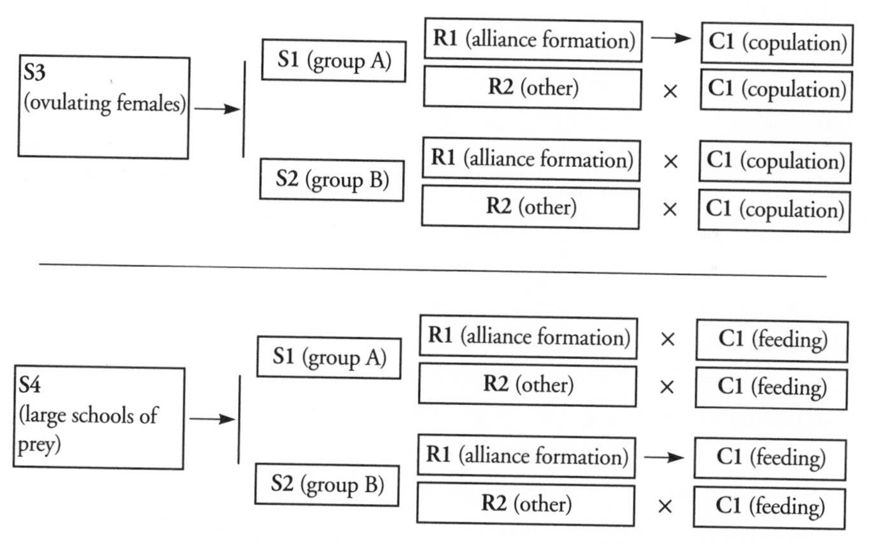

Figure 7.4. Four-term behavioral contingencies as applied to Connor et al.'s (1999) descriptions of alliance formation in bottlenose dolphins. The presence of a conditional stimulus, related to a given discriminative stimulus, occasions an appropriate response that leads to a specific reinforcing consequence. In each of the two cases depicted, the conditional and discriminative stimuli, appropriate response, and contingent reinforcer become members of an equivalence class.

stimulus 1), the male may form an alliance (response 1) with the first group (discriminative stimulus 1), leading to copulation (reinforcer 1). If schools of prey are present (conditional stimulus 2), the male may form an alliance (response 1) with the second group (discriminative stimulus 2), leading to successful foraging (reinforcer 2). Other factors, specifically the motivational state of the male dolphin, will enter the contingency as additional conditional cues and determine which stimuli and responses are appropriate for a particular context. When the concept of conditional control is considered, equivalence becomes a powerful and flexible paradigm within which a variety of very complex social relations in animals may be studied. Further, if an understanding of ordinal rank in social settings is superimposed as an additional contextual control, hierarchical preferences of association can emerge, dramatically increasing the number of pairwise or dyadic relations possible. As discussed earlier, sea lions and monkeys readily understand ordinal position in laboratory tasks, suggesting that they can form concepts of rank relations among individuals in social settings. These concepts allow an understanding of where another animal fits in with respect to all other members of the group.

Another criticism of the equivalence model is that it fails to account for "learning without reinforcement" (Seyfarth & Cheney in press). This criticism appears especially valid given the incorporation of behavioral units into the definition of equivalence, and Sidman's position that "The establishment of equivalence relations is, then, one of the outcomes of reinforcement contingencies" (1994, p. 387). Thus, distinctions between behaviors demonstrated in the laboratory, under controlled conditions of food reinforcement, and those that occur in the field, with no observable or explicit reinforcement, seem to be valid. However, this argument is countered by the examples of classical conditioning leading to emergent relations that satisfy the revised definition of equivalence (reviewed in Zentall 1998). Classical conditioning establishes associations between biologically neutral events and biologically significant events. Consider a vervet monkey observing an unknown, unrelated monkey who is grooming a third monkey, with whom the first monkey has recently had an aggressive encounter. The spatiotemporal contiguity of the unknown monkey, the known enemy, and the aversion, aggression, or stress response of the first may be sufficient to condition a similar response to the unknown monkey in the absence of explicit reinforcement. Thus, equivalence classification readily explains how relations between many pairs of individuals can be learned and remembered.

Experiments with humans and nonhuman animals, and theoretical treatments of equivalence (e.g., Sidman 2000) have spawned novel ideas resulting in a broader view of what equivalence classification means. As equivalence theory continues to develop, the reinterpretation of previous laboratory and field studies and the design of future experiments will likely generate a more parsimonious approach to a wide array of behavioral phenomena, including complex social behavior by nonhuman animals.

Approaches to the Study of Social Knowledge

The model of equivalence relations was developed to better understand a variety of behavioral phenomena, including 1) how symbols acquire meaning, 2) how novel behavior occurs in the absence of explicit reinforcement, and 3) how an individual classifies perceptually dissimilar items into the same category. We have applied the model of equivalence in order to better understand how animals living in social groups may benefit by forming categories, which include not only recognizable individuals but also relationships between individuals.

Primatologists have often observed that the behavior of individual animals may be influenced by the behavior of other individuals with whom they have previously associated; these interactions can be further influenced by the participants' mutual interactions with other individuals. For example, individuals may be more likely to interact positively with their siblings, in part because of their previous positive interactions and in part through their mutual interactions with their mothers. More complicated cases of social knowledge occur when the information acquired by an individual is gained through observations of others interacting, rather than by direct interactions. Thus, the attitudes and behaviors of animal A in relation to animals B, C, or D may be affected by A's observation and evaluation of the interactions and outcomes involving B and C, B and D, C and D, and so on. This type of social knowledge, which requires an animal to observe, classify, and evaluate interactions between other recognizable individuals, has been termed "triadic awareness" by de Waal (1982; Chapter 9), "non-egocentric social knowledge" by Cheney and Seyfarth (1990; see also Chapter 8), and "knowledge of third party relationships" by Tomasello and Call (1997). The latter investigators have hypothesized that this is an exclusive ability of simian primates (Tomasello & Call 1997; Tomasello 2000).

The ability of primates to perceive triadic relationships among classes or groups of conspecifics was first noted in the 1950s by Japanese primatologists and European ethologists. For example, Kawai (1958) observed that a juvenile Japanese macaque might be aggressed against by nonrelatives when it was alone, but that such behavior was rarely observed when its mother, who was dominant to the aggressive individuals, was present. However, the presence of other dominant adults near the juvenile did not have the same inhibiting effect. Kawai concluded that the potential aggressors recognized that only the dominant mother, and not the dominant nonrelatives, would protect the juvenile offspring. Similarly, Kummer (1967) observed that hamadryas baboons displayed knowledge about third-party relationships in the context of dominance interactions. He describes a situation in which a male is dominant to two females, A and B. Female A threatens female B, blocking the latter's access to the male while she simultaneously appeases the male by sex presenting. Female A then keeps herself between the male and female B, so that any return threat by B would also be directed toward the male. According to Kummer, this demonstration of a triangular or tripartite relationship, or what he also terms the "tactic of protected threatening" (1995, p. 41), indicates that the threatening female evaluates the relative ranks of the individual that is threatening as well as the one that she is soliciting. Further, experimental evidence shows that an adult male hamadryas baboon will not challenge a rival male for access to a female if he has previously observed the rival interacting positively with the female (e.g., in a grooming interaction), but this same male will frequently fight the rival male if he has not seen them together (Bachman & Kummer 1980). This suggests that the male's decision to engage his rival in an aggressive encounter is based on his evaluation of the strength of the social bond between the rival and the female.

Tomasello and Call (1997) list a number of complex interactions showing that different primates have an understanding of third-party social relationships. One of these interactions was originally described by Cheney and Seyfarth (1986, 1989) as "complex redirected aggression." An individual from one kin group witnesses a member of its matriline in an aggressive encounter with an individual from another matriline. After observing this encounter, the focal animal is likely to act aggressively toward a relative of its kin's opponent. Not only did Cheney and Seyfarth (1989) find that relatives of two opponents were significantly more likely to display aggression toward each other following a fight than during a matched control period, but they also noted that the vervet monkeys reconciled following fights, not only with their former opponents, but with the kin of their former opponents as well.

Cheney and Seyfarth (1999) documented similar trends in aggressive behavior from free-ranging baboons that "observed" fights between relatives and control subjects by listening to sequences of prerecorded calls that simulated aggressive interactions. Thus, these field observations suggest that members of a kin group, or even more generally perhaps, the members of a nongenetic alliance (see Connor et al. 2001), recognize others as equivalent members of the same category (friends) or not (foes) as hypothesized by Schusterman and colleagues (2000). This hypothesis suggests that equivalence classification constitutes a general learning process that is the basis of the coding of social relationships in many species. In this manner, an animal may learn to combine individuals of the same matriline because they have a common history of shared spatiotemporal interactions and functional relations. These equivalence judgments may be made on the basis of one's direct past interactions with other individuals or by observing the social encounters of others. Thus, when an individual reconciles after a fight, not with a former opponent, but with the kin of a former opponent, it does so because members of the same matrilineal group or alliance have become functionally interchangeable.

Recently, Seyfarth and Cheney (in press; Chapter 8) have proposed that the social knowledge of monkeys and baboons is more complex and hierarchically structured than either conventional associative theories of learning or our current version of equivalence theory can account for. Rather, Seyfarth and Cheney propose that simian primates organize information about their social companions into two-level nested hierarchies based on matrilineal kinship and rank; they emphasize important contextual or evaluative aspects of social knowledge, a point also made by de Waal (1982). It is important to note that the presence of nested hierarchies does not preclude an analysis based on the equivalence model. In this paper and elsewhere (Schusterman & Kastak in press) we have shown how context and the affective state of the perceiver can control equivalence classification. Unlike the many descriptive behavioral terms discussed in this section, this mechanism can explain how animals form flexible categories as reflected in complex social behavior.

Social Knowledge in Sea Lions

As reviewed earlier, the best evidence of equivalence classification in the laboratory comes not from simian primates but from sea lions. Sea lions are pinnipeds, a group of flipper-footed mammals that give birth on land or ice and forage at sea. They comprise three taxonomic families derived from a bear-like ancestor: the true seals, or phocids (for example, elephant seals and harbor seals), the eared seals, or otariids (fur seals and sea lions), and the odobenids (walruses). In contrast to both simian primates and fully aquatic dolphins, the pinnipeds are smaller-brained, shorter-lived, and show less complex social and coordinated behaviors. Nevertheless, there is good observational evidence showing that most if not all pinnipeds are capable of using social knowledge in ways that are somewhat comparable to highly social taxa such as simian primates, cetaceans, and terrestrial carnivores.

Our laboratory experiments with sea lions suggest that their cognitive skills allow them to organize streams of sensory input from several modalities into categories of meaningfully related items. Similar mechanisms are probably at work when sea lions try to solve social problems in nature.

Sea lions and fur seals, the otariid pinnipeds, are colonial breeders whose social structure is usually characterized by moderate to extreme polygyny with groups of females densely aggregated and with males highly territorial. In contrast to the brief but intense maternal care and attendance in phocids, the maternal behavior of fur seals and sea lions more closely resembles that of terrestrial mammals; that is, female otariids routinely leave their pups to forage at sea during a relatively protracted lactation period in which the pup is not weaned for about 6 to 24 months of age (Reidman 1990). Prior to weaning, females and their pups repeatedly reunite by calling to one another over some distance and then confirming their mutual recognition by visual, olfactory, and tactile cues at short range. Indeed, it is likely that older pups recognize their mothers by voice, smell, gait, posture, and facial expression. Thus, early in the pup's development, it acquires multimodal input and begins forming meaningful representations. These repeated mother-pup reunions undoubtedly play a key role in both parties learning to recognize each other's vocal characteristics at great distances and under a wide variety of environmental and behavioral conditions. Furthermore, females apparently learn about the gradual and subtle changes that must occur during the maturation of their pups' voices. Indeed, Insley (2000) has shown, in playback experiments to northern fur seals on St. Paul Island in Alaska, that despite not interacting with one another for nearly eight months, females and their pups were still responsive to one another's vocal playback. Moreover, Insley tested a few mature females returning four years after weaning and they still remained responsive to the playbacks of their mother's calls, which had been recorded while they were still pups. This study provides strong evidence that there is the potential for long-term social relationships in otariid pinnipeds.

There is a good deal of other evidence that suggests that there are long-term interactions among related female otariid pinnipeds. Some of this evidence stems from their tendency toward natal philopatry. Although offspring tend to return to the same breeding sites as their parents, there is little known about whether parents and offspring meet and interact throughout their lifetimes. However, given several factors relating to their known social behavior, it seems likely that maternal kin do establish social relationships with one another. These factors include 1) strong site fidelity, 2) very slow and gradual weakening of bonds between mothers and their offspring, 3) the overlap in maternal care sometimes observed between offspring born in successive years, 4) the fact that many otariids remain at their breeding grounds the majority of the year or even all year, and 5) observational and experimental evidence showing individual recognition (for reviews see Trillmich 1996; Insley 2001).

Two additional pieces of evidence collected in captive settings provide further support for the idea that these pinnipeds may establish long-term social bonds. With respect to the bonds between mothers and their pups, experimental demonstrations of filial imprinting with captive sea lion pups show that sea lions form strong attachments to human surrogate mothers that last for three or more years, well beyond weaning (Schusterman et al. 1992a,b). In terms of relationships with other maternal kin, observations of association patterns within a colony of captive California sea lions showed females interacting affiliatively with their kin and more aggressively with nonrelatives (Hanggi & Schusterman 1990). These relationships lasted over several years and involved mature breeding daughters associating preferentially with their mothers and each other. On the basis of these findings with California sea lions, Trillmich (1996) suggests that interactions between matrilineal kin may be a major factor influencing the social organization of these animals in the field. Mature daughters, for example, may potentially benefit from their interactions with their mothers. Such interactions may provide them with some degree of protection as they rear their young. Indeed, in an often overlooked study, Sandegren (1970) reports that on Lewis Island, Alaska, some Steller sea lion females maintain a social bond with their subadult offspring that is just as strong as the bonds they form with their newborn pups. He also reports that despite aggressive competition for preferred nursing sites, some females with their own pups are extremely affiliative toward other females with young pups. Are these adult individuals from the same matriline? If so, then perhaps pinnipeds aggregating in a particular site may be more closely related than expected by chance. Given this possibility, female social relationships and the role of female mate choice may be far more important in structuring pinniped sociality than had hitherto been realized (Trillmich 1996).

There is also increasing evidence that males, like females, have long-term social interactions spanning three or more years. For example, it has been found that the most reproductively successful Steller sea lion males were those that had at least three seasons of experience maintaining the same territorial site and interacting with many of the same neighboring opponents (Gisiner 1985). The classifying of rivals into "familiar" and "novel" groups appears critical to the success of experienced territorial males. Such males do not waste energy interacting aggressively with familiar rivals, but these established males will aggress against newcomers with alacrity and authority. The "dear enemy" effect in the context of habituation is not sufficient to account for these patterns of aggression because the decrease in aggressive sensitivity toward a territorial neighbor persists over successive breeding seasons without any apparent recovery effects.

Although territorial male otariids may not act in a cohesive and well-coordinated fashion to drive out newcomers, several territorial males will frequently come together where their territorial borders meet and simultaneously attack any rival that is situated at the area of territorial boundary convergence, as shown in Figure 7.5a. During such skirmishes, all males in the vicinity are aroused, and will cross territorial boundaries on their way to the fight (Peterson 1968; Gisiner 1985). These multimale fights (Gentry 1970) frequently gain the attention of females who may be resting nearby (see Figure 7.5b). The motivational mechanism for the convergence of territorial males attacking a newcomer is consistent with the classic idea of social facilitation (Zajonc 1972) and the more recent notion of social enhancement (Galef 1996). In this case, the presence of the original combatants in a particular area may energize or prime nearby territorial males to exhibit similar aggressive behavior toward the intruder. Although observing the fights of neighboring males may spur an observer to join in the aggression, male observers sometimes seem to simply monitor the outcomes of these interactions. Following their observation of a rival in combat, Steller sea lions are more likely to attack while their territorial rival is still fatigued. In this way, experienced territorial males may capitalize on "windows of opportunity" by initiating fights when they have the greatest chance of success (Gisiner 1985).

During the breeding season of California sea lions, nonterritorial subadult males also exhibit behaviors indicating that they are sensitive to how their own aggressive actions might affect a territorial bull (R. Schusterman, personal observation). For example, if a sexually motivated subadult moves alone through a territory, he will invariably move stealthily and silently; sometimes the young male will even hide behind a boulder, looking out in the direction of the resident male in a very surreptitious fashion. Once the resident male has been alerted to the presence of the intruder (frequently by the vocalizations of nearby females), the young male, remaining silent, immediately shows submissive behavior toward the bull and usually escapes into the water, and on at least one occasion, into the blind of the observers. Sometimes two or more nonterritorial males will slink through a bull's territory in a coordinated fashion. Once detected by the bull, they race off together, charging side by side out of the territory (Figure 7.5c). Similar observations of males modulating their behavior in the presence of territorial bulls have been made in captivity (Schusterman & Dawson 1968).

In the context of triadic relations, male sea lions frequently respond to females squabbling on their territories (Eibl-Eibesfeldt 1955; Peterson & Bartholomew 1967; Sandegren 1976; Gisiner 1985). From the standpoint of proximate and ultimate causes, why do they do this? The male invariably intervenes during these female fights by vocalizing and then moving between the females, thereby separating them. Such "pacifying" interventions by these males in the fights of females appear similar, at least superficially, to those observed in chimpanzees, when a third party breaks up a fight by pulling the combatants apart and standing between them to prevent further fighting (de Waal 1982). In chimpanzees, and of course in humans as well, such a tactic is understood to relate to coalition formation and reciprocity (de Waal 1992). However, in the case of male sea lions intervening in female fights on their territories, the motivation for the behavior can be described in a different way. The proximal cause of the bulls' intervention behavior likely depends on loud female vocalizations that capture the attention of the sexually motivated male and direct his behavior to the female interactions. The ultimate cause of the male interventions is most likely related to factors involved in reproductive success. Gisiner (1985) suggests that breaking up female fights puts males into contact with pre-estrous females more often than random ola.

b.

c.

Figure 7.5. Territorial male California sea lions driving out newcomers. *a)* Two males converge upon their territorial boundaries to simultaneously bark at and bite an intruder *(far right)* who had entered the rookery a bit earlier. *b)* One of the territorial males *(left),* in the act of attacking the newcomer *(right),* is in a "dominance" posture; the newcomer is displaying a submissive posture and facial display and is sidling away to retreat from his adversary. Neither sea lion is vocalizing and the females in the background are observing the fight. *c)* A territorial male discovers that two subadult males have entered his territory; he begins barking and the two young males gallop away without producing any vocalizations. Note the resting females on the male's territory.

factory inspection of females normally would. This finding is further supported by Gisiner's data showing that, in one year at Año Nuevo Island, California, male intervention in female fighting preceded nearly a third of 61 copulations observed. Clearly, the intervention behavior of sea lions, unlike that of chimpanzees, has little to do with peacemaking per se. Instead, this behavior appears related to the fact that sexually motivated males are sensitive and hyper-vigilant to female aggressive behaviors that are correlated with sexual receptivity. If this is true, and social cognition plays a role, then one might expect to find that more experienced territorial males might show more intervention behavior than less experienced territorial males. However, it is as yet unknown if this is the case.

The question arises as to whether male otariids, like nonhuman primates, dolphins, and some terrestrial carnivores, form coalitions during the breeding season in order to increase their reproductive success. Campagña and colleagues (1988) observed that subadult male Southern sea lions come together in groups to make what has been termed "group raids" on breeding areas containing resident territorial males and aggregations of females and pups in order to secure females for copulation. Although these raids are frequent and sometimes successful, Campagña and colleagues suggest that, rather than being coordinated activity among group members, the apparent synchrony of the raiding group is due to social facilitation among sexually motivated males. However, the fact that the composition of these raiding parties is frequently similar suggests that there is some degree of durable relationships within the group.

Like males, female sea lions rely on their experiences and observations to categorize individuals in their environment. Sexually mature females appear to divide males into one of two groups: experienced and familiar males, or inexperienced and novel males. Both Steller sea lion and California sea lion females, while avoiding territorial sites occupied by males in their first or second year of tenure, tend to approach and copulate on similar territorial sites that have been occupied by the same male for three or more seasons (Gisiner 1985; Heath 1989). Thus, the reproductive success of males and mate choice by females likely depends on learning to recognize individuals and on placing those individuals into meaningful categories or groups.

Although these observations of sea lions fall short of the myriad of complex behaviors produced by nonhuman primates and dolphins, they do suggest a level of social complexity that depends on monitoring and categorizing their own relationships with others. At this time, it is unclear whether these cognitive processes extend to interactions that sea lions observe occurring between other individuals. However, it is important to note that the social behavior and social complexity of otariid pinnipeds has not as yet been investigated to the same extent as that of larger-brained, longer-lived species.

Social Knowledge in Other Animals

Alarm and Food Calling

As we pointed out earlier, there is a growing body of evidence suggesting that nonhuman primates, marine and terrestrial mammals, and birds are capable of equivalence classification in natural as well as laboratory settings. Since the original laboratory experiments on mediated generalization and equivalence were spurred by attempts to analyze symbolic meaning in humans, it is not surprising that when animal language research peaked in the late 1970s and early 1980s there were numerous examples suggestive of equivalence formation. The evidence came from studies in which great apes and parrots were taught to label things symbolically, either with gestures, icons of various types, or vocally (for reviews see Pepperberg 1986; Schusterman & Kastak 1993; Dube et al. 1993; Cerutti & Rumbaugh 1993; Sidman 1994). Similarly, there have been classic field studies analyzing the symbolic nature of the alarm calls responded to by vervet monkeys (see Cheney & Seyfarth 1990 for a review). These studies showed that individual monkeys respond to acoustically discriminative calls with different types of anti-predator responses, even when the predator is not visually present. As the upper panel of Figure 7.6 indicates, the monkeys that perceive these calls place the structurally different alarm vocalizations of conspecifics and extraspecifics into the same category of predator. Thus, the meaning of a neighbor's call to a predator bird and a starling's call to a raptor are the same; in other words, the calls, although structurally different, are functionally equivalent. The critical evidence that the vervet monkeys are responding to the symbolic meaning or common referent of the alarm calls and not to the emotional intensity or other structural characteristics of the sound comes from playback experiments. For example, a monkey that has habituated to the playback of an alarm call dishabituates only when the meaning of the call is changed and not when the structure of the alarm call is changed (Cheney & Seyfarth 1990).

Recently, Zuberbühler's (2000a,b; Case Study 13A) research with Diana monkeys can be closely compared to that just described for vervet monkeys. This study is particularly interesting from the standpoint of applying the equivalence model to social knowledge because it brings into play the idea that categories are subject to contextual control, that is, the composition of a class can shift with the motivational or emotional state of the individual doing the classifying. As shown in the lower panel of Figure 7.6, Diana monkeys learn to classify at least five different events into a "leopard" equivalence class. Two of these events include alarm calls, "SOS" screams, and "waa" barks given by chimpanzees, which are also subjected to leopard predation. Thus, in one context, that of defense against predation by leopards, Diana monkeys respond to chimpanzee calls with conspicuous alarm-calling behavior that functions to warn conspecifics and to signal the leopard predator that it has been detected. On the other hand, chimpanzees also prey on Di-

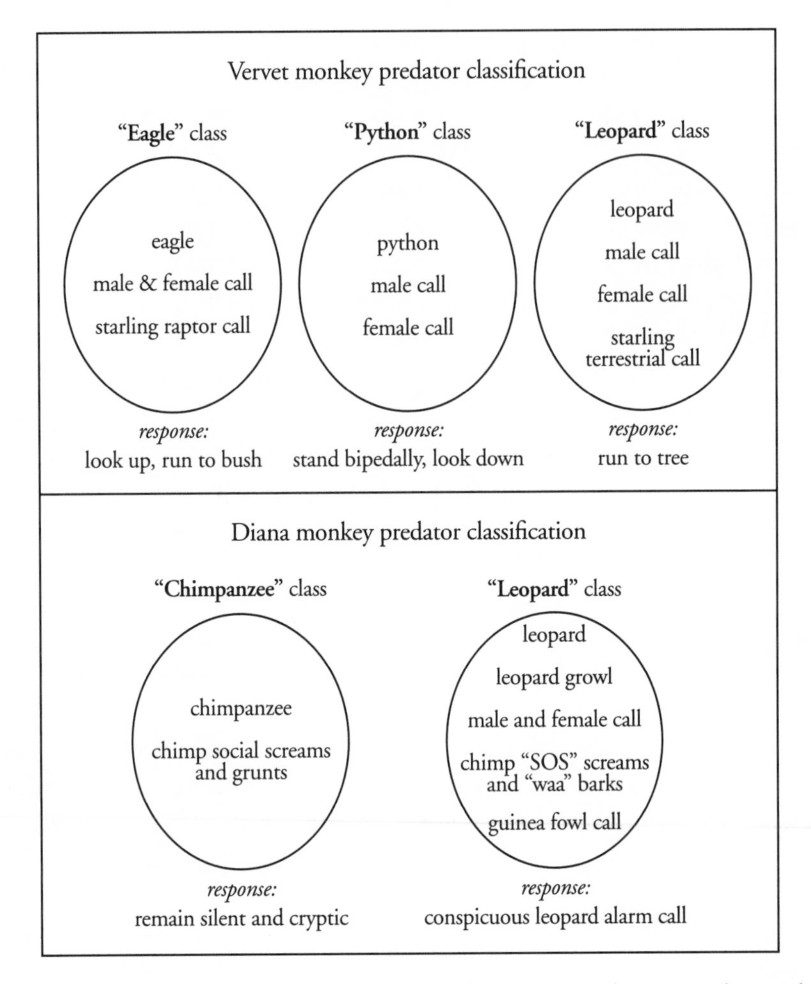

Figure 7.6. Examples showing the cues and alarm calls representing the same predator and eliciting the same behavioral response in vervet monkeys *(upper panel)* and Diana monkeys *(lower panel).* It is likely that the cues comprising each predator class and their associated response are related by equivalence.

ana monkeys. In the context of defense against predation by chimpanzees, two other chimpanzee vocalizations, social screams and grunts, receive a response from the Diana monkeys as an equivalence class; upon hearing the social calls of chimpanzees, the monkeys respond cryptically rather than conspicuously. Alarm call playback experiments involving vervet monkeys and Diana monkeys show how dissimilar events can take on the same meaning and elicit a similar, appropriate, and context-dependent response. Thus, in a manner similar to the linguistic performance described earlier, in which a boy connects a printed word with a corresponding object on the basis of their mutual relationship with a spoken word (the nodal link), the monkeys connect conspecific alarm calls and extraspecific calls to the sound and sight of a predator (the nodal link).

Playback experiments in the field dealing with food calling in rhesus monkeys in Cayo Santiago, Panama, are comparable to those just described concerning alarm calling in other simian primates. Briefly, Hauser (1998), by means of habituation experiments, found that rhesus monkeys place three distinctive calls (warbles, harmonic arches, and chirps) into the class "high quality, rare food" and two other distinctive calls (coos and grunts) into the "low quality, common food" class. Hauser found that his monkeys responded to the meaning of the food calls rather than the structure of the individual calls, and therefore the calls within each class were interchangeable in terms of function.

Mothers and Their Offspring

The most basic experimental observation of triadic relationships in simian primates is one in which the scream of a juvenile vervet monkey is played through a concealed speaker to a group of females; the typical result is that the control females immediately look to the mother of the juvenile, sometimes before she reacts to the cry of her offspring (Cheney & Seyfarth 1980). Thus, the adult female monkey formed an equivalence class comprising the *juvenile,* its *scream,* and its *mother.* Although some primatologists view this behavior as "uniquely primate" in nature (Tomasello & Call 1997) the scant experimental evidence from similar playback experiments conducted with nonprimate species has generated conflicting findings. In a rarely cited study, Porter (1979) observed that when the cries of an infant leaf-nosed bat were played back to its creche, harem males and some neighbor males behaved as if they recognized the relation between the *infant,* its *cries,* and its *mother.* Upon hearing the playback of an infant's cries, harem males frequently crawled directly to the mother of that particular infant and pestered her until she reacted to the voice of her offspring by flying to the speaker. Conversely, in a field playback experiment with spotted hyenas, Holekamp and colleagues (1999) found that whereas hyena mothers were likely to orient toward the speaker from which the "whoop" calls of their pups was being played, control females were no more likely to look to the mother than to any other control females.

Coalitions

Undoubtedly, the best example of complex social behavior in animals comes from Jane Goodall's (1986) observations of male common chimpanzees involved in territorial wars in Gombe Stream National Park in Tanzania (Goodall 1986). Over a period of many years, she and her colleagues observed the development of two large male cohesive groups, one from the Kasekela community and one from the Kahama community. These groups had strong identities that were cemented through mutually reinforcing events such as reciprocal grooming, food sharing, and reconciliation behavior. Aggressive interactions between the two groups escalated over time, and resulted in violent behavior being directed by the stronger Kasekela community toward the weaker Kahama community. This violent behavior was directed not only toward the individual males allied with the Kahama community, but also toward the females and offspring associated with them, and even their abandoned nests. The end result of these "chimpanzee wars" was the extermination of every male individual in the Kahama community (see Figure 7.7).

Goodall's observations suggest that the members of these groups, whose internal relationships were unequal with respect to dominance, age, status, kin relatedness, and friendships, were able to reconcile their individual differences within the context of a broader category, that is, their community identity. Thus, although each group was composed of very different individuals, the members of each group were virtually interchangeable with respect to their aggressive attitudes toward their adversaries. An equivalence model explains why empty nests were viciously attacked as if they were the members of the opposing group, as well as why females and infants were attacked during patrols. Additionally, this model explains why the patrolling Kasekela males hunted down *any* individual belonging to the Kahama group, suggesting that the "search image" of the Kasekela males was for the Kahama group as a category and not for the individual males being hunted. This view is supported by the fact that the killing of Kahama males continued well past the point of any reasonable retaliation. Sniff, the last Kahama male killed, posed little threat to the coalition of eight aggressive Kasekela males; al-

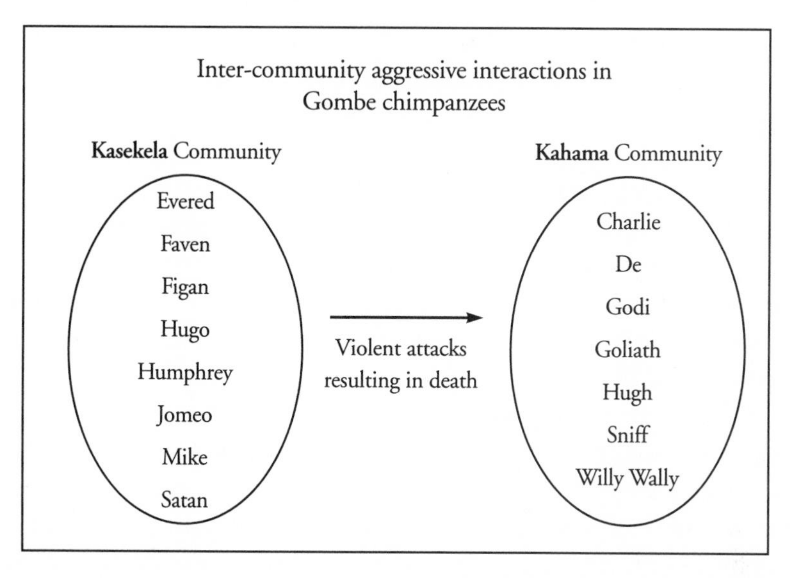

Figure 7.7. Schematic depiction of inter-community aggression between groups of male chimpanzees observed by Goodall (1986). There is a strong sense of identity within each of the communities and individuals clearly distinguish between those who "belong" to their group and those who do not. Before the division of the Gombe chimpanzees into two communities, the members of the Kahama community had enjoyed close and friendly relations with members of the Kasekela community. Once the Kahama community formed, however, the individuals in the Kasekela community no longer treated them like group members; rather they treated the Kahama males like prey animals, who were systematically hunted and killed.

though he had played with many of the Kasekela males as a youngster, he was nonetheless killed, presumably on the basis of his former association with the Kahama group.

Conclusions

There is a growing body of evidence indicating that many avian and mammalian species maintain some degree of group stability and individual recognition. These characteristics likely result in individuals forming such simple social categories as "friend" and "foe" (Schusterman et al. 2000). Behavioral interactions in the context of kinship, dominance, friendship, coalition formation, mate choice, and referential alarm calling may reveal even greater social complexity. What are the learning skills that allow an animal to organize social and communicative information into meaningful categories? Laboratory studies suggest these skills depend on a type of relational learning that has been specifically presented here in the framework of equivalence relations.

Whereas some primatologists (Tomasello & Call 1997; Tomasello 2000) have hypothesized that relational learning is a cognitive skill restricted to primates, there is a good deal of experimental evidence showing that marine mammals such as California sea lions and bottlenose dolphins can also understand categories of object relationships (for a review see Schusterman & Kastak in press). Do these laboratory studies on physical cognition, in which sea lions, dolphins, and simians clearly show an understanding of categories of object relations, hold true for their understanding and use of categories of social relations? Based on our current understanding of social knowledge in these animals, the answer is a very definite yes for simians and probably yes for dolphins. The answer is much more problematic for sea lions and fur seals. Although species may possess comparable cognitive abilities as demonstrated in the laboratory, these physical cognitive skills may not be evident in their behavior in social settings; rather, these skills may be more obviously expressed in ecological contexts such as foraging. The extent to which individuals acquire an understanding of social relationships, and the extent to which they act on that knowledge, appears to be more strongly correlated with degree of sociality. Nevertheless, a model of equivalence relations provides a behavioral learning mechanism that is basic to the formation of flexible categories when an animal's social environment requires referential communication and dyadic or triadic relationships. These same relational skills likely extend to other contexts. No other current approaches to animal social knowledge that we know of provide an alternative process capable of explaining these behavioral phenomena.

8

The Structure of Social Knowledge in Monkeys

ROBERT M. SEYFARTH AND DOROTHY L. CHENEY

The chapters in this volume richly document the complex societies found in many animal species. In this chapter we explore the kinds of intelligence that might underlie social complexity.

Animal societies have been described as "socially complex" for many different reasons. Some, like the colonies of bees or wasps, involve thousands or even tens of thousands of individuals that behave differently toward others depending on colony membership, age, sex, caste, or reproductive state. The sheer size of these groups, and the intricate pattern of interaction among their members, have led some to describe them as "superorganisms" (Seeley 1995). Other animal societies, like those of the dolphins, whales, elephants, and primates described in this book, involve tens or hundreds of animals that recognize each other as individuals. The members of these societies behave differently toward one another depending on kinship, sex, dominance rank, reproductive condition, and their previous history of interaction. Categories cut across one another, and the social calculus required to survive and reproduce seems formidable indeed. Finally, the social behavior of monogamous and even solitary species has been described as complex because of the sophisticated ways in which individuals locate one another, and the intricate coordination and timing involved in mating and rearing offspring.

By itself, however, social complexity does not necessarily imply sophisticated intelligence. Data from insects, together with experiments conducted by Schusterman and colleagues on sea lions (see Chapter 7) as well as many others, have shown clearly that complex societies and social relationships can be generated by relatively simple mechanisms. Indeed, if we have learned anything from decades of debate between behaviorists and cognitive scientists, it is that the same bit of behavior can be explained equally well in many different ways—some cognitively complex, others less so. Does a baboon that apparently knows the matrilineal kin relations of others in her group have a "social concept," as some have argued (e.g., Dasser 1988), or has the baboon simply learned to link individual A_1 with individual A_2 through a relatively simple process like associative conditioning, as others believe (e.g., Thompson 1995)? At present, the preferred explanation depends more upon the scientist's mind than upon any objective understanding of the baboon's.

As a benchmark for comparison with nonhuman species, human social intelligence is always lurking in the background. Humans have sophisticated ways of classifying each other and using these classifications to predict what others are likely to do. Somehow, during the course of human evolution, these cognitive skills were favored by natural selection. How did this occur? How might we have gotten from the relatively simple learning mechanisms proposed by Schusterman and colleagues (Chapter 7) and others to the sort of complex social intelligence we see in humans today?

We begin with a brief review of data on the recognition of social relations in monkeys. We ask: What must a monkey know, and how must its knowledge be structured, in order to account for its behavior? We then review two hypotheses derived from studies of laboratory animals—one offered to explain primate social knowledge, another not—and consider their strengths and limitations. Our goal is to uncover a model of social intelligence that both accounts for existing behavior and explains why, during the course of evolution, some cognitive strategies gained an evolutionary advantage over others.

Knowledge of Other Animals' Kin Relations

East African vervet monkeys *(Cercopithecus aethiops)* live in groups of 10–30 individuals. Each group occupies a territory that is surrounded by the territories of other vervet groups. A typical group contains three to seven adult males, together with five to eight adult females and their offspring. When young males reach adult size between five and six years of age, they emigrate to a neighboring group. Females remain in the group where they were born throughout their lives and form close, long-lasting bonds with their matrilineal relatives. Adult female vervets and their offspring can be arranged in a linear dominance hierarchy in which offspring rank immediately below their mothers. The stable core of a vervet social group, then, is a hierarchy of matrilineal families (Cheney & Seyfarth 1990).

Most affinitive social interactions, such as grooming, mutual tolerance at feeding sites, and the formation of aggressive alliances, occur within families (Seyfarth 1980; Whiten 1983; reviewed in Cheney & Seyfarth 1990). Clearly, individuals distinguish their own close matrilineal relatives from all others because their behavior toward them is so different. For a monkey to achieve a complete understanding of her society, however, she must be able to step outside her own sphere of interactions and recognize the relations that exist among others (Cheney & Seyfarth 1986; Harcourt 1988). Such knowledge can only be obtained by observing interactions in which oneself is not involved and making the appropriate inferences (Cheney & Seyfarth 1990). There is, in fact, growing evidence that monkeys do possess knowledge of other animals' social relationships and that such knowledge affects their behavior.

Evidence that vervet monkeys recognize other animals' social relations first emerged as part of a relatively simple playback experiment designed to document individual recognition by voice (Cheney & Seyfarth 1980). We had noticed that mothers often ran to support their juvenile and infant offspring when these individuals screamed during rough play. This observation, like many other studies (e.g., Hansen 1976; see also Gouzoules et al. 1984), suggested that mothers recognized the calls of their offspring. To test this hypothesis, we designed a playback experiment in which we played the distress scream of a two-year-old juvenile to a group of three adult females, one of whom was the juvenile's mother. As expected, mothers consistently looked toward or approached the loudspeaker for longer durations than did control females. Even before she had responded, however, a significant number of control females looked at the mother. They behaved as if they recognized the close social bonds that existed between particular juveniles and particular adult females (Cheney & Seyfarth 1980, 1982).

In an attempt to replicate these results, we recently carried out a similar set of experiments on free-ranging baboons *(Papio cynocephalus ursinus)* in the Okavango Delta of Botswana (for details of the study area and subjects, see Hamilton et al. 1976; Cheney et al. 1995a; Silk et al. 1999). The social organization of baboons is similar to that of vervets. In these experiments, two unrelated females were played a sequence of calls that mimicked a fight between each of their close relatives. The females' immediate responses to the playback were videotaped, and both subjects were also followed for 15 minutes after the playback to determine whether their behavior was affected by the calls they had heard. In separate trials, the same two subjects also heard two control sequences of calls. The first sequence mimicked a fight involving the dominant subject's relative and an individual unrelated to either female; the second mimicked a fight involving two individuals who were both unrelated to either female (for details see Cheney & Seyfarth 1999).

After hearing the test sequence, a significant number of subjects looked toward the other female, suggesting that they recognized not just the calls of unrelated individuals, but also those individuals' kin (or close associates). Moreover, in the minutes following playback, dominant subjects were significantly more likely to supplant subordinate subjects, suggesting that the dominant female's behavior toward others was influenced by her perception of whether one of her own relatives and another individual's relative had recently been involved in a fight. Females' responses following the test sequence differed significantly from their responses following control sequences. Following the first control sequence, when only the dominant subject's relative appeared to be involved in the fight, only the subordinate subject looked at her partner. Following the second control sequence, when neither of the subjects' relatives was involved, neither subject looked at the other. Finally, following both control sequences, the two subjects were significantly more likely to approach each other and interact in a friendly manner than following the test sequence.

Taken together, these experiments argue that baboons and vervet monkeys recognize the individual identities of even unrelated group members. Moreover, they appear to view their social groups not just in terms of the individuals that comprise them but also in terms of a web of social relationships in which certain individuals are linked with several others. Their behavior is influenced not only by their own recent interactions with others but also by the interactions of their close associates with other individuals' close associates.

Other studies provide additional evidence of monkeys' abilities to distinguish both their own and other individuals' close associates. For example, in a playback study using the contact calls of rhesus macaques *(Macaca mulatta),* Rendall and colleagues (1996) found that females not only distinguish the identities of different signalers but also categorize signalers according to matrilineal kinship. Similarly, in an experiment performed on captive long-tailed macaques *(Macaca fascicularis),* Dasser (1988) trained a female subject to choose between slides of one mother-offspring pair and slides of two unrelated individuals. Having been trained to respond to one mother-offspring pair, the subject was then tested with 14 novel slides of different mothers and offspring paired with an equal number of novel pairs of unrelated animals. In all tests, she correctly selected the mother-offspring pair. In so doing, she appeared not to rely on physical similarity (compare Parr & de Waal 1999), but instead, Dasser argued, classified individuals on the basis of an abstract category analogous to our concept of "mother-child affiliation."

In each of these studies, animals that were grouped into familial associations nonetheless retained their individual identities: a mother and her offspring, for example, were judged to be alike in belonging to the same family but still recognized as distinct individuals.

Finally, in many species of monkeys, an individual who has just threatened or been threatened by another animal will often redirect aggression by threatening a third, previously uninvolved, individual. Judge (1982) was the first to note that redirected aggression in rhesus macaques does not always occur at random; rather than simply threatening any nearby individual, animals will instead specifically target a close matrilineal relative of their recent opponent. Similar kin-biased redirected aggression occurs in Japanese macaques *(Macaca fuscata)* (Aureli et al. 1992) and vervets (Cheney & Seyfarth 1986, 1989).

Kin-biased redirected aggression also appears in more complex forms. In two different vervet groups studied over two different time periods, we found that a female was more likely to threaten another individual if one of her own close relatives and one of her opponent's close relatives had recently been involved in a fight (Figure 8.1; Cheney & Seyfarth 1986, 1989). These results support Dasser's (1988) contention that monkeys recognize that certain types of social relationships share similar characteristics. When a vervet monkey (A_2 in Figure 8.1) threatens B_2 following a fight between one of her own relatives (A_1) and one of her opponent's relatives (B_1), A_2 acts as if she recognizes that the relationship between B_2 and B_1 is in some way similar to her own relationship with A_1 (Cheney & Seyfarth 1990). In a similar manner, when a baboon female hears a playback sequence mimicking a fight be-

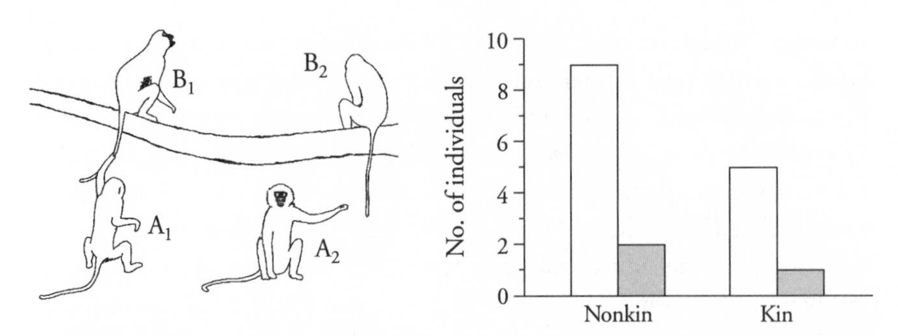

Figure 8.1. Redirected aggression in vervet monkeys. Open histograms show the number of individuals who behaved more aggressively toward an opponent after a fight between one of their own relatives and one of their opponent's relatives than during a matched control period. Shaded histograms show the number of individuals who were as likely to act aggressively after a fight as during the matched control period. Drawing by John Watanabe. From Cheney & Seyfarth (1990), with permission.

tween her own relative and the relative of another female, this temporarily increases the likelihood that her subsequent interactions with that female will be antagonistic (Cheney & Seyfarth 1999).

Knowledge of Other Animals' Dominance Ranks

Along with matrilineal kinship, linear, transitive dominance relations are a pervasive feature of social behavior in groups of Old World monkeys. And like matrilineal kin relations, dominance relations provide human observers with an opportunity to explore what monkeys know about their companions. A linear, transitive rank order might emerge because individuals can recognize the transitive dominance relations that exist among others: A is dominant to B and B is dominant to C, therefore A must be dominant to C. Alternatively, monkeys might simply recognize who is dominant or subordinate to themselves. In the latter case, a transitive, linear hierarchy would emerge as the incidental outcome of paired interactions. The hierarchy would be a product of the human mind, but not exist in the minds of the monkeys themselves.

There is evidence, however, that monkeys do recognize the rank relations that exist among others in their group. For example, dominant female baboons often grunt to mothers with infants as they approach the mothers and attempt to handle or touch their infants. The grunts seem to function to facilitate social interactions by appeasing anxious mothers, because an approach accompanied by a grunt is significantly more likely to lead to subsequent friendly interaction than is an approach without a grunt (Cheney et al. 1995b). Occasionally, however, a mother will utter a submissive call, or "fear bark," as a dominant female approaches. Fear barks are an unambiguous indicator of subordination; they are never given to lower-ranking females. To test whether baboons recognize that only a more dominant animal can cause another individual to give a fear bark, we designed a playback experiment in which adult female subjects were played a causally inconsistent call sequence in which a lower-ranking female apparently grunted to a higher-ranking female and the higher-ranking female apparently responded with fear barks. As a control, the same subjects heard the same sequence of grunts and fear barks made causally consistent by the inclusion of additional grunts from a third female who was dominant to both of the others. For example, if the inconsistent sequence was composed of female 6's grunts followed by female 2's fear barks, the corresponding consistent sequence might begin with female 1's grunts, followed by female 6's grunts and ending with female 2's fear barks. Subjects responded to all playbacks by looking toward the speaker, but did so for significantly longer durations to the causally inconsistent sequences. This consistent difference in response suggests that they recognize not only the identities of different signalers, but also the rank relations that exist among others in their group (Cheney et al. 1995a).

Further evidence that monkeys recognize other individuals' ranks comes from cases in which adult female vervet monkeys compete with one another for access to a grooming partner (Seyfarth 1980). Such competition occurs whenever one female approaches two that are grooming, supplants one of them, and then grooms with the female that remains. In a small proportion of cases, this competition takes a form that is especially interesting for our present purposes. As shown in Figure 8.2, a high-ranking female (ranked 2, for example) approaches two groomers who are both subordinate to herself (say, females ranked 4 and 5). Though 4 and 5 both rank lower than 2, they are not equally likely to depart. In a significant number of cases, the higher-ranking of the two females remains seated, while the lower-ranking of the two moves away (Cheney & Seyfarth 1990).

In so doing, the higher-ranking of the two females acts as if she recognizes that, although she is lower-ranking than the approaching female, her grooming partner is even more subordinate. In order to accomplish this ranking, a female must know not only her own status relative to other individuals but

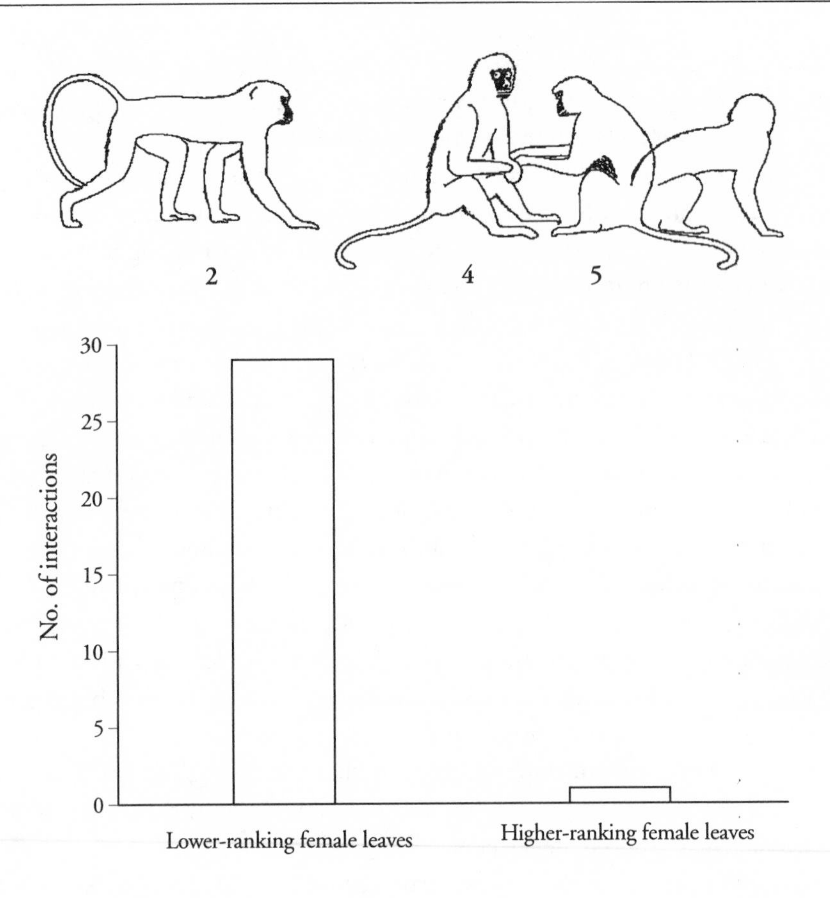

Figure 8.2. Competition over access to a grooming partner in vervet monkeys. Data are taken from all cases during 1985 and 1986 in which a high-ranking female (for example, female 2) approached two lower-ranking females (for example, females 4 and 5) and supplanted one individual and groomed the other. Drawing by John Watanabe. From Cheney & Seyfarth (1990), with permission.

also other individuals' status relative to each other. In other words, she must recognize a rank hierarchy (Cheney & Seyfarth 1990).

The ability to rank other group members is perhaps not surprising, given the evidence that captive monkeys and apes can be taught to rank objects according to an arbitrary sequential order (D'Amato & Colombo 1988), the amount of food contained within a container (Gillan 1981), their size, or the number of objects contained within an array (e.g., Hauser et al. 1996;

Brannon & Terrace 1998). What distinguishes the social example, however, is the fact that even in the absence of human training female monkeys seem able to construct a rank hierarchy and then place themselves at the appropriate location within it.

More Transient, Complicated Social Relations

Although male vervets, baboons, and macaques emigrate from their natal group at about the time of sexual maturity, female kin remain closely bonded throughout their lives. Similarly, dominance ranks among females and immatures are relatively unchanging, though an occasional upheaval may result in all of the members of one matriline rising in rank together (Samuels et al. 1987; Chapais 1988; Cheney & Seyfarth 1990; Gygax et al. 1997). It might seem, therefore, that an individual baboon, vervet, or macaque could simply learn her relative dominance rank early in life and thereafter navigate easily through familiar social terrain. This relatively stable social network is complicated, however, by many short-term, transient social relations that change often. These more temporary relationships cannot be ignored if an individual is to predict the behavior of others.

Male vervets, baboons, and macaques also form linear dominance hierarchies. Because dominance is determined primarily by age and fighting ability, however, rank relations are considerably less stable than they are among females (Walters & Seyfarth 1987). Nevertheless, males appear to recognize other individuals' relative ranks. For example, in a study of captive male bonnet macaques *(Macaca radiata),* Silk (1993, 1999) found that males formed linear, transitive dominance hierarchies that remained stable for only short periods of time. As in other primate species, males occasionally attempted to recruit alliance support during aggressive interactions (roughly 12 percent of all aggressive encounters). Silk found that males consistently solicited allies that outranked both themselves and their opponents. Silk's analysis ruled out simpler explanations based on the hypotheses that males chose allies that outranked themselves, or that males chose the highest-ranking individual in the group. Instead, soliciting males seemed to recognize not only their own rank relative to a potential ally but also the rank relation between the ally and their opponent. If dominance ranks remained stable this would not be a difficult task. However, during Silk's year-long study of 16 males, roughly half of the males changed dominance rank each month (data from Silk 1993, Table 3).

As a second example, consider the close bonds formed by lactating female baboons with resident adult males (Seyfarth 1978; Smuts 1985). Such "friendships" are particularly likely to occur after a new adult male has joined the group and rapidly risen to alpha status, and they typically endure until the female resumes sexual cycling. During this period, the female's friend intervenes on her behalf during aggressive encounters and also carries and protects her infant. Indirect evidence suggests that the male friend is often the infant's father (Bulger & Hamilton 1988; Palombit et al. 1997). One clear function of male-female friendships is to protect the female's infant from infanticide (Palombit et al. 1997).

Other group members seem to recognize the friendships that exist between particular females and particular males. Having been threatened by a more dominant male, for example, subordinate males will sometimes redirect aggression toward that male's female friend (Smuts 1985). In so doing, they act as if they recognize the close bond that exists between the two individuals.

Female members of the same matriline do not necessarily form friendships with the same male. A male often maintains simultaneous friendships with two females of different ranks and from different matrilines (Palombit et al. 1997, 2000). Although the definitive experiments have not yet been conducted, other baboons appear to recognize these patterns of association. A baboon who recognizes that females A_1 and A_2 associate at high rates, and that female A_2 and male X associate at high rates, does not act as if she expects female A_1 and male X also to associate at high rates. Instead, she identifies female A_1 with an entirely different male. Females A_1 and A_2, in other words, are not treated as interchangeable just because they belong to the same matrilineal kin group; instead, baboons seem to recognize that the same individual can simultaneously belong to more than one class.

Underlying Mechanisms

If the data just reviewed had come from a study of children in a nursery school, we would not hesitate to conclude that individuals divide their social companions into groups, that groups have a nested, hierarchical structure, and that the understanding of such relations is both complex and abstract.

Adult humans, for example, know that all the members of a matriarchy form a group, in the sense that they are more closely linked to one another than any one individual is to those outside the family. In their relations with others, the members of a matriarchy are in some respects interchangeable, but they also retain their own individual identities. Taken together, human family members exhibit a hierarchical structure in the sense that a parent, A_1, has a different relationship with her children $A_2 \ldots A_n$ than they do with each other, yet all parent-child relations share common properties that distinguish them from all sibling relations.

Human understanding of social relations is complex because it is not based on any one—or even a few—behavioral measures. Young children learn quickly that X and Y are friends even if they interact only rarely. And finally, human understanding of social relationships is abstract because we give names to types of relationships and can compare one type of relationship to another in a manner that is independent of the particular individuals involved. If someone mentions a sister, friend, or enemy, we immediately have some idea of her relationship with that person even if we have never met the individual involved.

Although human knowledge of social relationships is structured, complex, and abstract, there is no a priori reason to believe that the same mental operations underlie the social knowledge of monkeys and apes. In recent years, several authors (e.g., Heyes 1994; Thompson 1995; Schusterman & Kastak 1998) have argued that the complex behavior of nonhuman primates can be explained by relatively simple processes of associative learning and conditioning. Below we consider these arguments in light of the data just reviewed.

Equivalence Classes

Laboratory studies of equivalence class formation suggest that many animal species can be taught to place dissimilar stimuli into the same functional class. For example, Schusterman & Kastak (1993, 1998) taught a California sea lion, Rio, to group seemingly arbitrary stimuli into "equivalence classes." The stimuli were grouped together based on prior association, not physical similarity (for details see Chapter 7). Next, Rio was presented with one member from the A stimulus class and one from the B stimulus class, allowed to select one of the stimuli, and rewarded with food. Assuming Rio selected A_1 over B_1, she then received repeated presentations of the same stimuli, with A_1 rewarded and B_1 not rewarded, until she achieved a performance of 90 percent correct in a block of 10 consecutive trials (Schusterman &

Kastak 1993). Then Rio was tested with, for example, A_2 and B_2 (transfer test 1) or A_3 and B_3 (transfer test 2), to determine whether she had begun to treat all A stimuli as equivalent to each other and all B stimuli as equivalent to each other, at least insofar as they followed the rule "If $A_1 > B_1$ then $A_n > B_n$." Rio performed correctly on 28 of 30 transfer tests, significantly above chance (Schusterman & Kastak 1998; see also Chapter 7).

The authors suggest that the kind of equivalence judgments demonstrated by Rio constitute a general learning process that underlies much of the social behavior of animals, including the recognition of social relationships. According to Schusterman and Kastak (1998) "both social and non-social features of the environment can become related through behavioral contingencies, becoming mutually substitutable even when sharing few or no perceptual similarities" (1998, p. 1088; see also Dube et al. 1993; Fields 1993; Heyes 1994; Sidman 1994; Wasserman & Astley 1994). Thus, for example, a baboon or vervet monkey learns to group members of the same matriline together because they share a history of common association and functional relations. And when one monkey threatens the close relative of a recent opponent, she does so because members of the same matriline have effectively become "interchangeable" (Schusterman & Kastak 1998, p. 1094). As a result of equivalence class formation, members of the same class are not only "mutually substitutable" but also exhibit "transitivity": if $A_1 > B_1$ and A_2 is a member of the same class as A_1 and B_2 is a member of the same class as B_1, then $A_2 > B_2$.

There is no doubt that associative processes provide a powerful and often accurate means for animals to assess the relationships that exist among different stimuli, including members of their own species. However, before rushing to conclude that nonhuman primate social knowledge can be explained entirely on the basis of learned behavioral contingencies (e.g., Heyes 1994), it seems worth pointing out several ways in which social relations among nonhuman primates do not conform to equivalence class relations.

No Single Behavioral Measure Underlies the Associations between Individuals

It is, of course, a truism that monkeys can learn which other individuals share a close social relationship by attending to patterns of association. Matrilineal kin, for example, almost always associate at higher rates than nonkin. But no single behavioral measure is either necessary or sufficient to recognize such associations. For example, aggression often occurs at as high a rate within families as it does between families, and different family members may groom each other and associate with each other at different rates (Cheney & Seyfarth 1986). To recognize that two individuals are closely bonded despite relatively high rates of aggression or relatively low rates of grooming, a monkey must take note of a variety of different patterns of aggression, reconciliation, grooming, and proximity. There is no threshold or defining criterion for a "close" social bond.

By contrast, the equivalence classes in Schusterman's & Kastak's (1998) experiments were established by repeatedly presenting arbitrary visual stimuli aligned in groups of three. Either spatial or temporal juxtaposition would therefore suffice as a basis for the formation of association of stimuli within an equivalence class.

As yet, we do not know if monkeys distinguish between matrilineal kin bonds and the equally strong bonds that may form between unrelated animals who interact at high rates (for example, male and female baboon "friends"). If monkeys do make such distinctions, this would argue that they assess and compare social relationships using a metric that is based on more than just patterns of association.

Class Members Are Sometimes Mutually Substitutable, Sometimes Not

In discussing the experiment in which an adult female vervet hears a juvenile's scream and then looks at the juvenile's mother (Cheney & Seyfarth 1980), Schusterman & Kastak (1998) argue "that the existing relation between the scream, the juvenile itself, and the frequent association between the infant [sic] and its mother resulted in a three-member equivalence class" (1998, p. 1093). As a result, they are treated the same—not because they have become indistinguishable (Schusterman and colleagues, Chapter 7) but because one stimulus can be substituted for another without violating the associative link that has been formed between them. But in fact the call, the juvenile, and the mother are not interchangeable in this manner. Although listeners may place these stimuli in the same class under some circumstances, the call itself is linked primarily to the juvenile and only secondarily to the mother. In habituation/dishabituation experiments on rhesus macaques, monkeys both distinguished calls from different matrilines and distinguished among the calls of different individuals within the same matriline (Rendall et al. 1996).

Some Social Relationships Are Transitive, Others Are Not

If infant A_1 and juvenile A_2 both associate at high rates with a particular adult female, it is usually correct to infer that the juvenile and infant are also closely bonded and will support one another in an aggressive dispute (e.g., Altmann et al. 1996). Similarly, if A is dominant to B and B is dominant to C, it is usually correct to infer that A is dominant to C (Cheney & Seyfarth 1990). By contrast, if infant baboon A_1 and juvenile baboon A_2 both associate at high rates with the same adult female and she associates with an adult male friend, it would be correct to assume that the male is closely allied to the infant but incorrect to assume that he is equally closely allied to the juvenile. Male baboon friends form close bonds with their female's infant but not with any other of her older offspring (Seyfarth 1978; Smuts 1985; Palombit et al. 1997). To cite another example mentioned earlier, female members of the same matriline often form friendships with different males, and, conversely, the same male may form simultaneous friendships with females from two different matrilines. In the former case, a close bond between female A_1 and female A_2, and between female A_1 and male X, does not imply that A_2 and X are closely linked. In the latter case, the existence of close bonds between male X and females A_1 and C_1 do not predict a close bond between the two females. In fact, their relationship is more likely to be competitive than friendly (Palombit et al. 2000).

Individuals May Belong to Multiple Classes Simultaneously

As the previous examples make clear, at any one time an individual monkey belongs simultaneously to many different classes. An adult female, for example, belongs to a matrilineal kin group, associates with one or more adult males, holds a particular dominance rank, and may be weakly or strongly linked to other females outside her matriline. Again, the natural situation is considerably more complex than that in Schusterman's and Kastak's experiment.

Class Membership Changes Often

Whereas female dominance rank and membership in a kin group constitute relatively stable, predictable behavioral associations, other social relationships change often and unpredictably. For example, we know from field experiments that closely related vervet monkeys groom and support one another in alliances at rates much higher than those for nonkin. Unrelated animals, however, are more attentive to another individual's recruitment call if they have recently engaged in a grooming interaction than if they have not (Seyfarth & Cheney 1984). Thus the social relations among nonkin wax and wane throughout the day, with transient periods when they resemble the bonds among kin and many other times when they do not.

Considering a slightly longer time scale, when female vervets or baboons give birth, they often become extremely attractive to other females, who groom them at high rates and attempt to interact with their infants (Seyfarth 1976, 1977, 1980; Altmann 1980; Silk et al. 1999). The change in the rate of grooming received is particularly pronounced for females of low rank, who otherwise receive little attention from higher-ranking individuals. The low-ranking mother's dominance rank, however, does not change. Moreover, as her infant matures, her attractiveness to others diminishes. Similarly, when a low-ranking female forms a close friendship with a dominant male, she gains access to feeding sites from which she might normally be excluded by higher-ranking females. This preferential access disappears, however, when the friendship is terminated (Seyfarth 1978; Smuts 1985; Palombit et al. 2000).

Finally, consider the problem faced by a male bonnet macaque who, in order to recruit the most useful allies, must keep track of transitive rank relations among 16 male companions in a group in which half of the males change rank each month (Silk 1993).

Training May Distort an Animal's Natural Method of Classification

In Schusterman's and Kastak's experiment, as in many other studies, the subject was first presented with stimuli that had links to one another (A_1, A_2, A_3) and then rewarded for choosing stimuli from one class over stimuli from another ($A_1 > B_1$). Thus trained, the subject generalized her knowledge such that, when presented with any other AB stimulus pair, she always chose A. Speaking conservatively, these results tell us only that, when presented with certain stimuli and rewarded for following a particular rule with a subset, a sea lion will generalize the rule and apply it to all of the other members of that subset. The experiment does not tell us whether, in the absence of training and reward, the sea lion would naturally recognize this particular rule, or, if she did recognize it, whether she would apply it generally beyond her immediate experience.

The distinction between learning that is rewarded in the laboratory and learning that occurs in the wild is important, because any intervention by humans that selectively rewards one kind of learning over another potentially distorts an animal's natural method of acquiring and storing information. For example, pigeons trained to match to sample with just a few stimuli are not able to transfer to novel stimuli, although monkeys and chimpanzees do so easily. However, pigeons do learn to match similar stimuli if they are trained with hundreds of exemplars over hundreds of trials. Apparently, although pigeons can acquire the abstract concept same/different, they seem predisposed to attend to absolute stimulus properties and to form item-specific associations (Wright et al. 1988; see also Wasserman et al. 1995; Shettleworth 1998). Extensive human training, therefore, leads to qualitative changes in the ways in which pigeons classify stimuli. Similarly, Tomasello and colleagues compared the performance of chimpanzees raised by humans (but without language training), chimpanzees raised by their own mothers, and two-year-old children. Human-reared chimpanzees showed more imitation (Tomasello et al. 1993), more joint attention, and were more likely to use gestures to direct the demonstrator's attention (Carpenter et al. 1995) than did chimpanzees raised by their own mothers. In another study, chimpanzees that had been trained to use tokens as symbols were able to solve match-to-sample tasks that required them to judge relations between relations. Naïve chimpanzees could perceive these relations but their knowledge seemed to remain tacit (Premack 1983; Thompson & Oden 1995). Mere exposure to humans, therefore, enhances chimpanzees' problem-solving skills, either by changing their attention and motivation or altering their cognitive abilities.

The Magnitude of the Problem

In Schusterman's and Kastak's experiment, Rio was confronted with a total of 180 dyadic comparisons. This is roughly equivalent to the number of different dyadic comparisons—but not the number of triads—that confront a monkey in a group of 18 individuals. As shown in Figure 8.3, however, the number of possible dyads and triads increases rapidly as group size increases.

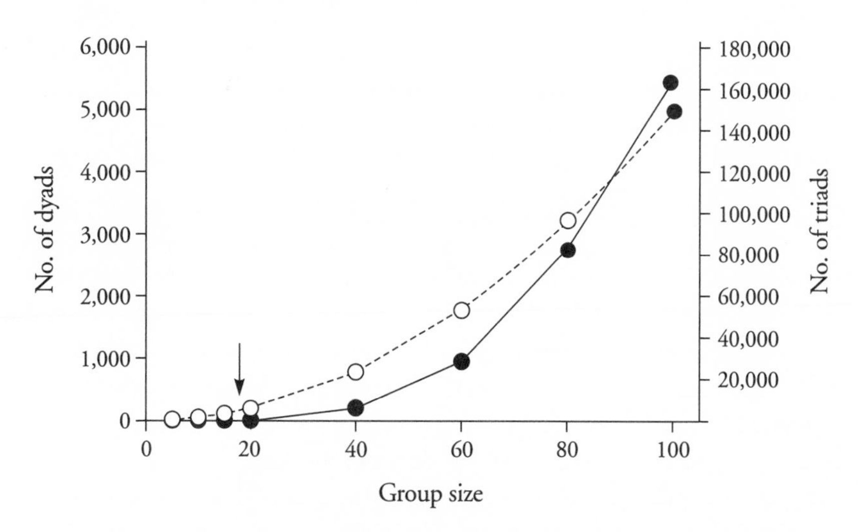

Figure 8.3. The number of different dyadic (two-individual) and triadic (three-individual) combinations possible in groups of different sizes.

In a group of 80 animals (not an unusual group size for baboons, macaques, or mangabeys), each individual confronts 3,160 different possible dyadic combinations and 82,160 different triadic combinations. Under these conditions, it seems likely that free-ranging monkeys and apes face problems in learning and memory that are not just quantitatively but also qualitatively different from those presented in the typical laboratory experiment.

Summary

In several respects the "equivalence classes" that make up nonhuman primate groups exhibit complexities not present in Schusterman's and Kastak's experiments. This is not to say that Schusterman's and Kastak's experiments, or the similar arguments offered by Heyes (1994), are completely erroneous, nor do we mean to suggest that associative learning plays no role in the development of primate social knowledge. In fact, it seems unlikely that a monkey could form a concept such as "closely bonded" without attending to social interactions and forming associations between one individual and another. To some extent, learning about other individuals' social relationships is by definition dependent on some form of conditioning. However, in order to conclude that all primate social knowledge results only from the kind of associative processes discussed by these authors, we need empirical evidence that associative mechanisms can account for behavior as complex as that known to occur in free-ranging primate groups, not a simplified surrogate.

Chunking in Human and Animal Memory

To survive and reproduce successfully, nonhuman primates must be able to predict the behavior of others. Predicting other animals' behavior demands, in turn, that individuals memorize information about all of the dyadic and triadic relations in their group. And, as already noted, memory loads will be enormous in species with large social groups because increases in group size lead to an explosive increase in the number of dyads and triads.

Faced with the problem of remembering long strings of letters, words, or numbers, human subjects learn the string faster and remember it better if some kind of "rule" allows them to group items into "chunks" that conform to a particular rule. Chunking in humans is an adaptive strategy because it increases the capacity of short-term memory (Miller 1956; Simon 1974). It is facilitated if the stimuli to be remembered are segregated by some kind of "phrasing," like spatial or temporal separation, that corresponds to or reinforces the higher-order rule governing the formation of chunks (Restle 1972; Fountain et al. 1984). Finally, even when a chunked structure is not obvious, human subjects will work to discover one. People presented with randomly ordered lists of words will learn to remember them according to semantic categories, like food, clothing, or animals (Bousfield 1953), and in the absence of any obvious categories subjects will invent idiosyncratic relations between words to facilitate chunking and thereby improve recall (Tulving 1962; Macuda & Roberts 1995). Humans thus bring to problems in learning and memory a predisposition to search for statistical regularities in the data.

Chunking in human memory might, of course, be entirely the result of language, since humans can attach verbal labels to the categories they detect or the rules they use to identify patterns. Several recent studies involving animals, however, indicate that "chunking is a more primitive and biologically pervasive cognitive process than has been recognized previously" (Terrace 1987).

In a study that is directly relevant to research on primate social knowledge, Dallal and Meck (1990) observed the foraging behavior of rats in a 12-arm radial maze. Three different food types were hidden in four arms each, and the baiting configuration remained constant from trial to trial. For example, arms 3, 5, 6, and 9 always contained one food type, arms 2, 7, 10, and 12 always contained another, and so on. Performance was measured by calculating the number of arms an individual needed to visit in order to obtain food from all 12 locations.

Subjects initially moved at random from arm to arm as they foraged, but they soon diverged from this pattern to one in which food was selected in a highly predictable order; specifically seed, seed, seed, seed, pellet, pellet, pellet, pellet, and rice, rice, rice, rice. Dallal and Meck (1990) argue that "Rather than manage the 12 locations independently in memory, rats . . . could organize the stable arrangement of differentiable food types/outcomes in a hierarchical fashion" and this mental organization led to improved performance.

Macuda and Roberts (1995) replicated Dallal's and Meck's results and proposed two contrasting models of spatial memory representation for a 12-arm radial maze containing three food types. In the nonchunked representation (Figure 8.4), the organization of memories is not hierarchical, chunking is absent, and each arm and its food type is independently represented in memory. In the chunked representation (Figure 8.5), organization is hierarchical. Food types now serve as higher-order nodes in a memory tree, and each food node prompts retrieval of a spatial map of the arms containing

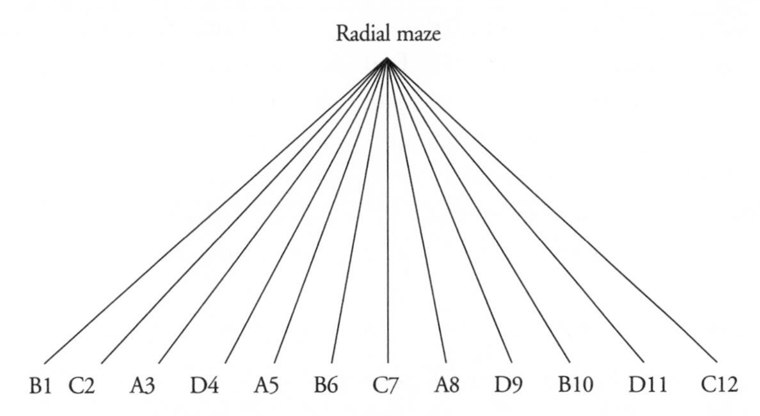

Figure 8.4. A representation of a rat's knowledge of the location of food in a 12-arm radial maze where there are four different food types (represented by the letters A–D) placed randomly among 12 arms (represented by the numbers 1–12). The drawing assumes that each arm–food type combination is remembered independently. Modified from Macuda & Roberts (1995).

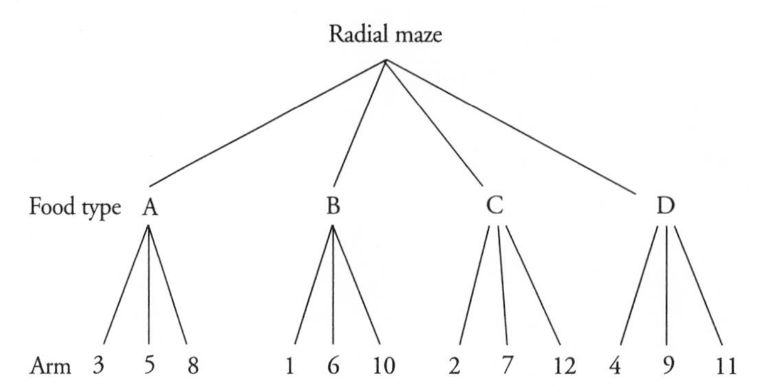

Figure 8.5. A representation of a rat's knowledge of the location of food in a 12-arm radial maze, assuming that the rat groups together, or "chunks," arms that contain similar types of food. Modified from Macuda & Roberts (1995).

that food. Hierarchical chunking reduces the load on working memory, promotes a lower error rate, and leads to the clustering of visits to arms containing the same food (Macuda & Roberts 1995, p. 21). In two experiments designed to test between these models, rats provided strong evidence that their knowledge was organized in a chunked, hierarchical fashion.

The earliest experiments on chunking in animal memory compared animals' performance on tests in which the stimuli had or had not been previously segregated by the experimenter into groups. Reviewing his results, Terrace (1987, p. 351) concluded that pigeons "can impose a self-generated organizational scheme on a series of arbitrary elements" when stimuli have already been segregated. The studies summarized above go an important step further by suggesting that rats actively organize stimuli into groups even in cases in which the experimenter has not provided any structure. They are comparable to studies of free-recall in humans, in which subjects imposed their own idiosyncratic mental structure on an otherwise unstructured dataset (Tulving 1962).

The Link with Primate Social Knowledge

Data on the social knowledge of nonhuman primates are compatible with models of knowledge acquisition based on either the formation of equivalence sets or hierarchical chunking. Both models assume that stimuli—in this case other individuals—are linked in memory through elementary associative processes, and that associative links are structured according to one or more rules. For two reasons, however, the model based on chunking seems more relevant to studies of primate social knowledge: chunking emerges naturally, without human intervention, and chunking has known adaptive advantages over other mental strategies.

Free-Recall versus Training

In Schusterman's and Kastak's study, the subject was rewarded for behaving in a manner consistent with the formation of equivalence classes; alternative choices were not rewarded. By contrast, in the chunking studies reviewed above, subjects received no reward other than food, and this food would have been obtained even if they had not formed chunks. The fact that the rats were not specifically trained suggests that the mechanisms underlying their actions were the same as the mechanisms underlying their behavior under more natural conditions.

Perhaps more important, rats in the radial maze were presented with a world in which there was no inherent structure imposed by the experimenter, yet they imposed a structure of their own. Even in the earliest minutes after they were released into the maze, the animals appeared to search actively for a rule that would allow them to organize and remember the locations of different food (see also Fountain & Annau 1984; Fountain et al. 1984). If rats, without explicit reinforcement, organize randomly placed food locations into chunks, how much more likely is it that monkeys, confronted with a society in which there are already statistical regularities in behavior, will organize their knowledge of social companions into hierarchical groups? The hierarchical chunking model predicts that monkeys who were asked to remember the order in which they had seen pictures of their social companions would perform poorly if these pictures were presented at random. By contrast, if the pictures occurred in groups of family members and in the order of the current dominance hierarchy, the monkeys should perform well (see Swartz et al. 1991 for the beginnings of such a study).

The Adaptive Value of Structured Associations

The adaptive value of chunking is directly relevant to the problem confronting primates like baboons, macaques, or mangabeys that live in large social

Figure 8.6. Vervet monkeys grooming.

groups. Faced with the need to remember hundreds or even thousands of dyadic and triadic relations in order to monitor and predict other animals' behavior, it is logical to assume that monkeys, like rats in a radial maze, will actively search for any rule that decreases memory load. A monkey could organize others into groups by noting patterns of grooming, alliances, and spatial proximity (Figure 8.6). The basis for chunking could be quantitative, in the sense that other individuals would be grouped together only if they had been seen grooming for a certain minimal length of time. Alternatively, it could follow an either/or classification—a single, pivotal alliance could cause animals to be grouped together regardless of whatever else they did.

Whatever metric is used, the adaptive value of chunking as a cognitive strategy is clear, particularly when hierarchical chunking is compared with models of knowledge representation like equivalence sets, which rely on associations between stimuli but lack a hierarchical structure. As group size gets larger, equivalence sets become increasingly implausible: there are simply too many dyads and triads to store in memory (see above, Figure 8.3). By contrast, as group size gets larger a model based on chunking into hierarchical groups becomes increasingly *more* plausible because it offers a means by which individuals can overcome the limits imposed by working memory.

Conclusion

To survive and reproduce, a monkey must predict the behavior of others. In nonhuman primate groups, where alliances are common, prediction demands that a monkey learn and remember all of its opponents' dyadic and triadic relations. The task is similar to the problems faced by human and animal subjects in memory experiments.

In response to these pressures, we suggest that nonhuman primates are innately predisposed to group other individuals into hierarchical classes. They actively search for ways to arrange their companions into rule-governed clusters. Once such groups are formed they somehow label the groups as higher-order nodes in a memory tree, both for ease of recall and to facilitate predictions of behavior. The formation of hierarchical classes is an adaptive social strategy, shaped by natural selection.

9

Social Syntax: The If-Then Structure of Social Problem Solving

FRANS B. M. DE WAAL

Long-lived animals in individualized societies, such as most species treated in this volume, face a range of similar social problems because of life with competitors whose collaboration is needed for survival (or, to turn it around: because of life with collaborators with divergent interests). Many of these animals build societies that encompass multiple reproductive units, meaning that not all members of the society are genetically related. The observed cooperation is therefore incompletely explained by kinship.

Primates: First but Not Unique

The scientific study of individualized societies requires the identification and naming of individuals, the longitudinal documentation of genealogies and life histories, and attention to the dynamics underlying social transaction, such as the network of enduring friendships and rivalries. It is this network that sets these societies apart from anonymous herds of grazers or shoals of fish as well as from simple family units, such as pairs of birds, even those with helpers at the nest. Friendships and rivalries also set these societies apart from colonies of eusocial insects, in which individuals contribute to the shared interests of a genetic clan. Human society has been compared with all sorts of animal groupings, yet its social dynamics most resemble those of the individualized societies of long-lived animals, such as dolphins, hyenas, and chimpanzees.

Primates are by no means unique in this regard, but it is in the study of this taxonomic group that the first steps toward a thorough understanding of individualized societies were made. Perhaps because monkeys and apes remind us so much of ourselves, scientists were prepared to treat them as individual actors, whereas they regarded other animals as interchangeable representatives of their species. When, in the 1950s, Kinji Imanishi and his students broke with the traditional focus on species-typical behavior and decided to individually identify the macaques native to Japan, they were criticized and mocked by Western scientists for humanizing their research subjects (de Waal 2001; see also Chapters 14 and 15 of this volume). Goodall (1990) reveals a glimpse of the resistance to individual identification when Western scientists adopted the same habit: one respected colleague told her that even if individuals of a species were different it might be better to sweep such knowledge under the carpet.

We now agree that individual identification is a prerequisite for the analysis of complex societies, and that the interindividual relationship is the unit of choice to describe social dynamics. This tool, previously employed in a narrow sense by students of dominance-subordination relations, received scientific credibility only in the 1970s through the complementary efforts of Hinde (1976) and Kummer (1978).

Hinde (1976) described social organization as a pattern of relationships emerging from repeated interaction among group members who individually recognize one another. Interactions do not only have immediate consequences but also serve long-term functions: interactions shape social relationships. From an evolutionary perspective, benefits to interactants are most evident in relationships with a reproductive function, such as male-female and mother-offspring relationships. Relationships may also produce nonreproductive benefits, however, such as when two individuals protect one another against attack, tolerate one another around food resources, provide vigilance against predators, or cooperate during intra- or intergroup competition. With regards to the evolution of cooperation it is, of course, assumed that it pays off in the long run in terms of survival and reproduction.

Kummer (1978) considered the benefits that individual A provides to B as A's value to B. Any individual will try to improve this value: B will select the best available A, predict A's behavior, and modify A's behavior to its own advantage. In other words, B will invest in the relationship with A. Whereas most of B's investments may not lead to quick profits, such as immediately useful actions by A, they may help cultivate a relationship that is beneficial to both A and B over the long haul. A good example of such investment is social grooming. One primate may groom another for over one hour without any immediate reciprocation or return of the favor; after the session the two may separate, each going their own way. Why would individuals provide such services to others—grooming is one of the most common social activities in primate groups—if not to foster future beneficial exchange?

Thus, whereas Hinde focused on descriptive aspects of social relationships and how social organization is defined by the way relationships are structured, Kummer was more concerned with the functional side, regarding relationships as commodities. These were the beginnings of the modern view of primate society as a place where costs and benefits are traded between individual actors, a view inspired by Trivers's (1971) theory of reciprocal altruism and highlighted by my (1982) depiction of chimpanzee *(Pan troglodytes)* society as a politically regulated "marketplace of services."

Parallel with the development of this thoroughly social perspective on behavior, there was the even more influential rise of cognitive approaches. This rise was by no means limited to nonhuman primates, yet these animals were part of this movement from the start (e.g., Köhler 1925; Yerkes 1943). Cognitive studies on primates took on a distinctly social flavor. The current fascination with self-awareness, theory of mind, imitation, and deception—a fascination visible in both the animal behavior and child development literatures—found its origin in primate studies by Gallup (1970), Menzel (1974), Premack & Woodruff (1978), and others. Students of primate behavior described social life as a set of problems to be solved by various means, including social tactics requiring intentionality and foresight, or at least a sound knowledge of the social relationships among group mates (e.g., Goodall 1971; de Waal 1982; Cheney & Seyfarth 1990).

Shared Social Dilemmas

Traditionally, primate research has operated somewhat apart from other areas of animal behavior as reflected in the early establishment of separate primatological societies, conferences, and journals, as well as the unique emphasis by primatologists on social context. There is no good reason to maintain this separation, however: most of the concepts of social relationships, reciprocal exchange, and strategizing may apply equally well to societies of other large-brained, long-lived animals. Outside primatology interest in these issues is growing rapidly.

If we consider the basic challenges facing individual players in large multimale, multifemale primate societies, such as those of baboons or chimpanzees, we can easily see a range of challenges that probably apply to societies of unrelated individuals in general. Individual members need to balance several priorities:

Exploiting others: obtaining services, making friends in the right places, getting the most out of profitable partnerships;
Making relationships worthwhile: maintaining and fostering the bonds of friendships by engaging in mutualism, reciprocity, and social tolerance;
Coexisting peacefully: getting along and protecting valued partnerships despite the competition that is inevitable when interests collide.

Individuals capable only of exploiting others will never manage to form the cohesive aggregations observed: they will fight over every scrap of food and fail to support each other in any endeavors. A situation of unlimited tolerance is equally unimaginable among parties with divergent reproductive interests. What we see instead is a fine balance between competitive and cooperative tendencies: ritualized fighting techniques and powerful peacemaking mechanisms serve to maintain cooperation and keep the society together. Inasmuch as groups are formed whenever the advantages of living together outweigh those of competing over the same resources (van Schaik & van Hooff 1983), there is a high premium on maintaining group cohesion. The result is a complex and dynamic interplay between the pursuit of private interests and protection of the common good.

A Syntactical Structure

Here I will discuss the structure of problem solving in such societies, focusing on coalition politics, conflict resolution, and cooperative arrangements, such as reciprocal altruism. In each case, I will first introduce some essential background knowledge after which I will argue that occurrence of the behavior is conditional, constrained by highly specific necessary conditions. In the more complex cases, this conditionality resembles a syntax of hierarchically organized if-then routines similar to the structures characterizing the ordering of symbols in human language. Whereas in protolanguage each communicative unit stands alone without being structurally integrated into the rest of the sentence, true language imposes a grammatical structure that constrains the interpretation of each unit (Calvin & Bickerton 2000; Figure 9.1). Even if not everyone may agree with Bickerton's (2000, p. 20) claim that "there is widespread agreement that the origins of language are to be sought in primate social intelligence, rather than in foraging, tool-making, and the like," his statement does drive home that the term "syntax" is perhaps not a mere metaphor when applied to the social domain.

Similar to the difference between protolanguage and language, the syntax of problem solving ranges from disconnected actions to tree-like structures. Chimpanzees, in particular, seem marked by free arrangements among elements that together lead to a solution. Following Köhler's (1925) thinking on this matter, Rumbaugh and colleagues (1996) declared the creative assembly of pre-existing elements of knowledge a hallmark of Hominoid intelligence. A chimpanzee knows how to stack objects, how to climb on a box, how to handle sticks, how to reach for fruits, and so on, and combines all of this knowledge in a single well-structured solution when faced with all of these elements at once. Rumbaugh and colleagues (1996) speak of *emergents* to set such insightful problem solving apart from the better-known trial-and-error learning.

Matsuzawa (1996; Chapter 14) applied a tree-like structure to technical problem solving by wild chimpanzees in West Africa. He argued that tool-use sequences run from the simple one-on-one level, in which, for example, a nut is picked up by hand (level 0) or a wand is used to catch ants (level 1), to

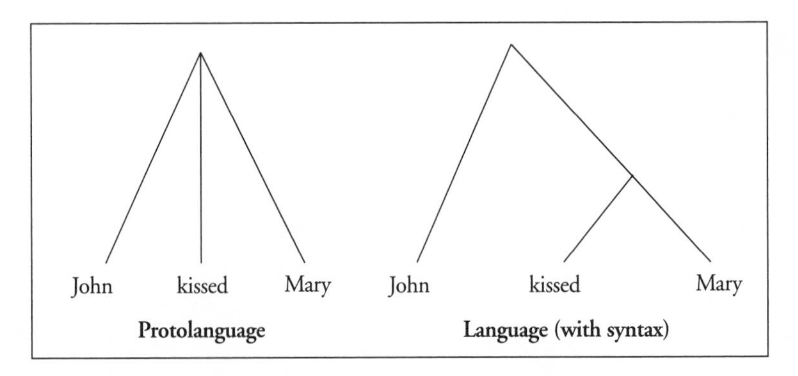

Figure 9.1. Syntax makes the difference between protolanguage and language. After Calvin & Bickerton 2000.

hierarchical structures with a variety of nodes: the more nodes in the tree, the more complex the sequence. For example, in nut cracking, the chimpanzee places a nut on an anvil stone, then lifts a hammer stone to strike the nut (level 2). When the chimpanzee uses an additional stone to wedge and level the anvil stone, the tree structure becomes even more complex, involving three nodes (level 3; Figure 9.2a). Each step is conditional on others: e.g., there is no purpose in hitting the nut with a hammer if it hasn't been placed on a level anvil. Westergaard (1999) subjected the remarkable tool-use abilities of capuchin monkeys (*Cebus apella;* see also Case Study 16A) to the same kind of hierarchical analysis, and a somewhat less formalized approach has been applied to the plant-food processing of wild mountain gorillas *(Gorilla g. beringei)* by Byrne (1996). Matsuzawa (1996) himself compared his problem-solving trees with the way chimpanzees in the laboratory recognize mul-

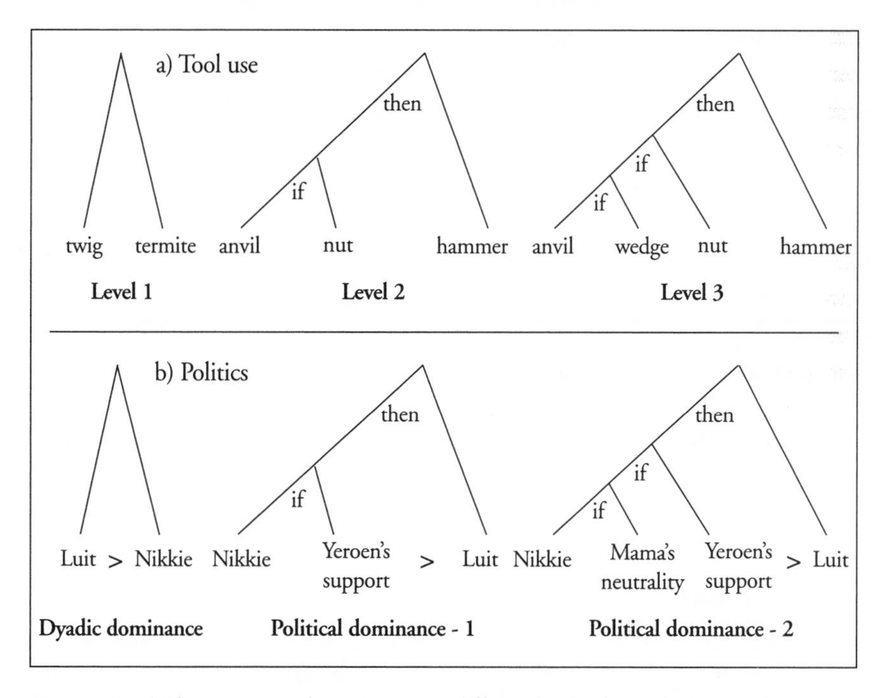

Figure 9.2. *a)* Chimpanzee tool use operates at different levels of complexity, as reflected in the number of nodes in a hierarchical structure of decision making (modified after Matsuzawa 1996). *b)* Political maneuvering of chimpanzees follows the same hierarchical if-then syntax as tool use, here expressing the specific conditions under which Nikkie, who by himself is incapable of dominating Luit, is able to express his top position. Based on de Waal 1982.

tifaceted symbols as well as with the structure of human action sequences (compare Greenfield 1991).

I will apply the same syntactical approach to social problem solving, thus returning to an earlier suggestion in relation to chimpanzee politics, namely that emergents in the technical or symbolic domain are equivalent to those in the social domain: "The extra faculty which makes chimpanzee behavior so flexible is their ability to *combine* separate bits of knowledge. Because their knowledge is not limited to familiar situations, they do not need to feel their way blindly when confronted with new problems . . . In their *social* application of reason and thought, chimpanzees are truly remarkable. Technically their inventiveness is clearly inferior to that of human beings, but socially I would hesitate to make such a claim" (de Waal 1982, p. 51, emphasis in original).

The tree structures seen in social life are essentially the same as those applicable to the technical domain: they have an if-then structure, such that one first needs to do *x* before one can do *y*. The contingencies are similar to those observed in nut cracking. A strong opponent shouldn't be challenged unless one's main ally is at hand to help out. There are many indications that nonhuman primates possess an exquisite knowledge of the contingencies underlying their social position relative to others. Kummer (1971) spoke in this context of knowledge of the "social field," and of the need of nonhuman primates to process large amounts of information for successful social action: "In turning from one to another context at a rapid rate, the individual primate constantly adapts to the equally versatile activities of the group members around him. Such a society requires two qualities in its members: a highly developed capacity for releasing or suppressing their own motivations according to what the situation permits and forbids; and an ability to evaluate complex social situations, that is, to respond not to specific social stimuli but to a social field" (Kummer 1971, p. 36).

The use of a linguistic concept to describe the structure of such social evaluations is deliberate. I deem it possible, even likely, that when our species evolved its unique language capacity and began ordering symbols in syntactical structures, it built upon the reiterative, hierarchical arrangements characterizing more ancient forms of action-sequence, the oldest ones being in the social domain. Exploring the same connection between language and social intelligence, Calvin and Bickerton (2000) specifically emphasized reciprocal altruism. Reciprocal altruism implies an if-then routine *between* individuals. The same cross-individual dialogue applies to other examples discussed here, which differ fundamentally from action chains or tool-use solutions in that the unfolding of events is not under a single individual's control. In a social encounter, each participant follows its own agenda, hence its outcome is not fully predictable to any participant.

One could of course argue that animals may perform complex sequences of tool use or social interaction in a manner that *looks* hierarchically organized, whereas their interaction with the environment only follows a simple stimulus-response scheme. Furthermore, similar outcomes in different species do not necessarily rest on similar processes. Thus, the members of two separate taxonomic groups may build matrilineal hierarchies, yet Cercopithecine monkeys may do so based on triadic awareness—or knowledge of third-party relationships (see Chapter 7)—whereas spotted hyenas *(Crocuta crocuta)* may do so without this ability (see Case Study 5A and Chapter 8). To uncover the precise cognitive capacities underlying behavior is a major task that hopefully will be undertaken with regard to a great variety of socially complex animals. For the moment, all we can do is focus on surface structures, looking for hierarchical if-then rules to describe these structures regardless of the precise cognitive and social mechanisms underpinning them.

Coalition Politics

In many species the dominance hierarchy and the network of affiliative ties appear to exist side by side without much interplay (but this impression may be due to lack of investigation into the connection). The remarkable social complexity of primates is brought about by their capacity to a) alter competitive outcomes and dominance positions through collaboration, and b) establish interindividual relationships for this very reason. Alliance formation links the vertical and horizontal components of social organization by making an individual's dominance position dependent on its place in the affiliative network. Consequently, this network becomes an arena of dominance-related strategy.

The first studies demonstrating that a monkey's position may depend on its affiliative ties appeared in the 1950s in Japanese and in the 1960s in English. One primatologist, Masao Kawai (1958), carried out simple field experiments, such as throwing a sweet potato between two juvenile monkeys and recording the outcome of the competition. The monkey taking the food or winning the ensuing aggressive encounter was regarded as dominant. Kawai conducted many such tests and found that some dominance relationships depended on the distance of the two youngsters to their respective mothers. For instance, juvenile A dominated its peer B if their mothers were far away, yet the dominance relationship was reversed if their mothers were nearby. Such reversals occurred if B's mother dominated A's mother. Thus, the offspring of a high-ranking mother benefits from her presence. Kawai termed the dominance relationship between the juveniles in their mothers' absence the *basic rank,* and in their mothers' presence the *dependent rank.*

Dependent rank is a triadic phenomenon, that is, although it manifests at the level of dyadic (two-animal) interaction, it can be explained only by the relationships of these parties with third individuals and the interrelationships among these third ones. This integration between dyadic and triadic levels is likely to stem from repeated interactions involving three or more individuals. Dependent rank may be caused, for example, by a mother's inclination to defend her offspring against individuals that she is able to dominate. The result is the so-called matrilineal hierarchy in which daughters occupy stable ranks adjacent to but below their mothers (Kawamura 1958; Sade 1967).

Another early example of dependent rank came from Hall and DeVore's (1965) study of wild savannah baboons *(Papio cynocephalus).* The rank of an adult male baboon seems to depend on both individual fighting ability and mutual cooperation. Hall and DeVore observed how an entire baboon troop was controlled jointly by three adult males, which formed the so-called central hierarchy. Individually, without support from the others, none of the central males had much clout. They needed each other to make a common front against individually more powerful rivals.

In the late 1970s, investigators first ventured to describe apparently intelligent tactics of primates, such as seeking affiliative relationships with strategically important individuals; intervening in fights to fortify their own positions; testing the reactions of potential supporters; undermining hostile alliances through disruptive behavior, and so forth (e.g., Chance et al. 1977; Riss & Goodall 1977; de Waal 1978; Walker Leonard 1979). If we follow Lasswell's (1936) classical definition of politics as "who gets what, when, and how," there is no reason why dominance strategies among nonhuman primates should not be labeled as such: the behavior clearly affects access to resources. Some primate studies explicitly stressed the connection with politics, even Machiavellianism, by judging each individual's objectives and decision making from a history of social maneuvers and by adopting an intentionalistic vocabulary to describe these processes (e.g., de Waal 1982). Opening of this area of inquiry was stimulated by the general thesis that social problem solving is the original function of primate (including human) higher mental faculties from which, evolutionarily speaking, all other applications of higher cognition derive (e.g., Humphrey 1976). This "social intelligence hypothesis," according to which the social milieu provided the main selection impetus for increased intelligence, remains popular to explain the relatively large brains found in the primate order (Byrne & Whiten 1997; but see Chapter 1 for a broader perspective).

If coalitions exemplify intelligence applied in the social domain, primates seem far from unique (reviewed in Harcourt & de Waal 1992), even if there remains disagreement about how primate coalitions compare in complexity to those of animals such as dolphins (Harcourt 1992; Connor et al. 1992; see also Case Study 4B). With respect to dependent rank, for example, there is good evidence that the matrilineal hierarchy of macaques and baboons, which is based on a female kin support system, is found also in spotted hyenas (see Case Study 5A).

In the Arnhem chimpanzee colony complex patterns of coalition formation have been described, both anecdotally and systematically, that may serve to illustrate the contingencies of dependent rank in terms of an if-then hierarchical structure. For example, the basic relationship between two males, Luit and Nikkie, would have Luit being dominant, because it was known that in one-on-one encounters away from the rest of the group (e.g., in an isolated indoor cage), Luit was self-assured and dominant. However, Nikkie had formed a partnership with an older male, named Yeroen, in whose presence he had no trouble dominating Luit (Figure 9.2b).

This is only what happened within the top male trio, however. The situation was more complex, because females tended to support Luit. The alpha female of the group, Mama, was a power to be reckoned with: she was able to mobilize all other females. It was not unusual for Nikkie, before he would even approach Luit and initiate a bluff display against him, to first embrace and kiss Mama as if to predispose her favorably toward himself, or at least to have her stay out of the ensuing confrontation. Often Nikkie also embraced Yeroen, or stood next to him to hoot together with him in Luit's direction before going over to his rival. Nikkie would thus put two necessary conditions in place before performing an act of dominance over Luit, the conditions being Yeroen's support and Mama's neutrality. These agonistic sequences, described in detail by de Waal (1982), are every bit as intricate and conditional as those described for chimpanzee tool use with the added complexity of lack of control over the "tools" being employed.

Conflict Resolution

As recounted by de Waal (2000a), the discovery of reconciliation behavior in chimpanzees occurred at a time of increased appreciation of the role of cooperative relationships in primate societies. The behavioral mechanism of reconciliation is relatively simple, consisting of a friendly reunion, such as kissing or embracing, following a conflict between two parties (de Waal & van Roosmalen 1979; Figure 9.3).

The reconciliation concept applies to animals a familiar human interpretation, which comes with connotations of rapprochement, conflict settlement, and even forgiveness. Reconciliation is best regarded as a heuristic concept

Figure 9.3. Chimpanzees typically reconcile after fights with a kiss and embrace, here by a female *(right)* to the alpha male. Photograph by Frans de Waal.

capable of generating testable predictions regarding the problem of relationship maintenance. One central assumption is that a motivational state can be replaced relatively rapidly by its opposite: hostility and fear make way for a positive inclination. Another assumption is that this motivational shift serves to restore relationships. Since the introduction of the reconciliation concept, over one hundred reports on 27 different primate species have been published, mostly in support of predictions derived from the concept (reviewed in Aureli & de Waal 2000).

The first aim of research in this area has been to compare different expectations regarding the social consequences of aggression. The traditional notion that aggression serves as a spacing function would predict decreased contact between individuals following open conflict. The reconciliation hypothesis, in contrast, predicts that individuals try to "undo" the social damage inflicted by aggression, hence that they will actively seek contact, specifically with former opponents. To test these predictions requires a comparison with baseline data. The standard procedure is a controlled design known as the PC-MC method (de Waal & Yoshihara 1983). One of the participants in a spontaneous fight is followed for a given time window, say ten minutes, to collect post-conflict (PC) data, which is then compared with baseline information on the contact tendencies of the same individual in the absence of previous aggression (matched-control, or MC). These two sets of data allow for a division of opponent pairs into "attracted" pairs (i.e., contacting each other earlier in the PC than the MC) and "dispersed" pairs (i.e., contacting each other earlier in the MC than the PC). According to Veenema and colleagues (1994), the conciliatory tendency (CT) after observed fights can then be expressed as CT = (attracted − dispersed pairs) / (all pairs).

This measure has a built-in correction for normal contact rates, such that a CT of 0 percent means that the rate of friendly interaction between any two individuals is unaffected by previous aggression. Studies adopting this paradigm for primates have almost universally demonstrated positive CT values (for some species exceeding 50 percent), meaning that former opponents systematically contact each other more than expected (Aureli & de Waal 2000). That former opponents frequently engage in friendly interaction flies in the face of earlier assumptions about the dispersive impact of aggression, which should have resulted in negative CT values. Moreover, some species show behavioral specificity, that is, their post-conflict reunions stand out by special gestures, vocalizations, or body contacts. Dependent on the species, post-conflict reunions may include mouth-to-mouth kissing, embracing, sexual intercourse, clasping the other's hips, grooming, grunting, and holding hands.

With regard to the pacifying function implied by the reconciliation label, several studies have confirmed that the chance of renewed aggression is reduced and tolerance is restored following post-conflict reunion. For example, when conflict was experimentally induced in pairs of monkeys, individuals permitted to reconcile were more tolerant of each other around a juice dispenser than individuals that had been prevented from reconciling, thus suggesting that reconciliation reduces aggression in the dominant and fear in the subordinate (Cords 1992).

Whereas all of these findings support the specific function suggested by the reconciliation label, which is to repair damaged relationships, it is hard to measure lasting effects. Could it be that the effects concern merely the immediate future (Silk 1996)? This has become a point of debate, and careful data collection on long-term consequences remains needed. It has been argued, though, that since long-term social relationships are an emergent property of short-term interactions a distinction between the two is artificial (Cords & Aureli 1996). Moreover, in virtually all primates studied reconciliation is typical of partners with close ties even after controlling for their high level of interaction. Thus, in macaques, which form matrilineal societies with kin-based alliances, fights among kin are more often reconciled than those among nonkin (Veenema et al. 1994). And chimpanzee males, which band together in territorial aggression against neighbors, counter the disruptive impact of status competition within the group with a conciliatory tendency that far exceeds that of females (de Waal 1986). These cases support the notion that long-term cooperative arrangements are associated with frequent relationship repair.

One of the most powerful generalizations to come out of reconciliation research on nonhuman primates is the valuable relationship hypothesis (Kappeler & van Schaik 1992; Cords & Aureli 2000), according to which reconciliation will typically occur after conflict between parties that represent a high social or reproductive value to each other. In other words, social relationships are commodities the deterioration of which needs to be prevented. Apart from the above-mentioned observational data, experimental evidence comes from a study that manipulated the degree of cooperation among longtail macaques *(Macaca fascicularis)*. Pairs of monkeys were trained to obtain rewards by acting in a coordinated fashion: the only way to obtain popcorn would be for two monkeys to sit side by side at a dispenser, a procedure attaching significant benefits to their relationship. After this training, subjects showed a three times greater tendency to reconcile an induced fight than subjects that had not been trained to cooperate (Cords & Thurnheer 1993).

If reconciliation indeed preserves cooperative arrangements in the face of competition, many animals other than primates would obviously stand to gain from such a mechanism so long as they possess the cognitive capacities to permit it. These capacities, mainly individual recognition and memory of previous fights, don't seem prohibitive. From the beginning, therefore, there have been calls to look beyond the primate order. Yet only recently has post-conflict behavior become a topic of systematic research in such disparate mammals as spotted hyenas (Hofer & East 2000; Case Study 9A), domestic goats (*Capra hircus;* Schino 1998), and bottlenose dolphins (*Tursiops spp.;* Samuels & Flaherty 2000). The results have been positive, suggesting that conflict resolution may be widespread indeed.

Like dependent rank, reconciliation is a behavior that will occur only if certain conditions are met. Even in the simplest case, when one former opponent approaches another, there is great advantage to gauging the emotional state of the other before physical contact is made. This explains why sometimes several approaches are made before the actual reunion, and why apes tend to establish eye contact before proceeding any further. The risks associated with approaching a dominant former aggressor are visible in anxiety indicators, such as self-scratching, which indeed drop in frequency following reconciliation (Aureli & van Schaik 1991). Apart from this conditionality, resting on behavioral indicators of residual hostility, there are the more complex evaluations of the nature of the relationship between the former opponents, and even the relationship with third parties. As explained above, reconciliation takes into account the value of the disturbed relationship. It is not based entirely on what happened in the preceding agonistic encounter, but also on the overall nature of the relationship between the two combatants. Reconciliation may even be directed at or received from kin of the opponent, thus taking into account the relations between entire matrilines (e.g., Judge 1991). There is also evidence for reconciliation between different groups of macaques after intergroup hostilities, which may depend on even more complex social evaluations (Judge & de Waal, 1994).

In perhaps the most complex pattern, thus far known only of chimpanzees, a reconciliation is brought about by a third party: a female acts as

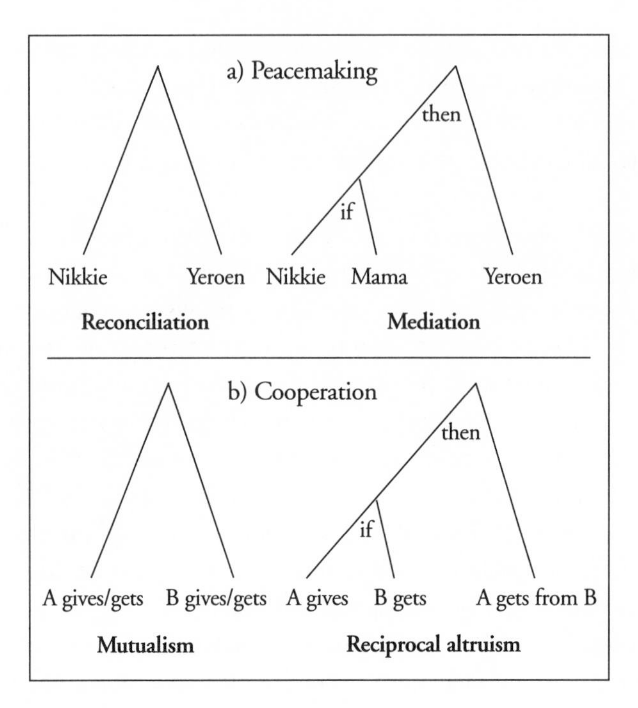

Figure 9.4. *a)* Peacemaking reaches a higher level of complexity when mediation takes place, i.e., when a third party (Mama) facilitates the establishment of contact between two others (Nikkie and Yeroen). *b)* Reciprocal altruism differs from mutualism by an extra step: first A needs to help B before B will help A. This contingency across time doesn't exist in mutualism, since both parties reap benefits at the same time.

catalyst by bringing male rivals together. After a fight, males may remain oriented toward each other, staying close but without initiating an actual reunion. Females have been observed to break the deadlock by grooming one male, then the other, until she has brought the two of them together, after which she withdraws. Thus, the female has created a condition that appears conducive to the reconciliation process (de Waal & van Roosmalen 1979; Figure 9.4a).

Reciprocal Exchange

Reciprocal altruism (RA) has a built-in contingency. The usual definitions, refined following Trivers's (1971) pioneering article, require RA to involve a) repeated dealings between individuals, b) a time delay between given and received helping behavior, c) an obligatory contingency between giving and receiving, and d) discouragement or punishment of cheaters (Rothstein & Pierotti 1988). RA is contrasted with "mutualism," in which cooperating parties benefit simultaneously, such as when a group of wild dogs brings down a wildebeest (Dugatkin 1997). Mutualism lacks all of the above requirements for RA, which stipulate an interindividual if-then rule known as tit-for-tat: "If you help me—I will help you" (Figure 9.4b).

But what is the actual evidence for RA in animals? Over the years, my team has examined the food-sharing tendencies of both brown capuchin monkeys and chimpanzees to determine whether food is exchanged reciprocally over time or is shared in return for other "currencies" in the marketplace of services. The capuchins' food-sharing tendencies were examined in a dyadic context rather than in the entire group in which they lived.

Two adult monkeys of the same sex were placed in a test chamber divided into two sections by a mesh partition. One monkey was allowed access to a bucket full of attractive food, and free to either monopolize it or move close to the mesh and share actively or passively by allowing its counterpart access to dropped pieces. The situation was then reversed so that the second individual had access to attractive food of a different type, while the first did not. The rate of food transfer in both directions through the mesh was found to be reciprocal both in its distribution over different dyads, and over separate testing days within each dyad (de Waal 1997a, 2000b). Although this study examined food sharing in an artificial setting, the results were not anomalous—spontaneous food sharing among unrelated adult capuchins has been observed both in a colony at the Yerkes Primate Center and in the field (Perry & Rose 1994; de Waal 1997a; Rose 1997).

In order to study how chimpanzees share food, a situation was created in which a monopolizable food source was available to individuals in an entire social group. The group was provided with branches and leaves that were tightly bundled together so that the possibility existed for some group members to keep all of the food for themselves. Based on an analysis of nearly seven thousand interactions over food, de Waal (1989) found that exchanges among nine adult group members were balanced per dyad so that on average, individuals A and B shared similar amounts of food with each other. In a subsequent study in which sequences of interaction were compared within the same day, it was found that grooming affected the subsequent likelihood of food sharing: individual A was more likely to share with B if B had groomed A earlier that day. The effect of grooming on food sharing did not extend to individuals other than the grooming partner, making this the first evidence for time-delayed partner-specific exchange in any nonhuman animal (de Waal 1997b).

These food-for-grooming sequences among chimpanzees seem to fit the definition of calculated reciprocity, which requires mental scorekeeping of services rendered and received (de Waal & Luttrell 1988). It seems hard to explain these sequences without the involvement of memory and perhaps even a psychological mechanism such as "gratitude" (compare Trivers 1971). Such a cognitively demanding explanation may not be needed for the food sharing among capuchin monkeys. The exchange in our test paradigm did not involve a substantial time delay, and it involved only a single partner at a time. Possibly, the observed reciprocity rested on what we have termed *attitudinal reciprocity,* that is, a mirroring of the partner's attitude. This would still involve an if-then contingency rule ("If you are nice, I'll be nice"), but one that is played out immediately, without a mediating role of memory (de Waal 2000b).

To further test this mechanism, a more complex exchange paradigm was developed in which capuchin monkeys could share food obtained with the help of a cooperation partner. Two monkeys had to work together to pull in a counterweighted tray, at which point one or both of them would be rewarded (compare Crawford 1937). They were again separated from each other by a mesh partition, which offered them the option to share food or not. Each monkey had its own bar to pull in the tray, although these bars could be removed for control tests. Food was placed in transparent bowls so that each monkey could see which one of them would receive the food (Figure 9.5). There were three relevant conditions:

Mutualism: two monkeys were required to pull together and both cups were baited.

Cooperation: two monkeys were required to pull together, but only one food cup was baited. This represented altruism on the part of the helper.

Solo effort: only one monkey had a pull-bar and only this individual received food, although both monkeys were present in the test chamber. This required no cooperation.

As expected, the success rate of cooperative trials was significantly lower than that of mutualistic tests or solo efforts. The monkeys successfully pulled the tray in 85.4 percent and 88.9 percent of the trials for solo effort and mutualism tests, respectively, but only in 39.2 percent of the trials in cooper-

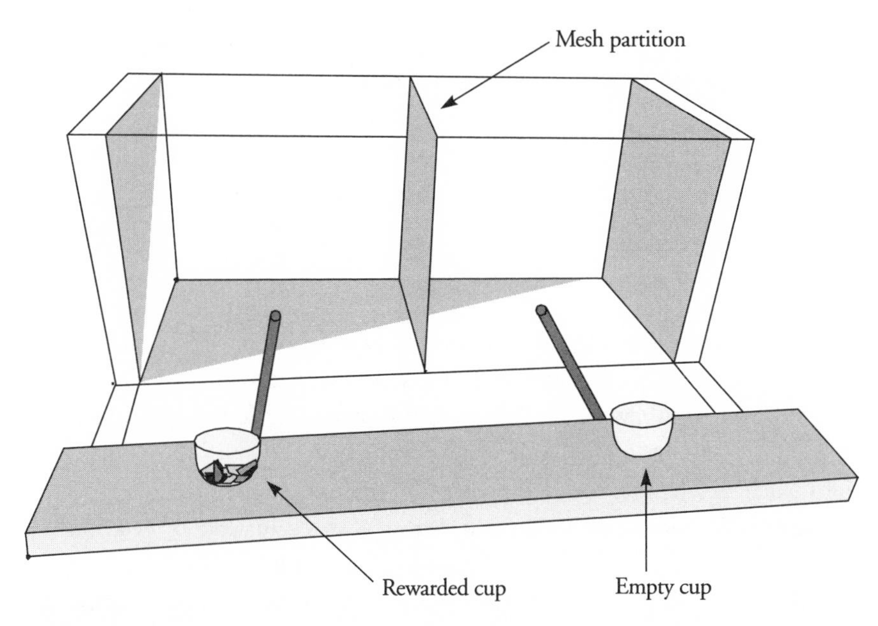

Figure 9.5. The apparatus used to test for attitudinal reciprocity in capuchin monkeys. Two monkeys are situated in adjacent sections of a test chamber, separated by a mesh partition. The apparatus consists of a counter-weighted tray with two pull bars, with each monkey having access to one bar. In cooperation tests, both monkeys are required to pull, but only one individual's food cup is baited. In mutualism tests, both monkeys are required to pull their respective pull bars, and both food cups are baited. In solo effort tests, finally, two monkeys are in the test chamber, but only one monkey has a bar and only this individual's food cup is baited. From de Waal & Berger 2000.

ation tests. This difference confirms that mutualism is easier to achieve than reciprocity as also indicated by the successful mutualistic cooperation in a similar paradigm among hyenas (Chapter 5).

The central question underlying this project, however, was if food sharing would increase in the context of a cooperative enterprise. Our analyses of the amount of sharing indicated that capuchins share significantly more after successful cooperative trials than after solo efforts (de Waal & Berger 2000). The most cognitively demanding interpretation of this result is that the food possessor understands that its partner has helped and that the partner must be rewarded for cooperation to continue. This would represent calculated reciprocity, in which the exchange of favors on a one-on-one basis drives reciprocal altruism. However, until we are compelled by evidence to the contrary, a sufficient explanation seems to be attitudinal reciprocity, according to which the mutual attention and coordination that the cooperative trial entails induces a positive attitude toward the partner, which is expressed in social tolerance, mutual attraction, and ultimately, the sharing of food.

As indicated in Figure 9.4b, we can represent reciprocal altruism as a more complex level of cooperation than mutualism, one involving an if-then hierarchy of events in which first A has to help B before B will help A.

Conclusion

It is obvious from the above examples that primates operate in a social environment in which the expression of their behavior is subject to constraints dictated by what they expect others to do, or not do, under particular circumstances. An individual can only dominate another if the situation is right, such as when its own supporters are present, or the other's supporters are absent. Similarly, reconciliation requires an evaluation of relationship value, and sometimes mediation by a third party. Or, cooperation may not happen in the absence of a history of previous cooperation. This conditionality, which third-party influences can render quite complex, requires primates to take multiple factors into account before they make a move. Undoubtedly, the contingencies are learned, based on experience with particular outcomes under particular circumstances. Yet, the same knowledge can be extended to novel situations through a mental process of creative assembly that is particularly well-documented for the great apes (Köhler 1925; de Waal 1982; Rumbaugh et al. 1996), yet perhaps not limited to this taxonomic group.

The basic message here is that the actions in a social chain are as constrained as word usage in a sentence. One cannot say "I dog the walk" any more than a capuchin monkey can count on much cooperation from a partner with whom in the past it has repeatedly refused to share abundant food. Hence animals can be said to follow a syntax of interaction, in which certain probabilistic rules are to be followed in order to achieve success. Only in the last couple of decades have primatological researchers attempted to document these rules in detail. Similar rules can be expected to apply to the societies of other large-brained, long-lived animals. My attempt at outlining these rules in this purely exploratory chapter has been highly informal. We can envision more systematic efforts, however, that not only address the decision trees in the social lives of animals, but also the underlying mechanisms. Only then will we know whether the observable structural similarities in decision making reflect similarities in cognitive processing.

CASE STUDY 9A

Conflict Resolution in the Spotted Hyena

SOFIA A. WAHAJ AND KAY E. HOLEKAMP

Gregarious animals that live in permanent social groups experience intragroup competition, and are therefore expected to exhibit nondispersive forms of conflict resolution (de Waal 1986). Behavioral mechanisms are often needed to repair social relationships damaged by aggression, and reconciliation behavior represents one such mechanism operating in the complex societies of many primates (Aureli & de Waal 2000). Defining reconciliation as "any friendly reunion between former opponents occurring soon after an agonistic conflict" (de Waal & van Roosmalen 1979), we recently documented reconciliation behavior in free-living spotted hyenas (*Crocuta crocuta,* see Wahaj et al. 2001). These are gregarious carnivores whose complex social lives have much in common with those of many Cercopithecine primates (Holekamp et al. 1999; Chapter 5). Both spotted hyenas and certain Old World primates are long-lived mammals that live in permanent social groups, the members of which cooperate to acquire and defend resources (Kruuk 1972; Harcourt 1992). Like certain primates, spotted hyenas also often depend on help from other group members during formation of coalitions that are important in both the acquisition and maintenance of social rank (Zabel et al. 1992; Engh et al. 2000). Thus, as occurs in primates, the enduring social relationships found among these large carnivores affect survival and reproduction of individual group members.

Like those of many primates, social groups of spotted hyenas usually contain multiple adult males and several matrilines of adult females with offspring, including individuals from several overlapping generations. In social groups of *Crocuta,* as in those of many Old World monkeys, adults can be ranked in a linear dominance hierarchy, members of the same matriline occupy adjacent rank positions, and female dominance relations are stable for extended periods and across a variety of contexts (Andelman 1985; Frank 1986). In both taxa, an individual's position in the group's hierarchy may determine its priority of access to food (Frank 1986). Reconciliation is important in regulating social relationships and reducing social tension in primate societies (Aureli & de Waal 2000). Given that primates and hyenas live in similarly complex societies, we sought to determine whether spotted hyenas exhibit patterns of reconciliation comparable to those found in primates.

In a large group of free-living hyenas in Kenya, we used focal animal observations of 160 individuals to monitor rates at which various affiliative behaviors occurred during five-minute intervals before and after each of 698 dyadic aggressive interactions. To facilitate comparison with primates, we also sampled 14 hyena dyads for which matched controls were available within 30 days of an observed conflict (de Waal & Yoshihara 1983). An affiliative behavior was only identified as having a conciliatory function if it occurred more frequently after than before fights, and if it was also followed by reduced rates of aggression between former opponents during the post-conflict interval. Of several different affiliative behaviors examined, only two satisfied both these criteria: greeting behavior and friendly approach. Hyenas exhibited one or both of these behaviors after 14.6 percent of fights, a finding consistent with that obtained recently in *Crocuta* by Hofer & East (2000). We used the formula developed by Veneema and colleagues (1994) to calculate a conciliatory tendency (CT) for each individual hyena. This measure controls for variable baseline levels of contact among pairs of animals. Overall mean CT in our study population was 11.3 percent. Conciliatory tendencies calculated for primates have ranged from 3.1 percent in Japanese macaques (*Macaca fuscata:* Chaffin et al. 1995) to 51.4 percent in crested macaques (*M. nigra:* Petit et al. 1997), so conciliatory behavior in hyenas clearly falls near the low end of this continuum.

Hyenas initiated higher CTs when they were recipients of aggression than when they were aggressors (Figure 9A.1a), as is also true in many primates (reviewed in Aureli & de Waal 2000). Hyenas also showed higher CTs in interactions with nonkin than with matrilineal kin (Figure 9A.1b). In fact, hy-

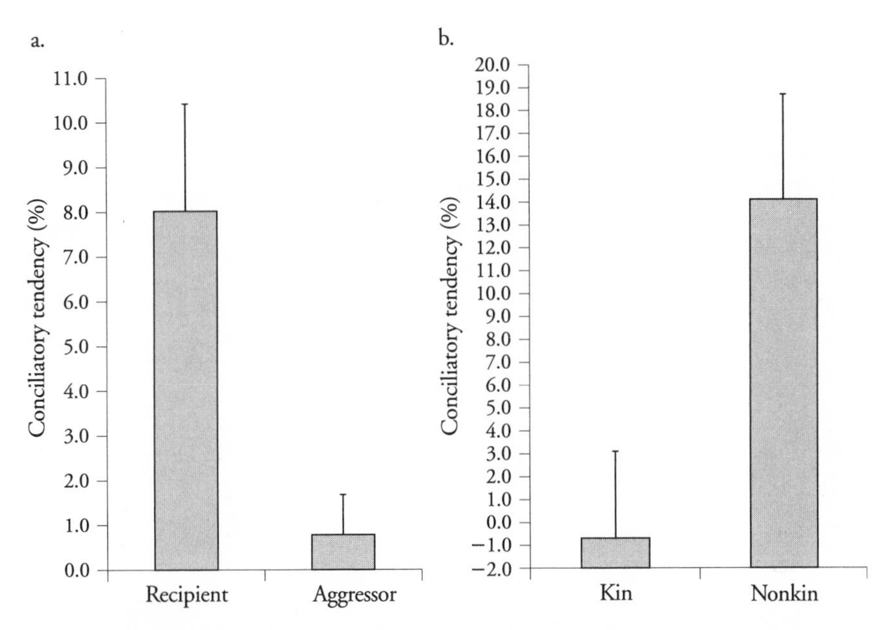

Figure 9A.1. Percent conciliatory tendency of *a)* 83 individual hyenas observed to play roles in different fights of both recipient and aggressor, *b)* 28 individual recipients of aggression observed to interact in different fights with both kin and nonkin.

ena kin seldom reconciled after fights. However, only 12 percent of 698 fights occurred between kin, suggesting that kin are more tolerant of each other than are nonkin. Conciliatory tendencies in hyenas did not vary with age-sex classes of opponents, the context in which aggression occurred, attack intensity, or rank distance between opponents.

Variables expected to affect rates of conciliatory behavior in animal societies include the extent of social cohesion necessary to survive in the wild, intensity of local resource competition, potential short-term costs of continued or escalated fighting, opportunities for dispersive conflict resolution, the value of the relationships between particular opponents, and the security of those relationships (Kummer 1978; Cords 1988; Cords & Aureli 2000). Because individual hyenas can often hunt and avoid predators on their own, the social cohesion evident in the fission-fusion society of the spotted hyena is far less than that observed in social groups of most Old World primates. Feeding competition among *Crocuta* is extraordinarily intense, with group members using their formidable weaponry on a daily basis in fights with conspecifics over ungulate carcasses. Thus, short-term costs of conflicts might promote unusually heavy reliance in this species on dispersive rather than nondispersive mechanisms of conflict resolution.

The value of a relationship reflects the magnitude of social or ecological benefits likely to accrue from it, whereas the security of a relationship reflects its predictability or resilience. The tendency to reconcile should be lowest when a relationship of low value is highly secure, and highest when a highly valued relationship is insecure (Cords 1988). As is also true in various primates (e.g., Cheney & Seyfarth 1989), hyena kin associate more closely than do nonkin, and kin also serve as frequent hunting partners and allies during coalition formation (Smale et al. 1995; Holekamp et al. 1997). Thus relationships with relatives should be more valuable than relationships with nonkin, yet hyenas have higher CTs with nonkin than with kin. Value of re-

Figure 9A.2. A greeting between two spotted hyenas, during which individuals sniff or lick their partner's ano-genital region. This behavior can reconcile former opponents. Photograph by Micaela Szykman.

lationships with nonkin may be relatively high in this species, since nonkin often join forces to repel lions and conspecifics from ungulate carcasses, and to defend territorial boundaries (Henschel & Skinner 1991). Thus, keeping relationships with nonkin in good repair (Figure 9A.2) should enhance the fitness of individual clan members. Relationships with relatives appear to be more predictable and resilient than do those with nonkin in this species, suggesting that relationship security better predicts CT than does relationship value.

Our results are consistent with the hypothesis that reconciliation may be widespread among gregarious mammals (Schino 2000). In addition to the spotted hyena, post-conflict behavior has now been evaluated in three other nonprimate species: bottlenose dolphins (Samuels & Flaherty 2000), domestic goats (Schino 2000), and domestic cats (van den Bos 1997). Like most primates, all of these species except cats live in stable social groups, and all except cats exhibit nondispersive mechanisms of conflict regulation (Schino 2000). Both cats and hyenas belong to the same monophyletic superfamily (Feloidea) of mammalian carnivores, and are thus closely related (Flynn 1996). However, hyenas reconcile, whereas cats do not. This suggests that taxonomic affiliations are less important determinants of post-conflict behavior than selection pressures associated with living in groups.

IV

Communication

Communication is closely linked to the topic of our book—it is the facet of social behavior that emphasizes the signals used by animals when they interact. We are interested in a broad comparative perspective on how communication fits into the evolution of intelligence, culture, and complex individualized societies. Several disparate approaches have been used to study how open and complex the communication of nonhuman animals can be. The contributions to this section exemplify these eclectic approaches from: (1) a broad comparison of emotional displays in primates to (2) close looks at the behavioral ecology of communication in parrots, bats, and dolphins, to (3) careful experiments to evaluate the information transmitted in a specific call type. One approach that we do not discuss in detail are the so-called "animal language" experiments (such as those involving apes who have learned American Sign Language), which have been extensively presented in other books. Since we are primarily interested in the evolutionary origins of communication skills from a broad comparative perspective, we prefer not to focus exclusively on the combination of skills that are used in human language. Rather than training animals to use an artificial language, the experiments we discuss use playback of natural sounds to explore what animals understand about their own communication system.

The case studies by Klaus Zuberbühler and Karen I. Hallberg, Douglas A. Nelson, and Sarah T. Boysen use playback of natural sounds to address the question of whether the natural vocalizations of nonhuman primates carry information about external referents. Zuberbühler played back natural sounds to cercopithecine primates in the wild to show not only that these primates respond to the sounds of a predator with the appropriate alarm call, but also that they can categorize their calls not just in terms of how they sound, but at a semantic level. Hallberg, Nelson, and Boysen tested the ability of trained chimpanzees to select pictures of foods after hearing recordings of spontaneous barks produced by chimpanzees in the presence of those same foods. Not only could the chimpanzees do so, but a discriminant analysis also could discriminate the calls to food type based solely upon acoustic parameters.

Chapter 10, by Jan A. R. A. M. van Hooff and Signe Preuschoft, explores the evolutionary origins of smiling and laughter in primates. These emotional expressions have clear homologues among other primates, and van Hooff and Preuschoft use the comparative method to explore the functional contexts of these expressions in order to understand their evolutionary origin. Although the smile is often considered to be an attenuated laugh, van Hooff and Preuschoft show that smiling and laughter have completely different origins, judging by the behavioral contexts in which their homologues are used. Laughter is related to the play face used by primates (and other social mammals) to signal a playful intent. By contrast, the smile is related to a display used as a submissive gesture in agonistic interactions. Van Hooff and Preuschoft thus argue that laughter evolved out of intention movements of gnawing during play, while the smile evolved from baring of the teeth signaling terror and fear.

If laughter and smiling have such different origins, how can humans have so misjudged their similarity? Van Hooff and Preuschoft compare different primate species and find that the two homologous displays are used in very distinct contexts in species with strong dominance hierarchies and asymmetric power relations, whereas the distinction between the two expressions is blurred in species with more symmetric, egalitarian relations. They suggest that hierarchical societies require an unambiguous distinction between status signals versus signaling a playful release from these status rules, but more egalitarian societies may not require such a clear distinction. Since the human species lies on the more egalitarian end of the continuum, we too have blurred the distinction between these displays.

Van Hooff and Preuschoft use a classic ethological method of comparison and reach some provocative cognitive conclusions. In most nonhuman primates, the play face functions as a message of a playful attitude directed to a potential play partner. Yet there is evidence that some chimpanzees may produce the play face while engaged in play without a partner. There is no evidence that monkeys do the same. Van Hooff and Preuschoft argue that these private emotional expressions suggest a capacity for self-reflection or self-awareness. In her case study, Lisa A. Parr also argues that skills in understanding emotional expressions are critical precursors to more complex forms of social cognition, such as empathy. Testing responses of chimpanzees to video clips of emotionally charged scenes, she showed that they understand the positive or negative valence of the scenes. Without specific training, the apes matched the scenes with images of facial expressions showing a similar valence.

The remaining chapters in Part IV discuss vocal communication. We are particularly interested in the evolution of complex open systems of communication in which animals can learn to produce new vocalizations. One of the puzzles regarding the evolution of vocal learning in humans concerns the lack of evidence that other primates can modify their vocal signals based upon what they hear. Even drastic operations like deafening or cross-fostering monkeys with other species have shown little effect on the species-specific vocal repertoire. However, several other animal groups, including birds, bats, and marine mammals, have independently evolved sophisticated skills of vocal learning.

Jack W. Bradbury discusses recent studies of vocal communication in parrots. It has long been known that parrots are excellent mimics of sounds, including human speech, and recent studies have shown that African grey parrots can be trained to associate meanings with spoken words, and to understand different combinations of words. Little was known prior to Bradbury's work about vocal communication in wild parrots. Bradbury suggests that social learning is an important part of the ecological adaptation of parrots, and that they learn from others how to select, handle, and find food.

Parrots produce a variety of different calls, but Bradbury selects three as likely to be learned. The best understood are loud contact calls, which are individually distinctive. Parrots appear to modify these calls so that they converge with birds with whom they share a bond—either the mate or members of a larger flock. Parrots that share communal night roosts often show interroost differences in contact calls. These contact calls are the most important signal used to recruit parrots to the roost, and there is clear evidence for call recognition. Bradbury suggests that parrots may retain the ability to modify their calls throughout their life span to learn new group-specific contact calls when they switch flocks to gain new habitat lore.

Gerald S. Wilkinson discusses the correlation between social and vocal complexity in bats. He argues that a critical issue for social complexity in individualized societies is the size of the group within which an animal can discriminate different individuals and remember their history of interaction. This is not the same thing as colony size, which is the group size measured for most bats. Many different functional call categories have been identified in bats. Wilkinson suggests that there is little general association between social and vocal complexity in bats, given the crude and inadequate measures available for most species, but he reports two specific correlations. In one bat species, females may select males with the most diverse repertoires, and Wilkinson suggests that sexual selection has driven the evolution of complex advertisements in these bats as in some bird and marine mammal species. Wilkinson also reports a strong correlation between colony size and complexity of the isolation calls produced by infants when separated from their mothers. These isolation calls vary between individuals, which suggests the presence of a vocal signature, and playback experiments in some species demonstrate that mothers are attracted to calls of their young. Wilkinson compared the total information contained in the isolation calls of eight bat species to the size of their colonies, which varied from one to 10 million. There was a strong correlation between call complexity and colony size, suggesting that the increased demands on call discrimination within larger colonies have selected for more complex isolation calls.

Wilkinson also conducted a comparative neuroanatomical analysis on bats similar to the one conducted on primates by van Schaik and Deaner in Chapter 1. Wilkinson found that the relative size of the neocortex was independent of colony size, mating system, and echolocation ability, but did correlate with group stability and feeding on vertebrates rather than fruit or insects. These features are linked with a specific kind of social complexity in which individuals within these groups share food and groom one another. In at least one species, this food sharing involves a complex system of reciprocation.

Peter L. Tyack opens his chapter on dolphin communication emphasizing that animals with individualized societies require a system for individual recognition. Because of their marine habitat, cetaceans require an acoustic recognition signal, and they may require vocal learning to maintain a stereotyped signature signal as the sound production apparatus changes shape during a dive. Paradoxically, dolphins appear to develop their distinctive signature signals, called signature whistles, by matching sounds present in their natal environment. The signature whistles of females remain stable, but adult males may gradually change their whistles to become more similar to animals with whom they share a strong bond. Similar patterns are seen in birds, bats, and humans, where convergence has been viewed as a marker of membership in a group and as an affiliative signal. In addition, dolphins exchange signature whistles especially when isolated from one another and may imitate the signature whistles of animals with which they are interacting. Tyack raises the issue of reference, also addressed in the case studies by Zuberbühler and Hallberg, Nelson, and Boysen, suggesting that dolphins may imitate the signature of another in order to "name" that particular individual. Captive dolphins have been trained to associate new vocalizations with arbitrary objects, but it remains unknown if or how this skill of vocal labeling is used in the wild.

All three chapters on vocal communication discuss contact calls used for individual recognition. Individual recognition is a prerequisite for the kind of individualized societies that form the focus of this book. Primates have specialized abilities to recognize distinctive visual cues of different individuals, especially facial cues. Face recognition need not involve active signaling, so may not be obvious to an observer. However, bats, birds, and marine mammals live in environments that favor vocal signatures. The distinctiveness of signature vocalizations highlights the critical need to identify individuals for species that maintain individualized societies. Wilkinson's analysis of bat isolation calls suggests that the complexity of these signatures may correlate with the number of individuals an animal must keep track of.

As Hal Whitehead suggests in Chapter 17, the vocal traditions of animals that learn their signals from conspecifics are a form of culture. Discussions of culture in nonhuman primates have thus far focused on manual tasks, at which primates excel, and not on vocal learning, of which primates are largely incapable. Many other animal groups, however, share with humans a talent for vocal imitation. Birds and marine mammals, in particular, have remarkably open systems of communication with vocal traditions that can be thought of as the vocal aspects of their cultures. Current research would suggest that the primary function of this communication is to mediate social relationships between individuals or members of groups. This reinforces from a communication perspective the main point of this book: the cognitive demands of social relationships in individualized societies may have driven the evolution of certain forms of animal intelligence.

10

Laughter and Smiling: The Intertwining of Nature and Culture

JAN A. R. A. M. VAN HOOFF AND SIGNE PREUSCHOFT

To a detached observer, human laughter presents itself as a remarkably instinctive and reflexive behavior, complex in the coordination of its components and the activity patterns of muscular and glandular effectors. The salvos of expirations, given in a series of exhalatory barks, end in a state of deep expiration, in which the laugher may choke and, face red, gasp for breath. Subsequently there is an inspiration howl, after which a new series of expiration barks may follow. The mouth is wide open, the mouth corners are retracted, the teeth bared, and sometimes there is shedding of tears. The laugher makes boisterous, seemingly uncontrollable movements with his body and arms. These are without the tension and rigidity characteristic of aggression or fear. On the contrary, in true whole-hearted laughter there is a general relaxation of the musculature, to the extent even that the laugher may lose his equilibrium, seeking support from a neighbor while slapping him on the shoulders.

Obviously the term "reflex" for this almost orgasmic discharge is a misnomer because this behavior is highly motivation- and context-dependent. We immediately recognize its emotional content and interpret it as an expression of mirth. It clearly indicates that we are enjoying what we are doing and in what is happening, and that we interpret the situation as amusing. Laughter is thus associated with appreciation of the comic and with humor, which are among the most sophisticated cognitive performances of humans.

A sense of humor traditionally has been viewed as a uniquely human attribute, based on a multifaceted interpretation of situations. But then, what a peculiar way to express this! One of our most esteemed intellectual faculties, the sense of humor, finds expression in a "fixed action pattern." Although there clearly is cultural variation, there is a growing conviction that the basic pattern is universal (e.g., Ekman 1973, 1989, 1994) and develops early in infancy in a process of maturation that takes place even in infants who lack major sensory input, namely the blind and deaf (Eibl-Eibesfeldt 1973, 1997). Intense laughter often "erupts" in an almost uncontrollable and compulsive way, and is both very contagious and difficult to fake—nonspontaneous and posed laughter is easily uncovered as such. A puzzling question is then why we should express our emotional appreciation in this particular way? This is a question about the evolution of laughter. A comparative approach is indicated if we want to get an insight into the possible evolutionary origin of such a complex species-specific behavior. We shall review evidence showing that in other animals, particularly in the primates, the zoological taxon to which we ourselves belong, forms of behavior exist that may be homologous to the human laugh. These behavior patterns resemble in their morphology and in their motivational and functional context both the intense forms of human laughter and more subdued expressions that are often referred to as smiling. The comparative approach also reveals considerable differences between species, and we will argue that these can be understood as consequences of differences in the social structure of the respective species.

Expression of Emotion or Communicative Display

The first comparative study of emotion was Darwin's. In 1872 he published his classic book *The Expression of the Emotions in Man and Animals.* For Darwin it was very important to show that the theory of evolution was relevant for the development not only of the physical features of organisms, but of all aspects of life, even mental phenomena such as emotions. The expression of emotions appeared particularly relevant to him because he was convinced to have found similar and, by implication, homologous expressions of such emotions in both humans and nonhumans.

Figure 10.1a. Darwin interpreted this silent bared-teeth grin of a Sulawesi macaque *(Macaca nigra)* as an expression of being pleased.

Figure 10.1b. A silent bared-teeth display in another Sulawesi macaque, *Macaca tonkeana* (from Preuschoft & van Hooff 1999). A quantitative message–meaning analysis shows that the silent bared-teeth display often marks a transition to a more affiliative interaction between the sender and the addressee, and therefore reflects a positive social attitude (Preuschoft 1995).

Darwin gave detailed and precise descriptions of many expressions, particularly of dogs, cats, and primates. He interpreted these in terms of emotions. This he did with a certain amount of anthropomorphic empathy (Figure 10.1a). Such an approach, which relies on the spontaneous intuition with which we judge the behavior of our conspecifics, comes almost naturally when we study species closely related to us. A more recent example is the pioneer study by Nadia Kohts (1937), in which she compared the expressions of a young chimpanzee and a human child. This "psychological" approach contrasts strongly with that of the early ethologists. Because they worked with species that often were distant from humans, such as invertebrates, fishes, and birds, early ethologists were not tempted into intuitive interpretations in terms of emotions. On the contrary, they were suspicious about such an approach, because it could not be objectively operationalized. Instead they noted that displays should be understood as social signals, evolved devices for communication with conspecifics (Figure 10.1b). The signals were seen to reflect action tendencies that were thwarted, by conflicting impulses or otherwise, and, because of this, manifested themselves in "intention," "ambivalence," or "displacement" movements or postures (compare Frijda 1986). Thus they could act as messages informing conspecifics of the likely courses of action of the "sender" (Figure 10.2). In the traditional ethological approach the message and meaning of the signal (as defined by Smith 1965) is deduced from its context and the way the signal is embedded in changes within the stream of behavior of the sender as well as the recipient.

Ritualization: The Evidence for a Communicative Function

Movements evolved for a signaling function are exaggerated in their form and dynamics compared to the original movement. They are often accentuated by ornaments that seem to have been specifically developed for signaling. In the course of their evolution signal behaviors were thus stylized, which led to a "typical intensity" or stereotypy of the performance (Morris 1957). This process by which the signals evolve into conspicuous and unam-

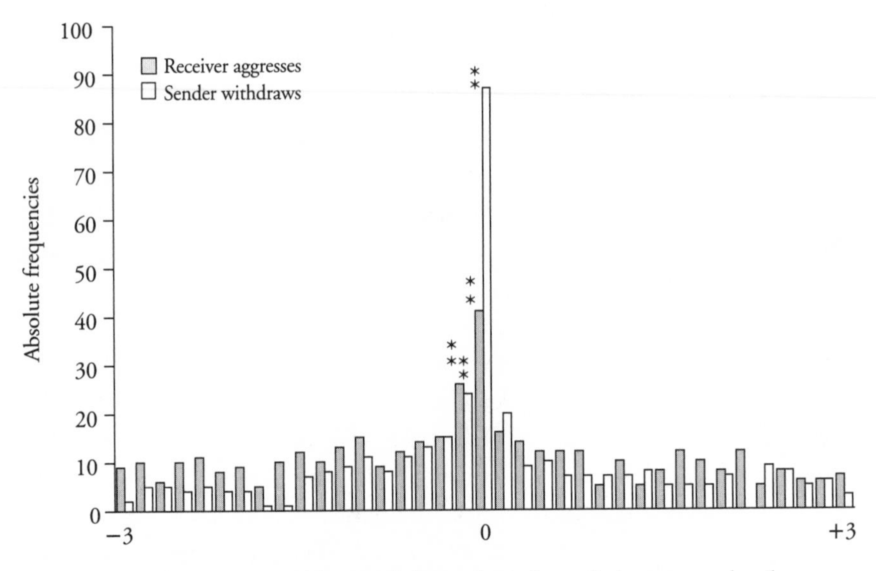

Figure 10.2. A "pre/post-event histogram" shows that in longtailed macaques the silent bared-teeth display (at interval 0) is embedded in agonistic behavior: The receivers show aggression *(gray)*, and the senders show withdrawal *(white)*. The sequences cover six minutes, from three minutes before (−3) to three minutes after (+3) the occurrence of the silent bared-teeth display. (From Preuschoft 1995.)

biguous communicative displays has been termed ritualization (Tinbergen 1952).

Only if the receiver's reaction to an accidental cue is beneficial for the sender can this cue evolve into a communication signal (Tinbergen 1952). Since evolution is a process, signals differ in how much ritualization they have undergone, and, consequently, in their degree of stereotypy. A misunderstanding of these classical ethological concepts can lead to confusion. For instance, Bachorowski and Owren (2001) question that human laughter is a stereotyped signal because they were able to demonstrate considerable variation in the acoustic properties of different laughs. Their reason to include these various types of laughs (which differs, e.g., from Provine 1993), however, was that all these variants were perceived as laughter by conspecific listeners. Thus, obviously the conspecific recipients must have detected commonalities that demanded this classification. Stereotypy does not imply absence of variation; it only means that there is a recognizable pattern or structure that enables receivers to classify the stimulus as signal *X*.

Early attempts to inventory the display repertoires of species were led by the expectation that discrete fixed action patterns would be found. This was indeed the expectation of Schenkel (1947) in his now classic study on the expressive behavior of wolves. This expectation was based on studies, mainly of insects and birds, in which displays were perceived as rather rigid. Soon, however, it was realized that, especially in mammals, signals are more often than not of a graded nature (Green 1975; Bradbury & Vehrencamp 1998). In response to Schenkel, Lorenz (1953) argued convincingly that, for instance, the seemingly endless variation in the facial threat expressions of the dog could be plotted on two main behavioral tendencies, namely the tendencies to attack and to flee from an opponent.

In sum, ritualized communication signals can be classified just like different handwriting, to the extent that they evince a certain stereotypy. Yet stereotypy does not imply that there is no meaningful, i.e., functionally relevant, variation among instances of a specific signal type. The categorization of stimuli, at least by primates, may in fact follow a fuzzy logic system (Lakoff 1987; Gouzoules et al. 1998).

In Search of the Homologue of Our Human Laughter

A number of studies have been performed in which the repertoire of primate displays, in particular their facial expressions, has been described and motivationally analyzed (e.g., Andrew 1963; van Hooff 1967; Redican

1975). Depending on the extent to which one wants to lump or split facial displays into categories, these studies reveal that there are about eight major types of facial expression. Among these there is indeed a clear candidate for the homologue of our human laughter, namely the "relaxed open-mouth" display or "play face" (van Hooff 1972, 1976; Preuschoft & van Hooff 1995).

Play and the Display of a Playful Attitude

Mammals, and to some extent birds (Ortega & Bekoff 1987), share a characteristic not clearly present in other taxa. That is play. Especially in juvenile individuals elements of behavior that normally are integrated in adaptive, functionally directed patterns, such as obtaining food, hunting, mating, fleeing, fighting, or even simple locomotion (jumping, sliding, pirouetting) can be performed "for their own sake" in a way that appears to be self-rewarding (e.g., Fagen 1981). In play these actions are not followed by the consequences that provide the functional goal or reward for the "serious" sequence. Instead they are performed, so to speak, for the joy, for the fun of it. Proverbial examples are the cat playing with a mouse or with a knot of wool, the dog running after a stick thrown away by its owner, the otter performing swimming and sliding acrobatics. The testing of one's own performance, the so-called competence motive or self-assessment, may offer the essential satisfaction and thus be the motivating factor (Thompson 1998). In the ultimate sense the exercise and sense of control it generates no doubt is an important adaptive consequence (Fagen 1981; Bekoff & Byers 1985; Martin & Caro 1985).

Play can occur in various forms. Both in birds and in mammals locomotory play and object play have been observed (e.g., Ficken 1977; Ortega & Bekoff 1987). In social mammals, such as primates, developing young playfully perform movement patterns normally involved in fighting, fleeing, and sexual contact. Thus primates, and especially juveniles, spend long hours in mock fighting. They chase, grasp, tackle, and tickle one another and engage in bouts of boisterous but at the same time tensionless wrestling, in which the role of the chaser and the chased, the "attacker" and the "attacked" are changed in often unpredictable ways. The interaction appears to be subtly coordinated between play partners. In this respect the communication of playful intent and the appreciation of a social interaction as playful is certainly remarkable, because it implies "a cognitive appreciation of the distinction between reality and pretense" (Bekoff & Allen 1998, p. 109). This is an element that seems characteristic for many animal species—not only primates—and which is also the essence of our humor (see also below).

In all primates social play is accompanied by a typical expression, the relaxed open-mouth display or play face (e.g., van Hooff 1967, 1972). The mouth is opened wide and the mouth corners may be slightly retracted. In most (but not all!) primate species the lips are not retracted but still cover the teeth. In many species this facial posture is often accompanied by a rhythmic staccato shallow breathing (play chuckles) and by vehement but supple body movements. The posture and movements, both of the face and of the body as a whole, lack the tension, rigidity, and brusqueness that is characteristic of expressions of aggression, threat, and fear. The relaxed open-mouth face has been interpreted as a ritualized version of the biting intention movement that precedes the playful gnawing on one another's body parts, so prominent in play-wrestling (Figures 10.3, 10.4 and 10.5).

Figure 10.3. Two juvenile chimpanzees play-wrestling. While one partner gnaws the other's arms with lips covering the teeth, the recipient of this play bite responds with the ritualized relaxed open-mouth face that is derived from the gnaw-intention face. Photograph courtesy of Yerkes Primate Center.

Figure 10.4. A juvenile gorilla playfully punches his father on the nose, and the silverback responds with a relaxed open-mouth play face. Photograph courtesy of Frank Kiernan.

Both partners in play usually give the relaxed open-mouth display. It has been interpreted as a meta-communicative signal, indicating that the behavior that it accompanies is to be understood differently from what it looks like at first sight; so, it should *not* be mistaken for real aggression, but should be taken as mock aggression. And so it is regarded by the receiver (Figure 10.4).

Laugh Variants

The relaxed open-mouth displays of the primates, humans included, are much the same. Still there are some interesting differences of detail. Many primates "laugh" without really baring their teeth. This is the case in several species of macaques (Figure 10.5) and also in our nearest relative, the chimpanzee (Preuschoft & van Hooff 1997). In other species, however, the teeth are fully bared. This is the case in the Sulawesi macaques (Figure 10.6), liontailed macaques, and gelada baboons (Preuschoft 1995; Duecker 1996), and apparently also in sooty mangabeys (personal observation, Signe

Preuschoft). The play face of these species is therefore intuitively interpreted by humans as laughter. As in humans this bared-teeth version of the open-mouth laughter in macaques is ontogenetically preceded by the classical play face, in which the upper teeth are covered. What might be the evolutionary reason for this difference will be discussed below.

It is the recognizability (i.e., stereotypy) of ritualized signals that allows us to compare signals between species. Similarity judgments concerning the form of signals are based on criteria such as similarity in details of performance and in the underlying body structures (Preuschoft & van Hooff 1995). To reconstruct the phylogenetic development of signals it is necessary to assess the signal's distribution in the phylogenetic tree, because this permits us to date evolutionary changes and to distinguish between ancestral and derived homologies (Preuschoft & van Hooff 1995, Figure 10.7). The distribution of relaxed open-mouth (rom) and relaxed open-mouth bared-

Figure 10.5. Playful tickling and grappling among juvenile Barbary macaques. While reaching for his brother, the older juvenile on the left makes gaze contact and shows a low-intensity relaxed open-mouth display. Waiting for his next move, the younger male returns a full-fledged relaxed open-mouth play face. Photograph by Signe Preuschoft.

Figure 10.6. Tonkean macaques at play. With an open-mouth bared-teeth laugh the juvenile to the right surrenders her arm to the female on the left. Laughing as well, the female makes gaze contact with the juvenile before she applies a playful bite. Photograph from Preuschoft & van Hooff 1999.

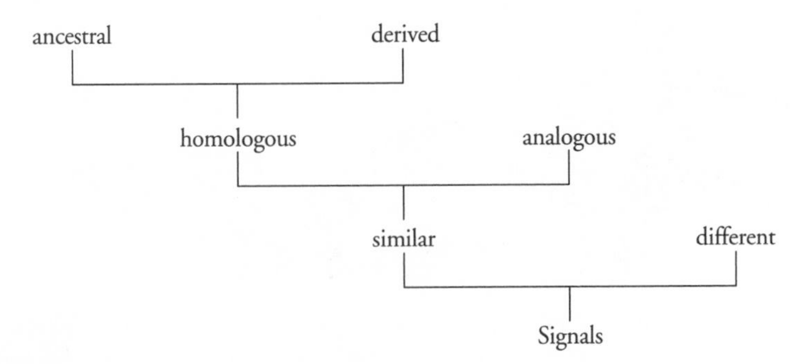

Figure 10.7. Decisions required for reconstructing the phylogenetic relationships between signals of different species. (After Martin 1990.)

teeth (rombt) play faces in the genus *Macaca* shows that the relaxed open-mouth display is more widespread. This would suggest that the relaxed open-mouth face is the ancestral signal, and that some but not all species in the silenus group have derived the open-mouth bared-teeth face from the original signal. However, this may be an incorrect inference. It is also possible—in principle—that the species that use the rombt have evolved at a slow rate, and that those lineages that use rom have evolved rather fast and experienced many speciation events. To distinguish between these two alternative scenarios an "outgroup comparison" is required (Hennig 1979). This means, for our example, comparing the macaques with other, more distantly related taxa. Although the documentation of play faces in primates is rather patchy, it appears that the rom version of the play face is the rule among colobines (the other large group of Old World monkeys, separated from the cercopithecines around 22 million years ago), and also among the phylogenetically more distant groups of the New World monkeys (split off around 40 million years ago) and the prosimians (Preuschoft & van Hooff 1997).

The frontispiece of the fourth edition of the well-known textbook on animal behavior by A. Manning and M. Stamp Dawkins (1992) shows a spectacular photograph of two polar bears standing face to face and displaying wide play gapes. Fox (1970) has described a well-developed play signal in the canidae. It consists of a play-bite intention, in the form of a jaw gape that strongly resembles the mouth posture of the aggressive jaw gape. However, it is accompanied by boisterous, frolicsome body movements, the "play dance." In the wolf and the dog it is accompanied, moreover, by rapid panting—an increase in the frequency but not the depth of respiration. The similarity to the play panting of primates (see below) is striking. The documentation of gape-mouthed play faces in which the upper teeth remain covered in canids and bears (ursinae) suggests that the relaxed open-mouth play face pre-dates the evolution of the primates and was perhaps already used by primordial insectivores romping about more than 65 million years ago.

Rombt displays, however, have been observed in many cercopithecoid species, but in a pattern apparently random to phylogenetic affiliation (Preuschoft & van Hooff 1997): Most macaque species in the silenus group use the rombt, but pigtailed macaques *(Macaca nemestrina)* use the rom display (Preuschoft 2000); Geladas use the rombt, but their closest relatives, the papionines, use the rom display (Duecker 1996; Preuschoft & van Hooff 1995); humans and bonobos habitually use the rombt version, whereas common chimpanzees, gorillas, and siamangs use primarily the classical rom, and the situation in orangutans and gibbons is poorly documented (Chevalier-Skolnikoff 1982; Preuschoft & van Hooff 1997).

In summary, it is most likely that the rom display represents the ancestral version of the play face. The rombt display seems to be derived easily from this classical version, because it must have evolved independently in at least three different lineages.

In macaques, both forms of the play face are usually accompanied by a subtle vocalization, a rhythmical voiced breathing (Preuschoft 1992; Preuschoft & van Hooff 1999). In chimpanzees this vocalization is more conspicuous, and takes the form of a staccato, throaty panting (van Hooff 1972, 1973; Haanstra et al. 1982). The laughter of humans and our nearest relatives, the chimpanzees, thus differs in a noteworthy respect: the vocalization. Our human laugh consists of a series of loud, staccato barking vocalizations given during one exhalation phase. After this there is a long deep inhalation before the next series of barks is given. The chimpanzee expression is differ-

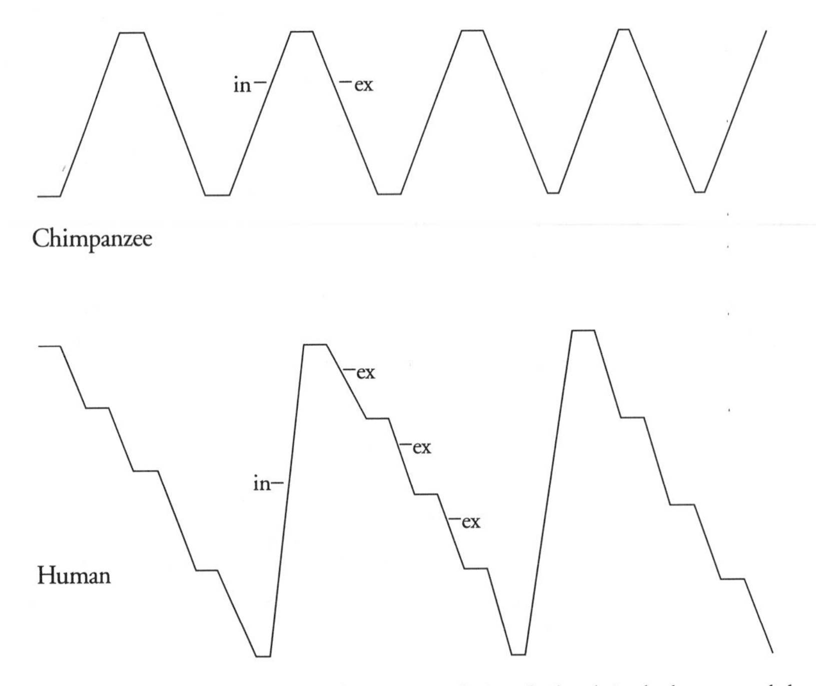

Figure 10.8. Pattern of inspirations and expirations during the laugh in the human and the chimpanzee. In the human the laugh consists of salvos of interrupted expirations (-ex) followed by a deep inspiration howl (in-); in the chimpanzee it consists of alternating shallow in- and expirations.

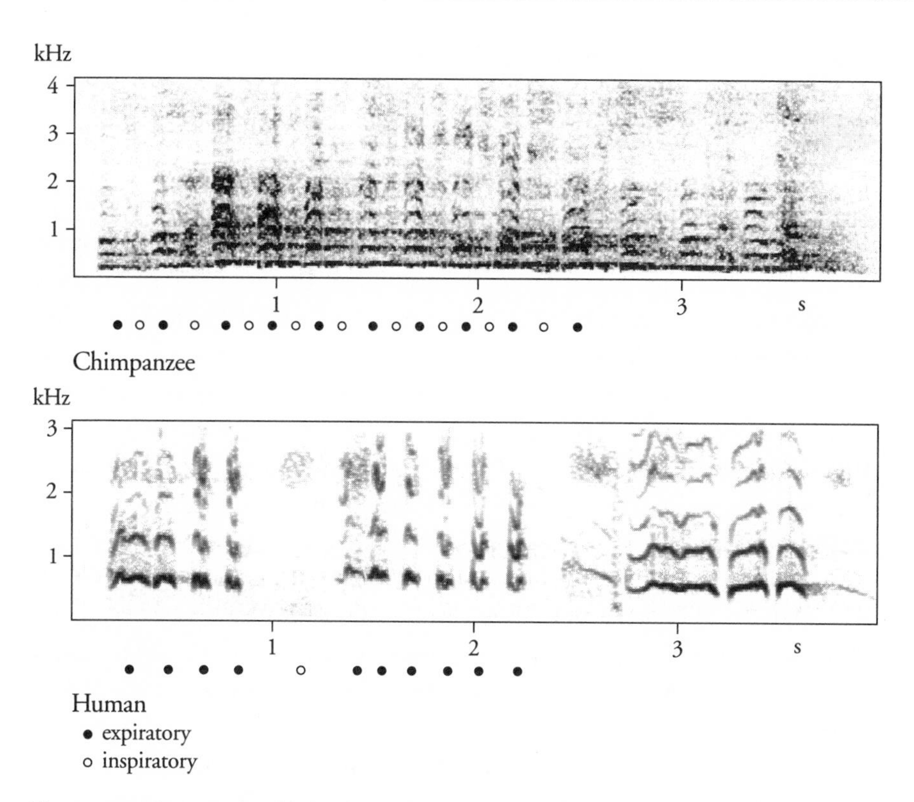

Figure 10.9. Sonographs of human and chimpanzee laughter. Black circles mark exhalation and white circles mark inhalation. (From Vettin et al. 1999.)

ent in that barks or chuckles are much quieter than in humans, and, more important, they are not given during one long exhalation that is interrupted at every bark. Instead there is a short inspiration after each brief expiratory bark (van Hooff 1973). This means that the average level of inspiration over the whole of the laugh bout remains about constant. This has been represented schematically in Figure 10.8. A precise spectrographic representation is given in Figure 10.9. The "during in- and exhalation" vocalization pattern has been noted in a number of primates. However, since sonographic analyses of other nonhuman primate laughter vocalizations are lacking it is currently impossible to know whether the human "during exhalation only" version of laughter is typical for humans alone, or if it may occur in other primate species as well. Clearly, the acoustic documentation of snort- and grunt-like laughs in humans shows that the leap from the chimpanzees' throaty panting chuckles to the tonal song-like versions of human laughter may not be as abrupt as earlier investigations made it appear (Provine 1996; Bachorowski & Owren 2001; Bachorowski et al. in press).

The Primate Play Face: When Is It Shown?

It is revealing to look in detail at the way in which the play face is embedded in the play. There are two ways to do this: One is to observe when precisely the relaxed open-mouth face occurs in play sequences of spontaneously interacting primates. The other is to play with a primate. This is particularly illuminating because one can steer the interaction in various ways. Jan van Hooff has often engaged in tickling games with young chimpanzees born in the study colony at the Arnhem Zoo in the Netherlands.

When you engage in play, certain variations in the relaxed open-mouth display become obvious. If the chimpanzee takes the active role, he shows the wide-open mouth. Especially if he takes the passive role, that is when he offers himself and is being tickled, we notice the rapid play chuckles. Similar observations have been made for the gorilla by Schenkel (1964). The chimpanzee reacts with its typical "ah ah" chuckles especially when the tickler changes his grip in unpredictable ways, that is when there is an element of surprise in the playful interaction. Obviously this makes all the difference between a dull and an exciting game. The resemblance with the interactions of very young children who release bouts of laughter is unmistakable. Here also the excitement of tickle and peek-a-boo games is in the "expected unexpected." Another striking aspect is that there must be a relaxed and trusting social atmosphere to get a young chimpanzee to play with you at all. Mason (1965) found that if young chimpanzees were given a choice between a play partner and a protective cuddling partner their decision was dependent on the familiarity of the surroundings: In an unfamiliar setting the young chimpanzee would choose the protective partner to huddle with, but in a familiar setting it would opt for the playmate.

Observations of spontaneous play among conspecifics reveal that play faces and the voiced breathing laughter typically occur during rough-and-tumble play-wrestling and play-chasing. Macaques "laugh" when they jump on, grasp, lunge, or bite their playmates, but also when they are being pounced on, stalked, tugged, or wrestled (Preuschoft 1995). When one partner shows a play face, the other answers in kind. Playing primates coordinate their actions: Two juveniles may play chase, circling a bush, one pursuing the other, each ready to change roles. When the pursued wheels round both face each other for an instant with mutual play faces before the pursuer changes direction as well and runs off, round and round the bush, now as the pursued. The king of the castle defends a tree stump against a group of would-be invaders creeping up from all directions. When the "king" is busy defending the front and guarding the side then it is time to mount an attack from the rear. Tug-of-war can be played with any branch or liana. Here also the timing of the expected surprise is the challenge, and the appreciation of the excitement is expressed in the breathing display. Playing primates seem to seek the adventurous: they climb up the high wall, and dangle before they finally dare to jump down; they play at the water hole, ducking and splashing. A playing primate often seeks out a big, potentially dangerous partner; when these are apt at self-handicapping and tickle, tug, and bite with restraint, the fun can't be greater. In our own species the nature and contents of playful interactions gradually broaden as individuals mature; it is no longer only physical play that is associated with laughter, but also the intellectual and symbolic romping about with one another in verbal joking and humor.

In other words, it is very plausible to regard this expression of joyful mirth and humorous intellectual wrestling as a present-day end point in a trajectory that reaches far back in the evolutionary history of our ancestors. This must be so because it has its clear homologues in the other end points of primate evolution, the existing primates, and also in some other mammalian taxa.

The Motivational Basis of Laughter

The question of the origin of laughter has long fascinated philosophers, psychologists, and cultural anthropologists. Two main categories of hypotheses can be distinguished: those who see laughter as an original expression of happiness, joy, and high spirits (e.g., Darwin 1872; Berlyne 1969), and those who seek its origin in aggressive and intimidating barks used by our forebears in the collective threatening of a common enemy (e.g., Hayworth 1928; Eibl-Eibesfeldt 1997). The "laughter as aggression" theory supposes that whoops of triumph, given when the members of a group succeed in defeating the enemy, would have evolved into a signal by which they communicate their superiority, and which at the same time fosters coherence among the group members. When the opponent no longer forms a serious threat, the signal turns into jeering and ridiculing, and thus into a laugh of derision. The next step would be that it expresses the collective appreciation that something awesome, grandiose, or impressive turns out to be a trifle. This demasking of the pompous has been seen by many as an essential and original aspect of the comic.

The problems with this "laughter as aggression" theory are twofold. First of all it finds no support in the comparative approach. Second, the ontogenetic evidence points in another direction. The first laughter of very young human children occurs in the context of play, in the physical interactions of tickling and peek-a-boo games.

The Laugh and the Smile: Different Origins

Most people consider the laugh and the smile as expressions of different intensity of the same motivation, that of merriment and happiness. The smile is regarded as a subdued, low-intensity form of the laugh. This is reflected in the etymology of the words in many languages. In French there is *rire* and *sourire* (an "under-laugh"); in German there is *Lachen* and *Lächeln* (a little laugh); in Dutch there is *lachen* and *glimlachen* (a glimmer of laughter). This has also been the traditional scientific view. For instance, Darwin (1872), Hayworth (1928), and Koestler (1949) believed that in the history of our species laughter preceded smiling, which was the diminutive. However, a comparative evolutionary approach suggests a different view.

The Primate Silent Bared-Teeth Display

Two long-tailed macaques meet one another near the same bunch of fruit in a tree. One shows the tense open-mouth stare, a threatening warning to the other. This smaller one responds by ducking away and by showing the silent bared-teeth face or grimace. The animal retracts its mouth corners and lifts its lips, thus fully baring the teeth, while the mouth is kept more or less closed. The eyes are wide open and the glance is fleeting and slightly oblique (Figure 10.10). In such a situation the animal may subsequently withdraw. However, it may also be permitted to stay around and be left alone by its more dominant companion. Systematic context analysis, in which the behavior of the dominant animal is related to the preceding and following behavior both of his adversary and of himself (see Figure 10.2) shows that his adversary's behavior should be interpreted as a submissive gesture. It is an expression of fear, which may be paraphrased: "Please, do not attack me. I am afraid of you and I shall not oppose. I shall give in and lend priority to you." This message thus is a negation of aggressive intention and can have an ap-

Figure 10.10. A male longtailed macaque responds with a submissive silent bared-teeth display to the approach of a dominant. (From Preuschoft & van Hooff 1999.)

peasing effect on the (potential) aggressor. This silent bared-teeth display may even function as a status signal by which subordinates express their acceptance of the subordinate position.

The silent bared-teeth display clearly can be regarded as a ritualized low-intensity version of the "bared-teeth scream display," by which animals express strong fear and frustration. The latter is found in almost all species of primates.

In some primate species the meaning of this facial expression has been broadened. In Sulawesi macaques we may see this facial gesture being given by a confident animal to a subordinate conspecific, who behaves in a nervous and fearful manner. Again this expression is one of nonhostile intent, but now from a self-assured position: "Don't worry, I am not going to attack you. Don't flee." It may allow the more confident animal to reassure the subordinate and to approach him. It even may lead to a social contact (see Figure 10.1b). For instance, a male may thus persuade a female to let him approach her. And then it is only a little step toward a friendly meaning: "Don't worry,

Figure 10.11. Zebra silent bared-teeth faces: *(above)* submission face and *(below)* greeting face. (Courtesy of Matthijs Schilder.)

I am not hostile. On the contrary, I am friendly—I like you." The distribution of this display over the primates suggests that this emancipation of the submissive and fearful grin into a confident reassurance gesture and subsequently into a friendly gesture must have taken place at various places in the phylogenetic tree leading to present-day primates (Preuschoft & van Hooff 1997).

A similar silent bared-teeth display has also been described for other mammalian taxa, in particular for such widely different taxa as the canidae (e.g., Schenkel 1947; Fox 1970) and the equidae (Trumler 1959; Schilder et al. 1984). In the canidae there is a submissive facial expression in which the mouth is not or only slightly opened while the mouth corners are withdrawn and the lips are lifted a bit, thus baring the teeth partly. At the same time the ears are folded back. The expression clearly differs from aggressive teeth-baring. There the mouth corners are pulled forward and the nose is wrinkled because the lips are strongly lifted. In the equidae a similar expression of submission has been described (the "Unterlegenheitsgebärde" of Trumler [1959]). Here the teeth and the gums are well exposed. In this expression the ears are folded back. In the latter respect it differs from the "Begrüßungsgesicht," or greeting face, a more confident expression given when meeting a conspecific. In this face the ears are kept upright and the ear opening is turned forward (Figure 10.11).

From the distribution and function of the silent bared-teeth face in many taxa we infer that its original meaning and function is that of submission. The silent bared-teeth face, especially the confident, friendly form found in some primate species, bears great resemblance to the emphatic greeting smile, the "broad smile," which Anglophones call the "Say 'cheese'–smile" and which characterizes the emphatic manifestation of friendliness, but which is also prominent when people excuse themselves (Goldenthal et al. 1981). The latter is clearly a reflection of the submissive, anxious meaning it must have started with in its evolutionary development.

Society Shapes the Meaning of Displays: The Power Asymmetry Hypothesis

The comparative evidence thus suggests that laughter and bared-teeth smiling have different origins. Laughter has evolved out of the intention movement of biting or gnawing that is such a prominent part of playful wrestling. The broad smile has its origins in the baring of the teeth, which signals terror and fear. Smiling can, therefore, no longer be regarded as a subdued low-intensity form of laughter. Both expressions can no longer be seen as the extremes on one continuum of variation; they both have their own characteristic variation.

The smile can vary from a faint retraction and lifting of the mouth corners, with the lips closed, to the "say-cheese" smile with fully bared teeth. The laugh can vary from a few soft chuckles given with the mouth almost closed to the wide-open-mouth burst of laughter. It is worth noting that we have come to fully realize this only after the results of the comparative evolutionary approach had been obtained (van Hooff 1972, 1989). This view has since been confirmed by ethological studies of human laughter and smiling (e.g., Kraut & Johnston 1979; Fernández-Dols & Ruiz-Belda 1995). The broad teeth-baring smile is indeed characteristic of an apologizing, appeasing attitude (Goldenthal et al. 1981).

Still, the misunderstanding that "rire" and "sourire" are one of a kind can easily be understood. Although we can clearly distinguish the full-fledged forms of both laughter and smiling, it is also obvious that they often occur closely together in the stream of ongoing behavior. Laughter may go over into smiling and vice-versa. There are also intermediate forms. This means that in human social contact the spheres of friendly and playful interaction have come quite close together. We often (but not always) can assume a joking attitude to "break the ice" when we establish social contacts and thus create a relaxed atmosphere (van Hooff 1976, 1989). According to Provine (1998), only 20 percent of the laughter in human interaction is associated with jokes and puns; in most cases it is used to create and maintain an atmosphere of congeniality. A number of studies have found correlations as low as .3 or .4 between funniness and laughter and smiling (e.g., see articles in Ekman & Rosenberg 1997). But, as Ruch (1997) suggests, these low values may also be attributed to insufficient attention paid to the existence of different types of laughter and smiles.

The first indication that variants of laughs may convey different meanings (as defined by Smith 1965) to recipients comes from a series of experiments conducted by Bachorowski and colleagues. These studies show that tonal, song-like laughs were found to be more likeable, attractive, sexy, and contagious than noisier variants, especially when the sender was female. Male laughs were more numerous, lasted longer, were higher pitched, and covered a larger spectrum when their interaction partner was a friend, especially a male friend, rather than a stranger. The same features characterized female laughs, but here the decisive factor was that the interaction partner was male not female, and friendship played a lesser role (Bachorowski & Owren 2001; Bachorowski et al. submitted). It thus appears that there are functionally significant variants within the class of human laughter vocalizations. These variants differentially influence a receiver's readiness to interact positively with the laughers; hence variants have different meanings for listeners. Human laughter can also vary with respect to the message emitted by the sender. For instance, the desire to interact affiliatively, as measured by verbal self-report, was reflected in a synchronization of a person's laughter with the partner's. Strangers' laughs were not synchronized (Smoski & Bachorowski in press).

How the Smile Came to Be Regarded as a Weak Laugh

In the extent of overlap between laughter and smile humans appear to differ from at least some other primates, for example, from chimpanzees and a number of monkeys, such as rhesus macaques and long-tailed macaques. There the play face and the greeting grin are clearly distinct expressions. They do not show overlap either in their appearance or in the context in which they are used.

However, there are a number of species in which the distinction between the two expressions is blurry, as it is in humans (Thierry et al. 1989; Preuschoft & van Hooff 1997). After comparing various species we have suggested that the degree to which the relaxed open-mouth play face and the silent bared-teeth smile are distinct is related to the nature of the social relationships of the respective species (Preuschoft 1995; Preuschoft & van Hooff 1997). Primate species appear to differ in the degree to which the relationships between individuals are "despotic" or "egalitarian." In the first case there is a strong dominance hierarchy with asymmetric power relationships; in the second case the hierarchy is shallow and relationships are more symmetric. The latter should be expected when there is a greater emphasis on cooperation, because this requires that both parties remain motivated to invest in the common goals (Vehrencamp 1983; Hand 1986). If one individual were to exploit the other, this would reduce the willingness of the other to make his part of the investment. The skew in the hierarchy depends also on the alternative options that are open for the subordinate members in a community. If strong outside pressures prevent an animal from leaving a group—for instance, if there is a high risk of predation—then a dominant animal can enforce his priority in the competition with subordinates without risking attrition from the group.

When, however, a subordinate has the option to move to another group, and when his presence in the group is also of some value to the dominant—for instance, as an alliance partner—then the subordinate has some leverage. If the dominant animal exploits the subordinate to the extent that his benefit-cost balance of maintaining the relationship becomes negative, the subordinate can leave. The dominant has to be sufficiently tolerant to keep the balance for his subordinate positive. Our *power asymmetry hypothesis* suggests that in a system with strongly hierarchical relationships, there is a greater need for formal and unambiguous status recognition signals, whereas an equally distinct signal is needed to earmark playful transgression of rules. Studies that have investigated the relationships among these different characteristics are still limited in numbers, but the available data certainly are suggestive. We have recently tested this prediction with an additional species of *Macaca,* the pig-tailed macaque, which is closely related to the tolerant liontailed and Sulawesi macaques (Preuschoft 1995; Fa 1989; Melnick & Hoelzer 1996; Thierry 2000). Pig-tailed macaque society is in many ways similar to the rather hierarchical society of long-tailed macaques (Kaufmann & Rosenblum 1966; Castles et al. 1996). As expected, we found that pig-tailed macaques use the classical play face in which the upper teeth remain covered, and a distinct silent-bared teeth face. Moreover, pig-tail play faces are strictly confined to the context of play, whereas silent bared-teeth smiles indicate formal subordination and function to appease hostile dominants.

It is, of course, fascinating to wonder whether similar differences exist between or even within human cultures, namely dependent on the degree of authoritarianism versus egalitarianism in a society or in segments of a society. In other words: Do formalized smiles without chuckles of laughter prevail where hierarchical distances are emphasized? And conversely, do relaxed laughs prevail where hierarchical aspects are absent or subdued? Everyday experience would suggest that this is so. Unfortunately, we still lack explicit systematic studies in which different cultures and social contexts are compared ethologically in this respect.

Cognitive Aspects and the Sense of Humor

The "Comment Laugh"

Although laughter is part of our primate heritage, humans differ in its use in one respect, which undoubtedly is correlated with the development of referential communication in our species. We have developed verbal speech as an important tool to communicate about the world around us and to comment on it. In this respect our communication differs at least gradually from that of our nonhuman fellows. Although distinctions may not be as sharp and discrete as we once thought, it is fair to say that animals mostly communicate about their relations and their relational attitudes (whether they like one another or not, and so on). In addition to this, we humans use our language to share our impressions and knowledge about the world around us.

This referential aspect is also characteristic of our humor, our "play with the world." We do not laugh only at a person to tell him that we approach him with playful intentions and that we interpret his acts as playful, in other words as a specification of our relationship, as do monkeys and apes. We also laugh about things around us, thus expressing that we regard them as funny. All right, we do this in company, and it also presupposes a certain "joking" relationship. Yet the comment is not about ourselves but about the world we share. This "comment laugh" we use continuously in our verbal exchanges. It then modifies and colors the content of our verbal interchange. There is no evidence that monkeys and apes do anything like this, although in our closest relative, the chimpanzee, we may observe something that comes very close.

A couple of years ago a group of ethologists was visiting Burgers Zoo in Arnhem, and looked at the chimpanzee colony, which, by that time, had become famous by the studies of Frans de Waal, Otto Adang, Charlotte Hemelrijk, and others. We had decided to do a little experiment to demonstrate cooperative aggression toward a threat from outside. A collaborator, Dr. Matthijs Schilder, had been conducting experiments at the Arnhem Zoo on the defense by zebra stallions of their herd against predators. For this he used models and masks of lions and leopards. So we decided to play a little trick on the chimpanzees. Matthijs put on a leopard mask and kept hidden behind a bush just across the ditch that surrounds the chimpanzee forest. Then he suddenly appeared from behind the bush. The chimps noticed him at once and got completely upset and aroused. They started screaming, embracing and kissing one another, then took sticks and branches and threw them toward the "leopard." At that moment Matthijs took off his mask and showed his true face, that of Matthijs, whom the chimps knew all along. To our amazement, Mama, one of the elder females in the group, now looked at Matthijs and us, showing a full-fledged play face, as if she were saying, "Gosh, you got me there! This was a good joke!"

In the human the vocalized play face or laughter is closely associated with verbal expression. It becomes part of the verbal exchanges that display wit. Thus the human laugh becomes part of tournaments of wit that are evident when groups of people congregate in a congenial atmosphere, where impressing the companions with one's versatility of mind, and especially, it seems, the companions of the other sex, has become a major motivation.

The Expression of Joy: An Expression of Self-Awareness?

In nonhuman primates the play face is part of playful interaction. It is clearly directed at the play partner, and, insofar as there are "chuckles" or breathing coughs, these are given particularly in response to playful twists by the playing partner. Thus it clearly functions as a message that can be interpreted by the addressee as a signal of a playful attitude. It is in this context of regulating social relationships between individuals that this signal must have been ritualized into a display or expression movement. When using the term "expression movement" we emphasize the psychological background of the display, a particular mood, or emotion that is expressed in the display. As we noted above, ethologists, in contrast to psychologists, tend to see the latter not as an essential evolutionary cause of the display's existence but as a phenomenon complementary to the communicative function. The element of emotional expression comes to the fore when the display is performed in the absence of an audience (Miller et al. 1967; Preuschoft 2000).

A few cases of apparently "pure" expression have been nicely documented in the film by Haanstra and colleagues (1982), "The Family of Chimps." It shows young chimpanzees engaging in "funny" movements, such as pirouetting, or walking and looking backward through the legs, head dangling, at the world upside-down. During these performances the chimpanzees show fully fledged "laughing faces," but without addressing them to anyone in particular. Obviously, here we have pure expressions of joy, of fun in "clowning" with its unusual experiential consequences. The possible relevance of such performances for the existence of self-reflection—second-order intentionality where one is one's own audience—and thus of self-awareness is very cogent and casts doubt on the idea that the expression of emotion is necessarily a simpler function than is the communication of cognition. It is also noteworthy that in many hours of observing macaque play behavior we have never seen a monkey privately laughing to itself during solitary play.

On the other hand, it is clear that, also in chimpanzees, there is an audience effect on the performance of play faces. Juvenile chimpanzees laugh more when the mother of a playmate that is younger than themselves is close by (Jeanotte & de Waal 1996). Like other signals, facial expressions seem to communicate conditional action tendencies (Hinde 1985; Preuschoft & Preuschoft 1994). In other words, facial expressions are used in processes of negotiation between interaction partners. Thus, even though facial expressions may be tightly coupled to emotional states, the information content of a display may greatly exceed the mere expression of an acute emotion. It is likely that the silent bared-teeth smile is associated with fear in the "despotic" species. However, at the same time the smile may symbolize the sender's subordination under the receiver. This is by no means trivial as it implies that primates are capable of assessing the quality of relationships in which they are engaged and communicating this assessment, potentially independent of being in a state of acute emotional arousal (Preuschoft 1999).

In chimpanzees the smile can often be understood as an expression of fear, and seems to appease the receiver, irrespective of the interactants' rank positions (van Hooff 1972). However, the chimpanzee smile also indicates that the sender is anxious to achieve a certain result, for instance to take over a desired resting place from a subordinate without fighting, or to establish peaceful contact with an attractive partner. When conducting experiments at the Yerkes Field Station Signe Preuschoft could use only certain individuals at a time, but sometimes other group members were eager to participate also. It often happened that when she slightly opened the door to admit the next subject, she found another group member anxiously grinning at her in an obvious attempt to gain admission to the experimental cages. In these cases the smile seems both the expression of anxious anticipation and an overture in a process of negotiation between the experimenter and the chimpanzee who is not scheduled to be the next subject.

This "wooing" function of the smile is quite similar to the affiliative smile as found in the tolerant species of macaques, in which adults approach infants and juveniles with smiles. Here, the smile clearly induces trust and toleration of intensive body contact, involving nibbling, nudging, and patting. Such smiles are obviously not motivated by fear; instead senders express their anxious affection that may develop into frolicsome playfulness if the infant can be enticed to perform a play chase or some wrestling. It is in these species that smiling and laughing merge, and co-occur in the same social contexts (Preuschoft 1995). These are also the species in which not only juveniles, but also adults can regularly be observed playing and wrestling. While the general participation in play further highlights the pleasant social manners of these species, it is also probable that individuals use play to assess and probe each other. Obtaining information at low cost about each other's competitive potential and behavioral inclinations is arguably particularly useful in tolerant species with their tight-knit social networks and their symmetrical power relationships (for similar reasoning see Pellis & Iwaniuk 2000). The use of play for the double purposes of joyful entertainment and strategic probing shows the limited usefulness of the traditional distinction between emotional and cognitive underpinnings of animal behavior.

Conclusion

Convincing examples illustrating that human behavior is also a product of evolutionary development are found most easily in our expressions of emotions. The respective behavior patterns are species-specific in their form, motivations, and functions. The expression movements are, therefore, the best candidates for a comparative study that can lead to a reconstruction of their evolutionary origins and development. Laughter and smiling are excellent examples.

The comparative approach clearly leads to insights that would not have been gained easily by looking at human behavior in isolation, as has mostly been done until recently. An anthropocentric perspective cannot reveal that the laugh and the broad smile have different evolutionary origins. The comparative approach brings strong arguments that laughter has evolved in the context of joyful play, and that the broad smile has evolved as an expression of nonhostility and friendliness, taking its origin in the expression of fearful submission. The approach also leads to a clear choice between earlier theories on the origin of laughter. Theories that seek its origin in a context of aggression and regard it as an intimidating derogatory display are not supported. First, there is no equivalent of a derisory, jeering expression in our nearest relatives, whereas there is the expression of nonserious playfulness. Second, the first appearances of laughter in very young children occur in the context of relaxed playful interaction. There the signal operates as a very strong rewarding stimulus reinforcing playful interaction of elders with the infant. So both the comparative evidence and the early development of laughter support the idea that social play is the original context.

This, however, is far from saying that there are no forms of laughter or smiling with a strong aggressive component. Mocking laughter, laughter of derision, and mean smiles do exist, as do various kinds of laughter that betray elements of fear and uneasiness, such as nervous giggling. These forms are better regarded as the result of mixed motivations; they are secondary intermediate forms, where elements of other behavioral attitudes, such as aggression or fear, have been mixed in.

CASE STUDY 10A

Emotional Recognition by Chimpanzees

LISA A. PARR

Numerous studies have demonstrated that human facial expressions represent universally recognized emotions, including anger, disgust, fear, happiness, sadness, and surprise (Ekman et al. 1972; Izard 1971). The perception of these expressions in others can provide accurate information about that individual's emotional state. In addition, viewing emotional behavior such as facial expressions in conspecifics is capable of producing physiological changes in the observer that may be sufficient to induce similar emotions, providing a mechanism for the evolution of shared emotional experiences, or empathy, among socially living animals that is independent of conscious awareness (Ekman et al. 1983; Hatfield et al. 1994; Kappas et al. 1993; Levenson & Ruef 1992). Dolphins and elephants, for example, may accomplish similar feats using tactile and auditory contacts, such as contact between dolphin mothers and offspring, and elephants may even show empathic reactions when caressing the bones of conspecifics (see Case Study 4B and Chapter 3).

Previous research has suggested that chimpanzees, like other social animals, are highly emotional, capable of showing strong conciliatory tendencies (de Waal 1996; Flack & de Waal 2000; Goodall 1968). They also have great social intelligence, demonstrating skillful categorization of conspecifics' faces and facial expressions, even basic kin relationships (Parr et al. 1998; Parr & de Waal 1999; Parr et al. 2000). No studies to date have examined whether chimpanzees understand the relationship between facial expressions and their underlying emotion meaning, despite the overwhelming evidence for biologically determined facial emotion in humans, and evidence that some facial expressions are homologous in humans and nonhuman primates (Ekman et al. 1972; Izard 1971; Preuschoft & van Hooff 1995). These studies would not only provide valuable insight into the function of facial expressions, their evolution as emotional signals in humans, and their phylogenetic continuity, but they would also significantly increase our understanding of the mechanisms by which chimpanzees acquire and process emotional information about conspecifics. Such abilities are likely to have played a pivotal role in the evolution of empathy in our living ancestors, in addition to being a mechanism that could facilitate conciliatory behaviors (Davis 1994).

This study examined the ability of three chimpanzees *(Pan troglodytes),* two males and one female 12 years of age, to associate emotionally positive and negative facial expressions with other video images that share a similar emotional valence. Subjects were required to capitalize on the inherent emotional meaning of facial expressions to successfully perform the task. Subjects had previously been trained to use a computerized joystick testing system (Parr et al. 1998; Parr & de Waal 1999; Parr et al. 2000). This paradigm involves positioning a computer in front of the subjects' home cage and attaching a joystick that they may use to control the movements of a cursor on the computer screen. Subjects are required to match an initially presented sample stimulus to one of two comparison stimuli, only one of which is similar to the sample. In this study, chimpanzees were first presented with a short video depicting a positive, negative, or neutral scene. They were then shown two facial expressions, one representing an emotion similar to that depicted in the video, the other representing either a different or a neutral emotion. The correct choice, rewarded with a squirt of juice, was to select the facial expression that was emotionally similar to that present in the video. Unlike traditional matching-to-sample tasks (MTS), these discriminations were not aided by perceptual similarities between the sample and correct comparison images, but instead required an understanding of their shared emotional meaning. Hence, this task is a variation on the MTS hereafter referred to as matching-to-meaning, MTM. Readers are directed to Parr (2001) for a detailed description of this experiment.

Four examples of six videos from seven categories were presented. The first was a control (CON1) that included videos of conspecifics engaged in severe aggression and intense play. The correct choice for these scenes was the facial expression that occurred most often in those social contexts, i.e., screaming and play faces, respectively. Subjects were pre-trained on this category so that their performance exceeded >85 percent, making CON1 a control for whether subjects were motivated to perform the task on a given testing session. A random control (CON2) showed scenes of sleeping chimpanzees. These were arbitrarily paired with facial expressions that would not normally occur in this context, e.g., play faces, screams, hoots, and whimpers, making CON2 a control for learning by reinforcement history.

Five experimental categories showed videos of emotionally positive and negative scenes. Negative videos included scenes of veterinarians performing routine medical procedures (CHASE), anaesthetized chimpanzees (KD), injection needles (DART), and conspecifics receiving injections (INJECT). Correct comparisons for these were negative facial expressions, e.g., screams and bared-teeth displays, versus hoots, play faces, relaxed-lip faces, and neutral portraits. Positive emotion videos (POS) included scenes of preferred food and objects paired with play faces as the correct expression. The ability of chimpanzees to categorize these facial expressions is reported elsewhere (Parr et al. 1998).

Even though not specifically trained on these "emotional" discriminations, subjects performed significantly above chance on the very first testing day (5 categories = 66 percent, $z = 1.87$, $p < 0.05$), regardless of whether the videos included social information, like the presence of socio-emotional information (SOCIAL = CHASE and INJ scenes, 68.8 percent, $z = 2.60$, $p < 0.005$), or were largely nonsocial (food, objects, needles, and anesthetized chimpanzees, STIM = DART, KD, and POS scenes, 63.9 percent, $z = 2.35$, $p < 0.01$). As expected, subjects performed best on the previously learned control trials (CON1, 83.3 percent, $z = 3.27$, $p < 0.001$), indicating that they were motivated to perform the task, while performance did not exceed chance on the random control (CON2, 50.0 percent, $z = 0$, $p = 0.50$), indicating that their performance was not learned through reinforcement. The second testing session saw above chance performance for each individual stimulus category, again except for CON2. Figure 10A.1 shows the performance on both testing sessions where each subject was presented with two exposures to each of the 28 novel video discriminations on each day.

These data confirm that chimpanzees have a basic understanding of the emotional valence, positive or negative, that is communicated by their facial expressions, and are able to use this information to categorize emotionally meaningful social and nonsocial scenes. Additionally, these data support the

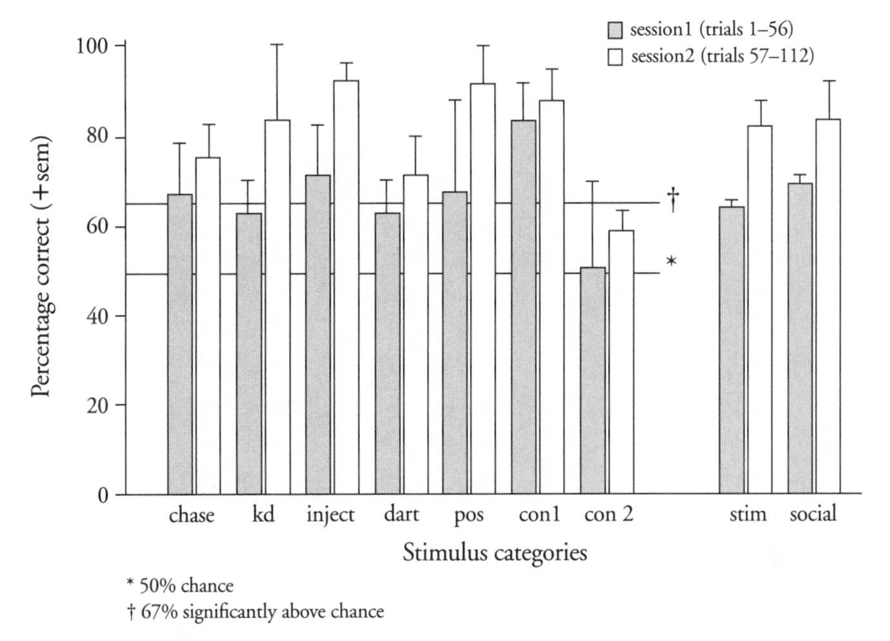

Figure 10A.1. The mean performance (+sem) on the first (trials 1–56) and second (trials 57–112) testing sessions, including the grouped categories of STIM (DART + KD + POS) and SOCIAL (INJECT + CHASE). Subjects are given two exposures to each of the 28 novel videos per session. An average of 205 trials was required before subjects reached the final performance criterion of >85 percent over two daily sessions (112 trials), fewer than eight exposures to each of the 28 stimulus discriminations in the task.

claim that emotional facial expressions in humans and nonhuman primates share a biological, as opposed to cultural, continuity (Brothers 1989; Ekman 1973), and that emotional valence may be a natural way for chimpanzees to organize social information (Ortony et al. 1988). Discrimination performance did not appear to be affected by the presence of social or emotional information in the videos, like the presence of specific emotional vocalizations. This would suggest that discrimination was based on prior knowledge about the emotional meaning of the scenes that were presented. This would favor the interpretation that subjects used their facial expressions representationally to categorize social and nonsocial videos according to their basic emotional meaning, i.e., positive or negative emotional valence.

Although previous studies have demonstrated that chimpanzees exhibit human-like emotional behavior, including consolation (de Waal 1996; de Waal & van Roosmalen 1979), empirical evidence for emotion processing in this and other species of mammals has remained elusive. This study provides much-needed empirical support to conclude that chimpanzees not only have a keen social awareness (de Waal 1996; Parr & de Waal 1999), but they also have a basic emotional awareness demonstrated in their capacity to perform discriminations involving emotional information in the absence of extensive testing. This confers distinct advantages to chimpanzees in that the observation of specific facial expressions in conspecifics may lead to a general understanding of that individual's emotional state, not just an understanding of how that expression is related to a specific social context, i.e., as in the CON1 trials. Thus, individuals may come to learn about the emotional states of others regardless of unfamiliarity with precipitating events. Such an understanding may provide bystanders with the motivation to console distressed individuals. It should be noted that it is highly unlikely that other social mammals would not possess similar levels of emotional awareness, although this remains to be demonstrated. This type of emotional awareness is important for comparative studies of social cognition because it is a likely precursor to more cognitive forms of empathy, including emotional perspective taking. Although such mentalistic forms of empathy have yet to be empirically demonstrated in nonhuman species, these data suggest that empathy as a shared emotional feeling, or basic level of emotional awareness, may indeed be something we have in common with the chimpanzees.

11

Vocal Communication in Wild Parrots

JACK W. BRADBURY

Captive parrots mimic an astonishing number of exotic sounds, including human speech. The vocabularies of pet African grey and Amazon parrots can number hundreds of such sounds. Recent research by Irene Pepperberg with a hand-raised African grey parrot (Alex) showed that this species can learn to associate meanings with spoken words and to interpret complex combinations of adjectives, nouns, and verbs (Pepperberg 1987, 1988, 1990, 1994, 1999). Alex correctly interprets novel combinations of known terms, and can reply "none" when asked to search an array that lacks a requested item. All of these facts raise the question of how parrots use their vocal learning abilities in the wild. Since vocal learning is limited to only a few groups of birds, why has it evolved in parrots? And is it possible that wild parrots use vocal signals in a semantic way, like Alex, to exchange environmental information?

One immediately looks to better-studied bird taxa for insights. The three groups of birds that show extensive vocal learning are the songbirds, the hummingbirds, and the parrots. Although little is yet known about vocal learning in hummingbirds, there is an extensive literature on the development, evolutionary functions, production, and central processing of songbird vocalizations (Kroodsma & Miller 1982, 1996; Catchpole & Slater 1995). This body of information suggests that vocal learning in songbirds has minimal parallels with that in parrots. Specifically:

1. Parrots use different parts of the syrinx to generate sounds (Nottebohm 1976; Gaunt & Gaunt 1985) and nonhomologous parts of the brain to control sound production (Streidter 1994; Durand et al. 1997; Jarvis & Mello 2000; Nespor 2000).
2. Whereas songbird vocal learning is largely limited to males and to a limited period of their lives (Catchpole & Slater 1995; Kroodsma & Miller 1996), most parrots show vocal learning by both sexes throughout life.
3. Male songbirds use their learned vocalizations to defend territories and attract mates. Parrots appear to use their learned vocalizations in a wide variety of contexts other than, or in addition to, nest site defense and mate attraction (see evidence below).

If typical songbirds fail as models for parrot vocal learning, we can only turn to the parrots themselves. Obtaining data for parrots equivalent to that accumulated for songbirds is challenging because parrots are notoriously difficult to study. They are rarely found in fixed territories like songbirds, but instead fly rapidly over large areas. Because of poaching for the pet trade, most populations are extremely wary of humans, making it difficult to capture and follow known birds. Once captured, few bands or transmitters can long withstand their powerful bills. Finally, the remarkable learning abilities of parrots should make us wary about inferring functions of vocal learning in wild populations from observations made on captive parrots or domesticated strains.

Despite these limitations, new technologies, access to protected and less timid populations, and systematic observations at sites where parrots aggregate are generating a reasonably concordant picture of parrot biology and the role of vocal signals in their lives. In this chapter, I shall illustrate patterns emerging from the work by my research group in Costa Rica and compare it to other studies worldwide. Our Costa Rica project has focused on four sympatric species of the Guanacaste dry forest (listed in order of decreasing body size): yellow-naped Amazons *(Amazona auropalliata),* white-fronted Amazons (*Amazona albifrons),* orange-fronted conures *(Aratinga canicularis),* and orange-chinned parakeets *(Brotogeris jugularis).* Although we do not yet have a complete story for any of these species, together they provide a useful cross-section of parrot biology and vocal communication.

A Précis of Parrot Biology

Diet

Most parrots are seed predators (Saunders 1980; Wyndham 1980; Snyder et al. 1987; Clout 1989; Forshaw 1989; Stiles & Skutch 1989; Rowley 1980; Acedo 1992; Waltman & Beissinger 1992; Galetti 1993; Martuscelli 1995; Gilardi 1996; Norconk et al. 1997; Wermundsen 1997; Greene 1998; Juniper & Parr 1998). Although many will also eat ripe fruits, flowers, nectar, and even insects, the original function of the curved parrot bill is surely the extraction and ingestion of seeds. Most plants encourage the visits of seed-dispersing frugivores by making fruit conspicuous and accessible. In contrast, plants discourage parrot visits using crypticity, toxicity, hard capsules, spines, and other seed defenses. When parrots eat toxic seeds, they must do so in moderation and may require ingestion of clay or other materials to absorb toxins (Gilardi et al. 1999). Plant defenses against seed predation force parrots to have diverse diets and to acquire a correspondingly diverse set of handling and extraction skills. Orange-fronted conures use widely divergent techniques for extracting seeds from the long pods of *Inga* fruit, removing the thin pericarp on *Cordia* seeds, obtaining purchase on the spherical fruit of *Byrsonima* (Figure 11.1), and decimating the ovaries of *Gliricidia* flowers. Some parrots avoid plant defenses by specializing on the seeds of fruit designed for seed-dispersing frugivores. Thus orange-chinned parakeets crack and digest the thousands of tiny seeds in wild fig fruit that most frugivores ingest whole and disperse in their feces (Janzen 1981). The benefit to the parrots is that these fruits have no defenses against seed predators. The costs are that they must crack every tiny seed and find fruit crops before the hordes of frugivores that also specialize in figs.

Ranging

Parrots have much larger home ranges than similarly sized birds of other taxa, and they move throughout these ranges rapidly and frequently. Large home ranges are surely related to their diets. Species eating toxic seeds will have to limit intake at any one location even when additional food is present and move to other food types. Nontoxic foods often have phenological patterns that force parrots to adopt large ranges. The primary foods of orange-

Figure 11.1. Sharing of *Byrsonima crassifolia* fruit by a mated pair of orange-fronted conures (*Aratinga canicularis*) in Costa Rica. Photograph courtesy of Erica Spotswood.

fronted conures are often widespread and patchy. Many species mature only a few fruit per day, requiring the parrots to move from plant to plant and to move between patches if another flock has already harvested the site. Range lengths (longest axis of cumulative home range) for these conures during the rainy season average six to nine kilometers. Fig specialists, such as orange-chinned parakeets, nearly always have large home ranges because fig reproductive biology requires trees to fruit asynchronously and randomly. Birds must hunt a large area to find a currently fruiting tree. We have measured range lengths for orange-chinned parakeets of at least 10–15 kilometers. Open-country parrots, such as wild budgerigars in Australia, are often nomadic and have enormous annual ranges (Wyndham 1983).

Daily Cycle

Most parrots follow a similar daily cycle. Large night-roosting groups divide into smaller foraging flocks shortly after dawn. Typical flock sizes for conures and white-fronted Amazons are 10–20 birds; parakeets tend to have two to eight birds per group. Flocks depart the night roost in different directions and forage independently of each other. Conure flights are typically swift and direct with many short foraging stops. Radio tracking indicates that different foraging flocks have highly overlapping ranges. A single fruiting patch may be visited by six to eight different foraging conure flocks in a morning. Parakeets use vocal exchanges to recruit passing flocks of conspecifics to a discovered fruiting tree. Conures do not show active recruitment but overlapping visits by multiple flocks at the same site often result in temporary fusion of the groups while foraging and separation before departure. None of the three smaller species exhibit defense of food sites.

After several hours of foraging, flocks move to resting sites. Several flocks may fuse into a single aggregation at this time. Birds then preen, vocalize, play, and sleep. Play is particularly conspicuous during the early and late parts of the resting period. Midday is a period of minimal activity in most parrots. In early afternoon, flocks leave separately for further foraging, but usually visit fewer sites than in the morning. By late afternoon, the flocks begin to aggregate into sleeping groups.

The large yellow-naped Amazon is relatively asocial when compared to sympatric species. Although these birds interact noisily at sleeping sites, pairs are often widely separated during the day. Counter-duetting may occur shortly after arrival on the foraging grounds, but even then pairs remain at a distance. On occasion, several pairs may share a rich feeding site. They then feed silently with none of the vocal exchanges or coordinated movements of the smaller species.

Night Roosts

Most parrots sleep at a communal night roost. This is preceded by late afternoon staging near to the eventual sleeping trees. Conures, parakeets, and white-fronted Amazons all engage in active recruitment of conspecifics to staging aggregations. These grow over a period of one to two hours and then the entire assembly moves frenetically into and out of a series of candidate sleeping trees. Eventually they nestle densely into the foliage of one or a few nearby trees and become totally silent. All three species shift communal roost locations at intervals. Orange-fronted conures do so nightly, and radio-tracked individuals never sleep in the same tree two nights in a row. Successive night roosts for a given bird can be kilometers apart. Sleeping aggregations range from 15 to hundreds of birds. One typically finds many different conure sleeping groups all within a few kilometers of each other. Orange-chinned parakeets move communal roosts at intervals of two to five weeks, and white-fronted Amazons shift roosts at four- to six-week intervals. In both species, there is usually only one such aggregation of 50–100 birds within a one or two kilometer radius. Different sites are favored even at the same time of year in successive years.

This shifting of night roosts with little interannual predictability puts a premium on the recruitment and staging process. Birds often do not know where the next night roost will be and can find a new site only if it is advertised to overflying birds. The large yellow-naped Amazon again differs from the three smaller species because its widely separated night roosts remain unchanged for decades. Although these aggregations are extremely noisy, one sees none of the interactive recruitment of the smaller species, and pairs often delay return to the night roost until just before sundown. Whereas the smaller species always nestle into dense foliage and become cryptic, yellow-naped Amazons favor large bare trees and thus are highly conspicuous.

Nests and Breeding

The majority of parrots are obligate hole-nesters. The two *Amazona* nest in tree cavities excavated by other taxa such as woodpeckers and insects. Because they cannot create cavities, there is intense competition for the limited number of suitable holes, and nesting pairs actively defend their nest sites. Conures and parakeets both excavate nest cavities in arboreal termiteria. As with most parrots, females of our study species do all the incubation, although both sexes may sleep in the nest site at night. Males usually regurgitate food to the female during this period. Both parents feed the young both before and for at least a considerable period after fledging. Although subadults remain with parents for up to a year in some cockatoos (Saunders 1982, 1983; Rowley 1980), our species resemble budgerigars (*Melopsittacus undulatus;* Wyndham 1980), parrotlets (*Forpus passerinus;* Waltman & Beissinger 1992), and galahs (*Cacatua roseicapilla;* Rowley 1980) by being independent several months after fledging. We have found no evidence for retention of young as breeding helpers in any of our four species.

Predation

Because of their heavy musculature, parrots are a favored prey of raptors (Saunders 1982; Snyder et al. 1987; Forshaw 1989; Stiles & Skutch 1989; Juniper & Parr 1998). The green coloration of many species makes them nearly invisible to over-flying hawks when nestled in the foliage. On the other hand, parrots are highly visible in flight and this has presumably favored high speed, an obsessive directionality, and tight flock coordination (Hardy 1965; Garnetzke-Stollman & Franck 1991). While the large yellow-naped Amazons appear relatively blasé about passing hawks, parakeets panic and seek cover even at the sighting of distant turkey vultures, whose rocking glides are often mimicked by a major parakeet predator, the zone-tailed hawk *(Buteo albonotatus).* Nocturnal predators of parrots include barn owls *(Tyto alba),* which rush at night roosts just before dawn to flush hidden parakeets, and the large carnivorous bat, *Vampyrum spectrum,* for which conure is a favored prey (Vehrencamp et al. 1977). The daily moving of night roosts in conures may be an adaptation to reduce bat predation. Nest predation in our study area is most often inflicted by humans, coatis, snakes, and capuchin monkeys. There is little the adult birds can do about any of these predators.

Social Structure

The fundamental social unit in most parrot societies is the mated pair (Forshaw 1989; Juniper & Parr 1998). Pairs remain together throughout the year and act as coordinated units in agonistic interactions with other pairs. Allopreening, food sharing, play, and contact sleeping all tend to be pair-specific. Parrots also retain mates across years (Serpell 1981; Saunders 1982, 1983; Rowley 1980; Garnetzke-Stollman & Franck 1991; Waltman & Beissinger 1992), although some reassortment due to death or mate switching has been reported. Saunders (1982) estimates remating between years at 10 percent for females and 29 percent for males in black cockatoos *(Calyptorhynchus funereus).* Rowley (1980) recorded mate-change rates in galahs as high as 50 percent per year. Paternity has not been measured genetically in any wild parrot, but Rowley (1980) notes that extrapair copulations are not uncommon in galahs.

Compositional stability of higher-order associations varies. Fledged siblings of spectacled parrotlets *(Forpus conspicillatus)* remain together until pair bonds are formed (Garnetzke-Stollman & Franck 1991; Wanker et al. 1996). Kin associations at older ages have yet to be reported in any species. Galahs favor nearby nesting pairs when forming foraging flocks (Rowley 1980). Many species show daily fission/fusion of foraging flocks (Brereton & Pidgeon 1966; Wyndham 1980; Juniper & Parr 1998). Individually radio-tracked orange-fronted conures may be found together in the same foraging flock on successive days, in different flocks a week later, and back together again several weeks after that (Bradbury et al. in press). There is little evidence that conures have any social units other than mated pairs that are closed and stable. Foraging aggregations, resting groups, and night roosts provide many occasions in which individual pairs can switch group affiliations. Observed levels of eventual re-association suggest that birds may be selective in making these shifts. This would require some mechanism for recognizing individuals, pairs, or higher-level subsets of the population.

Dominance hierarchies are common in captive parrot colonies (Hardy 1965; Power 1966a; Arrowood 1988; Garnetzke-Stollman & Franck 1991), and we see indications of such dominance in wild populations. Conflicts between pairs over preferred perches are common in all species. Most disputes are settled vocally. The occasional escalations into physical attacks rarely result in injury in the smaller species, but can be fatal in captive yellow-naped Amazons. As noted below, this species may use vocal duets to assert dominance over localized foraging areas.

Reliance on Learning

A typical parrot clearly must have access to extensive "habitat lore" to survive (Saunders 1982). Birds must know what is safe to eat in each season and how to handle an immense variety of food types. Because of their status as raptor prey, parrots must move between foraging sites swiftly and directly. They can only do so if they already know when and where to find food. They must also learn how to handle diverse foods quickly and efficiently, or be left behind by their rapidly moving flock. Finding suitable food sites is complicated by the need to search a succession of sites before finding one not yet harvested by other flocks. If each individual bird had to acquire this information by trial-and-error learning, we ought to see more sampling and reconnaissance than we do. In fact, reconnoitering is rare in our study species. It thus seems highly likely that our parrots learn the requisite lore by following other birds to feeding sites and by imitating their handling techniques and ingestion limits.

There is good evidence that habitat lore is learned socially from other parrots. A natural experiment in which Major Mitchell cockatoos *(Cacatua leadbeateri)* accidentally fostered galah offspring produced adult galahs that had learned the full repertoire of food types and handling techniques of their foster parents (Rowley & Chapman 1986). Even at full body size and three months of age, conure fledglings do not yet know how to handle many of the foods their foraging flocks encounter. Parents still regurgitate to them while the youngsters make fumbling attempts to handle new food types. It is widely reported that wild and captive parrots are neophobic about new foods. This is certainly true with wild-caught conures held for several weeks in our aviary. Individuals are extremely variable in their willingness to test new foods and their abilities to solve the handling problems. However, once one bird in a captive flock solves the problem, the others generally learn it within days.

The Vocal Repertoires of Parrots

Most researchers studying parrots in the wild have listed 10–15 separate call types in their species' repertoire. The type definitions are based on auditory distinctiveness to human listeners and usage in a definable context. Despite the potential subjectivity in such classification schemes, there is remarkable concordance in the results. Below, I list the nine call types that appear to be most widespread in parrot repertoires:

Loud Contact Call

Nearly all species of parrots have a call that is given in flight, exchanged by perched and flying birds, and exchanged by separated members of a pair. Because it typically elicits a like response from conspecifics, the loud contact call appears to establish a vocal connection between specific birds. It is often the loudest call in the repertoire. Examples include the "wa-wa" of yellow-naped Amazons (Wright 1996), the "ack-ack" of white-fronted Amazons, the "tinkle" of orange-chinned parakeets, the "chee" of orange-fronted conures (= the "intrapair" call of Hardy 1963), the "bugle" of the Caribbean Amazon (Synder et al. 1987), the "S" or "chet" call of the galah (Pidgeon 1981; Rowley 1980), the "whylah" of black cockatoos (Saunders 1983), and the "contact call" of sulphur-crested cockatoos *(Cacatua galerita)* (Noske 1980), budgerigars (Wyndham 1980; Farabaugh & Dooling 1996), parrotlets (Wanker et al. 1996, 1998), and monk parakeets (Martella & Bucher 1990).

Soft Contact Call

This is a call that coordinates the movement of flock members when moving through vegetation as a group. It tends to be very low in amplitude and repeated often with or without responses by other flock members. Examples include the "zip" of orange-fronted conures, the "soft buzz" of orange-chinned parakeets (or, "low intensity squawks" of Power 1966a), the "mumble" of yellow-naped Amazons, the "chet-it" of the galah (Rowley 1980), and the "chatter" of monk parakeets (Martella & Bucher 1990).

Preflight Call

Most species have a specific call that is given by flock members prior to taking flight. It is often a relatively loud and harsh call. Examples have been described in orange-fronted conures (the "peach" call of Hardy 1963), sulphur-crested cockatoos (Noske 1980), galahs (Rowley 1980), budgerigars (Wyndham 1980) and rosellas *(Platycercus eximius)* (Brereton & Pidgeon 1966).

Begging Call

All parrots studied to date have a special call used by begging offspring (Hardy 1963; Pidgeon 1980; Rowley 1980; Saunders 1983; Wyndham 1980; Martella & Bucher 1990). In budgerigars, begging calls morph into the first loud contact call as fledglings mature (Brittan-Powell et al. 1997).

Pair Duets

Some, but not all, parrot species have pair duets. Both members may vocalize simultaneously, but more often they call antiphonally or sequentially. Examples have been described in several species of parakeets (Power 1966a,b; Arrowood 1988), peachfaced lovebirds *(Agapornis roseicollis)* (Mebes 1978), sulphur-crested cockatoos (Noske 1980), galahs (Pidgeon 1981), several species of lorikeets *(Trichoglossus)* (Serpell 1981), and a number of *Amazona* species (Nottebohm 1972; Snyder et al. 1987; Wright & Dorin 2001).

Warbles

Many species have long rambling vocalizations with highly variable note types that are produced during rest periods, around nesting sites, or during late-afternoon staging. In budgerigars, males use these calls to stimulate females for reproduction (Brockway 1965). Their function is not yet clear in other species as they are often produced both in and out of the breeding season. Warbles appear to be present in all four of our study species.

Agonistic Protest

Nearly all species produce some form of loud "squawk" during fights (Hardy 1963; Power 1966a; Wyndham 1980; Pidgeon 1981; Saunders 1983; Martella & Bucher 1990).

Distress Call

This is given by injured or threatened birds of certain species (Pidgeon 1970; Wyndham 1980; Saunders 1983; Martella & Bucher 1990; Rowley 1980).

Alarm Calls

These are given when an approaching predator is sighted. They usually elicit evasive action by listeners (Brereton & Pidgeon 1966; Power 1966a; Noske 1980; Saunders 1983; Martella & Bucher 1990).

A variety of less widely shared call types have also been noted. Some species have calls they give when a bird finds itself isolated (Saunders 1983; Martella & Bucher 1990; Wright 1997). Others have specific appeasement calls to de-escalate a conflict (Pidgeon 1981). At least one cockatoo has a male pre-copulation call (Saunders 1983), the white-fronted Amazon has a female postcopulatory call (Skeate 1984), and some species make soft sounds just prior to becoming silent in the night roost (Wyndham 1980). The exceptional lek-mating kakapo *(Strigops habroptilus)* produces a loud drumming sound to attract potential mates (Merton et al. 1984).

Different species of parrots draw on the general list in different ways. The basic vocal repertoire for wild orange-fronted conures in Costa Rica is shown in Figure 11.2. It includes loud contact calls, soft contact calls, preflight calls, offspring begging calls, agonistic protests, and warbles. We have found no evidence that this species duets in the wild, and we have found no specific calls used as predator alarms or isolation calls. In contrast, sympatric orange-chinned parakeets have two kinds of duets but seem to use the pair-specific components of duets as their contact calls. Yellow-naped Amazons appear to have a distinct example of each of the nine call types with the exception of alarm calls.

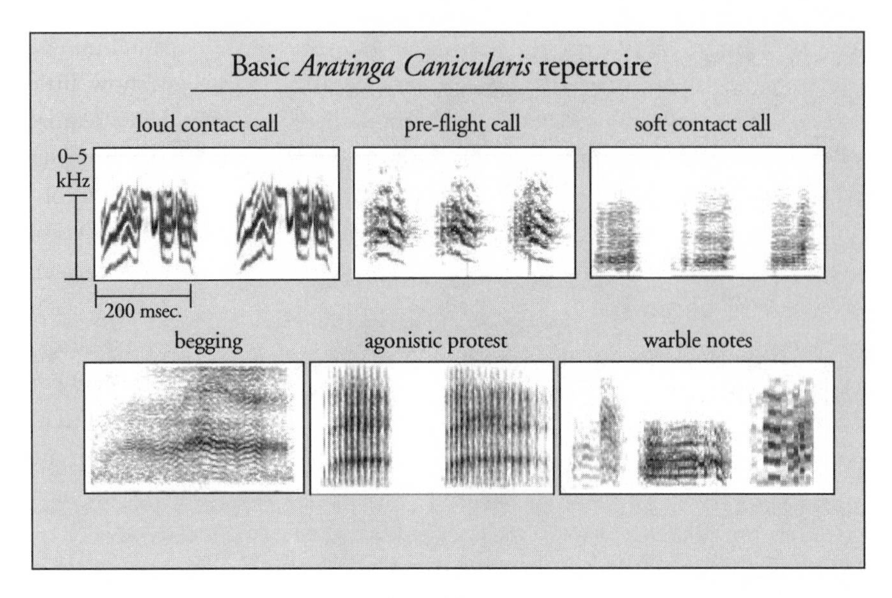

Figure 11.2. Samples from each of six major call types in the vocal repertoire of orange-fronted conures in Guanacaste, Costa Rica.

Likely Foci of Vocal Learning in Parrots

Given that one can define a core repertoire common to many parrot species, which parts of that repertoire seem most reliant on lifelong vocal learning? One tack is to see which parts of the repertoire are learned and which not. Unfortunately, this fails to help. Where offspring have been monitored in wild parrots, it appears that youngsters must learn nearly the entire vocal repertoire from conspecifics. The galahs cross-fostered by Major Mitchell cockatoos acquired most of the foster parent vocal repertoire and used the calls in the proper (foster parent) contexts (Rowley & Chapman 1986). The exceptions were their alarm calls, which remained species-specific, and their earliest begging calls, which were initially like galah chicks but quickly shifted to mimic those of their foster siblings. Pidgeon (1970) compared vocalizations of a hand-raised galah nestling with those from similarly aged wild nestlings. While all youngsters stopped making nestling begging sounds at the same age, the captive bird never replaced its begging calls with adult sounds. Similar observations are reported for other hand-raised parrot species.

The question is thus not which parrot call types are learned, but which ones may be relearned throughout life. Because this is difficult to determine with wild birds, we have to resort to more indirect cues. We can eliminate some candidates by examining call variability. Call types that show little structural variation between and within populations probably do not require adult relearning. Were there relearning, it would have to be synchronous across populations to retain low variability and this would be obvious to observers. Although call variability is one likely outcome of continued vocal learning, it is not sufficient evidence as there may be other causes of such variation. Hence we must use this cue warily. A second indicator is an identifiable adaptive function for adult vocal learning, given parrot ecology. For which call types might learning of a new variant benefit the learner? For which call types might adult learning not be valuable? Finally, with persistent fieldwork, we hope to accumulate examples of adult vocal learning that are concordant with patterns of variability and identifiable benefits of such learning.

Our review of current information suggests three candidate call types that fit the above criteria. These are the loud contact calls, pair duets, and warbles. The remaining categories of calls so far fail to show one or more of the criteria. This does not mean they may not turn out to be important with subsequent work, but at this stage, they seem less likely foci for vocal learning.

Loud Contact Calls

These calls show consistently high patterns of individual and social group variability (Pidgeon 1970; Noske 1980; Saunders 1983; Martella & Bucher 1990; Farabaugh & Dooling 1996). Variation begins at the level of individual differences. Although there is a common structure shared by all birds, the loud contact calls of orange-fronted conures are individually distinctive (Figure 11.3). Where individuals produce more than one variant, there is a dominant variant that is used far more frequently than any others (Figure 11.4). A similar pattern occurs in budgerigars (Farabaugh et al. 1994). Contact calls of orange-chinned parakeets also show individual variation within a common template, as do those of yellow-naped Amazons (Wright 1996) and captive budgerigars (Farabaugh & Dooling 1996). Individuality of loud contact calls has also been inferred for several other species from the ability of nestlings or breeding females to respond to mate and parental calls but ignore calls of other individuals (Rowley 1980; Saunders 1983).

Individual distinctiveness does not require adult learning, and in fact, it

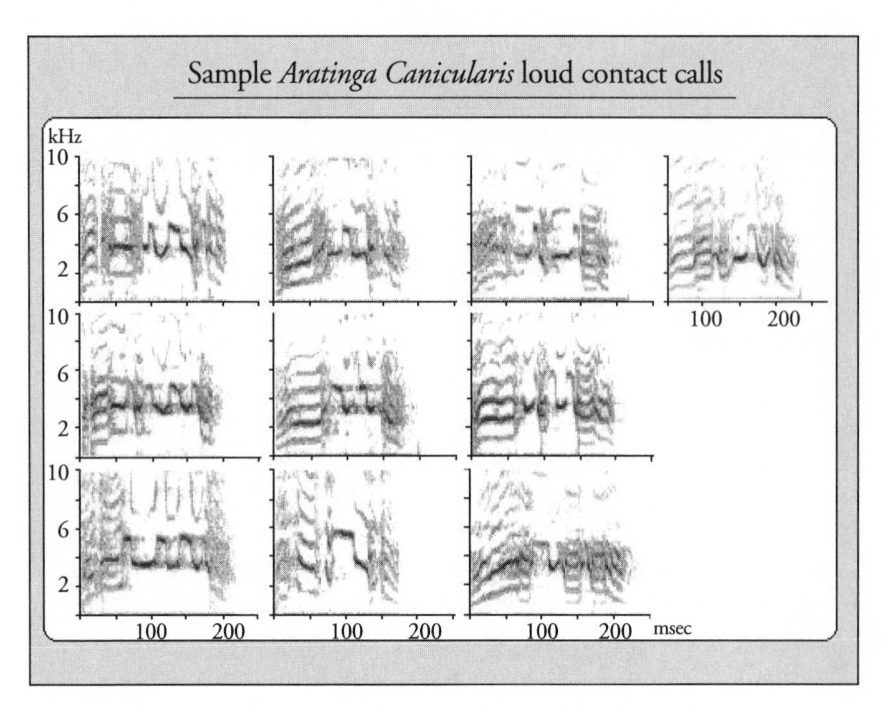

Figure 11.3. Sample loud contact calls from 10 separate individual orange-fronted conures in Guanacaste, Costa Rica. Note the common template with a rising harmonic series in the first phase, a rectangularly modulated middle section, and a falling harmonic series at the end.

may require no learning at any stage. What does suggest some form of lifelong learning is evidence for secondary modification of individual contact call structure as a function of social affiliation. The lowest level of such convergence involves call convergence of mated pairs. Domesticated budgerigar males have been shown to modify their own loud contact call to more closely match that of their female mate (Hile et al. 2000). Yellow-naped Amazon pairs also show significant structural convergence in their contact calls (Wright 1996). We have inconsistent evidence for such convergence in contact calls of wild pairs of conures: some appear strikingly convergent, while others do not. Given that mates die and birds occasionally shift to new mates, those species showing within-pair convergence of contact calls would require lifelong learning so as to allow occasional remating.

The next highest social level in which contact call convergence has been reported is at the level of the flock. The primary evidence is from domesticated budgerigars. Birds placed simultaneously into a new group showed detectable convergence in their contact calls after six to eight weeks (Farabaugh

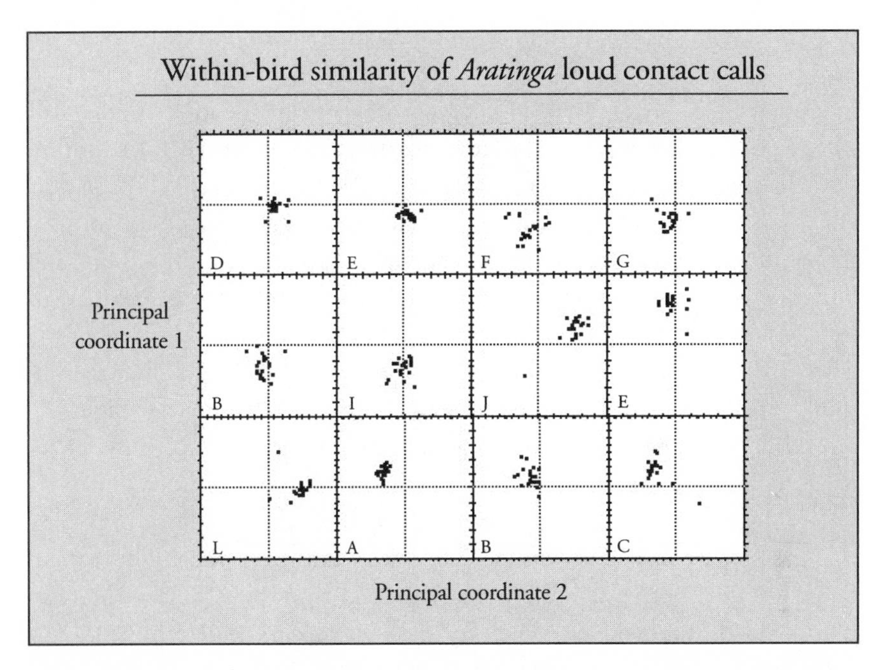

Figure 11.4. Similarity plots for 20 sample loud contact calls from each of 12 individual orange-fronted conures in Costa Rica. The axes in each plot are the first two principal coordinates extracted from a similarity matrix in which every call has been compared to every other call using spectrographic cross correlation (Cortopassi & Bradbury 2000). The center of each plot corresponds to the origin (0,0) of the principal coordinate axes. Note that for most birds, their own calls cluster in a small area of principal coordinate space with occasional variant outliers. Note also that the main clusters for different birds occupy different regions of principal coordinate space. Multivariate analyses of these plots show that there are highly significant between-bird differences in loud contact calls even given the obvious within-bird variation.

et al. 1994; Farabaugh & Dooling 1996). Another study added single individuals to existing flocks and found the new member modified its contact call so as to match better the common call pattern of the group (Bartlett & Slater 1999). Whereas budgerigars can discriminate between over 30 individuals with different contact call patterns (Dooling 1986; Ali et al. 1993), discrimination between individuals with a convergent contact call pattern is only achieved by those birds who live in the same flock and have learned the subtle differences among individuals (Brown et al. 1988). *Forpus conspicillatus* are selectively responsive to their mate's and siblings' contact calls, but do not distinguish between other group members (Wanker et al. 1998).

Based on their captive studies, Farabaugh and Dooling (1996) suggested that vocal learning might have arisen to facilitate flock-specific contact call convergence in the wild. We have so far been unable to find evidence of significant within- versus between-flock convergence in the loud contact calls of orange-fronted conures. Although Hardy (1965) reported flock compositional stability in captive colonies of this species, the frequent shifts in flock affiliation that we have seen with radio-tracked birds in the wild do not seem to support the existence of closed and stable groups. On the other hand, both orange-fronted conures and orange-chinned parakeets are extremely selective in their responsiveness to contact calls of over-flying flocks of conspecifics. Perched birds may totally ignore over-flight by one flock, but then respond with a cacophony of reciprocated contact calls to the over-flight of a second flock minutes later. Loud contact calls are the usual recruitment signal at night roost staging sites. One over-flying flock will continue to pass the staging aggregation without a single reply, whereas the next flock to pass exchanges contact calls with the perched birds and immediately joins them. Some form of contact call discrimination is clearly at work here. Whether it is at the level of individuals, pairs, or flocks remains to be determined.

Several species of wild parrots show significant divergence in contact call structure among different night roosts. Saunders (1983) describes easily distinguishable differences between contact calls at different night roosts of black cockatoos, and Wright (1996, 1997) provides detailed quantitative comparisons showing interroost differences in yellow-naped Amazons. Both species use fixed communal roost sites for long periods. This contrasts with the frequent shifting of night roost locations and compositions seen in smaller taxa such as conures, parakeets, and white-fronted Amazons. Not surprisingly, we have found no detectable differences in contact calls from conures using adjacent night roosts.

Finally, many species of parrots show contact call convergence within and divergence between different geographical regions. In orange-fronted conures, geographical variation in loud contact calls is relatively continuous over a 120-kilometer study area with calls becoming structurally more distinct the farther apart the two sampling localities (Bradbury et al. in press). Playbacks to foraging birds and short-term captives show significantly greater responses to contact calls from the subject's own region than to calls from more distant regions. The zone within which calls are statistically homogeneous in structure and elicit playback responses is very similar to the range lengths measured for radio-tracked individuals (six to eight kilometers). Over exactly the same sampling area in Costa Rica, the contact calls of sympatric yellow-naped Amazons divide up into three contiguous but discrete dialects (Wright 1996). Each dialect incorporates tens of night roosts, and each night roost can harbor hundreds of birds. Although there is interroost variation in contact calls within a dialect, it is subtle compared to the striking differences between dialects. Naive observers are invariably surprised to learn that birds using different dialects are from the same species! Birds living near dialect boundaries are often bilingual, producing either type of contact call as needed, but never melding them. Recent studies indicate that there are no genetic boundaries commensurate with dialect boundaries (Wright & Wilkinson 2001). This implies that there is free migration between dialects, as long as immigrants learn the new dialect. It is not clear why they should have to do so. These birds are relatively unresponsive to contact call playbacks, and showed no differential responses to calls of different dialects (Wright 1997). Why orange-fronted conures and yellow-naped Amazons show such different patterns of geographical convergence remains unclear. There are certainly candidate differences in their patterns of night roosting, daytime dispersion, and sociality, any of which could be linked to the different patterns of call variation.

What are the advantages of reducing individual differences in loud contact call diversity as a function of flock, night roost, and local geographical affiliations? As noted, fusion and fission of social units are common. Creating unit-specific tags before fusing would clearly facilitate later fission into the original units. This could easily be favored to minimize foraging path overlaps. Consider two conure flocks that fuse to pool predator risks at a foraging patch. Because flocks may have overlapping ranges, a pair that shifted groups during fission would risk having to follow the new flock to foraging sites that the pair had already visited earlier that day. Tags could also facilitate fusion by allowing birds to be choosy about which groups they joined. This might be important during the formation of midday resting and night roosting aggregations. Selective aggregation with familiar flocks might ensure that dominance relationships were already established and thus reduce agonistic conflicts over preferred perches.

Neither the fission nor the fusion advantages of flock tags require a lifelong ability to learn to produce new vocal signals unless the composition of units is labile. If tagged contact calls only served to ensure group compositional stability, then once a group had a distinctive tag, the only additional learning required would be that by new recruits. There is a solution to this conundrum. Our data suggest that conure flocks are not closed, and pairs move between flocks on occasion. If socially acquired habitat lore is as important to parrots as we think, and if the aggregate lore for all flocks in a local area is greater than that within any given flock, it would pay birds to move between flocks. Such transfer cannot be too frequent, or path overlap and dominance conflicts would undermine the benefits of new lore. Occasional transfers between flocks and the existence of flock-specific contact call tags together could thus explain continued adult vocal learning. The observed ability of conures to discriminate between local and distant loud contact calls fits this scenario. Conures from quite distant areas have been found to mingle together at bottleneck feeding sites. A pair that exploited a temporary feeding aggregation to switch foraging flocks would have to be very careful about which flock they joined. Were they to join a distant group, all of their current lore about suitable foraging locations and associated seasonalities would be useless, and they might have to learn a whole set of new food handling techniques. They would do better to join a new but local flock.

This hypothesis can explain many of our results on orange-fronted conures, orange-chinned parakeets, and white-fronted Amazons. Loud contact calls are the primary call type heard during the frequent fusions and fissions seen in these species. And birds of all three species appear highly selective in their responsiveness to contact calls of conspecifics. It is more difficult to envision fission/fusion tags as the reason for vocal learning in yellow-naped Amazons. Unlike the smaller species, pairs rarely fuse into larger flocks during the day, and they are markedly unresponsive to contact call playbacks. Still, there must be some reason why contact calls are convergent within pairs, within night roosts, and sufficiently so within regions to maintain discrete dialects despite persistent migration across boundaries. The apparent aggressiveness and daytime asociality of this species may hold some clues here. Perhaps birds cannot gain access to favored food trees or defend nest sites without adopting local dialect and tag patterns in their contact calls. This species remains an intriguing exception to the rule.

Pair Duets

Both members of mated pairs vocalize concurrently or antiphonally in all monogamous parrots. Where the pattern of mutual calling is at least somewhat stereotyped or syntactically predictable and rarely given by a single individual, the exchange has been called a "pair duet." Given this definition, duets have been reported in species of parakeets, Amazons, lorikeets, and lovebirds (Power 1966a,b; Mebes 1978; Serpell 1981; Arrowood 1988; Snyder et al. 1987; Wright & Dorin 2001). Structured duets have not been reported in orange-fronted conures (although warbling near nests may be a joint effort), white-billed black cockatoos (Saunders 1983), budgerigars (Wyndham 1980), and monk parakeets (Martella & Bucher 1990).

Captive orange-chinned parakeets exhibit two types of pair duets: one is sequential with a short time delay between each pair member's contribution, and the other consists of overlapping pair-member contributions (Power 1966a). Power also found that the male duet contribution always preceded that of the female and was slightly higher in mean frequency. We have confirmed that there are two types of pair duets in wild populations. Interestingly, the same basic element appears to be used by each bird as its contact call and its contribution to both kinds of duets. Arrowood (1988) also found two different duet types in her study of captive white-winged parakeets *(B. versicolurus).* However, one of these types may consist of many more notes than the short contributions of each partner seen in orange-chinned parakeets. Both Powers and Arrowood agree that pair duets in these species largely function to mediate agonistic conflicts between pairs. A similar function has been assigned to duets in the genera *Trichoglossus* (Serpell 1981) and *Amazona,* the latter counter-duetting with other pairs at nests or on the foraging grounds (Nottebohm 1972; Snyder et al. 1987; Wright & Dorin 2001). Remating by captive white-winged parakeets results in the creation of a new duet (Arrowood 1988). This is not developed gradually, but appears de novo with few timing errors the minute the birds begin to act as a pair. Timing is more refined the longer the pair remains together, but the basic duet is unchanged unless the birds remate.

Wild yellow-naped Amazons show a more complicated pattern (Wright 1997; Wright & Dorin 2001). Each pair has a number of duet motifs that it can produce. All of these follow the same three-phased syntax: a) contact call alternation; b) alternating female and male sex-specific notes called "screes"; and c) male "yodelling." Each individual has several scree variants that it can produce, and each pair produces a number of different duet motifs that vary with the number and type of screes used in the latter phases. Soon after arrival at the foraging grounds each morning, neighboring pairs can be heard singing duets. Pairs in close proximity may counter-duet and even repeat each other's motifs. Pairs also duet and counter-duet around active nest sites. Playbacks of duets near nests elicit active counter-duetting and aggressive approach to the speakers. Duet motifs, like loud contact calls in this species, show marked geographical variation and in fact have the same discrete dialect boundaries as do contact calls (Wright 1996). Birds respond more strongly to local dialects than to more distant ones. All of this suggests that the duets of yellow-naped Amazons, like those of other species, are largely used to mediate pair agonistic and dominance interactions.

Why might the large Amazons have so many different duets? Vehrencamp (2000) recently outlined an adaptive reason for multiple motifs in counter-singing songbirds. The ability to match or not match the last song of an opponent is a convenient way for combatants to regulate disputes using vocal signals. Exact matches evoke further escalation, singing of a shared but not matched motif maintains involvement without further escalation, and singing a motif not shared by the combatants signals retreat. These conventional rules appear to be present in several songbirds with multiple song repertoires. To acquire a territory in a given neighborhood, young must learn a minimal number of songs common to that area. Extrapolating this model to parrots, it is possible that pairs of counter-duetting species must acquire a number of duet motifs that are shared with potential opponent pairs if they are to regulate conflicts over resources or nest sites. The high overlap of pair home ranges, even in the relatively asocial yellow-naped Amazon, must surely lead to much larger numbers of potential opponents than is the case for sedentary and territorial songbirds. This could place a very high premium on adult vocal learning. The caveat here, of course, is that structured duets only exist in about half the parrot species studied. Although learning is certainly required to play the counter-duet game, it seems more likely that vocal learning preceded duets than that duets were the original substrate favoring adult vocal learning.

Warble Songs

This is the candidate call type for vocal learning origins in parrots about which we currently know least. In orange-fronted conures, the rambling vocalizations that we call warbles can consist of a very high diversity of note types. Although notes do repeat at intervals, we have not yet discerned an underlying syntax as described for *Amazona* duets. We have recorded birds warbling at mid-morning play aggregations, as solitary pairs or even solitary birds during midday, in the midst of large night roost staging aggregations, and around nests during the breeding season. Unlike duets and loud contact calls, warbles usually need not be directed at particular listeners, and birds often warble when there is no apparent audience in the vicinity. On the other hand, we have observed conure pairs at nearby nest sites exchanging warbles in the same aggressive way that parakeets or yellow-naped Amazons exchange pair duets. We are not yet sure whether both or only one member of a pair warbles in conures. We have often heard a solitary bird warble, so pairs are not required. Conures never perform this call type while in captivity, which has made it challenging to study.

Warbling occurs in both *Amazona* species and in orange-chinned parakeets in relatively similar contexts to that observed for conures. In yellow-naped Amazons, pairs may warble as solos or duets (the "gurgle calls" of Wright 1997). Wild African grey parrots in Gabon also produce long warbling vocalizations at the same time of day as our Costa Rican species and without any other nearby audience (personal observations). In budgerigars, both sexes warble, but the male does so more often and for longer periods (Brockway 1962. 1969; Wyndham 1980; Farabaugh & Dooling 1996). Farabaugh and Dooling (1996) suggested that warbling might be absent in parrots that do not have unpredictable breeding episodes. Our findings suggest that it is much more widespread taxonomically and seasonally than they proposed, but is often missed in the field and absent from captive groups.

Among the parrot call types that we have examined, this is the only one that could conceivably encode complex semantic information about the environment. Unfortunately, warbles are not associated with either foraging or predator alarms, but instead given when birds are resting or forming night roost staging. They are also given frequently without an obvious audience. Given the costs of exchanging this kind of information (Bradbury & Vehrencamp 1998), it seems very unlikely that birds would broadcast environmental information to no one in particular. The warbling of one pair of wild African grey parrots recorded in Zaire were found to contain whole and piecemeal mimicry of local bird and bat species (Cruickshank et al. 1993). This is intriguing because several species of jays mimic raptor calls and it has been suggested that they do so to identify nearby predators (Löhrl 1958; Hope 1980). Most of the species mimicked by the parrots were not raptors but just conspicuous vocalists in the local forests.

Our recent observations of warble exchanges by conures with adjacent nest sites could imply that warbles here substitute for duets in site defense. This would also fit the observed broadcasting of warbles at midday with or without a focal audience present. On the other hand, the nest site is the only location in which we have seen any evidence of conure territoriality. In addition, the trees used by local birds for warbling out of the breeding season are almost never the same on successive days. An alternative explanation for all of these results is that warbling serves to attract new mates and defend existing ones. Perhaps the conure pairs at nearby nests warbled to discourage mate desertion or extrapair copulations. This interpretation would provide a better link with the use of warbling in budgerigars to coordinate breeding and mate attraction.

Parallels with Other Taxa

It is instructive at this point to look beyond parrots for parallels in other taxa. Dolphins, African elephants, and spider monkeys all exhibit periodic fission/fusion of social units (Moss 1988; Ballance 1990; Chapman 1990; Scott et al. 1990; Robbins et al. 1991; Norconk & Kinzey 1994). Although the composition of small foraging groups can be variable in these species, the larger aggregates from which they are drawn are relatively fixed and stable in composition. It is entirely possible that a similar organization exists in orange-fronted conures and other social parrots. This would mesh with their reliance on contact calls to provide both individual and local population information, and with their basic foraging and roosting biology. As described by Peter Tyack in Chapter 13, the signature calls of dolphins function in many ways like the loud contact calls of parrots. There is similar evidence of individual recognition and troop affiliation in the calls of spider monkeys (Chapman & Weary 1990; Teixidor & Byrne 1997, 1999) and elephants (see Chapter 3). Although I dismissed most songbirds as poor analogues for parrot vocal learning, there are interesting exceptions among the parids and the carduelines. In nonbreeding chickadees and house finches, individual vocal signatures are modified with flock-specific tags that are acquired when adults join new groups (Mundinger 1970, 1979; Ficken et al. 1978; Mammen & Nowicki 1981; Nowicki 1983). Although there are other groups of birds that have flock-specific calls, most of these are worse analogues for parrots in that they defend year-round group territories (e.g., Brown & Farabaugh 1991) or colonial nest sites (Feekes 1977; Trainer 1989).

The functions of pair duets in other avian and mammalian taxa remain controversial (Seibt & Wickler 1977; Wiley & Wiley 1977; Wickler 1980; Farabaugh 1982; Sonnenschein & Reyer 1983; Morton 1996; Levin 1996a,b; Langmore 1998; Hall 2000). Popular hypotheses include joint defense of territories against all intruders, defense and reinforcement of pair bonds, and same-sex exclusion of competitors from a territory. The tropical monogamous birds and gibbons that have been the main focus of duet studies all have long-term pair territories. In this sense, they are poor analogues to parrots, which at most defend a nesting hole for a short period. On the other hand, the facts that duets are the calls of choice when duetting parrots defend their nest sites, and that they counter-sing with intruders just like territorial songbirds, argue for convergent functions. Parakeets' use of duets to settle daily squabbles over preferred perches may simply be an extension of the same principles to mobile sites. There is yet no evidence that pair duets in parrots serve to reinforce pair bonds or effect mate defense. They are not often used in suitable contexts for this function and the absence of duets in species with equally fanatical pair-bonding suggests they are not a necessary tool for mate defense. It remains possible that species differ in extrapair paternity risks and that duets tend to evolve in those with the higher rates.

What parallels exist for the complex warbled vocalizations of parrots? If these are truly used to attract new mates, defend current ones, or obtain sneak copulations, then there are many avian and mammalian examples. Promiscuous species such as bowerbirds, lyrebirds, and humpback whales are inappropriate analogues to the monogamous mainstream parrots (Nottebohm 1972). A better fit is achieved with male starlings, thrashers, and mockingbirds, which are ostensibly monogamous and continue to learn new songs throughout their lives. All of these species, including the promiscuous birds cited, are good mimics of other animal sounds. As much as a third of a starling's song repertoire consists of mimicked sounds (Hausberger et al. 1991; see also Chapter 18). As mentioned earlier, the only interspecific mimicry observed in any wild parrot to date is found in the warbles of African grey parrots. Although this suggests interesting parallels with starlings or mockingbirds, and provides a tantalizing reason for the fabled mimicry of captive parrots, it remains to be seen whether warble mimicry is very widespread in other parrot species.

Conclusion

The evidence to date suggests that vocal learning in parrots has evolved to manage complex and rapidly changing social affiliations and not for the transfer of environmental information. When one considers their ecology, social organization, and vocal communication, parrots appear to be more convergent with dolphins than they are with other birds. The degree to which such parallels prove useful will only be known with further work. However, the perspectives afforded by a volume such as this one force us to look more widely for basic patterns and cannot help but provide new insights and novel approaches for subsequent research.

CASE STUDY IIA

Representational Vocal Signaling in the Chimpanzee

KAREN I. HALLBERG, DOUGLAS A. NELSON, AND SARAH T. BOYSEN

Evidence of the complex cognitive abilities of chimpanzees (*Pan troglodytes*) that allow them to understand and use human representational symbols raises the question of how such abilities might be manifest in their natural vocal behavior. That is, do chimpanzee vocal signals contain information that represents or refers to objects or events in the animal's external environment?

Upon discovery and eating of palatable foods, chimpanzees in both wild and captive environments produce a vocalization referred to as a "food bark" or "rough grunt" (Marler & Tenaza 1977; Goodall 1986). Because these signals are emitted only in food-related contexts and receivers respond by approaching the food source, we were interested in evaluating the extent to which these calls may serve as representational, or referential, signals.

To investigate the referential specificity of food bark vocalizations, five chimpanzees at the Ohio State University Chimpanzee Center were tested using a playback paradigm, wherein animals matched vocal stimuli to photographic images of different foods. The recorded food barks used in the playback test were collected from two adult male chimpanzees, Bobby and Darrell, who were isolated from the other animals and from each other during recording. One vocalization for each of nine food items was used as playback stimuli. The nine food items represented three food-quality categories: low-value foods (beans, carrots, and lettuce); moderate-value foods (apples, oranges, and peanuts); and high-value foods (grapes, lemon candies, and M&M candies). Note that the recorded vocalizations for grapes, apples, and lettuce used in Experiments 1 and 2 were collected from Bobby, whereas the recordings for the six remaining food items were collected from Darrell. By evaluating the images that the five chimpanzees matched to the vocalizations during this playback task, we were able to analyze the association between their choice and the food-quality category.

In both experiments described below, a trial consisted of the playback of a prerecorded vocalization through a speaker located in the testing cage. The presentation of the acoustic stimulus was immediately followed by the appearance of four photographic images of foods on the subject's computer touch-frame monitor. A choice was considered correct for "item" if it matched the vocalization for that specific food item and correct for "category" when it matched any other item of the same quality as the target.

Experiment 1

Each animal was tested with 16 presentations of each food vocalization, randomly presented within eight 18-trial sessions. A nondifferential schedule of reinforcement was employed, with the subjects receiving a reward after every trial. Data analyses revealed that all five chimpanzees selected the image of a food belonging to the same food-value category as the recorded vocalization at a level above chance, with combined group performance reaching significance ($p[(T \leq 5) \mid \text{Binomial}(.5,5)] = .031$). Additional individual analyses revealed that the chimpanzees most effectively discriminated between low- and high-value foods. Assigning numeric values of 1, 2, and 3 to the three food-value categories of low, moderate, and high, respectively, and calculating the mean values for the animals' responses by category, analysis of variance posthoc comparisons revealed that the mean values for low- and high-value foods differed significantly ($p < .01$). Although the other comparisons were nonsignificant, the mean response values increased systematically from low- to high-value foods for each animal.

Experiment 2

To address concerns regarding the animals' understanding of the cross-modal task demands of the approach used in Experiment 1, direct training with a limited number of stimuli was implemented in Experiment 2. Three of the nine foods were used as targets during these training sessions, with one food item representing each food-value category. Seventeen training sessions (12 trials each) were completed to acquaint the animals with the procedural demands of the task. Once all subjects had reached the criterion of at least 75 percent correct responses for two consecutive sessions, novel probe testing was initiated.

The chimpanzees were tested on novel probe trials similar to those in Experiment 1 using the six untrained foods as targets. Each of the six probe foods was presented 12 times over 10 sessions. To control for potential learning due to the differential reinforcement schedule employed in Experiment 2, only data from the first session of probe trials were analyzed. These analyses revealed that all five chimpanzees chose the correct food image by category and by item at a level above chance, with group performance reaching significance ($p[(T \leq 5) \mid \text{Binomial}(.5,5)] = .031$). Further, individual analyses revealed that three out of the five animals chose the correct category and the correct food item at a level significantly above chance.

Acoustic Analyses

In an effort to identify the acoustic features of the calls that might have allowed for the chimpanzees' cross-modal discriminations, a larger corpus of calls for the same nine food items was collected and analyzed. Food vocalization recordings were collected from three chimpanzees: one adult female (Sheba, age 18 years) and the two adult males whose vocalizations were used in Experiments 1 and 2 (Darrell, age 18.5 years, and Bobby, age 11.5 years). A total of 75 recordings were obtained: 30 recordings from Bobby, 36 from Darrell, and nine from Sheba.

Five acoustic features were measured for each call: 1) number of elements per call, 2) number of spectral peaks within elements, 3) dominant frequency, 4) time interval between element start times (periodicity), and 5) element duration. Based on these features, a cross-validated discriminant function analysis classified the three food-value categories at a rate of 62.7 percent correct. The greatest occurrence of misclassification was between low-value and moderate-value vocalizations, and between moderate- and high-value vocalizations, with low-value and high-value food vocalizations rarely (3.3 percent of cases) being misclassified as the other. The two variables most highly correlated with Function 1, which accounted for 95.2 per-

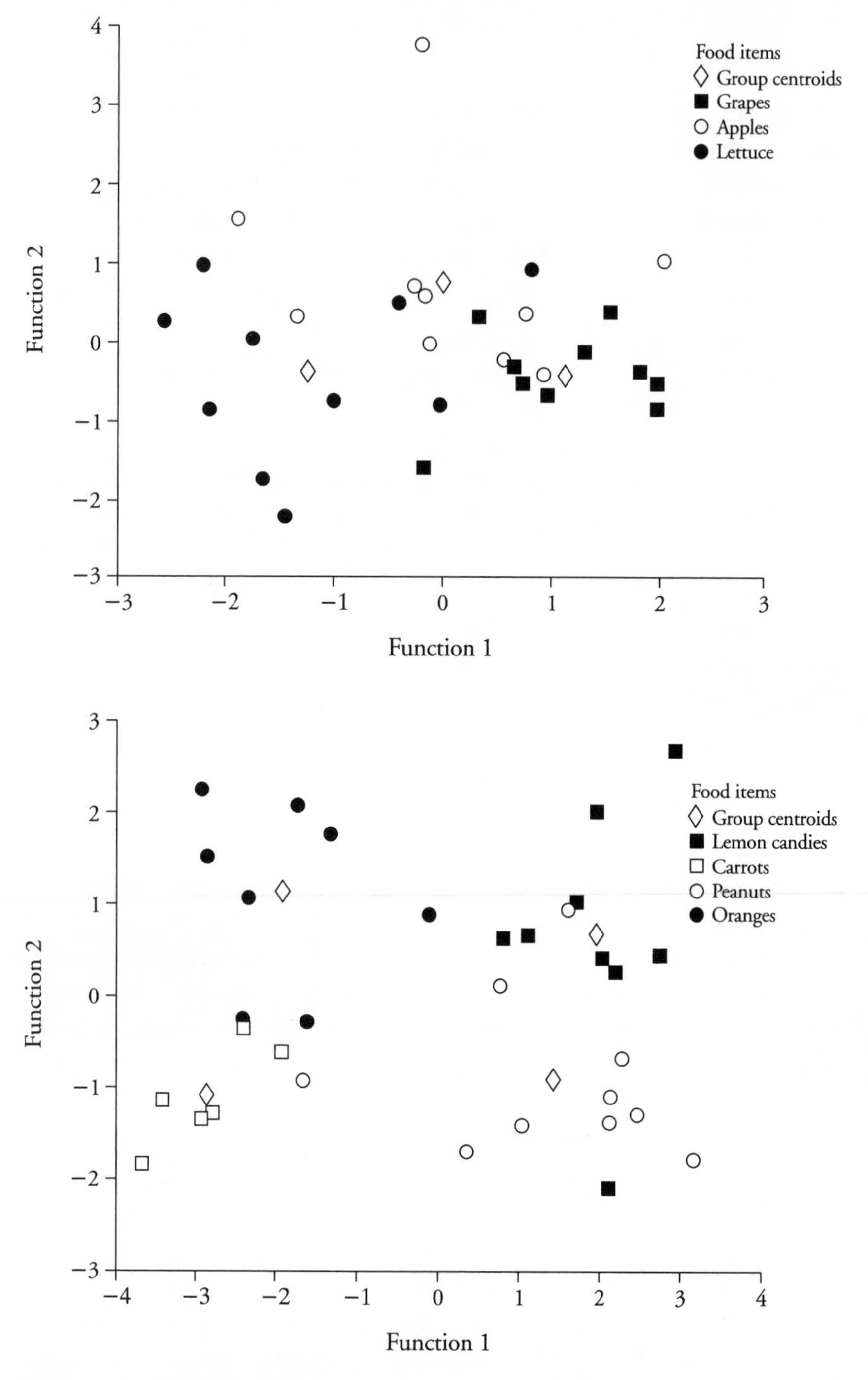

Figure 11A.1. Canonical discriminant function scatterplots illustrating the acoustic discriminability of food-bark vocalizations from two adult male chimpanzees produced in response to seven different food items.

cent of the variance, were the number of elements and number of spectral peaks (r = .608 and .558, respectively).

A similar cross-validated discriminant function analysis effectively discriminated between vocalizations for eight of the specific food stimuli at a rate of 66.25 percent correct classification. (The vocalization for beans was excluded from the analysis due to the small sample collected for this item; n = 1.) Again, most errors occurred in association with moderate-value calls. The number of spectral peaks within elements and the number of elements within calls contributed most to the accurate classification of signals (r = .929 and .748, respectively).

Since the majority of this larger corpus of vocalization recordings for these specific food items was collected from two animals, separate discriminant function analyses were run for Bobby's and Darrell's calls to control for classifications that may have been based on differences between the animals' voices. The resulting cross-validated confusion matrices indicated correct item classification of 53.3 percent of Bobby's calls and 81.8 percent of Darrell's calls, compared to 33 percent and 25 percent respective correct classification expected by chance (see Fig. 11A.1).

Discussion

The results of the playback experiments reported here provide support for the hypothesis that chimpanzee food barks contain referential information. Evidence from Experiments 1 and 2 indicated that the chimpanzees discriminated between food barks. Thereby, the food item, or some characteristic thereof, appears to be the referent of the call. In addition, the acoustic analyses of these vocalizations further supported the behavioral data: calls differed among the three food-value categories and among the specific food items.

The existence of such complexity in the chimpanzee's natural communication system further advances the hypothesis that these types of cross-modal mechanisms were "necessary prerequisites for the emergence of human language" (Marler 1985, p. 219). Although similar findings have been reported for other nonhuman primate species, these data from the chimpanzee, with whom we share a close evolutionary history, provide critical information toward furthering our understanding of the continuum between animal communication systems and human language.

12

Social and Vocal Complexity in Bats

GERALD S. WILKINSON

The Machiavellian intelligence (MI) hypothesis (Whiten & Byrne 1988; Whiten 1999) proposes that enhanced cognitive abilities permit an individual to exploit others in a group, thereby creating a social environment that favors development of counter-strategies, such as deception or mutual cooperation. Proponents of this idea suggest that as group size increased during hominid evolution, a cognitive arms race resulted in the evolution of larger brains. One of the most robust findings that has been offered in support of the MI hypothesis is that social group size in extant primates positively covaries with relative size of the neocortex—a proxy for enhanced cognitive abilities (Dunbar 1992, 1995). This result holds after controlling for phylogenetic effects (Dunbar & Bever 1998) and is not confounded by visual sensitivity since it occurs within diurnal taxa (Barton & Dunbar 1997). The MI hypothesis has also been applied to other groups. For example, relative size of the neocortex also covaries with group size in carnivores (Barton & Dunbar 1997) and insectivores (Dunbar & Bever 1998). Thus, these results suggest that a complex social environment should select for enhanced cognitive traits in other mammals.

Social interactions require communication. In many primates, social grooming is used to mediate social interactions and maintain social cohesion. Such tactile communication is, however, inadequate for large social groups. The evolution of language seems likely to have enabled the formation of larger hominid groups (Aiello & Dunbar 1993; Dunbar 1993). Although transformational grammar, which can create an infinite combination of phrases and sentences, is only found in humans, other features of human language are found in nonhuman mammals (Pinker & Bloom 1990). These include phonetic units (Richman 1976; Owren & Bernacki 1988; Snowdon 1990; Hauser & Fowler 1991), rhythmicity (Richman 1987), categorical perception (Kuhl & Miller 1975), and syntax-like combinations of discrete sounds that occur nonrandomly and recurrently (Kanwal et al. 1994). In bats and primates, such composite syllables elicit unique neuronal responses, suggesting they have cognitive salience (Esser et al. 1997; Wang 2000). Some bats, cetaceans, and primates can also modify their vocalizations to resemble those emitted by other individuals (Guinee et al. 1983; Boughman 1998; Mitani & Gros-Louis 1998; Smolker & Pepper 1999; Snowdon & Elowson 1999). Thus, social interactions in nonprimate groups may be mediated by vocal communication and, as a consequence, associated with vocal complexity.

In this chapter I define and then compare social complexity to vocal complexity in bats. Bats provide an interesting comparison to primates because these two mammalian orders do not have a recent common ancestor. All species in one suborder, the Microchiroptera, rely on the echoes of high-frequency vocalizations to perceive their world. Perhaps as a consequence, many bat species exhibit a rich repertoire of communication vocalizations (Fenton 1985). Bats also display a wide variety of social organizations that rival primates both in diversity and complexity (Bradbury 1977b). Although my goal in this chapter is to provide an independent test of the MI hypothesis, I also consider alternative scenarios that could give rise to associations among social complexity, relative neocortex volume, and vocal complexity in bats, as well as suggest promising areas for futher study.

Social Complexity

Group Size

Perhaps the most obvious dimension of social complexity is group size. With 936 described species (Findley 1993), bats exhibit unrivaled variation in aggregation size. At one extreme, species from five families have been reported to form colonies in excess of 500,000 individuals, including Mexican free-tailed bats (*Tadarida brasiliensis;* Barbour & Davis 1969), ghost-faced bats (*Mormoops megalophylla;* Barbour & Davis 1969), Sundeval's leaf-nosed bat (*Hipposideros caffer;* Nowak 1994), bent-winged bats (*Miniopterus schreibersii;* Smithers 1992), and the straw-colored fruit bat (*Eidolon helvum;* Fayenuwo & Halstead 1974). Colony size appears to be evolutionarily labile as other species in each family roost solitarily or in small groups. Important factors influencing colony size include diet, body size, and predation risk (Bradbury & Vehrencamp 1976b). All of the species listed above, except the straw-colored fruit bat, weigh 20 grams or less, capture abundant insect prey using echolocation, and roost in caves where they are inaccessible to most predators. Straw-colored fruit bats, in contrast, weigh 280 grams, eat fruit, do not echolocate, and roost in tall trees. Species that form small groups or roost solitarily often inhabit hollow trees or roost cryptically in foliage and forage on more dispersed food sources. Small group size is also typical of bat species that hunt and capture vertebrate prey (Norberg & Fenton 1988), including the false vampire bat (*Vampyrum spectrum*) and woolly false vampire bat (*Chrotopterus auritus*) in the family Phyllostomatidae and the yellow-winged bat (*Lavia frons*), lesser false vampire bat (*Megaderma spasma*), and ghost bat (*Macroderma gigas*) in the family Megadermatidae.

Colony size may not, however, be the same as social group size. Dunbar (1993) has argued that while group size is ultimately determined by ecological factors, the upper limit to social group size for any primate is set by cognitive constraints. Those constraints dictate the number of social relationships one animal can maintain by personal contact. Unfortunately, the average size of social groups in most colonial roosting bats is unknown. Where detailed studies have been conducted, colonies are often composed of multiple social groups, usually consisting of small groups (less than 30) of females that utilize and defend traditional roosting and foraging sites, i.e., common vampire bats (*Desmodus rotundus;* Wilkinson 1985a), greater spear-nosed bats (*Phyllostomus hastatus;* McCracken & Bradbury 1981), Bechstein's bats (*Myotis bechsteini;* Kerth & Konig 1999), and greater white-lined bats (*Saccopteryx bilineata;* Bradbury & Vehrencamp 1976a).

Social groups could be maintained by recognition and memory of individuals or by using a cue indicating group membership. Individual recognition must occur in female vampire bats to enable them to share blood reciprocally with roostmates (Wilkinson 1984). Individual differences in vocalizations could be used for recognition and have been described for echolocation calls of many species (Habersetzer 1981; Suga et al. 1987; Jones et al. 1992; Rydell 1993; Masters et al. 1995; Obrist 1995; Guillen et al. 2000). In contrast, female greater spear-nosed bats use group-specific contact calls to recognize groupmates while foraging (Boughman & Wilkinson 1998), much like parrots (see Chapter 11). Similarly, female fishing bats, *Noctilio leporinus,* scent-mark groupmates presumably to facilitate group recognition while foraging (Brooke 1997). Group membership tags may occur in other species because their use reduces the cognitive burden associated with recognizing group members in a large aggregation. However, to ensure honest signaling, group-specific cues need to be costly or difficult to acquire (Grafen 1990). Vocal learning in greater spear-nosed bats takes time, consistent with this requirement (Boughman 1998).

Group Stability

A second dimension likely to influence social complexity is group stability. The stability of bat roosting groups depends on individual longevity, roost fidelity, and affiliations among individuals. For their body size, bats are extraordinarily long-lived. Twenty of 41 longevity records for bats exceed 15 years (Tuttle & Stevenson 1982). The most long-lived species include greater long-eared bats *(Plecotus auritus)* and little brown bats *(Myotis lucifugus),* which have been recaptured 30 years after initial banding (Keen & Hitchcock 1980; Lehmann et al. 1992), and greater horseshoe bats *(Rhinolophus ferrumequinum),* which have survived 26 years (Ransome 1991). All three of these species weigh 25 grams or less. Most long-lived bat species are temperate insectivores, give birth to a single young each year and hibernate over winter. Whether temperate species live longer than tropical species is unclear since less information on lifespan is available for most tropical species. Some of the large flying foxes live more than 20 years (Tuttle & Stevenson 1982) and long lifespan could be characteristic of other *Pteropus* species that reproduce once per year (Racey & Entwhistle 2000). In the New World tropics, female common vampire bats and greater spear-nosed bats give birth once per year and have been recaptured after 15 years (Wilkinson & Boughman 1998; Tschapka & Wilkinson 1999) and 22 years (G. Wilkinson, personal observation), respectively. A female common vampire bat has even survived 30 years in captivity (U. Schmidt, personal communication). In contrast, neotropical fruit bats which give birth twice per year, such as the Jamaican fruit bat (*Artibeus jamaicensis*) and short-tailed fruit bat *(Carollia perspicillata),* live no more than nine years (Fleming 1988; Gardner et al. 1991).

A review of roost fidelity indicates that 14 of 43 species of bats rarely switch roosts whereas 25 frequently change roosts (Lewis 1995). High fidelity to a single roost is positively associated with the use of permanent roost sites, such as caves, mines, or buildings, and inversely associated with roost availability. Mark-recapture and molecular genetic studies have recently documented that females exhibit natal philopatry and roost fidelity in several temperate-region bat species. For example, female greater long-eared bats (Entwhistle et al. 2000; Burland et al. 2001), greater mouse-eared bats (*Myotis myotis;* Petri et al. 1997), evening bats (*Nycticeius humeralis;* Wilkinson 1992), and greater horseshoe bats (Rossiter et al. 2000) typically reproduce in the colonies where they were born. In each of these species, maternity colonies are persistently used for decades, suggesting that natal philopatry leads to long-term associations among individuals in a colony. Nevertheless, average pairwise relatedness among females in a colony does not differ from zero where it has been estimated (Wilkinson 1992; Kerth & Konig 1999; Burland et al. 2001). Low relatedness between females apparently results from male dispersal, extra-colony mating, high first-year mortality and low male reproductive skew (Watt & Fenton 1995; Rossiter et al. 2000; Burland et al. 2001).

Most bat species that frequently switch roosts exhibit low group stability. For example, even though pallid bats *(Antrozous pallidus)* often roost together in rocky cracks (Vaughan & O'Shea 1976), these groups do not persist through the reproductive season (Lewis 1996). In the tropics, many bats that switch roosts form female roosting groups, but these groups show relatively low compositional stability over time. Examples include the Jamaican fruit bat in Panama (Morrison 1987), tent-making fruit bat (*Uroderma bilobatum;* Lewis 1992), great fruit bat (*Artibeus literatus;* Morrison 1980), short-tailed fruit bats (Fleming 1988), and greater white-lined bats. Greater white-lined bats differ from all other bat species in that males exhibit natal philopatry to a colony whereas females disperse (Bradbury & Vehrencamp 1976a).

Stable aggregations of females that remain together for two or more years have been documented for greater spear-nosed bats (McCracken & Bradbury 1981), fishing bats (Brooke 1997), common vampire bats (Wilkinson 1985a), Jamaican fruit bats in Mexico (Ortega & Arita 1999), Bechstein's bats (Kerth & Konig 1999; Kerth et al. 2000), rufous hairy bats (*Myotis bocagei;* Brosset 1976), African sheath-tailed bats (*Coleura afra;* McWilliam 1987), little free-tailed bats (*Tadarida pumila;* McWilliam 1988), and greater short-nosed fruit bats (*Cynopterus sphinx;* Storz et al. 2000). Resightings of animals marked as infants indicates that both sexes disperse from their natal group in greater spear-nosed bats, fishing bats, and greater short-nosed fruit bats. In each of these species, females subsequently either join existing groups or assemble into age-structured cohorts, which then exhibit high levels of affiliation. In contrast, female matrilineal kin remain together in common vampire bats, Bechstein's bats, and little free-tailed bats. Two or more females have been recaptured in the same social group after 10 years in greater spear-nosed bats (McCracken & Wilkinson 2000) and after 12 years in common vampire bats (Wilkinson 1985a), suggesting that long-term roostmate affiliation can occur with and without matrilineal relatives present. In several species, notably common vampire bats, Bechstein's bats, and greater short-nosed fruit bats, females often switch roosts, but pairs or small groups of females can be identified that invariably roost together and occasionally join other subgroups. Roost affiliation varies independently of the level of relatedness between the bats (Wilkinson 1985b; Kerth & Konig 1999; Storz et al. 2000). Such fission-fusion movement patterns are characteristic of chimpanzees (Goodall 1986; Nishida 1990), dolphins (Shane et al. 1986; Bearzi et al. 1997; Chapter 2), elephants (Moss & Poole 1983), and some parrots (Bradbury, Chapter 11). More examples of fission-fusion social organization among bats may be revealed as more long-term studies of individually marked animals are conducted.

Mating

In primates, the presence of alliances and coalitions, involving pairs or trios of individuals, has been used to distinguish more complex from less complex societies. In polygamous primates, reproductive skew, i.e., the extent to which high-ranking males monopolize matings, exhibits negative correlated evolution with neocortical size independent of group size (Barton & Dunbar 1997). To the extent that reproductive skew is reduced by alliance formation, this result is consistent with coalition formation influencing social complexity. Bats exhibit considerable variation in mating systems (McCracken & Wilkinson 2000). The most common mating system in bats involves a single male mating with several females (52 percent of 66 species). This mating system can occur either when a male controls access to a group of females or mates sequentially with several females that visit a defended site. Because single males often control access to sites with females, this mating system offers little opportunity for coalitions to form among lower-ranking individuals. Alliances might, on the other hand, occur among individuals in species in which several males roost and mate with females. Such multimale and multifemale groups have been described for 18 percent of bat species (McCracken & Wilkinson 2000). Alliance formation has not been described for any bat species, but this may reflect the paucity of detailed observational studies of mating behavior in bats rather than absence of this type of behavior.

Cognition and Social Complexity

Group stability in bats has been reported to covary positively with relative neocortex size (Barton & Dunbar 1997), but only two species with purportedly stable groups, common vampire bats and lesser spear-nosed bats (*Phyllostomus discolor*), were used in the analysis. Here I reexamine this question using recent compilations of mating system, colony size, and stability of female groups (McCracken & Wilkinson 2000; Wilkinson & McCracken 2001) together with brain structure volume estimates (Baron et al. 1996) for bats. Because previous work has also found differences in brain-body allometry according to diet among mammals (Pagel & Harvey 1989), I also tested for an effect of diet (insects, vertebrates, fruit, or nectar) on brain size. In dolphins, increased brain size has been proposed to be a consequence of echolocation (Ridgway & Brownson 1984). Consequently, I also compare brain volumes between echolocating and nonecholocating bats. I follow the methods of Dunbar (1992) with the exception that I use species, rather than genera, as the unit of analysis because species within bat genera often differ in each of the social factors under consideration (e.g., Bradbury & Vehrencamp 1976a). To control for differences in body size, which scales allometrically with brain size, I regressed log volume of the neocortex on log volume of the rest of the brain ($F_{1,63} = 5705.2$, $p < 0.0001$), and then used residual neocortex volumes in subsequent analyses.

Residual neocortex volume was independent of colony size ($F_{1,42} = 1.92$, $p = 0.17$), type of mating system ($F_{2,37} = 0.3$, $p = 0.76$), and echolocation ability ($F_{1,63} = 0.08$, $p = 0.78$). In contrast, Figure 12.1 shows that residual neocortex volumes differed according to diet ($F_{3,61} = 4.6$, $p = 0.0061$) and group stability ($F_{1,24} = 8.5$, $p = 0.0075$). In a two-way ANOVA, group stability ($F_{1,21} = 2.4$, $p = 0.13$) explained more variation than diet ($F_{3,21} = 0.6$, $p = 0.61$), but neither factor was significant, indicating that diet and group stability are not independent. Bats that form stable groups or feed on vertebrates or nectar have larger relative neocortex size than bats without stable groups or those that feed on fruit or insects. A previous analysis indicated

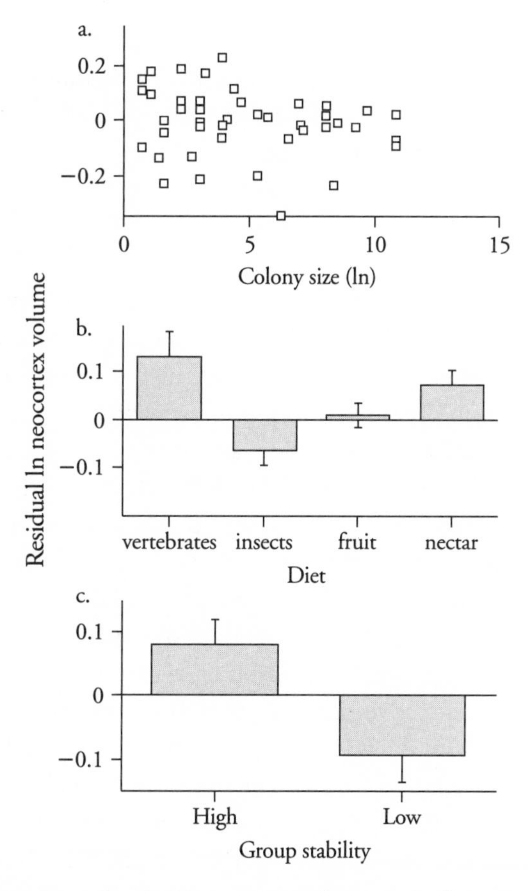

Figure 12.1. Comparison of residual ln neocortex volume to *a)* colony size, *b)* diet, and *c)* the stability of social groups. Mean ± SE shown in *b)* and *c)*. Species with more stable groups include *Coleura afra, Cardioderma cor, Lavia frons, Tadarida pumila, Noctilio leporinus, Artibeus jamaicensis* (Mexico), *Desmodus rotundus, Phyllostomus discolor, P. hastatus, Vampyrum spectrum, Cynopterus sphinx**, *Myotis bocagei*, and *Plecotus auritus*. Species with less stable groups include *Saccopteryx bilineata, Carollia perspcillata, Leptonycteris curasoae, Uroderma bilobatum, Hypsignathus monstrosus**, *Miniopterus australis, M. minor, M. schreibersi, Myotis adversus, M. albescens, M. lucifugus, Nyctalus noctula, P. nanus, P. pipistrellus, Tylonycteris pachypus*, and *T. robustula*. * indicates nonecholocating megachiropteran species.

that group stability predicted the occurrence of cooperative behaviors, such as allogrooming, across five species of phyllostomid bats (Wilkinson 1987). Group stability, as measured by an index of female association among day roosts, also predicted the frequency of food sharing (Wilkinson 1984) and social grooming among vampire bats (Wilkinson 1986). Taken together, these results suggest that group stability, and possibly social grooming, are key components of social complexity in bats. More information on group stability from additional species is needed to tease apart the effects of group stability and diet on neocortex volume.

Vocal Complexity

Complexity of vocalizations can be defined on at least four levels. First, some animals may have a larger lexicon than others, i.e., use more sounds to convey different meanings. To date, nine different functional categories of vocalizations have been described for bats, i.e., echolocation, infant isolation, maternal directive, mate advertisement, copulation, distress, alarm, contact, and defense (Fenton 1985; Wilkinson 1995). A second level of complexity involves the number and order of sound types, often referred to as syllables, emitted by an individual in a single context. If syllables can be combined in different orders to form composite syllables, then syllable order can contribute to syllable diversity. Composite syllables are produced by some bats (Kanwal et al. 1994; Davidson & Wilkinson 2002) and appear to be important for mate advertisement in at least one species (Davidson & Wilkinson 2001). A third level of complexity involves variation in the acoustic features of calls emitted by different individuals. Such variation is often associated with vocalizations that carry signature information, such as infant isolation calls (Gelfand & McCracken 1986; Jones et al. 1991; Scherrer & Wilkinson 1993), but also has been noted for maternal directive calls (Balcombe & McCracken 1992; Esser & Schubert 1998) and contact calls (Boughman 1997). Finally, the ability to modify vocalizations by learning provides a fourth potential source of complexity. Evidence for vocal learning in bats has been obtained for echolocation calls in greater horseshoe bats (Jones & Ransome 1993), infant isolation calls in lesser spear-nosed bats (Esser 1994), and contact calls in greater spear-nosed bats (Boughman 1998). Below I review the literature on several functional categories of bat vocalizations to determine if any aspect of vocal complexity is related to colony size, group stability, or mating behavior in bats.

Call Repertoires

Repertoires of vocalizations with some information on contextual association have been described for grey-headed flying foxes (*Pteropus poliocephalus;* Nelson 1964), short-tailed fruit bats (Porter 1979), little brown bats (Barclay et al. 1979), and greater white-lined bats (Bradbury & Emmons 1974). Insufficient information is available, however, to determine if the usage of different functional call types is associated with social complexity. For example, distress calls have only been described for little brown bats (Fenton et al. 1976), Jamaican fruit bats (August 1979), and pipistrelle bats (*Pipistrellus pipistrellus;* Russ et al. 1998), but also occur in many other microchiropteran and megachiropteran species (personal observation). Maternal directive calls, which are emitted by females when searching for pups, have only been described for species that roost in groups, e.g., pallid bats (Brown 1976), lesser bulldog bats (*Noctilio albiventris;* Brown et al. 1983), Mexican free-tailed bats (Balcombe & McCracken 1992), and lesser spear-nosed bats (Esser & Schmidt 1989). However, it is difficult to assess their distribution since they are not detectable without ultrasound recording equipment.

Recent work on the social calls of Parnell's mustached bats (*Pteronotus parnellii*), has revealed that these bats can combine calls in nonrandom orders to create composite syllables (Kanwal et al. 1994). Two-syllable composites represent a nonrandom subset of the 342 possible disyllabic combinations and constitute 30 percent of the sounds recorded (Kanwal et al. 1994). Playback studies have revealed that frequently used composite syllables elicit specific neuronal responses suggesting that syllable order, which the authors refer to as syntax, may have some behavioral salience (Esser et al. 1997). Unfortunately, the context in which each call variant was produced is unknown because observations were not conducted simultaneously with audio recordings. Observational and playback studies are, therefore, needed to assess syntax context and complexity among this and other species of bats, particularly those that differ in social complexity.

Mate Advertisement Calls

A variety of bat species emit calls that appear to attract females for mating. In most cases these are acoustically simple calls that are emitted repetitively. For example, male hammer-headed bats, *Hypsignathus monstrosus,* emit loud "honks" from a stationary location and increase call repetition rate as females approach (Bradbury 1977a). Similar types of calls are given by Wahlberg's fruit bats (*Epomophorus wahlbergi;* Wickler & Seibt 1976), Gambian epauletted fruit bats (*E. crypturus;* personal observation), and Franquet's fruit bats (*Epomops franqueti;* Bradbury 1981). These epomophorine bats exhibit lek or exploded lek mating systems in which males gather in groups to display (Bradbury 1981). Whether or not variation in any acoustic aspect of these calls is associated with female visitation rates has not yet been determined.

Several vespertilionid species, including the banana bat (*Pipistrellus nanus;* O'Shea 1980), both phonic types of pipistrelle bat (Gerell-Lundberg & Gerell 1994; Barlow & Jones 1997), Kuhl's pipistrelle (*P. kuhli;* Barak & Yom-Tov 1991), serotine bat and Daubenton's bat (*Eptesicus serotinus* and *Myotis daubentonii;* Miller & Degn 1981), greater mouse-eared bats (Zahn & Dipple 1997), and noctule bat (*Nyctalus noctula;* Sluiter & van Heedt 1966) also emit calls that appear to attract females to mating sites. None of these species are known to form stable female groups. Instead, where information is available, females visit males for short periods to mate (McCracken & Wilkinson 2000). Much as in epomophorine bats, the vespertilionid advertisement calls that have been described have simple acoustic structure, often consisting of repetitive frequency-modulated chirps. Acoustic differences occur in calls emitted by closely related species, but have not been described among individuals (Barlow & Jones 1997).

In contrast, male white-lined bats produce long series of complex vocalizations and defend small territories on the buttresses of large trees that contain up to eight females (Bradbury & Emmons 1974; Davidson & Wilkinson 2001). Calls are produced by males during the day throughout the year and differ from echolocation calls, in general, by being lower in frequency and longer in duration. Analysis of vocalizations from 16 individually marked males at four colonies in Trinidad, West Indies identified 21 simple (Figure 12.2) and 62 composite syllable types. Males differed in estimated repertoire size and in three out of six acoustic features measured from the most common syllable type (Davidson & Wilkinson 2002). The number of composite syllables in a male's repertoire, the number of times a particular element is repeated, and two other acoustic features of the common syllable significantly correlate with the number of females in a male's territory (Davidson & Wilkinson 2001). None of these variables covaries with male body size. Because females do not form stable groups in this species, these results suggest that females prefer to roost with males that produce more complex vocalizations. Although territorial males may not father all of the young in their territory, they do have higher mating success than males without females (Heckel et al. 1999). These results are consistent with some studies in birds that have shown female mating preferences for males with larger repertoires (Catchpole 1980; Heibert et al. 1989). Interestingly, the size of male

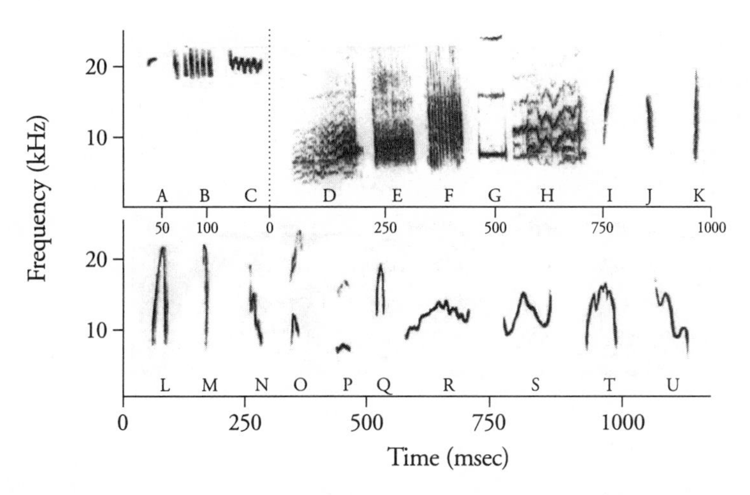

Figure 12.2. Sound spectrograms of representative syllables for 21 different types emitted by male greater white-lined bats (*Saccopteryx bilineata*) during vocal displays.

song repertoires in birds also correlates with the size of brain nuclei associated with song production, presumably as a consequence of past selection to enhance vocal learning (DeVoogd et al. 1993). Whether male repertoire size in greater white-lined bats is influenced by learning and possibly associated with the size of any brain structures remains to be determined. Neuroanatomical comparisons between greater white-lined bats and a congener, such as *S. leptura,* might be revealing since male *S. leptura* do not produce vocal displays. Also, playback studies on greater white-lined bats are needed to demonstrate that male vocalizations influence female roosting patterns.

Echolocation Calls

Echolocation calls are produced by megachiropteran bats in the genus *Rousettus* as well as by all microchiropteran bats. *Rousettus* emit short audible broad-band clicks that permit limited orientation in cave roosts (Suthers 1988). In contrast, microchiropteran bats emit specialized, often ultrasonic, calls that are used for orientation and prey capture. Two echolocation strategies have been recognized and are characterized by differences in signal bandwidth and in how outgoing calls are discriminated from incoming echoes (Fenton 1995). Rhinolophoid species in the Old World and Parnell's mustached bat (family Mormoopidae) in the New World separate pulse and echo by frequency. Echolocation calls emitted by these bats have narrow bandwidth (<1 kHz over >80 percent of a call), long duration (5–100 ms), high duty cycle (>50 percent), and consequently, extensive pulse-echo overlap. These bats are often referred to as constant frequency or CF bats. In contrast, all other microchiropteran species separate pulse and echo in time by producing calls of short duration (<5 ms), broad bandwidth (>40 kHz), low duty cycle (<20 percent), and no pulse-echo overlap. These bats are often referred to as frequency modulated or FM bats because they produce calls that sweep through many frequencies in a very short time. These two types of call design have been associated with differences in prey capture technique and auditory processing (Neuweiler 1984), with high duty cycle bats often hunting from perches and relying on Doppler-shift compensation to detect insect wing movements. Although dominant frequency is inversely related to body size in both groups of bats (Heller & Helverson 1989; Fenton et al. 1998; Bogdanowicz et al. 1999; Jones 1999), no evidence yet suggests that echolocation call design covaries with any dimension of social complexity.

One possibility worthy of consideration, however, is that the amount of variation in orientation call spectral characteristics may be greater in group-living species to avoid acoustic interference from conspecifics, especially during exodus from densely populated caves or mines. Individual variation in the dominant frequency of the echolocation call has been described for several colony-forming species, including Hardwicke's lesser mouse-tailed bats (*Rhinopoma hardwickei;* Habersetzer 1981), Parnell's mustached bats (Suga et al. 1987), big brown bats (*Eptesicus fuscus;* Masters et al. 1995), northern bats (*E. nilssoni;* Rydell 1993), lesser horseshoe bats (*Rhinolophus hipposideros;* Jones et al. 1992), and Sundeval's leaf-nosed bats (Guillen et al. 2000). Individual differences in call features have, though, also been reported for some solitary roosting species, such as spotted bats (*Euderma maculatum*), hoary bats (*Lasiurus cinereus*), and red bats (*L. borealis;* Obrist 1995). Because age can also influence echolocation call frequency (Jones & Ransome 1993), age-matched comparisons are needed to determine if any aspect of echolocation call variation is associated with colony size.

Isolation Calls

When isolated from their mothers, pups of most, if not all, species of bats produce loud, repetitive calls (Gould et al. 1973; Gould 1975, 1977). These isolation calls often exhibit variation in acoustic structure suggestive of a vocal signature (Brown 1976; Brown et al. 1983; Thomson et al. 1985;

Gelfand & McCracken 1986; Jones et al. 1991; Scherrer & Wilkinson 1993). Playback studies on Mexican free-tailed bats (Balcombe 1990) and little brown bats (Thomson et al. 1985) have demonstrated that isolation calls attract mothers to their young. Recordings of individuals over time show that frequency modulation patterns can be used to distinguish individuals even though frequencies typically increase while durations decrease as pups age (Esser & Schmidt 1989; Jones et al. 1991; Scherrer & Wilkinson 1993). Furthermore, comparison of calls between siblings indicates that many acoustic features are both repeatable and heritable (Scherrer & Wilkinson 1993). Thus, isolation call complexity is expected to increase with group size as the discrimination task for a female becomes more difficult (Beecher 1989).

To test if call complexity is greater for species that form larger colonies, I recorded and analyzed isolation calls from eight species of bats that differ in colony size. These species represent three phylogenetically related groups and include both high duty cycle CF echolocators and low duty cycle FM echolocators. One clade contains three closely related vespertilionid species: evening bats, Schlieffen's bats (*Nycticeinops schlieffenii*), and lesser yellow house bats (*Scotophilus viridis*). The evening bat forms maternity colonies containing up to 1,000 individuals in hollow trees or houses (Scherrer & Wilkinson 1993). Schlieffen's bats roost solitarily under bark (Merwe & Rautenbach 1987). Lesser yellow house bats form colonies in hollow trees or caves that contain up to 100 individuals (Fenton et al. 1977). Females of all three of these species typically produce twins, and pups of similar ages aggregate to form creches while their mothers are away feeding (Figure 12.3).

A second clade includes two molossid bats, Mexican free-tailed bats and little free-tailed bats, both of which produce a single young each year. Mexican free-tailed bats form large maternity colonies (McCracken 1984) with up to 20 million individuals in a single cave. Pup densities on the cave wall can reach 1,000 per square meter. Little free-tailed bats form colonies in attics that contain 20–500 individuals (McWilliam 1988).

The third clade includes one rhinolophid, the bushveld horseshoe bat (*Rhinolophus simulator*), and two hipposiderids, Sundeval's leaf-nosed bat and the short-eared trident bat (*Cleotis percivali*). All three of these species form colonies and use high duty cycle, narrow bandwidth echolocation calls. Colony size varies from several hundred in the bushveld horseshoe bat to many thousand in Sundeval's leaf-nosed bat. Little information is available on colony size for short-eared trident bats, but reports indicate colonies

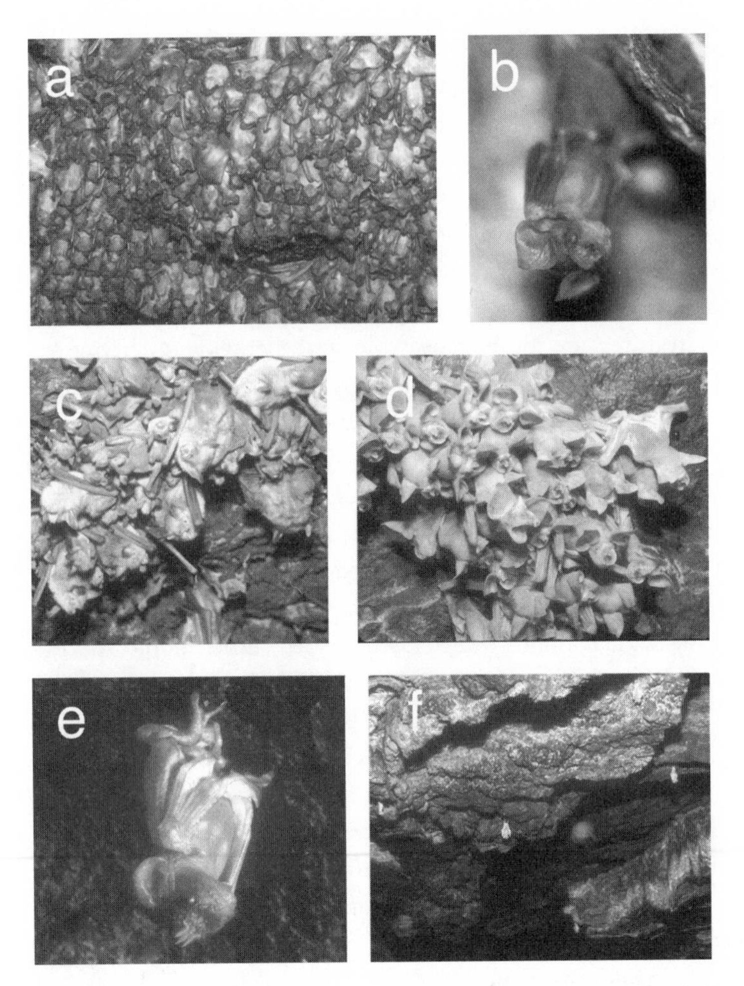

Figure 12.3. Infant aggregations for *a)* Mexican free-tailed bats (*Tadarida brasiliensis;* courtesy of G. McCracken), *b)* Sundeval's leaf-nosed bats (*Hipposideros caffer*), *c)* lesser yellow house bats (*Scotophilus borbonicus*), *d)* bushveld horseshoe bats (*Rhinolophus simulator*), *e)* and *f)* short-eared trident bats (*Cleotis percivali*).

range from tens to hundreds of bats. All three species give birth to singletons, but the spacing patterns of individuals within a colony differ. Bushveld horseshoe bats form dense aggregations and leave pups in creches (Figure 12.3d), whereas both Sundeval's leaf-nosed bats and short-eared trident bats leave pups in isolated locations, often a meter or more away from other pups (Figure 12.3b, e, f). Given these differences in roosting and echolocation behavior, I also compare vocal complexity to creching behavior and echolocation strategy (high duty cycle versus low duty cycle).

To compare call complexity across species I used comparable methods to quantify acoustic variation and when possible, measured 20 individuals per species. I obtained recordings of infant calls for Mexican free-tailed bats (Balcombe 1990) and evening bats (Scherrer & Wilkinson 1993) that were recorded on tape at 30 ips. All other species were recorded in South Africa in November, 1993, using a quarter-inch Bruel and Kjaer microphone connected to a Bruel and Kjaer sound level meter and a Portable Ultrasound Signal Processor (PUSP), which sampled 8 bits at 410 kHz. PUSP digital recordings were time-expanded 20:1 in the field and recorded with a Marantz PMD-430 cassette recorder. By playing back high-speed recordings at one-eighth speed or using time-expanded recordings for each bat we were able to digitize five nonsequential calls at 44 kHz using a PowerMac computer. Then, for each call we measured 10 time, frequency, and amplitude traits from the waveform, spectrogram, and power spectra, respectively (Figure 12.4), using the sound analysis program CANARY, v. 1.2.

Because call features often change with age as pups grow (Scherrer & Wilkinson 1993), for each species I removed effects of age statistically by computing residuals from regressions of forearm length on each of the 10 acoustic variables. Forearm length is linearly related to age during the first half of lactation in bats (Kunz & Stern 1995). To quantify the complexity in

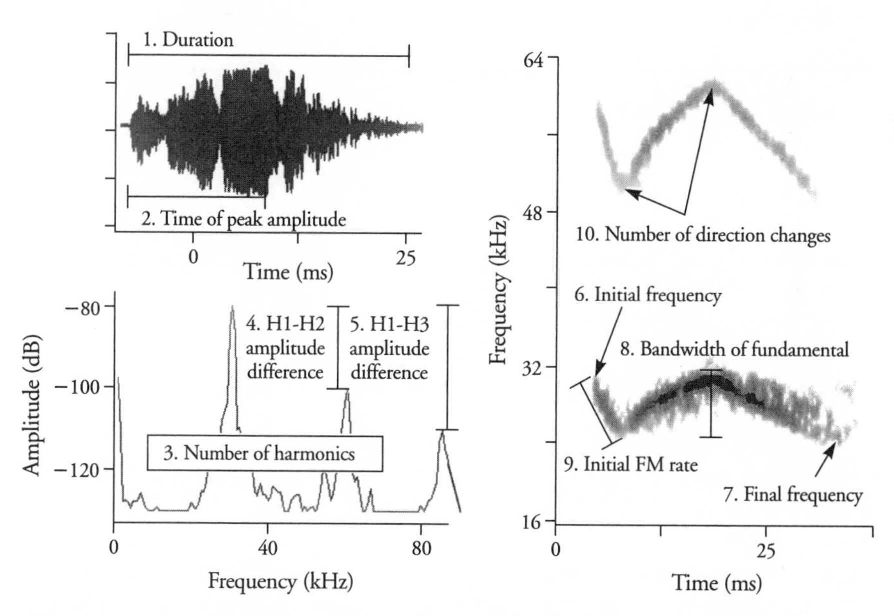

Figure 12.4. Description of 10 acoustic measurements taken on each isolation call.

each variable I used the Shannon-Wiener information statistic, H_s, which I calculated with a model II MANOVA on the regression residuals involving each acoustic variable (Beecher 1989). I then compared total complexity to visual count estimates of colony size obtained when the bats were captured and recorded.

Isolation calls can potentially function as vocal signatures given that there were highly significant differences between individuals in all eight species ($p < 0.0001$, MANOVA). In addition, the total information contained in the isolation calls of each species exhibited a significant positive relationship with colony size (Figure 12.5), as predicted if evolution has acted to increase call complexity. Neither echolocation type nor creching behavior showed significant associations with call complexity. Despite dramatically different forms of echolocation calls, infant isolation calls exhibit considerable acoustic similarity across species, genera, and families. Differences appear only in the degree to which individuals differ in acoustic dimensions, with CF bats showing more variation in the number and relative intensities of harmonics and little variation in duration compared to FM bats. Additional studies are

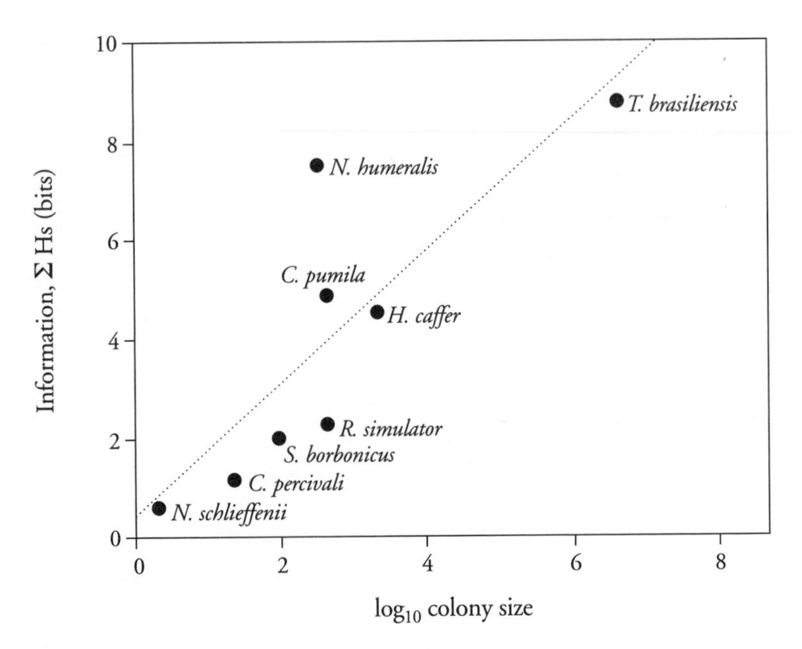

Figure 12.5. Infant isolation call complexity, as measured by the information contained in 10 acoustic variables (see text), plotted against log(10) colony size for eight species. A least-squares regression explains 69 percent of the variation in call complexity.

needed to determine if these acoustic differences reflect perceptual differences between species with different echolocation systems.

Discussion and Conclusions

The analysis of neocortical volume presented above is consistent with the hypothesis that group stability has been more important than colony size, mating behavior, or echolocation ability in shaping neocortical volume and presumably, therefore, the cognitive capabilities of bats. This result does not preclude the possibility that social group size has also influenced bat cognitive ability. As noted above, social group sizes are not available for most species of bats, and the MI hypothesis does not necessarily predict any association between colony size and cognitive ability. The association between female group stability and neocortical volume could reflect past social selection for cognitive ability if one assumes that individuals in stable social groups must remember many social transactions. At least one group-living species, the common vampire bat, exhibits evidence of reciprocal food sharing (Wilkinson 1984). A key requirement for the maintenance of reciprocal exchange systems is the presence of a mechanism for detecting cheaters, such as the ability to remember interactions with other members of a social group.

Alternatively, the association between group stability and neocortical volume may be caused by a third, unmeasured, variable. For example, those species that form stable social groups might also face difficult cognitive challenges related to obtaining food, such as finding vertebrate prey. The common vampire bat and the fishing bat are notable in this regard as these two species have some of the largest residual neocortical volumes in the order and face challenging foraging tasks. The relatively large neocortex among nectar-feeding bats might also reflect the cognitive difficulties associated with traveling long distances to food sources that change location frequently. Additional data on group stability as well as more specific comparisons involving aspects of social and cognitive complexity, such as the duration of social relationships or home range size and utilization, may help discriminate between these possibilities.

One interesting parallel between bats and primates is that social grooming appears to be confined to those bats that form stable social groups (Wilkinson 1987). Unfortunately, much less is known about the function of this behavior in bats than in primates. One notable difference is that some bats groom each other to obtain food. Nectar and pollen-feeding bats can become covered in pollen after visiting a flower. This pollen is typically ingested by a bat or its roostmates after returning to a communal roost. Nevertheless, not all nectar-feeding bats groom each other, for example, long-nosed bats (*Leptonycteris curasoae*) do not (T. Fleming, personal communication), whereas lesser spear-nosed bats do (personal observation). Long-nosed bats form large colonies and forage in groups (Howell 1979). Absence of social grooming suggests that foraging groups are not stable assemblages of individuals and recent observations on long-nosed bats are consistent with this inference (T. Fleming, personal communication). In contrast, analysis of social grooming among individually marked female vampire bats revealed grooming preferences for some individuals and an association between social grooming episodes and food sharing events (Wilkinson 1986). I hypothesized that bats inspected each other while grooming to determine who might be able to provide or be in need of a blood meal. Whether or not social grooming also functions to maintain social bonds, as has been proposed for primates (Dunbar 1991), remains to be determined.

In contrast to propositions for primates (Dunbar 1993), evidence available to date suggests that social complexity in bats has little general relationship to vocal complexity. This conclusion should, though, remain tentative until direct comparison of vocal complexity between species that form long-term social bonds, such as vampire bats and greater spear-nosed bats, to species with less stable social groups, has been made. On the other hand, repertoire diversity in greater white-lined bats is related to the number of females in a male's territory (Davidson & Wilkinson 2001). Thus, as in some birds and marine mammals (Janik & Slater 1997), sexual selection appears to provide the best explanation for this type of vocal complexity in bats. In addition, colony size appears to have influenced the evolution of infant isolation call complexity in bats. Presumably, similar patterns will be found between colony size and complexity of maternal directive calls in those species where females call to pups to facilitate reunions.

An essential adaptation for the evolution of language is the ability to learn to produce new vocalizations after hearing exemplars. An intriguing question to consider, therefore, is whether vocal learning contributes to call complexity in bats. Several answers are possible. Infant bats begin to produce isolation calls moments after birth. Even at that time, individuals differ in characteristic ways that persist for several subsequent weeks (Scherrer & Wilkinson 1993). Although some evidence indicates that pups may modify acoustic features of their calls in response to sounds they hear (Esser 1994), a consequence of such modication would typically be a reduction, not an increase, in call complexity as defined by the information available in the calls, according to Beecher (1989). However, if individuals have distinctive signature calls and, in addition, modify a call to match others in their social group, as greater spear-nosed bat females seem to do (Boughman 1998), then within-individual call diversity could increase. Vocal learning could also enhance call complexity if syllables can be acquired from others and syllable order is flexible. Learned vocalizations appear to play key roles in maintaining social cohesion within groups in other taxa, such as parrots (see Chapter 11) and cetaceans (see Chapter 13). Thus, additional studies on vocal learning in bats may prove particularly insightful at understanding how sociality influences vocal complexity in bats.

13

Dolphins Communicate about Individual-Specific Social Relationships

PETER L. TYACK

Most research and modeling on social complexity emphasizes the strategies employed in balancing competitive relations in societies in which animals must also depend upon one another (Byrne & Whiten 1988; Humphrey 1976). The study animals I discuss in this chapter, bottlenose dolphins, live in societies in which they rely upon one another for survival, feeding, and reproduction, including care of the young. The two areas where we can best understand the competing pressures of cooperation and competition involve adult females caring for their young and adult males that form coalitions to compete for access to females for mating. Current research suggests that dolphins depend upon one another for social defense from predators. As Wells discusses in Chapter 2 of this volume, social defense from predators is particularly important for young calves. Some dolphin mothers raise their calves in relative isolation from other dolphins, while other mothers associate in a group of other females with calves of similar age. Calves of mothers that live in larger and more stable groups have a lower rate of shark predation than do the calves of mothers who are seldom sighted in groups. Wells also points out more subtle benefits of group living for females. Young inexperienced mothers have poor reproductive success, and it may take years and several pregnancies for a mother to successfully raise a calf. The more experienced a mother is, the more likely she is to raise her calf in close association with other mother-calf pairs. Wells suggests that the improved reproductive success of social females may result from "calf exposure to other individuals for socialization, learning, and possibly allomaternal care." Yet allomaternal care may be a mixed blessing, offering significant benefits for cooperation along with significant potential costs. Mann and Whitehead (2000) describe how young calves are particularly attractive to inexperienced females. At least in captivity, females may steal a calf from the mother, preventing appropriate care. Although inexperienced females may benefit from making their mistakes with someone else's calf, a mother who allows this to happen with her own calf may face increased risks to her calf's survival. For a female to be successful raising her calves in a social setting, she must learn which females she can trust, and must be able to limit access of her calf to appropriate females.

Adult males face similarly tough choices in balancing cooperation and competition as do adult females. Both Wells and Connor and Krützen (see Case Study 4B) describe how pairs or trios of adult males may form coalitions, in which they are sighted together most of the time. The behavior of coalition males is remarkably synchronized and coordinated. The coordinated behaviors of coalition males are thought to enhance their ability to guard females and also to engage in unusually coordinated social feeding behaviors (Nowacek 1999). As Wells states in Chapter 2, the primary mating strategy of male bottlenose dolphins appears to involve guarding receptive females. Adult male bottlenose dolphins guard a receptive female by flanking her during her period of receptivity. Groups of two or three males are better able to control access to a female than a lone male, because they can swim in formation around her. In Wells's study site in Sarasota, Florida, most males pair around the age of sexual maturity, with a bond that may last for life. Most males pair at some point in their lives, but 42 percent of males do not have a partner. Genetic analysis of paternity suggests that paired and unpaired males have similar reproductive success, but that there is strong asymmetry in the reproductive success within a pair, with one member siring most of the calves. Although most males do not sire calves until more than 20 years of age, Wells has found that particularly large males may sire calves at ages as young as 13. As a male matures, it faces difficult decisions about whether to pair and who to pair with, decisions that are likely to affect reproductive success for the rest of his life. In Connor's study site in Western Australia, most males form coalitions with one or two other males. If one coalition finds another coalition with a female, it may fight for access to the female, and one coalition may ally with another in order to improve their competitive advantage. As Connor and Krützen describe in Case Study 4B, the competing demands of cooperation and competition within and between coalitions can be remarkably complex.

Individual Recognition

Implicit in this discussion of complexity in dolphin societies is the assumption that dolphins can recognize individuals over time and remember enough about past interactions to make strategic decisions about who to associate with. Most models of social complexity describe societies in which individuals recognize one another and remember the history of their past interaction. Modellers have gone into detail about the variety of strategies that may mediate interactions in these individualized societies (e.g., Axelrod & Hamilton 1984), but seldom have focused on the problem of individual recognition itself. Dolphins are particularly good animals for studying the role of individual recognition in communication, because they learn to develop particularly distinctive calls and because the process of learning to match calls appears to parallel the development of social bonds. However, individual recognition is a very common problem that can be addressed by simple mechanisms. Before exploring the complexity of individualized communication in dolphins, I will review the basics of individual recognition.

Most birds and mammals face problems requiring individual recognition. For example, both taxa provide resources to their young after birth. In some situations, such as when the young are in a nest or den, the parents may be able to provide resources to any young in their refuge, with little risk of misallocating resources. However, parental care is such a valuable resource that other animals may evolve special mechanisms to steal the care, as is seen in nest parasites of birds. This may involve interspecific parasites such as the cuckoo and the cowbird (Robert & Sorci 2001) or intraspecific nest parasites (Petrie & Møller 1991). This risk of parasitism selects for recognition mechanisms between parent and young so that parents do not misallocate resources. This kind of recognition could be based upon very simple discrimination. For example, the young of many interspecific nest parasites are often much bigger than the offspring of the parents; parents could follow a simple decision rule such as rejecting a nestling that is much bigger than the rest.

In more colonial species of birds and mammals, even if the young are born in a den or hatched in a nest with only sibs, the young may mix later with conspecifics before the period of parental care is over. The ability of different bird species to recognize individuals using these signals appears to correlate with the likelihood of misallocation of parental care. For example, barn swallows *(Hirundo rustica)* raise their young in nests that are far from other broods, whereas cliff swallows *(Petrochelidon pyrrhonota)* intermingle within a colony while still being fed by their parents. Barn swallow chicks make a begging call, but their parents do not distinguish between the calls of their own and unrelated chicks (Medvin & Beecher 1986). Cliff swallow parents can discriminate the begging calls of their own offspring from other young (Stoddard & Beecher 1983), and have evolved both a more distinctive begging call in the young and more rapid discrimination of begging calls by adults (Loesche et al. 1991). Some colonial species of seabirds switch from location cues to identifying the calls of their own offspring at the time when the young from different broods intermix (Beer 1970; Miller & Emlen 1975). These results suggest that evolution favors investment in parent-offspring recognition if the risk of misallocation of care outweighs the cost of the recognition system.

Individual recognition of close relatives within a colony need not involve learning. One of the most difficult mother-infant recognition problems is faced by mollossid bats, vast numbers of which may roost in the same cave. Up to 20 million Mexican free-tailed bats (*Tadarida brasiliensis*) may roost in one cave, with maternity colonies in which there are up to 1,000 pups per square meter (see Chapter 12). Imagine the job for a female to find her own pup among the cacophony. Bats solve this problem by individually distinctive isolation calls, which are produced by the pup, and which attract a mother to the call of her pup. In Chapter 12 Wilkinson shows that these isolation calls are individually distinctive, and that the acoustic complexity of a species' isolation call correlates with its typical colony size. This is to be expected if mothers must use the call to distinguish their pups from the rest of the colony. Even though this system appears closely tailored to the ecological setting, most acoustic features of the calls are heritable (Scherrer & Wilkinson 1993). Little is known about the mechanisms by which females distinguish the call of their pups, but mothers need not learn the calls. Since the calls are highly heritable, if she can remember her own call, she could compare pup calls to her own template.

Many animals have specialized mechanisms to learn to recognize individually distinctive signals. The imprinting process by which many birds learn to identify their caregiver may involve acoustic, olfactory, or visual cues. Some mammals such as seals also appear to have a similar imprinting process (Schusterman et al. 1992). These forms of individual recognition need not involve learning to *produce* a distinctive signal. Rather individually distinctive features of the signals may be generated by variability in development. Dolphins differ from the classical picture of imprinting by learning the acoustic structure of their recognition calls through imitation, and they retain this ability to imitate and learn calls throughout their lifetime (Tyack & Sayigh 1997).

Signature Whistles of Bottlenose Dolphins

Bottlenose dolphin calves *(Tursiops truncatus)* show an unusual combination of early mobility with prolonged dependence, which creates a strong need for early development of a mother-offspring recognition system. Dolphin calves are precocious in locomotory skills, and swim out of sight of the mother within the first few weeks of life (Smolker et al. 1993). Calves this young often associate with animals other than the mother during these separations. Yet as Wells discusses in his chapter, calves depend upon the mother for three to six years.

Dolphins have limited sensory modalities for individual recognition. Their sense of smell is atrophied—they have no olfactory bulb. Vision is limited to only a body length or so in most marine habitats. The most useful modality for rapid recognition for separated animals is acoustic. Dolphin mothers and young use frequency modulated tonal whistles as signals for individual recognition. Observations of captive dolphins suggest that whistles function to maintain contact between mothers and young (McBride & Kritzler 1951). When a dolphin mother and her young calf are forcibly separated in the wild, they whistle at high rates (Sayigh et al. 1990). This is similar to the acoustic structure and patterns of usage of isolation calls in a broad variety of mammals, from bats to ungulates to primates. Figure 13.1 illustrates an exchange of whistles between an adult female dolphin from the wild population near Sarasota, Florida, and her young calf. There is a relatively regular exchange of whistles, and in this interaction, the calf whistles appear to become more synchronized and more similar to those of the mother. When a mother and calf voluntarily separate in the wild, the whistle rate is

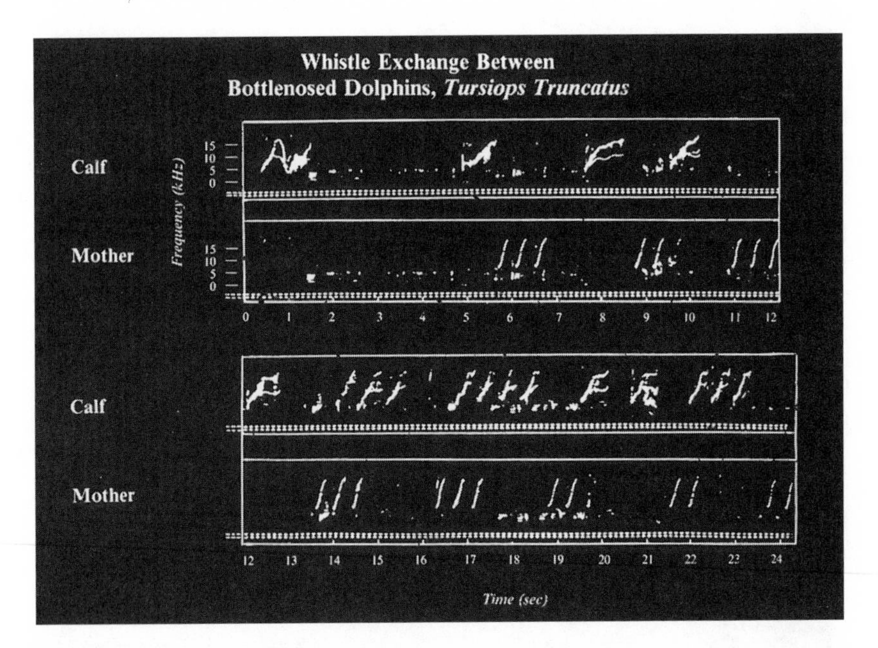

Figure 13.1. Spectrogram indicating an exchange of whistles between a wild bottlenose dolphin mother and her young calf. Recording was made while these wild dolphins were temporarily restrained (described in Chapter 2). As soon as the dolphins were held out of sight of one another, they started a regular exchange of whistles. Note how the calf whistles appear to become more synchronized and more similar to those of the mother.

usually lower, with less obvious exchanges. It is usually the calf that whistles to signal a reunion (Smolker et al. 1993).

Caldwell and Caldwell (1965, 1968) demonstrated that each dolphin within a captive group produced an individually distinctive whistle. The Caldwells called these "signature whistles" and postulated that signature whistles function to broadcast individual identity. Four independent research groups have studied signature whistles in a total of 132 captive (Caldwell et al. 1990; Janik et al. 1994; Janik & Slater 1998) and 90 wild dolphins (Sayigh et al. 1990, 1995, 1999; Smolker et al. 1993) for a total of 222 individuals. Terrestrial biologists are used to animal signals, such as bird song, that may give obvious and distinct cues for *species* recognition, but the *individually* distinctive cues in bird or primate calls are typically very subtle. Often it is easier to demonstrate individual recognition in playback experiments than it is for the investigator to identify the acoustic features used in recognition. In contrast to these other species, dolphin whistles are rated as highly

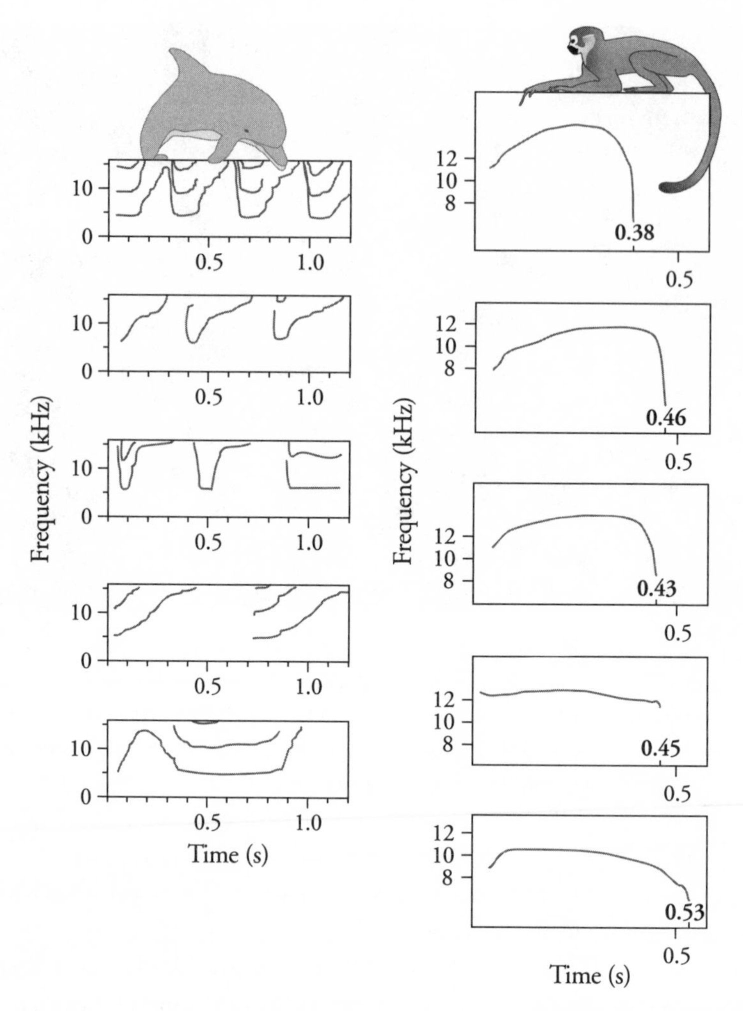

Figure 13.2. Spectrograms of signature signals from five bottlenose dolphins (whole figure from Tyack 2000; left side adapted from figures 2, 5, and 6 of Sayigh et al. 1990) and three squirrel monkeys (right side adapted from figure 2 of Symmes et al. 1979). Note how different the signature whistles of the dolphins are compared to the isolation peeps of the squirrel monkeys.

distinctive by human judges. Figure 13.2 illustrates how remarkably easy it is to distinguish the signature whistles of dolphins compared to the isolation peeps used by squirrel monkeys *(Saimiri)* for mother-infant recognition (Symmes et al. 1979).

Janik and Slater (1998) compared whistles from dolphins when they were swimming together in one large pool compared to when one dolphin swam into a side pool. When isolated in the side pool, each dolphin produced its own distinctive signature whistle but very few signature whistles were produced when the dolphins swam together. The tendency for signature whistles to be produced during isolation suggests that these whistles function as contact calls not just among mother-infant pairs, but also between adults. Experimental playbacks have demonstrated that mothers and offspring respond preferentially to each others' signature whistles even after calves become independent from their mothers (Sayigh et al. 1999).

Vocal Learning of Whistles

Tyack (2000) and Janik and Slater (1997) argue that the reason for the unusual distinctiveness of the signature whistle is that dolphins are unable to use voice cues, so must learn to produce a distinctive recognition signal. Individual recognition by voice cues appears to be common and has been demonstrated in many species of birds and mammals. As mentioned above, the signaler does not need to learn to produce a distinctive signal for voice cues—rather the receiver can learn to detect acoustic features correlated with slight differences in the vocal tract of the signaler compared to conspecifics. The diving adaptation of dolphins interferes with recognition by voice. The vocal tract of the dolphin is filled with air, and as the dolphin dives, the air is compressed, halving in volume for each 10 meters of depth. The need for individual recognition of vocal signals in diving dolphins may have driven the evolution of a communication system in which each individual dolphin must learn to produce a distinctive signal in addition to needing to learn to discriminate the calls of others.

In 1965, when David and Melba Caldwell initially introduced the signature whistle hypothesis, they believed that this was like a voice cue, and that each dolphin was physically capable of producing only one kind of whistle. During the next decade, though, they and several other research groups demonstrated remarkable abilities of dolphins to imitate manmade whistle-like sounds (Caldwell 1972; Richards et al. 1984). Figure 13.3 (from Richards et al. 1984) illustrates imitations made by dolphins (on the right half of each cell) of synthetic patterns of frequency modulation generated by a computer (shown on the left half of each cell). These imitation experiments demonstrate abilities of vocal learning that are unparalleled among nonhuman mammals.

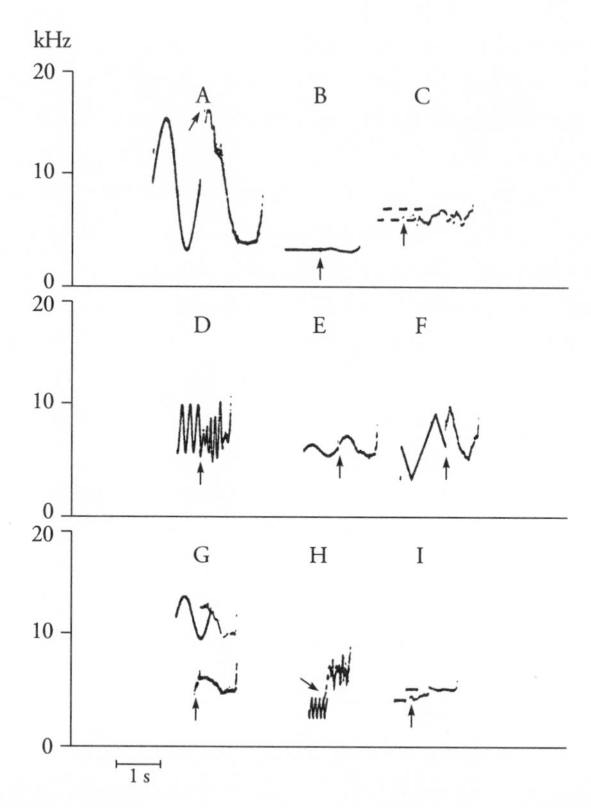

Figure 13.3. Vocal imitation by captive dolphins of computer-generated whistle-like sounds. The left part of each cell before the arrow shows synthetic whistles and to the right of the arrow shows imitation by a captive dolphin trained to imitate the model. From figure 2 of Richards et al. (1984).

Development of Signature Whistles

Studies on the development of signature whistles show that dolphins do not inherit whistles similar to those of their parents, but rather use vocal learning to copy acoustic models present in their natal environment. In a comparison of the similarity of whistles recorded from adult females in the wild and their offspring, Sayigh et al. (1995) report that 74 percent (31/42) produced signature whistles that were not judged similar to those of their mothers. The similarity of whistles was judged by humans who inspected spectrograms of the whistles, using a rating scale of 1 (not similar) to 5 (very similar). Figure 13.4 illustrates pairs of whistles given scores of 1 and 5. There was a pronounced sex difference in the tendency of offspring to have whistles similar

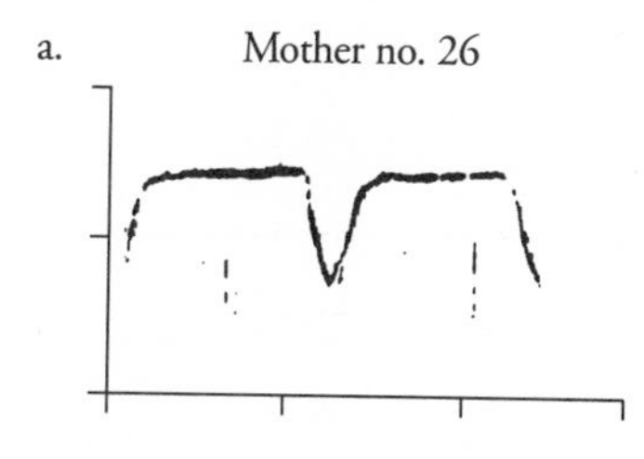

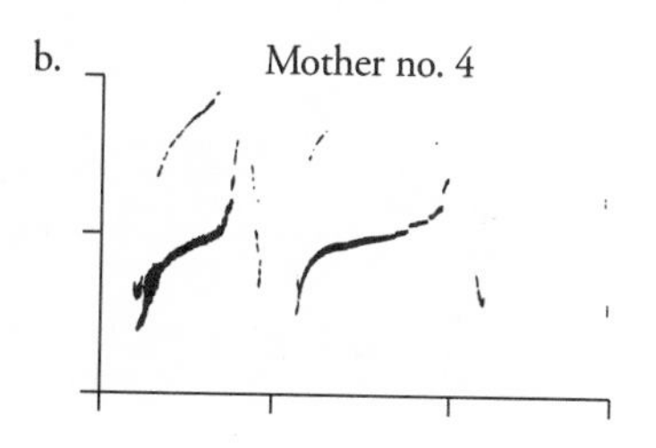

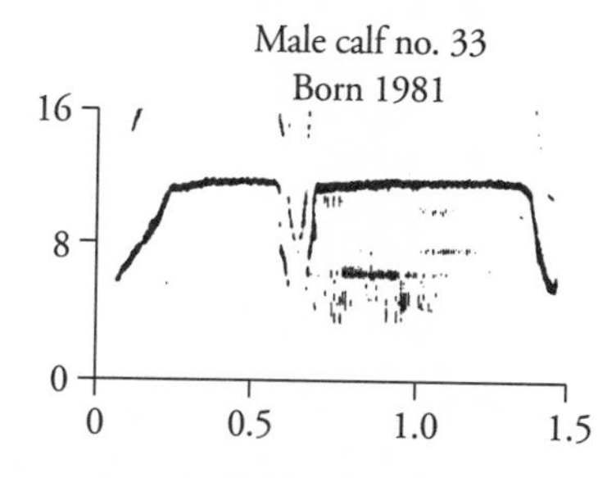

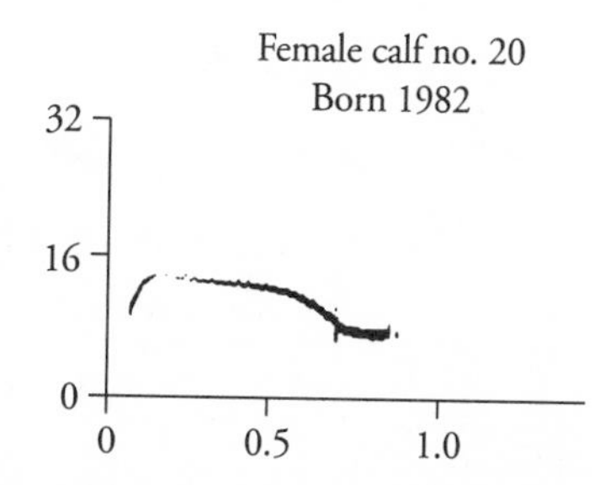

Figure 13.4. Spectrograms of pairs of whistles, judged by humans to be *(a)* very similar (similarity score = 5), or *(b)* very different (similarity score = 1). The whistle pair in *(a)* was recorded from a mother and her son and the pair in *(b)* from a mother and her daughter. Frequency (kHz) is on the *y* axis and time (in seconds) is on the *x* axis. From figures 2 and 3 of Sayigh et al. (1995).

to the mother: sons were more likely than daughters to produce whistles similar to those of their mothers (nine out of 21 sons as opposed to two out of 21 daughters sampled had similarity scores greater than 3.67), whereas daughters were more likely than sons to produce whistles highly distinct from those of their mothers (15 out of 21 daughters as opposed to seven out of 21 sons sampled had similarity scores less than 2.33). In two studies of captive dolphins, only two out of 23 calves developed a whistle similar to their mother (Caldwell & Caldwell 1979; Tyack & Sayigh 1997).

Tyack and Sayigh (1997) discuss evidence that bottlenose dolphins develop signature whistles that match acoustic models present in their natal environment. One bottlenose dolphin infant cross-fostered with a conspecific developed a whistle similar to that of the foster mother, and Caldwell and Caldwell (1979) report that one bottlenose dolphin infant developed a signature whistle similar to a Pacific white-sided dolphin, *Lagenorhynchus obliquidens.* Some of the most convincing data for vocal learning in whistle

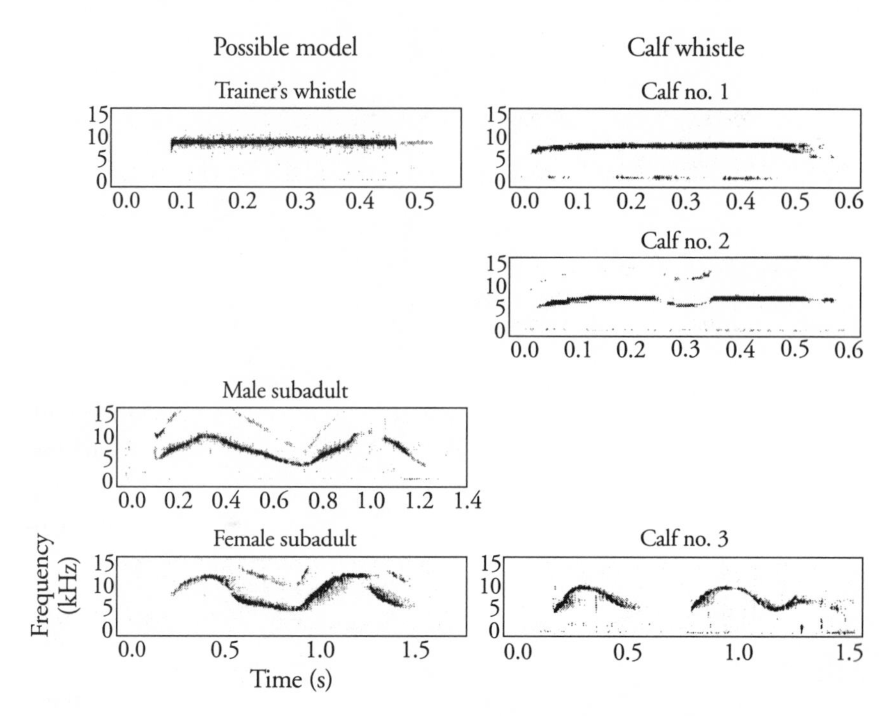

Figure 13.5. Comparisons of the signature whistles from three bottlenose dolphin calves born in captivity to the most similar acoustic models present in the natal pool. These three calves were selected because they were born within a three-month period in the same pool. Frequency (kHz) is on the *y* axis and time (in seconds) is on the *x* axis. From figure 11.10 of Tyack and Sayigh (1997).

ontogeny stems from comparisons of calf whistles to sounds present in their natal environment. Of three calves born in a colony pool, two developed whistles very similar to the whistle used by trainers to signal a dolphin that it can approach for food after performing a requested task, whereas another developed a whistle quite similar to the signature whistles of two subadults who themselves were raised together in the pool (Tyack & Sayigh 1997; Figure 13.5).

Stability and Modification of Signature Whistles in Adults

Bottlenose dolphins tend to develop a stereotyped signature whistle by one year of age. Once developed, the signature whistles of adult females are usually stable over the dolphin's lifespan. Stability of signature whistles has been determined by visually comparing spectrograms of whistles recorded from the same individuals over time. Caldwell and colleagues (1990) documented stability of signature whistles for periods of more than a decade in captive bottlenose dolphins and Sayigh and colleagues (1990) made similar analyses of wild dolphins. The combined data set of more than 200 individuals recorded either in captivity or in the wild indicates clearly that the signature whistles of most adult females, once developed, are stable throughout a dolphin's lifespan. Figure 13.6 shows signature whistles recorded from an adult female over 11 years and a calf at one and three years of age.

Not all dolphins retain a stable signature whistle; some maturing or adult male dolphins form strong bonds with other males in the wild, and the formation of this bond appears to be accompanied by a process of vocal convergence in which the signature whistles of the males become more similar. Co-

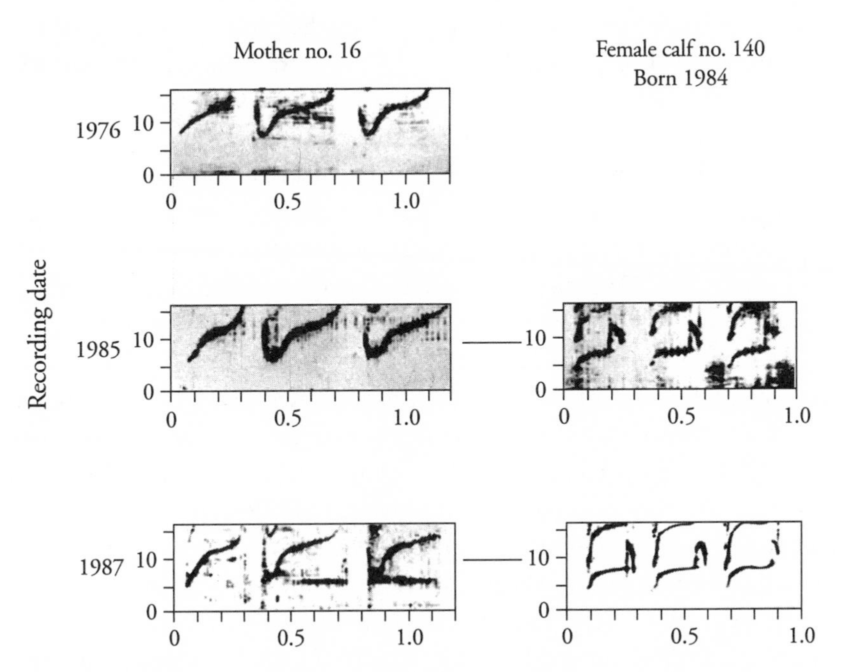

Figure 13.6. Spectrograms of signature whistles from one wild adult female recorded over a period of eleven years and from one of her calves at one and three years of age. Note the stability of both signature whistles. Frequency (kHz) is on the *y* axis and time (in seconds) is on the *x* axis. From figure 2 of Sayigh et al. (1990).

alitions of two or three adult males are reported from western Australia, where males in a coalition often gain advantage in competition with other males (see Case Study 4B; Connor et al. 1992). One coalition regularly came in to a beach to be fed by humans, and the whistles of each of these three adult males could be recorded and identified in this setting. Smolker and Pepper (1999) report convergence of whistles among these three males as the coalition formed. A whistle initially only produced by one member of the coalition gradually became the most common whistle for all three males over two years. These male coalitions are also common in Sarasota, Florida, as Wells describes in Chapter 2, but in Sarasota, all coalitions observed have involved pairs of males, not trios. No longitudinal data on changes in whistles are available from the Sarasota population during the period of coalition formation, but Watwood and colleagues (in revision) studied the similarity of male coalition partners in order to test for whistle convergence. Watwood and colleagues (in revision) studied whistles recorded from five pairs of males (Figure 13.7). Whistles were recorded when the males were involuntarily separated from their partners in a raft or net corral, although they usually could hear one another. Judges rated the partner's whistle to be more similar to whistles of the partner than of the nine nonpartners in the majority of these 10 comparisons, a rate that was much higher than predicted by chance.

Not only do adult male dolphins show vocal convergence, but dolphins also learn to imitate the signature whistles of animals with which they are interacting. Tyack (1986) used a simple telemetry device to determine which of two captive dolphins made each whistle while they were interacting. The most common whistle of each of these dolphins was categorized as a signature whistle, but each of these dolphins also imitated the signature whistle of the other—25 percent of all occurrences of each signature whistle were imitations produced by the other dolphin. Other studies reviewed in Tyack and Sayigh (1997) have reported rates of signature whistle imitation near 1 percent among captive dolphins that were in acoustic but not physical contact. These imitated signature whistles are not just produced immediately after the partner makes its signature whistle, but can become incorporated into a dolphin's whistle repertoire. For example, after a period of silence, dolphin A might produce a copy of dolphin B's signature whistle. The two animals in the Tyack (1986) study were first housed together at about five years of age, well after signature whistles are developed. Thus, dolphins maintain the ability well past infancy to add new whistles to their vocal repertoire through imitation of auditory models. However, dolphins tend to limit the production

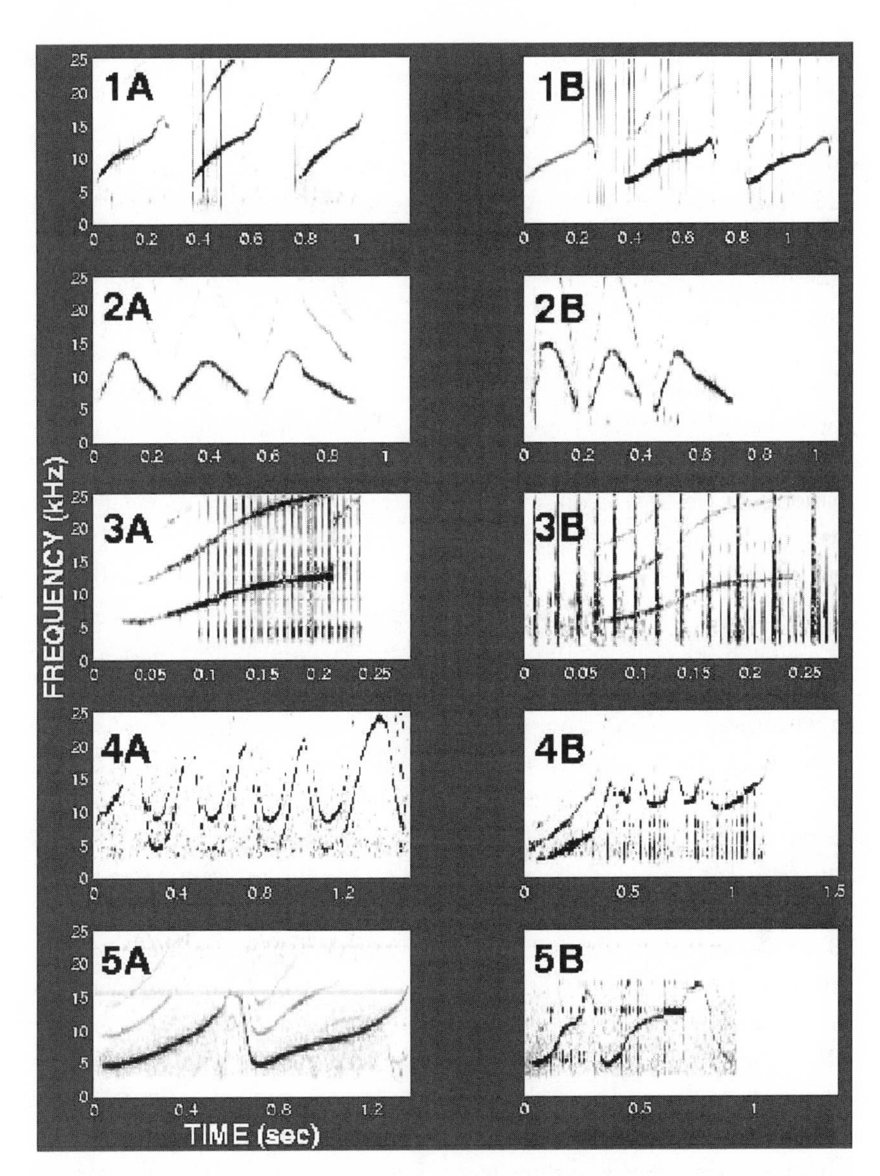

Figure 13.7. Male coalition partners have similar signature whistles. Spectrograms of five pairs of males, each pair of which forms a coalition. Each row represents one pair, with 1A showing a signature whistle from one member and 1B showing a signature whistle for the other member of the pair. Human judges compared the whistle of each dolphin to all nine others, without knowing which one was a partner. The coalition partner was more likely to have a similar whistle than expected by chance. This suggests that the whistles of males converge as they form a strong social bond. Figure from figure 1 of Watwood et al. (in revision).

of imitated whistles to dolphins with whom they are interacting. Recordings of isolated dolphins contain few imitated whistles. Two years after the death of one of the two dolphins in the Tyack (1986) study, the surviving dolphin, who was housed alone in the pool, produced only segments of his own signature whistle, with no imitations of the other dolphin's whistle (Tyack 1991).

Janik (2000) found that wild dolphins often immediately match the whistle of another dolphin that is within acoustic range but out of visual range. Janik observed the dolphins from shore and used an array of hydrophones previously installed in the study area to record the whistles and to determine their location. This method was unusual among dolphin studies in that the behavior of the dolphins could not have been affected by the observations. Janik defined a whistle interaction as occurring when one dolphin produced a whistle within three seconds of another dolphin. He could only determine that two whistles came from different dolphins if they were separated by more than 26 meters. About 10 percent of the whistles recorded with an average of 10 dolphins in the study area were classified as whistle interactions. Even though the criterion duration was three seconds, in over 80 percent of the whistle interactions, the whistles occurred within one second of one another, suggesting that one dolphin was whistling immediately in response to the other. Of the 176 whistle interactions, 39 involved matching whistles, a percentage significantly greater than expected if the dolphins were whistling independently of one another ($p = 0.001$). The mean distance between dolphins scored as matching whistles was 179 meters, a significantly smaller distance than for dolphins engaged in nonmatching whistle interactions ($p < 0.025$). These results show that undisturbed wild dolphins will exchange whistles when out of sight. Dolphins that are closer, and more likely to be able to interact directly, are more likely to match one another's whistles.

Vocal development for many species has been characterized as a progressive narrowing from a large and variable (overproduced) repertoire to a more fixed and mature repertoire. Although bottlenose dolphins develop a stable signature whistle, they also show increasing variability in the vocal repertoire with age. Sayigh and colleagues (1990) found that males produced a lower percentage of signature whistles at each age than females. Maturing male calves showed a particular increase in the number of nonsignature whistles after they separated from their mothers and presumably formed a broader network of social relationships. Some of these new whistles appeared to match those of other animals caught at the same time, and could be categorized as imitations of signature whistles. Caldwell and colleagues (1990) reviewed data on the percentage of signature whistles from 126 captive dolphins. There was a significant difference between the sexes, with males producing a larger proportion of variant whistles at each age. Males showed a steady increase in the percentage of variant whistles, whereas females showed a more complex pattern. Although infant females had the fewest variant whistles and adult females had the highest percentage of variant whistles of all female age classes, juvenile females had slightly more variant whistles than subadults.

Whistle Usage by Adult Dolphins

I have been emphasizing signature whistles so far in this chapter, but as the last paragraph indicates, dolphins do produce other whistles, even when they are isolated. Janik and Slater (1998) used a simple and ingenious design to demonstrate that the usage of signature whistles varies drastically depending upon relatively minor changes in the extent of isolation of each dolphin. They compared whistles from four captive dolphins when they were swimming together in one large pool compared to when one dolphin swam into a side pool. The most common whistle from each dolphin when in a separate pool from the other three was its signature whistle. However, when these captive dolphins were undisturbed and together in one pool, only 2.4 percent of their whistles were signature whistles.

When dolphins are swimming together and interacting, they tend to produce a repertoire of whistles in addition to the signature whistle. Tyack (1986) and Janik and colleagues (1994) defined several simple categories of nonsignature whistles such as upsweeps (rise whistles), downsweeps, flat or unmodulated whistles, and sinusoidal patterns of frequency modulation that differed from the signature whistle. Most dolphins seem to include these whistles in their repertoire. Rise whistles were the most common whistle reported by Janik and Slater (1998) when all the dolphins were swimming together in one group. In a study of an isolated dolphin engaged in a training task, Janik and colleagues (1994) found that rise whistles were the second most common whistle type (305/1743) after the signature whistle (1098/1743) of the subject. McCowan and Reiss (1995, 1997) also report upsweep whistles to be common among dolphins that are socially interacting. Some dolphin whistles fit neither these nonsignature categories nor match the individual's signature whistle. Some may represent imitated signature whistles, but others may represent production errors, improvisations (Kroodsma & Parker 1977), or types that are just too rare to be sampled regularly.

Evolution of Open Systems of Vocal Communication

Most animal species are thought to have a closed system of vocal development (as defined by Mayr 1976), in which animals tend to produce a relatively fixed repertoire of species- and sex-specific vocalizations in specific behavioral contexts. There is little evidence in terrestrial nonhuman mammals, even nonhuman primates, that auditory input affects acoustic features of vocalizations (Janik & Slater 1997). In many species of mammal, even as drastic an operation as deafening has little impact on the acoustic structure vocal repertoire. This stands in stark contrast to humans, dolphins, and some birds, which have an open system of vocal development in two senses. Auditory input has a dramatic impact on their vocal repertoire, for these animals copy the sounds they hear around them. Some of these species also have an open vocal repertoire, in that they can continue to add or create new vocalizations throughout their lifetime.

The ability to modify one's vocalizations based upon exposure to other animals is called vocal learning. Vocal learning is common among several orders of birds, especially among the Passeriformes (sparrow-like oscine songbirds: see Chapter 18) and the Psittaciformes (parrots: see Chapter 11). Among mammals, the only species for which there is strong evidence for vocal learning include: humans, *Homo sapiens;* the horseshoe bat, *Rhinolophus ferrumequinum;* the harbor seal, *Phoca vitulina;* the humpback whale, *Megaptera novaeangliae;* the beluga whale, *Delphinapterus leucas;* and bottlenose dolphins, *Tursiops truncatus* (Janik & Slater 1997). Most evidence on vocal learning in songbirds, whales, and seals involves reproductive advertisement displays called songs. Many reviewers have suggested that vocal learning may allow males to produce a more complex repertoire of songs, or to produce songs that are more attractive to females. In addition, males may use vocal learning to match the songs of other males, a behavior that is often associated with aggressive interactions between adult males (Payne & Payne 1997; Beecher et al. 2000; Vehrencamp 2001). These observations suggest that intra- and intersexual selection may have led to the evolution of vocally learned songs.

Although some bats also produce songs, as Wilkinson describes in Chapter 12, the best evidence for vocal learning in bats involves sounds used for echolocation. The echolocation calls of horseshoe bats rise in frequency over the first and second years of life and then decrease in frequency with increasing age. The echolocation calls of young bats with older mothers are lower in frequency than calls of young bats with younger mothers, suggesting that young horseshoe bats match the call of their mother (Jones & Ransome 1993). The echolocation of horseshoe bats inherently involves vocal learning, but of a sort that does not increase the diversity of signals, but that rather increases the stability of the call as heard by the bat. Horseshoe bats listen for slight shifts in frequency between a sonar signal and its echo induced by the relative motion between the bat and its target. In order to detect subtle changes in frequency, their ears are particularly sensitive to a very narrow frequency range. This frequency range is so narrow, that if the bat were to always make the same echolocation signal, the relative motion would often drive the echo frequency outside of this preferred frequency. The bat therefore modifies its outgoing signal so that the frequency-shifted echo will fall in the sweet spot.

Evidence on the stability and distinctiveness of signature whistles from adult female dolphins suggests that they too may use vocal learning to maintain a stable and distinctive signal. Tyack (2000) and Janik and Slater (1997) have also pointed out that diving marine mammals, like the frequency-shifting bats, may need to use auditory feedback to maintain a stable signal throughout the dive. As a mammal dives, the gas-filled cavities in its sound production apparatus change volume and shape. This is likely to change the sound produced when the animal makes the same motor pattern to vocalize. In order to maintain a stable signal throughout all these changes, diving mammals may require vocal learning that uses auditory feedback of one's own signal to modulate the motor patterns used to vocalize.

Communication Accommodation

Whereas many adult female dolphins appear to maintain a stable signature whistle from the first two years of age, there is evidence that adult males may modify their signature whistle as they form strong social bonds with another male. Brown and Farabaugh (1997) and Chapter 11 in this book argue that studying birds with long-term social relationships leads to a different picture of vocal learning than study of more migratory and seasonal singers, a picture with more parallels with dolphin whistles. For example, many birds learn to produce contact calls that are used in situations similar to those when dolphins whistle (Brown & Farabaugh 1997). The contact whistles of starlings, like those of dolphins, are very stable over time, even when other elements of the vocal repertoire change each year (Hausberger 1997; West et al. 1997). When birds form a group, their contact calls often converge in a process similar to that seen when dolphin males form a coalition (Nowicki 1989; Farabaugh et al. 1994). This is also the setting for which the best data is available on vocal learning in nonhuman primates, although the extent of vocal modification is much less for primates than for birds and dolphins (Elowson & Snowdon 1994). These studies all emphasize the role of vocal imitation in maintaining affiliative relationships.

There is a theory about communication in humans, called communication accommodation theory, which also emphasizes the idea that imitation or matching a signal may be an affiliative gesture. Giles (1973) observed that humans may modify the accent of their speech to match a partner as an affiliative signal and as a way to regulate the closeness of the interaction. Later work has shown that speakers may converge in a remarkable diversity of dimensions including loudness, rate of talking, and even more abstract linguistic features such as dialects (Shepard et al. 2001). As adult male dolphins form a coalition, they not only converge in their whistles, but develop an ability to synchronize complex movement patterns as well, suggesting a broader form of accommodation than just vocal, which parallels studies of accommodation in humans. This concept of vocal accommodation has also been expanded to suggest that speakers may retain differences in speech behavior in order to mark status differences or differences in group membership, a similar interpretation to the idea that vocal convergence may create a badge of membership within a group of birds or other animals. Vocal matching in dolphins provides another example of vocal accommodation as an affiliative signal that may be used to mark a social bond, or membership in a coalition. As with humans and many birds capable of vocal learning, dolphins show convergence of vocalizations used to maintain contact with animals with whom they share a strong social bond.

Imitation and Reference

The communication system of an animal species is structured to solve problems posed by its social life. The signature whistles of dolphins tell us that one of the most important things dolphins communicate about is each other—they use individually distinctive signals to mediate individual-specific social relationships. As Seyfarth and Cheney discuss in Chapter 8, many models of individual-specific social interaction among animals assume that each actor has an internal model of different individuals, along with the ability to link memories of the history of interaction with that individual. Individual-specific relationships are so important to dolphins, and signature whistles make up such an important part of their communication system, that I find it hard to believe that when a dolphin hears a signature whistle of an animal with which it shares a social bond, that this does not recruit a rich representation of that individual. This raises the question of whether when a dolphin imitates the signature whistle of another individual, it might in some sense be referring to that animal as if by name. Armstrong (1973) described vocal matching in birds as "naming an opponent," and Tyack (1993) and Thorpe and North (1966) have suggested that dolphins or duetting birds may use imitation of a signature signal to call a specific individual in order to initiate an interaction with that animal. However, what data there are suggest functional reference more than reference in the sense used by linguists. If an animal uses a signal to refer to something in the linguistic sense, then it must have an abstract representation of the referent divorced from the structural properties of the signal. Janik (2000) has shown that wild dolphins match the whistles of other dolphins nearby. We now need more controlled experimental tests of reference; for example, when one dolphin hears the signature whistle of another, can it select a visual representation of the same individual?

Analysis of the communication system of bottlenose dolphins in terms of the social problems posed by the social life of this species raises significant questions about the relationship between animal societies and cognition. Why have dolphins and a few other animal taxa evolved open systems of communication? Is communication accommodation a general pattern for animals that can learn to imitate? Is call matching used to label an individual whether as a threat among territorial songbird males, or as a contact call? How referential is the imitation of individually distinctive signals? Two basic themes of this book are that the study of animal social complexity and intelligence requires a broad comparative perspective, and a diversity of approaches from ecologically valid observations of wild animals to careful experiments of cognitive abilities. This approach is likely to continue to yield fascinating results for dolphins and other animals.

CASE STUDY 13A

Natural Semanticity in Wild Primates

KLAUS ZUBERBÜHLER

Two basic models have been distinguished to describe call comprehension in nonhuman primates. The first one assumes that primates respond to their own calls by simply attending to the calls' physical features. According to this idea, recipients possess only a working knowledge of how to best respond to a particular call but they do not really understand the associated referential situation that normally leads to the production of a particular call. For example, after hearing a conspecific's leopard alarm call a monkey might simply know that the best response is to run into a tree, but not that this particular alarm call is a semantic signal that indicates the presence of a leopard (e.g., Deacon 1997). Alternatively, the monkeys' processing of their own calls could be cognitively more complex, perhaps similar to the kind that underlies language perception (e.g., Cheney & Seyfarth 1990). According to this second model, the acoustic properties of a vocal stimulus are only relevant insofar as they refer to an associated mental structure. When hearing the call of a conspecific, a monkey generates a mental representation of the referential situation and it is this mental representation that is then driving the monkey's subsequent behavior (e.g., Tomasello & Zuberbühler, 2001). This chapter reviews some recent data that bear relevance for distinguishing these two models by specifically investigating the psychological processes underlying call comprehension in two forest primates, the West African Diana monkey *(Cercopithecus diana)* and the Campbell's monkey *(Cercopithecus campbelli).*

Alarm calls have been shown to be specifically productive in research on animal semanticity, mainly because the eliciting stimuli, and thus the associated mental representations, tend to be quite specific and identifiable to the researcher. This is especially true for forest species, such as the Diana monkey or the Campbell's monkey, where visual communication is often unreliable or restricted to very short distances. Using alarm calls as experimental stimuli, a number of studies have already shown interesting parallels between human linguistic behavior and nonhuman primates' vocal behavior in the wild. When East African vervet monkeys *(Cercopithecus aethiops),* for example, hear a playback of a conspecific's eagle alarm call, they look up into the sky and run for cover, apparently anticipating the attack of a predatory eagle even though no predator is present (Seyfarth et al. 1980). This and other studies suggested that nonhuman primates might have the cognitive capacities to treat their vocalizations as semantic signals, in the sense that they compared and responded to signals according to their meaning and not just the acoustic properties (Cheney & Seyfarth 1990).

Predator-specific alarm calls appear to be a common behavioral feature in various primate species. In the Taï forest of Western Ivory Coast, adult guenons readily give alarm calls to two of their main predators, the crowned-hawk eagle *(Stephanoaetus coronatus)* and the leopard *(Panthera pardus).* These alarm calls differ in their acoustic structure both between species and between predators. For instance, the Campbell's monkey eagle alarm calls are acoustically very different from Diana monkey eagle alarm calls, even though both calls refer to the same predator (Zuberbühler 2000a, 2001). Additionally, there is an interesting sexual dimorphism in the acoustic structure between male and female alarm calls (e.g., Zuberbühler et al. 1997; Zuberbühler 2000a). Playback experiments have shown that when hearing male alarm calls to either an eagle or a leopard, female Diana monkeys respond by giving their own predator-specific alarm calls, as if the corresponding predator were present (Zuberbühler et al. 1997).

However, as outlined before, this response could be based on a relatively simple cognitive process, which would make it less relevant for understanding the evolution of the cognitive abilities necessary for language. Complex cognitive processes can be distinguished from more simple ones if there is evidence of flexible behavioral adaptations, in which individuals make informed choices based on mental representations (Tomasello & Call 1997). Although mental representations are notoriously resistant to proper empirical analysis, some progress has been made using various experimental para-

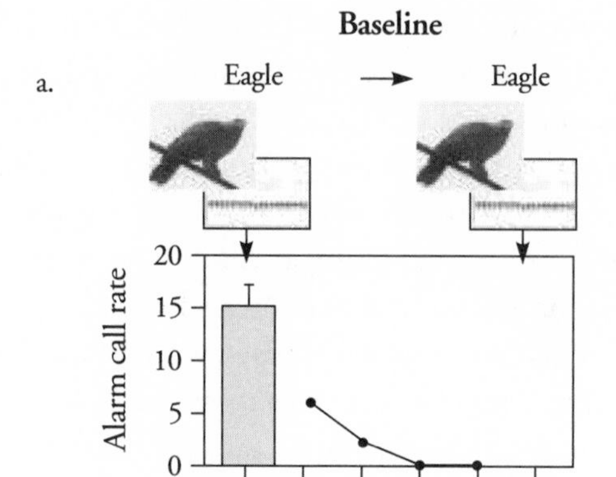
Baseline
a.
Eagle
Eagle
20
15
10
5
0
Alarm call rate
1 2 3 4 5 6
Time (min)

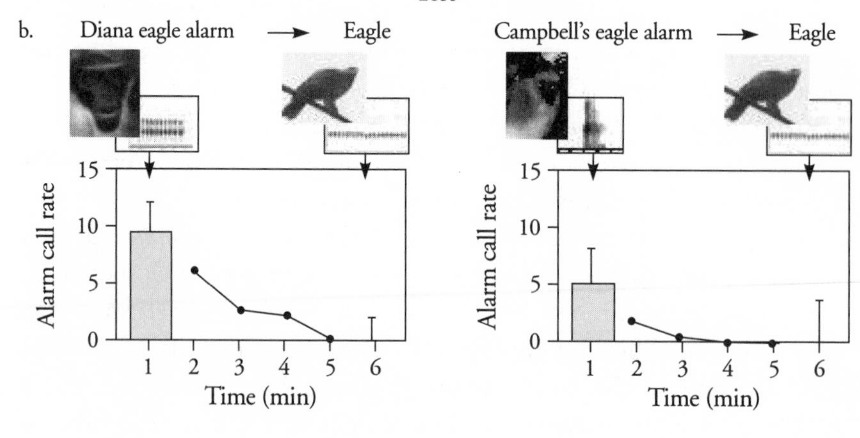
Test
b.
Diana eagle alarm
Eagle
Campbell's eagle alarm
Eagle
15
10
5
0
Alarm call rate
1 2 3 4 5 6
Time (min)

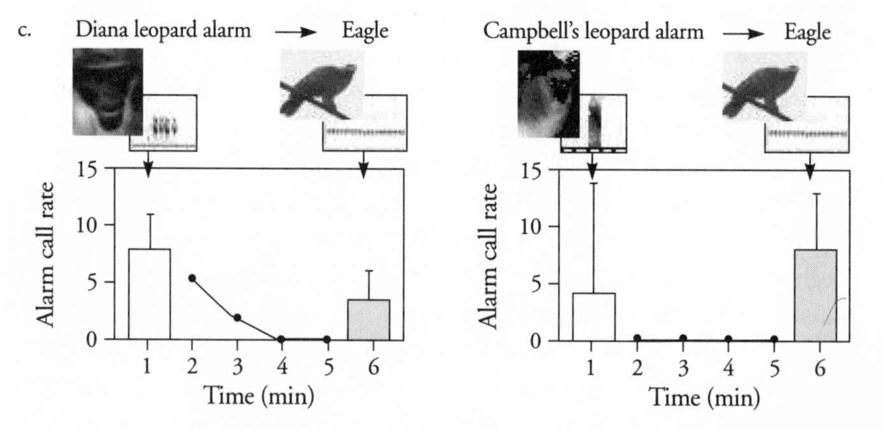
Control
c.
Diana leopard alarm
Eagle
Campbell's leopard alarm
Eagle
15
10
5
0
Alarm call rate
1 2 3 4 5 6
Time (min)

Figure 13A.1. Design and some characteristic results of the prime-probe experiment. Diana monkey groups are tested on two stimuli separated by five minutes of silence. Stimulus pairs differ in similarity of the acoustic and semantic features across conditions as follows: *(a)* Baseline condition: both the acoustic and the semantic features remain the same. *(b)* Test condition: the acoustic features change while the semantic features remain the same. *(c)* Control condition: both the acoustic and the semantic features change. Depicted are the females' responses to eagle probes as a function of priming history (primed with *(a)* baseline: eagle shrieks, *(b)* test: male eagle alarm calls, *(c)* control: male leopard alarm calls). The *x* axis represents time after beginning of the first playback stimulus. The *y* axis represents the median number of calls per minute. Error bars indicate the third quartile. Shaded bars are female eagle alarm calls; white bars are female leopard alarm calls.

digms, such as the match-to-sample technique (e.g., Premack & Dasser 1991). Under field conditions, the choice among these techniques is obviously rather limited and typically restricted to some variant of the classic habituation-dishabituation technique, initially developed for young children (Eimas et al. 1971; Cheney & Seyfarth 1988).

In the following, two studies are discussed that have applied one form of this technique, the prime-probe procedure, to primates living under undisturbed natural conditions. The prime-probe procedure is interesting because it does not expose the subject to a long habituation phase, in which stimuli are presented over and over again until the subject ceases to respond. Instead, it simply provides the animal with some critical information, in this case a predator vocalization or an alarm call. It then tests the effect of this priming experience on the animal's subsequent response to an experimental probe stimulus, in this case a predator vocalization. An animal capable of taking into account the semantic information of the prime stimulus is expected to respond differently to the subsequent probe stimulus than an animal that is not capable of doing so.

To address this issue, experiments have been carried out with various groups of wild Diana monkeys and Campbell's monkeys in the Taï forest of Western Ivory Coast. These groups occupy relatively small home ranges of less than one square kilometer and each group consists of one fully adult male and several adult females with their offspring. Females are philopatric and defend their home range against neighboring groups. Because of the high predation pressure in the Taï forest, both species frequently form mixed species associations to better protect themselves against predation, and some groups spend up to 90 percent of their time in association with each other. As mentioned before, the adult males of both species produce acoustically distinct alarm calls to crowned-hawk eagles and leopards, two of their main predators, and females respond to these alarm calls by producing their own predator-specific alarm calls.

In the first experiment, different groups of Diana monkeys were tested to see whether individuals attended to the acoustic or the semantic features of their alarm calls. The experiment consisted of baseline, test, and control trials. In each trial a Diana monkey group heard two playback stimuli, a prime and a probe, separated by an interval of five minutes of silence. For example, subjects first heard male Diana monkey leopard alarm calls followed, after five minutes of silence, by the shrieks of a crowned-hawk eagle. Across conditions, prime and probe stimuli varied with respect to their acoustic and semantic resemblance. In the baseline condition, both the acoustic and semantic features were alike (e.g., eagle shrieks were followed by the exact same eagle shrieks. See Figure 13A.1.). In the test condition, only the semantic features were alike (e.g., male eagle alarm calls were followed by eagle shrieks), whereas in the control condition both the acoustic and semantic features were different (e.g., male leopard alarm calls were followed by eagle shrieks).

Since both the acoustic and the semantic features were repeated in the baseline condition, a habituation effect was expected. Subjects were expected to produce many alarm calls to the prime, but to produce only few alarm calls to the probe. In the test condition, only the semantic features, but not the acoustic features, were repeated and subjects were expected to show a strong response to the prime and 1) a strong response again to the probe if they attended to the acoustic features, or 2) a weak response to the probe if they attended to the semantic features of the stimuli. In the control condition, finally, both the acoustic and the semantic features were different in the prime and probe, and subjects were expected to show a strong response to both the prime and the probe stimuli.

Results of this experiment clearly showed that the semantic content of the prime stimuli, but not the acoustic features of the playback stimuli alone, explained the response patterns of female Diana monkeys: both eagle shrieks and leopard growls, two normally very powerful stimuli, lost their effectiveness in eliciting alarm calls when subjects were primed with the semantically corresponding male alarm calls. However, if the referent changed between prime and probe, then subjects responded strongly. Figure 13A.1 illustrates the design and the females' responses to eagle probes; results using leopard growls as probes were analogous (Zuberbühler et al. 1999).

Results of a second experiment using Campbell's monkey alarm calls instead of Diana monkey alarm calls were analogous. Campbell's monkey eagle or leopard alarm calls were equally effective as prime stimuli for recipient Diana monkeys as the calls of the conspecific males (Figure 13A.1). Female Diana monkeys only responded to predator vocalizations in cases in which the probe stimuli were semantically novel, suggesting that the acoustic features of the calls were only relevant to refer to the underlying mental representations of the different referential situations (Zuberbühler 2000b). In sum, these results are consistent with the notion that nonhuman primates are capable of processing their calls at a semantic level, similar to how humans process words.

V

Cultural Transmission

The possibility of animal culture is currently about the hottest topic of debate among students of animal behavior. Any rendering of this topic will need to start with the influence of Kinji Imanishi, who put Japanese primatologists on the right track with a fictive debate between a layperson, a biologist, a monkey, and a wasp, which posed the question if animals could have something like culture. Imanishi was the first to answer the question in the affirmative.

It is appropriate, therefore, that this section on animal culture opens with chapters by Tetsuro Matsuzawa and Toshisada Nishida, who each in their own way have been influenced by Imanishi. In the 1950s, well before such studies were undertaken by Western primatologists, Imanishi initiated an extensive program of field studies on monkeys and apes that continues to this day. Apart from work on the one macaque species native to Japan, Imanishi and his colleagues studied primates elsewhere, as far away as India and Africa, always employing the trademark technique of recognized individuals and longitudinal data. Matsuzawa takes us on a historical review of this work, which started with the careful documentation of the spread of sweet-potato washing among monkeys on Koshima Island. This was to become the classic textbook example of an animal "tradition."

After this, Matsuzawa takes us to West Africa for a look at the various traditions established by chimpanzees in the forest of Bossou, Guinea. Here, the author's predecessor, Yukimaru Sugiyama, recorded the cracking of oil-palm nuts with stones, which is considered one of the most complex tool-use techniques known of any animal in the field. Matsuzawa, renowned for his laboratory experiments on numerosity, perception, imitation, and other cognitive abilities in chimpanzees, has taken up the task of finding out how nut cracking is acquired and transmitted in the field. An "outdoor laboratory" has been set up in Bossou, where the chimpanzees spontaneously display their skills while the human observers mark every phase of development of the young. Rather than seeing support for the contingency learning model favored by learning psychologists, the author believes in a comparison with the master-apprentice relationship of Japanese culture. He argues that watching and exposure are more important for cultural transmission than reinforcement contingencies.

Nishida also stresses Imanishi's pioneering influence, and how he inspired the study of African apes. He then takes us to the Mahale Mountains, Tanzania, where he began his research on chimpanzees in 1965, collecting a wealth of information on cultural patterns, from hand-clasp grooming to the leaf-clipping display, and from male bluff displays to the use of special plants for self-medication. Some of the most striking examples of chimpanzee culture come from Mahale, which is so close to Jane Goodall's study site in Gombe National Park (i.e., Mahale and Gombe have a similar ecology and the same subspecies of chimpanzees) that the striking behavioral differences between chimpanzees at the two sites have convinced many of the power of cultural learning.

Particularly intriguing is Nishida's discussion of the apparent lack of "fidelity" in the copying of cultural habits. The author argues that there are often multiple viable solutions to the same problem, hence that there is not always a need for precise behavioral copying. Cultural transmission in chimpanzees is imperfect and flexible, as it probably is in humans as well. Nishida gives as a striking example his own way of using chopsticks, which doesn't follow the standard pattern. There is great individual variability, even if the basic idea of bringing food to the mouth is the same for all chopstick users in the world.

Two case studies accompanying the chapters by Matsuzawa and Nishida illustrate attempts to understand the learning processes involved by means of rather contrasting techniques: one is laboratory experimentation, and the other is correlational treatment of field data. The study by Bernhard Voelkl and Ludwig Huber addresses the issue of imitation. One conclusion from the intense debate about social learning has been to restrict the term "imitation" (which used to refer to almost any kind of behavioral duplication) to cases in which the observer copies not only the location where the model obtains rewards, or the goal of the model's actions, but the precise technique employed. By exploiting the same individual variability noted by Nishida, Voelkl, and Huber demonstrated "true imitation" in marmoset monkeys—thus contradicting an entire literature that has questioned this possibility in most or all animals, including aping in apes.

The study by Stephanie Pandolfi, Carel P. van Schaik, and Anne E. Pusey, on the other hand, explores the social circumstances under which male and female chimpanzees fish for termites. It demonstrates strong effects of partner presence, which effect is quite different for the two sexes. When studying cultural transmission, such data help us understand the settings in which learning takes place, and why one sex may become more skilled than the other.

William McGrew takes up the larger issues surrounding cultural primatology. McGrew himself is one of the recognized founders of this field: he was one of the first to document different customs and technologies in chimpanzee communities at different sites in the field. He brings his background in anthropology to bear on the question of whether or not apes fit the cultural definitions of his discipline. These definitions run from explicitly exclusive ones, which state that culture is what makes us human, to inclusive ones, such as those that only require socially learned differences in behavior between groups or populations. The term "culture" obviously has many meanings to many scientists.

Should we make the mechanism of transmission our chief criterion of culture, as some have proposed? In such a definition, culture requires full-blown imitation, perhaps even active instruction and language. Or, should we accept all socially learned behavior as cultural? The first definition may be overly restrictive, the second overly inclusive. McGrew is positioned as no other to give us dispatches from the culture wars, as he has been in the middle of this heated debate for decades, and has witnessed the pendulum swing from "animal culture" as an absurd proposal to one that is taken seriously by many, both inside and outside the field of animal behavior. With humor and some distance, he questions the "obvious truths" about culture that social scientists still cling to unhampered by knowledge of recent developments in the field of animal behavior.

Case Study 16A by Eduardo Ottoni and Massimo Mannu highlights a semifree-ranging population of capuchins—considered the most dexterous monkeys—that has mastered the cracking of nuts with stones, thus matching what many consider the high point of chimpanzee culture. Their observations confirm that we have only scratched the surface of cultural habits in the animal world: many more group-specific behavior patterns no doubt remain to be discovered. And even if many of the most striking examples come from primates, there is absolutely no reason to believe cultural phenomena are limited to this taxonomic group. All that is needed for culture is good learning abilities and a social setting in which habits can be transmitted. As the remaining contributions to this section illustrate, this applies to a wide range of animals.

Hal Whitehead introduces us to his study animal, proudly announcing it as the species with the largest brain on Earth: the sperm whale. It is an animal that is hard to fathom for the human scientist since it lives in such a different environment, bears little resemblance to us, and will never let its behavior be brought under experimental control. The author then moves to the issue of animal culture, essentially taking the same position as other authors in this volume, namely that the definition of culture is not to be restricted to certain kinds of learning. What matters is that knowledge and habits are being socially transmitted: how this is done is hard to determine. It is also secondary when it comes to the survival value of learning from others.

Based on his analysis of whale mitochondrial DNA, Whitehead introduces the intriguing idea of cultural "hitchhiking." The co-evolution of cultural and genetic traits is discussed in the context of female whales passing on successful cultural features to their offspring, which features then make their offspring out-reproduce other whales, hence spread their mother's DNA. Cultural transmission may thus assist genetic dispersion, hence explain the low mtDNA diversity in matrilineal cetaceans. As an illustration of the behavioral variability in whales—even within the same species in essentially the same environment—Harald Yurk compares populations of killer whales. His review makes it clear that remarkable differences exist, but also that it will be hard to decide how precisely they have arisen.

In the final chapter, Meredith J. West, Andrew P. King, and David J. White rely on their extensive research experience with two bird species of "ill repute"—cowbirds and European starlings—to illuminate song development. Their work illustrates nicely why it would be foolish to define cultural learning too narrowly. Behavioral transmission can occur in rather unexpected ways. It would seem logical to think, for example, that the transmission of male song takes place from male to male, that is, that regional "dialects" are based on males from one region getting their song from males in the same region. The authors, however, show that this transmission is mediated by females: the female's preference for certain song patterns shapes the way males sing. This is a wonderful case of indirect transmission—quite a bit more complex, it seems, than straightforward imitation.

The authors conclude with a most inspiring notion of animal culture. Culture is not merely a set of discrete behavior patterns, knowledge, and skills transferred from models to observers, or from masters to apprentices, but rather cultural transmission is a totally interactive process in which attention is paid not only to what others do but also to how others react to what oneself does. Culture is thus an emergent property of social interaction. The authors even show how this idea may apply to language learning, thus linking what some consider the most complex learning process in our own species to the way the young cowbird learns to become a cowbird and the young starling to become a starling.

14

Koshima Monkeys and Bossou Chimpanzees: Long-Term Research on Culture in Nonhuman Primates

TETSURO MATSUZAWA

Different disciplines provide different definitions for the term "culture," which has become a hot topic of discussion in the study of nonhuman primates, and of animals in general (de Waal 2001). For our purposes here, bear in mind the following loose definition of culture: "a set of behaviors that are shared by the members of a community and are transmitted from one generation to the next through nongenetic channels" (Matsuzawa 1999). In this chapter I describe two long-running research projects: the study of Japanese monkeys at Koshima Island, Japan, and of chimpanzees at Bossou, Guinea, West Africa. By summarizing what has become clear about culture in monkeys and chimpanzees, I would like to suggest how primate origins underlie the emergence of human culture.

The existence of "culture" in nonhuman animals was first illustrated by the now well-known observations of Koshima monkeys. The idea of culture in nonhuman animals had been developed several years earlier by Kinji Imanishi (Imanishi 2001[1941]), the founder of primatological research in Japan. When research at Koshima first began, Japanese scientists gave each monkey a name and assumed that each had a different personality. They did not hesitate to use anthropomorphic terms when seeking to understand the monkeys' behavior—an approach that may partly be attributed to Japan's cultural traditions. Japan is unique among the highly industrialized nations of the world in having its own indigenous primate. There are no wild monkeys in North America and Europe. The Japanese are raised with a folklore rich in fairy tales in which monkeys feature prominently. Perhaps as a result, the Japanese worldview has little in common with the Western dualism of humans versus animals. According to Buddhist tradition, life is an eternal cycle, "Rinné": a human may be reborn as a dog or an insect, and any animal can have the prospect of existing as a human in its next life.

Recently, Westerners have come to acknowledge the existence of culture in nonhuman primates, mainly through the accumulation of knowledge about regional differences in chimpanzee behavioral repertoires (McGrew 1992; Wrangham et al. 1994; Whiten et al. 1999). The most recent review of chimpanzees in Africa listed 54 cases of reported tool use in a feeding context from 14 different study sites (Yamakoshi 2001). Each community has developed a unique set of behaviors in feeding, greeting, and communicative contexts. Among these, the chimpanzees of Bossou are well known for their versatility as tool makers. Through the studies carried out at Koshima and Bossou, the present paper aims to relate the findings to illuminate important aspects of culture in nonhuman primates.

Sweet-Potato Washing by Koshima Monkeys

It was December 3, 1948. World War II had been over for just three years. Kinji Imanishi (1902–1992) set off with his two students, Shunzo Kawamura (1924–) and Jun'ichiro Itani (1926–2001), to Koshima Island, where they were to begin a long-term study of wild Japanese macaques, *Macaca fuscata,* also known as "snow monkeys." Imanishi was 46 years old at the time, a lecturer at the Department of Zoology of Kyoto University. Kawamura and Itani were both undergraduate students. Imanishi, who began his career as a field entomologist, had in the 1930s discovered the phenomenon of "habitat segregation" in four species of mayfly larvae cohabiting a single stream. He expanded his idea to come to look upon evolution as the noncompetitive coexistence-as-a-whole of various kinds of organisms (Imanishi 2001[1941]; see also Chapter 15).

The three scientists began their fieldwork by observing the behavior of wild Japanese monkeys, giving names to all of the individuals in order to identify each one. Their goal was to unravel the social structure of the groups, as a way of seeking the evolutionary origins of human society. Imanishi organized a research team of ten, including himself, who visited 19 research sites in Japan—Takasakiyama and Yakushima among them—to study wild populations of Japanese monkeys. By 1955, in the first seven years of the study, they had sent out about 90 parties and spent a total of about 1,500 days observing wild monkeys.

Their attempt was so unusual that they succeeded in uncovering various hitherto unknown aspects of the monkeys' social behavior. Among their most important findings were the following:

1. The monkeys lived in communities called "troops," each of which consisted of multiple males and multiple females and moved from place to place as a group.
2. There is a breeding season. The monkeys copulate in autumn to winter and give birth in spring to early summer.
3. Society was matrilineal: males migrated to other communities whereas females remained in the troop.
4. Neighboring troops were separate, exclusive entities. Over the years, a troop could split into two different communities but not the reverse.
5. Members of the troop were ranked according to their dominance standing: some individuals were dominant, as were certain family lines, while within the same matrilineal family, the youngest individual was more dominant than older siblings—a phenomenon that became known as "Kawamura's law."
6. Itani (1963) recognized that the monkeys possessed a set of distinctive vocalizations. He identified six groups of 37 different vocal sounds, such as coo calls and alarm calls, emitted in different social contexts.

Among the numerous findings of Imanishi's team, the best known concerns the cultural propagation of sweet-potato washing. Sweet-potato washing (SPW) is a behavior in which monkeys carry a sweet potato to the shore and wash the sand off it with water (Figure 14.1). This behavior was first seen in the summer of 1953, performed by a female named Imo ("sweet potato" in Japanese), who was one-and-a-half years old at that time. The initial observation was made by Satsue Mito (1915–), a local collaborative researcher, and was later confirmed by Kawamura. Mito and her father had been dedicating themselves to the conservation of the wild monkeys in Koshima, now regarded as "national heritage" monkeys. The finding was first

Figure 14.1. Sweet-potato washing by a Japanese monkey at Koshima, Japan. Photograph by Satoshi Hirata.

reported by Kawamura in 1954 in Japanese. English-speaking readers will be more familiar with Masao Kawai's (1924–) widely cited follow-up paper, which described the propagation process in detail, along with reports of three other newly acquired behaviors: wheat placer mining behavior, snatching behavior, and bathing behavior (Kawai 1965). Cultural propagation among Koshima monkeys is a topic recently revisited by Kawai and his junior colleagues (Hirata et al. 2001b).

The SPW behavior gradually spread to other members of the community. The first five years' record shows that the acquisition rate in adults more than eight years old was only 18 percent, or two out of 11 (six males and five females), both of whom were females. The rate in young monkeys aged between two to seven years old was much higher: 79 percent, or 15 out of 19 (10 males and nine females) acquired the behavior. After that, most newborns began to show SPW. In sum, the younger generation was sensitive toward the new invention, whereas adults rarely adopted the behavior.

The propagation process was clear, involving two main channels: through kinship and through playmate relations. In the fourth year following the invention, 10 individuals including Imo performed the SPW behavior. One route of transmission was along the family line: after Imo's invention, her mother (Eba) adopted SPW behavior in the same year. Imo's brother, Ei, younger by two years, took up SPW at the age of one to one-and-a-half. The other route led to Imo's playmates. All seven remaining sweet-potato-washing individuals were Imo's playmates, no more than a year younger or older, and acquired the skill at the ages of two to five years. These data suggest a critical period for learning SPW.

The SPW behavior became fixed in the troop during the years 1958–1959. By this time almost all infants were acquiring the skill. The learning process seemed to consist of the following stages. First, infants strengthen their affinity to water from birth onward by being dipped in water or by having water splashed over them while the mothers engage in SPW. Second, infants at the age of six months begin eating fragments of potatoes that their mothers drop in the water. Finally, they acquire the skill at the age of one to two-and-a-half years old. In terms of learning mechanisms, there are no explicit observations that chart the process of acquisition, partly because of constraints on field observation. The prolonged exposure to the mother's SPW behavior may be important in providing a setting for learning. However, there is no evidence that the mothers' role extends beyond simply increasing opportunities for individual learning by the infants. (For debate among Western scientists about the mechanism of transmission most likely to have occurred at Koshima, see Galef 1990, and de Waal 2001.)

The Koshima troop also performs another cultural behavior called "wheat placer mining" (WPM, or "wheat washing"), which resembles gold mining. As part of their provisioning scheme, researchers scattered grains of wheat around the beach. At first, monkeys ate them by picking up one grain after the other—a time-consuming effort. Then, one monkey began to gather up the grains of wheat together with some sand, take them to the shore, and throw them into the water. The advantage of this method was clear: water easily separated the grains from the sand, and while the latter sank to the bottom, the grains floated to the surface.

The inventor of this behavior, first observed by Kawamura in 1957, was once again Imo. The propagation process for WPM was similar to that observed for SPW: through family lineage and playmate relations. The first followers were Ego (Imo's younger sister by five years, who began to display WPM at the age of one-and-a-half), Enoki (Imo's younger sister by four years, who began at the age of two-and-a-half), and Jugo (a male playmate of Imo, who began at the age of four). The WPM behavior was mostly acquired by monkeys aged two to four years: a little over the age at which SPW first appears, probably because of the relative complexity of the WPM.

Kawai and his colleagues conducted intensive follow-up studies. An interesting case was "pool-making," which was an especially efficient variant on the WPM behavior. When grains of wheat are scattered on the beach while it is still wet at low tide, some monkeys dig out the sand and make small pools from the water that oozes up. They then dip a piece of sweet potato or sweep nearby grains of wheat into the pool before they eat them. Building on previous innovations (a process labeled a "ratchet effect" by Tomasello et al. 1993), cultural development thus continues among the Koshima monkeys.

There are three key points to note when looking back at the origins of cultural behavior in Koshima monkeys. First, the natural environment had an important role to play. It provided the water and the sand, the key materials for the cultural behavior. The island is small, only 32 hectares, and provides poor nutritional conditions, reflected in high first-year mortality. Nearly 40 percent of the offspring die within their first year at Koshima (Watanabe 2001). Koshima monkeys were originally inhabitants of the mountain forests, but adapted to a new ecological environment by inventing new behaviors. Second, the cultural behaviors described above derived from the monkeys' foraging behavior and developed from their relation to human activity. That is, they came about as a result of sweet-potato and wheat-grain provisioning by humans. Such interaction between monkeys and humans facilitated the innovation. Provisioning was responsible for population expansion from 22 in 1952 (when provisioning first began) to about 120 in 1971, just before provisioning ceased (Watanabe et al. 1992). This inevitably increased competition for limited food resources. Third, there was an originator—Imo. Her role was a very significant one in the history of Koshima. All cultural behavior should have its origins.

Imo, the originator of SPW and WPM, died on May 21, 1972. She was 20 years old—the average life span of a Japanese monkey—and had given birth to six sons and three daughters. Long-term study at Koshima still continues. It is now in its sixth decade, and has so far chronicled the history of seven generations of wild monkeys.

Records at Koshima document the lives of a total of about 470 monkeys until the year 2001 (Watanabe 2001, personal communication). None of the monkeys who were present at the emergence of the cultural behaviors are still alive today. However, their descendants continue to dip sweet potatoes into the sea, and throw grains of wheat into the water. The behaviors have been transmitted over several generations.

Nut Cracking Using Stone Tools by Bossou Chimpanzees

On February 4, 1958, Imanishi and Itani arrived in Nairobi, Kenya. After ten years of accumulating knowledge of Japanese monkeys, they aimed to begin work on the African great apes in the wild. In search of a suitable research site, they visited Tanzania, Uganda, Zaire, Congo, and Cameroon. They first encountered wild mountain gorillas in Uganda and lowland gorillas in Cameroon. On the way back from Africa, the two visited various European countries as well as the United States, carrying with them a film of wild Japanese monkeys and the first volume of *Primates,* the international journal of primatology that they founded (which still exists today). They met with C. R. Carpenter and other distinguished scholars, and opened a window to the West.

Imanishi, and later Itani, continued to send graduate students to Tanzania, East Africa each year. Among them, Kosei Izawa (1939–) reached the Mahale Mountains in 1965, and Toshisada Nishida (1941–) succeeded in provisioning the Mahale chimpanzees in 1966. Research at Mahale in Tanzania—a site as well known as Gombe (Goodall 1986)—is now in its 36th year (Nishida 1990). This research site has revealed many interesting cultural variants in East African chimpanzee behavior (see Chapter 15). Another important research site is operated by Japanese scientists at Bossou, Guinea, West Africa. Dutch researchers were the first to arrive in Bossou in 1960, but they only carried out brief surveys lasting no more than a few weeks at a time in 1965, 1967, and 1969 (Kortlandt 1962, 1986). In 1976, Yukimaru Sugiyama (1935–) began his work at the same site, and since then research at Bossou has been running continuously. The past 25 years have provided a steady stream of discoveries from this unique community of Western chimpanzees (*Pan troglodytes verus*), starting with the first scientific report on nut cracking, i.e., behavioral observation of chimpanzees using a pair of stones as hammer and anvil to crack open oil-palm nuts (Sugiyama & Koman 1979). There are only six major research sites around Africa where long-term studies of chimpanzees in the forest have been carried out: Gombe, Mahale, Budongo, Bossou, Kibale, and Taï (Boesch & Boesch-Achermann 2000). Two of these are run by Japanese researchers, initiated largely by Imanishi and his colleagues.

I am Imanishi's junior by about 50 years, and belong to the second generation in the history of Japanese primatologists. The first generation, Imanishi and his students, struggled to establish research sites in the wild. As an intellectual descendant of Imanishi and Itani, I have been developing my own research niche: the study of the chimpanzee's mind rather than society (Matsuzawa 1985, 2001). I have been conducting long-term laboratory research on chimpanzee cognitive behavior, referred to as the "Ai-project," since 1978 (named after my first research subject). Through my research, and probably thanks to my own cultural background, I felt the need to explore the behavior and the ecological environment of chimpanzees by visiting their natural habitat. I joined the field study of Bossou chimpanzees in 1986 as the second researcher after Sugiyama, and have continued to visit Bossou once a year ever since. I have been interested in the chimpanzees' use of tools, focusing especially on the developmental processes underlying these skills, and the cultural variation that exists among neighboring communities (Matsuzawa 1994, 1999, 2001).

At Bossou, one can regularly observe five types of tool use: nut cracking with stones, pestle pounding of oil palms, algae scooping with a stick, ant dipping using a wand, and the use of leaves for drinking water (reviewed by Matsuzawa 1999). In addition to these, we are discovering unique traditions each year, thereby steadily extending the list of tools that are unique to this community. Using leaves as cushions (Hirata et al. 1998) and capturing and playing with hyraxes (Hirata et al. 2001a) are among these recent additions.

My colleagues and I have carried out a field experiment on tool use in an outdoor laboratory located on top of a hill that forms a core part of the chimpanzees' habitat. It is a place that the group regularly traverses and where they appear to be at ease. We provided nuts and stones for nut cracking, water in an artificial tree hollow for leaf sponging, and for ant dipping we attracted Safari ants to the area by laying out fresh oil palms or dead insects. Such experimental manipulations had minimal effects on the natural environment—for instance, in the absence of rainwater filling up the hollow, we provided fresh water ourselves every day—but succeeded to drastically increase our chances for close-range observation of tool use and for the video recording of tool-using episodes to be used in further, detailed analysis. This

Figure 14.2. Nut cracking by chimpanzees at Bossou, Guinea, West Africa. Photograph by Tetsuro Matsuzawa.

setup also gave us the opportunity to compare different types of tool use being performed at the same site at the same time (Figure 14.2).

In the following section, I will focus on some important aspects of the cultural tradition of tool use by chimpanzees at Bossou in relation to that of Koshima monkeys: ecological environment, critical period for learning, and education by master apprenticeship.

Ecological Environment

Chimpanzee tool use has been regarded as opportunistic, having a trivial effect on subsistence (Mann 1972). However, a year-round study revealed that the chimpanzees at Bossou spend 12 percent of their total annual feeding time on these tool-using behaviors (Yamakoshi 1998, 2001). There is a marked seasonal fluctuation in flowering and fruiting in tropical forests. A negative correlation exists between fruit availability and frequency of tool use at Bossou. When fruit becomes rather scarce in the summer rainy season, chimpanzees spend 32 percent of their feeding time on nut cracking and pestle pounding. Nut cracking gives them access to the nutritious kernel inside the nut's hard shell. Pestle pounding involves pulling out palm fronds and eating the base, followed by pounding of the apical meristem of the tree with the fronds in order to extract the juicy pith. Both nut cracking and pestle pounding concern products of the oil palm.

We can estimate energy intake obtained through tool use. Oil-palm nuts weigh 7.2 grams on average (SD = 2.2 g); of this, the kernel makes up about 2.0 grams (SD = 0.4 g). The nutritional energy of the kernel is 663 kcal/100 grams—close to that of walnuts—and rich in fat. Based on data from four years (1990–1993; 340 parties and 6,564 minutes of direct observation and video-recording time, over the course of 133 days), chimpanzees remain at the outdoor laboratory for about 30 to 60 minutes in a feeding bout (occasionally more than two hours) to crack open nuts. The adults crack open three nuts a minute on average, such that over the course of a single bout, they will consume 90 to 180 nuts in total. Total energy intake is thus about 1,200 to 2,400 kcal for each feeding bout.

It must be noted that although oil palms *(Elaeis guineensis)* occur naturally at Bossou, they are also important agricultural products for humans. In a sense, Bossou chimpanzees subsist on oil palms much like the local Manon people. In general, chimpanzees prefer the secondary forest over primary forest. The habitat of the Bossou community, in particular, is very small: at about six square kilometers, it is much smaller than habitats of other known chimpanzee communities across Africa. The area has been home to a group of 19 chimpanzees on average (with a minimum of 17 and maximum of 21 over the past 25 years). Bossou chimpanzees also derive some of their food from farm produce, such as rice (chewing on the grassy stem to obtain the juice within, rather than eating the grain), oranges, and mangos. Oil palms do not grow in the core area of the forest, but are distributed around the periphery. Bossou chimpanzees may have acquired various kinds of tool-using skills that specifically target these nutritious oil palms when they expanded their habitat because of population pressure.

Critical Period for Learning

Chimpanzees at Bossou begin to crack nuts at three-and-a-half to five years of age. In sum, acquisition takes place through a number of successive stages. First, infants start to manipulate a single object, such as a nut or a stone, at the age of one. They then proceed to manipulate two objects simultaneously or successively at the age of two. By this stage they can place a stone on another stone, or place a nut on a stone, but then instead of using a hammer, they stamp the nut with their foot. At the age of three, they eventually begin to manipulate three objects (a nut, a hammer stone, and an anvil stone) in some sort of sequence, but the order is still inappropriate. They sometimes hold in one hand a hammer stone with which they hit the anvil stone, but all the while holding the nut in the other hand! The use of leaves for drinking water first emerges at two-and-a-half years of age, one year earlier than nut cracking. The difference partly reflects the complexity of object manipulation that is involved in the skill, although in the case of tool-using activities other than nut cracking, essential data on acquisition is still lacking. For further details, see Inoue-Nakamura and Matsuzawa (1996) for the development of stone tool use from birth in the Bossou community, and Boesch and Boesch-Achermann (2000) for the acquisition age of nut cracking for a different nut species with different techniques in the Taï community.

According to the tree structure analysis of tool use in terms of "action grammar," leaf sponging and almost all the examples of chimpanzee tool use we know of represent "level 1" tool use—only a single relationship between objects exists, that is, one object is related to another (see Matsuzawa 1996 for the concept of "action grammar" and Chapter 9 for its application to the social domain). On the other hand, nut cracking is a rare example of a more complicated form of tool use. Chimpanzees use a hammer stone to crack open a nut that is placed on an anvil stone. This type of action is labeled "level 2" tool use, as there are two kinds of relationships among the objects involved: the nut is related to the anvil stone in the positioning phase, and the hammer stone is related to the nut in the cracking phase. Perhaps the most complex form of tool use ever encountered among nonhuman primates is what I have called the "metatool" of chimpanzees. Three individual chimpanzees utilized a third stone as a wedge to stabilize the anvil stone. Metatool use is rare partly because it is not necessarily the best and only solution to the problem: so far chimpanzees invented lots of other solutions such as rotating (upside-down or clockwise/counter-clockwise) the anvil stone on the ground to get the best surface, holding the anvil up with a supporting foot (a solution limited to adults), and precisely locating an oil-palm nut (oval like a rugby ball) along the slant surface (using friction), or placing it at the tip of the anvil stone.

The minimum age at which metatool use was exhibited was six-and-a-half years old. I carried out a comparative study with local Manon children and found that nut cracking started at the age of three whereas metatool use only emerged around six years of age—the results are thus comparable to the chimpanzee findings. It is impossible for chimpanzees more than six years old to master the nut-cracking skill. Within the Bossou group, there are two adults (Nina and Pama) and two juveniles (Yunro and Juru) who are unable to crack nuts. In the case of the Taï chimpanzees of the Ivory Coast, all members of the community crack nuts (Boesch, personal communication). This major difference might be a result of the critical complexity of the behaviors. Bossou chimpanzees' nut cracking represents true "level 2" tool use where two stones are used as a set of tools to pound a nut, whereas Taï chimpanzees' tool use is somewhere between "level 1" and "level 2": they often utilize roots of trees (a substrate) as the anvil. The two communities also differ in patterns of hand preference. Bossou chimpanzees show perfect laterality, with each individual always using one hand for manipulating nuts and another for handling a stone hammer ("right handers" make up about two-thirds of the population), whereas Taï chimpanzees often employ a wooden hammer and use both hands together for hammering. In sum, Bossou chimpanzees show extremely complex tool-use abilities, involving the coordination of an asymmetrical manual skill.

Mothers are the most important models for observation, but are not always the sole resource for the observer. Both of the non–nut-cracking mothers have given birth and succeeded in raising four offspring over the past 15 years. All four young mastered the nut-cracking skill. One of the daughters (Nto) who learned to crack nuts has even allowed her mother (Nina) to take kernels from her. The mothers of the two juvenile non–nut crackers are themselves skillful nut crackers.

Education by Master Apprenticeship

According to our analysis of observed behavior (Biro & Matsuzawa, in preparation), adults were not only the most likely to be the targets of observation by other members of the group, but also the least likely to be observers themselves. On the other hand, juveniles were rarely targets but often observers, both of adults and of juveniles. In sum, chimpanzees showed a strong tendency to pay attention to the stone-tool use of conspecifics in their own age group or older, but largely ignored younger models. This observation means that cultural innovations are more likely to spread horizontally or vertically/

orthogonally downward, but not upward as is sometimes observed in the case of humans. In this respect, the channels of social transmission and the learning mechanism involved should be different between the sweet-potato washing in Koshima monkeys and the stone-tool use in Bossou chimpanzees. The learning process underlying the acquisition of nut cracking has three features: long-term exposure from birth, high tolerance with no formal instruction from mothers, and intrinsic motivation of infants for imitation independent of direct food reward.

First, chimpanzee infants are exposed to the nut-cracking activities of others for a long time prior to making their own initial attempts. Most mothers practice nut cracking on an everyday basis—it really does comprise a tool for survival. Such long-term pre-exposure is likely to be essential for mastering a complex skill, much like long-term exposure to speech sounds precedes production of real speech in human infants.

Second, there is no active teaching, no molding, and no verbal instruction from the mother. However, chimpanzee mothers do show a high tolerance to the apprentice observers. Infants are allowed to steal nuts from their mothers. I have seen a mother allowing her infant to continuously steal the nuts over the course of seven consecutive bouts of cracking. Two further episodes may help to clarify the relationship between master and apprentice. In one, a four-year-old infant interrupted the mother's hammering, stole the nut to be cracked from her anvil, and went on to crack it on her own. In another, a three-year-old infant picked up a nut from the ground after having watched her mother's nut cracking for a long time. The infant moved toward the mother and placed the nut on her anvil stone; the mother briefly stopped in mid-motion, then proceeded to crack the nut. The infant removed the kernel from the anvil and ate it.

Third, learning does not depend on direct food reward. As described above, infants can easily obtain kernels from their mother. Moreover, all attempts by infants to manipulate nuts and stones fail to result in even a fraction of a kernel until the age of three-and-a-half to five years. In a sense, practicing the manipulatory behavior invariably leads to failure and no reward. However, youngsters continue to try their hand at nut cracking. Motivation thus seems to be intrinsic and drives infants to attempt to produce a copy of the mothers' (or the masters') behavior.

Taken together, the mother-infant interaction during nut cracking reminds me of the Sushi master-apprentice relationship long established in Japanese cultural tradition (Matsuzawa 2001). The master is the model who demonstrates in front of the apprentice, but provides no further guidance. The apprentice continues to observe the skill. Such prolonged exposure without formal instruction and reward or punishment may be essential for acquiring complex skills.

Conclusion

The study of Koshima monkeys showed us the three important phases of cultural tradition: emergence, transmission across generations, and modification. The study of Bossou chimpanzees revealed more sophisticated aspects of culture that are clearly different from that in monkeys. Culture in monkeys appears to be limited to feeding behavior, whereas culture in chimpanzees is not restricted to feeding, but includes tool use (even in the context of feeding), communication, and other habits and skills. Learning mechanisms responsible for acquisition are also different. The education by master apprenticeship described above is a clear feature of chimpanzee social learning, but is not evident in monkeys. Humans share many aspects of nonhuman primate culture, too. However, when we address "culture" in human societies, we refer to an extensive range of substances and levels, many of which are different from those in monkeys and chimpanzees. The word "culture" includes literature, art, music, religion, and so forth. It includes ethical values, too. In the case of human culture, those who do not follow the cultural tradition may become social outcasts. Further research is necessary to clarify the primate origins of human culture. The year 2002 marks the centenary of the birth of Kinji Imanishi, whose pioneering spirit of striving for new discoveries through fieldwork should be an inspiration to us all. It is already an important cultural tradition in the study of primates in Japan.

CASE STUDY 14A

Movement Imitation in Monkeys

BERNHARD VOELKL AND LUDWIG HUBER

Within social learning, imitation is perhaps one of the most complex learning mechanisms and plays an important role in the development of behavioral traditions (overview in Heyes & Galef 1996). Imitation consists of response learning by observation, i.e., learning how to move the body by observing the behavior of others. Other varieties of social learning consist of stimulus learning by observation; these are means of acquiring information about the static or dynamic properties of objects; about their value, location, or motion (Heyes & Ray 2000). Many observations of animal social learning that were described as imitation in the former literature do not fit this criterion and could also be explained by attentional or motivational effects. Robust evidence for imitation was only found in a few species, including great apes, parrots, and dolphins (Heyes & Galef 1996; Tayler & Saayman 1973). In monkeys only Bugnyar and Huber (1997) have claimed to show evidence for imitation, using marmosets *(Callithrix jacchus)* as experimental subjects. As it seemed implausible for us that parrots, dolphins, and apes do but monkeys do not imitate, we attempted to corroborate these earlier findings.

In Voelkl and Huber (2000) we permitted two groups of marmosets to observe a conspecific model using one of two alternative techniques to open plastic film canisters (pulling with a hand versus pulling with the mouth, Figure 14A.1). These different techniques arose spontaneously in the models, and were exploited by us for the experiment. Responses of the observers were compared with one another and with a control group of marmosets that

a.

b.

Figure 14A.1. The two opening techniques performed by *(a)* the hand-opening model and *(b)* the mouth-opening model.

were never given the opportunity to observe models. The particular requirement of the present experiment was that the two opening techniques had the same effect, that is, the lid to be opened would "behave" identically in both groups. Furthermore, one technique consisted of a behavioral peculiarity (mouth opening), that is, mouth opening was neither common in the animals under investigation nor necessary for lid removal. This requirement ensured that if the technique was performed by the observers, they had copied a novel behavior.

First, a male was trained to open the canisters with his hand. This subject and another animal that showed the spontaneous peculiarity of mouth opening were used as models. Then, six others were given the opportunity to observe the mouth-opening model (group Mouth) while five others observed the hand-opening model (group Hand). Both models were familiar to the observers. After five observation sessions all subjects were tested individually in two successive test sessions. The control experiment involved 11 naive individuals that had access to the canisters without previous opportunity to observe a social model (group Nonexposed Control). As we used the same canisters for all subjects we controlled for scent-mediated local enhancement, i.e., the influence of olfactory cues on the subjects' test performance, by conducting a second nonexposed control experiment. Fourteen naive subjects (group Olfactory Control) had access to the canisters that were previously opened (in visual isolation) by another animal by mouth.

In the first test session all subjects opened at least one canister with their hand. Interestingly opening the canisters with the mouth was shown only by four out of six subjects of group Mouth but by none of the five subjects of group Hand. Despite the small group size this difference was statistically significant (Fisher's exact test: $N = 11$, $P = 0.045$, one tailed).

In the second test session the canisters were shut tightly and could only be opened by mouth. This session was conducted to investigate if the subjects that used only their hands in the first session were also able to open canisters with the mouth. From the two subjects of group Mouth that did not use the mouth in the first session, one immediately began to open the canisters with its mouth, whereas none of the subjects from group Hand managed to open canisters with its mouth. Thus in the second test session five out of six subjects of group Mouth opened the canisters with the mouth whereas none of the five subjects of group Hand succeeded in opening a canister (Fisher's exact test: $N = 11$, $P = 0.013$, one tailed).

To summarize the main results, marmosets that had observed a mouth-opening model were more likely to use this technique to remove the lid from a film canister during a subsequent test session than marmosets that had observed a hand-opening model. The nonexposed control tests revealed that spontaneous mouth opening is quite rare, as it was observed only in two out of 30 subjects that had never seen a mouth-opening model or any opening demonstration before (group Hand, group Nonexposed Control, and group Olfactory Control). It is also unlikely that olfactory cues influenced the direction and strength of the observers' approach behavior. The olfactory control subjects opened the canisters significantly more often with their hands than with the mouth. In sum, these results cannot be accounted for by any nonimitative learning effect known in the literature as local or stimulus enhancement (Spence 1937; Thorpe 1963), object movement re-enactment (Custance et al. 1999), affordance learning (Gibson 1979), or contagion (Thorpe 1963). Therefore these results confirm that marmosets are capable of learning simple movement patterns from a conspecific by imitation.

This study implies that learning by imitation is more widespread in the animal kingdom than previously assumed. However, the considerable number of studies that failed to demonstrate imitation in monkeys as well as our own experiences from other studies on social learning suggest that imitation may be a robust but rarely used mechanism. More experiments and observations in the field are required to determine how often and under what circumstances monkeys learn a behavior by imitation.

15

Individuality and Flexibility of Cultural Behavior Patterns in Chimpanzees

TOSHISADA NISHIDA

Japanese studies of nonhuman primate culture started with Kinji Imanishi's (1952) prediction that not only humans but also other animals, in particular nonhuman primates, have cultures, or "acquired behavior that is socially acknowledged." Imanishi was the founder of Japanese primatology as well as a pioneer in international primatology after World War II. In 1937 he wrote a book in Japanese, *The World of Living Things* (1974[1941]), in which he expressed his concern about the structure of the living world and its continuity. He was also interested in the origins of societies, ranging all the way from insects to humans.

Imanishi's contributions to biosociology were based on three core convictions. First, he believed that human beings should be considered animals, discarding the conventional human-animal dichotomy. Second, he argued that "species society," which he called "specia," exists in all creatures, both so-called social and nonsocial animals and even plants. Third, he asserted that the concept of "culture" should be applied to animals as well as to humans and called his theoretical framework "cultural biology."

Inspired by Imanishi's approach, Shunzo Kawamura, Jun'ichiro Itani, Masao Kawai, and other students began to look for "cultures" among Japanese monkeys (de Waal 2001). This was not difficult because Imanishi's team had been traveling throughout Japan to habituate many troops of Japanese macaques. They did so in response to the demand for the monkeys as experimental animals in medical research and the desire of local communities to earn revenue by establishing monkey parks.

They had found interesting behavioral differences between monkeys at different localities. An intensive study of the potato-washing and wheat-washing traditions of the monkeys on Koshima Island (e.g., Kawamura 1959; Kawai 1965) had a tremendous influence on subsequent worldwide studies of primate culture and learning (Nishida 1987). Thus, by the mid-1960s, various aspects of Japanese monkey cultures had already been elucidated. These were mostly published in Japanese (e.g., Kawamura 1965; Kawai 1964). Tetsuro Matsuzawa provides further details in Chapter 14.

Imanishi's original interest in monkey culture was to find an explanation for what he called "troop-oriented behavior." Specific troop-oriented behaviors included alarm calling, vigilance, and leadership behavior, which is related to altruistic behavior in modern terms. It was difficult to explain such behavior by conventional learning theory, and Imanishi felt it imperative to instead consider cultural origin theory, such as Freudian identification theory (Imanishi 1957). One should remember that this was before the development of aspects of modern evolutionary theory such as kin selection (Hamilton 1964). When Imanishi's students began to concentrate on the culture of feeding behavior, such as potato washing, he himself gradually lost interest because his major concern was not to explain food customs but rather to understand more complex cultural characteristics, such as leadership, self-sacrificing vigilance, and social institutions.

I suspect that this, in addition to his desire to elucidate human evolution, is one of the reasons why Imanishi's target changed from Japanese macaques to the African great apes (Imanishi 1961). He expected great apes to have social structures and cultures more like those of humans. In 1958, Imanishi and Itani began their survey of great apes in Africa, and in 1961 they decided to begin long-term fieldwork on chimpanzees in Tanzania. In 1965, I joined their team and began to study chimpanzees in the foothills of the Mahale Mountains, which was one of three field sites (Nishida 1990).

In the 1980s, after field studies of chimpanzees were established in various parts of Africa, it was proven that local behavioral differences among chimpanzees were more extensive than those among Japanese macaques. These studies of chimpanzee behavioral variations added several new dimensions to the previous research. First, differences in tool use, which is not in the behavioral repertoire of macaques, were extensively documented. Second, there was a more systematic examination of whether local behavioral differences could be explained by ecological variables (e.g., McGrew & Tutin 1978). Third, social communication became a conspicuous element of focal comparative studies after documentation of the grooming hand-clasp (McGrew & Tutin 1978) and the leaf-clipping courtship display (Nishida 1980a).

Some psychologists have asserted that chimpanzee and human cultures are very different because the major channels of information transmission used by humans do not effectively develop among chimpanzees (Premack & Premack 1994; Tomasello & Call 1997). To summarize their premises, humans have outstanding imitative ability as well as languages and great teaching capabilities. Furthermore, Tomasello asserts that the underdeveloped imitative ability of chimpanzees suggests that they lack relevant cognitive ability.

Although I do not deny that humans and chimpanzees differ in these aspects, I would stress that their superb behavioral flexibility, rather than the lack of some cognitive abilities, explains the within-group heterogeneity of the cultural patterns of chimpanzees. Furthermore, this viewpoint is supported by the various patterns having quasi-equivalence in their functional values. Therefore, the lack of standardization in chimpanzee cultural patterns may reflect their high intelligence rather than cognitive inferiority.

The purpose of this paper is twofold. First, to report innovative and locality-specific behaviors among the chimpanzees of Mahale, with an emphasis on social behavior, since there have already been many reports on the variations in chimpanzee subsistence behavior (e.g., McGrew 1992; Whiten et al. 1999). Second, to highlight the individual variability and flexibility in the cultural behavior patterns of a chimpanzee community. I will try to explain how and why chimpanzees differ in their cultural patterns within a single social group. Culture is here defined as information shared by many members of a society and transmitted from generation to generation via social learning, hence not the direct result of environmental differences (Nishida 1987). "Many" means at least most members of an age/sex class of a unit group or community.

The observational data reported here come mainly from my recent trips to the Mahale Mountains National Park from 1991 to 2000. I focus on the contrast between populations of Mahale and Gombe since they not only belong to the same subspecies but also live in broadly similar environments. Readers unfamiliar with chimpanzee behavior are referred to Nishida (1990) for the Mahale population and Goodall (1986) for Gombe, which are situated alongside Lake Tanganyika and separated from each other by only 135 kilometers. Just to illustrate what we mean by locally specific behavior, of the many different plant species common to both Mahale and Gombe, there are 11 species that are consumed by chimpanzees at only one of the two sites (Nishida et al. 1983). Most of the behavior described below is either characteristic of Mahale or appeared in a temporally constrained wave (a fashion) suggesting social learning, as in the changes in male bluff displays.

Locality-Specific Behavior Patterns in Mahale

Certain behavior patterns appear to be specific to Mahale chimpanzees. I divide them into eating behavior, play, intimidation display, courtship, and grooming.

Eating Behavior

Ant Fishing

The chimpanzees of Mahale fish for arboreal carpenter ants, a technique similar to the termite fishing observed at Gombe. A one-year-old infant begins to hold a vine like a tool between his lips without even trying to insert it into the nest hole while its mother is fishing for ants. This is the first stage of tool manipulation. The youngest age of successful ant fishing was four years, although youngsters around this age use a shorter probe and fish less efficiently than do adults. However, youngsters of seven years can fish as skillfully as their adult counterparts.

Eating Crematogaster Ants

The chimpanzees of Mahale, on a daily basis, eat small ants of the genus *Crematogaster* that live inside hollow branches. Making quick sweeping movements with one hand while holding a branch with the other hand (Figure 15.1) is a technique used to remove adult ants so that the chimpanzee can eat the eggs or pupae. There has been no report of eating this kind of ant elsewhere.

Licking Old Wood

Mahale chimpanzees lick the old wood of fruit trees such as *Garcinia huillensis, Ficus capensis, Ficus vallis-choudae,* and *Pycnanthus angolensis* (Figure 15.2). Chemical analysis of the *Garcinia* wood extracted a mannitol, a

Figure 15.1. An adolescent male eating *Crematogaster* ants.

kind of sugar alcohol whose sweetness is one-sixth that of sugar. However, it is not certain whether this was what chimpanzees sought in licking the wood. Although similar types of old wood are available, the chimpanzees of Gombe do not lick them (Goodall 1986).

Licking Rocks

Mahale chimpanzees also extensively lick the rocks along the shore of Lake Tanganyika and some rivers (Figure 15.3). Although similar rocks are available, the chimpanzees of Gombe do not lick them (Goodall 1986).

Swallowing Leaves

The chimpanzees of Mahale swallow the leaves of *Aspilia mossambicensis, Commelina* spp, *Ficus exasperata, Lippia pulicata,* and *Trema orintalis,* probably for self-medication (see Huffman 1997 for a review). Following their mothers' lead, two-year-old infants were observed to put such leaves into their mouths, but they only chewed and spat them out instead of swallowing. The chimpanzee brings its lips to a single leaf, bites off the leaf-blade, appears to rotate the leaf with the tongue, and then swallows it. The whole

Figure 15.2. An adult female licking the wood of a dead *Garcinia* tree.

Figure 15.3. The chimpanzees of M group licking rocks on the shore of Lake Tanganyika.

process is very slow and deliberate. Seven-year-old juveniles can swallow these leaves in an adult style.

Most eating behavior, including the above, begins at less than one year of age, and infants likely learn their food repertoire and self-medication from their mothers. This is because infants intently watch the behavior of their mothers, who are typically at closest proximity at this age. Youngsters begin to swallow leaves for self-medication at a later age than eating normal adult food.

Play Behavior

Another area of interest is juvenile play. Let me describe one distinct self-play pattern as a possible juvenile culture of Mahale. Older juveniles and younger adolescents ranging from six to nine years of age will slowly push or draw a pile of dry leaves as they move backward on all fours, and sometimes forward, holding the dry leaves with both hands and producing a conspicuous sound on the ground. This play continues for 5–10 seconds. The behavior can be detected, even if not directly observed, by lines of naked ground that typically extend more than five meters. Youngsters often engage in this type of self-play when traveling in a procession along a slope. The youngest individual to show this pattern was a two-year-old male, while his mother was sitting nearby. Male chimpanzees virtually cease this behavior when they attain late adolescence (from 12 years onward). Therefore, youngsters must learn it from older playmates.

Intimidation Displays

Drumming Walls of Metal Houses

Metal "Uniport" houses were first installed in our camp in 1976. Chimpanzees simply passed by the metal houses for some years after they were built (K. Norikoshi, personal communication). By August, 1979 (Y. Takahata, personal communication), many adult males began to use the metal walls of the houses as drumming objects (Nishida 1994). Since the camp was in the center of the range of our main study group, M group, they regularly hit the wall when traveling from south to north and from north to south. This behavior was most likely transmitted via social learning (as defined by Heyes 1993) because it began suddenly, with most adult males doing it, after a long period of the chimpanzees ignoring the houses.

Apparently, the males drummed the wall as a type of intimidation display.

Drumming produces a conspicuous sound and has intimidation effects on onlookers (the same has been noted by Goodall (1971) in relation to a Gombe male, Mike, who learned to incorporate empty kerosene cans in his displays). At Mahale, male chimpanzees begin to drum metal walls much later than they drum tree buttresses or trunks. Even four-year-old males occasionally hit or kick such natural objects. However, young males do not drum the metal walls. The youngest male that was recorded touching a metal wall was a six-year-old male, Chopin, who was once seen to hit the wall with his fist softly. The second youngest recording concerned a ten-year-old male, Bonobo, who only touched the wall with one hand. An adolescent male, Masudi, did not even touch the metal wall until fourteen years of age, when he kissed the wall instead of hitting it! All of these examples were observed when a large party of chimpanzees passed in a procession through the camp. Another fourteen-year-old male, Carter, showed impressive charging displays, including drumming tree buttresses outside of the camp. He even dominated two fully adult males. However, this male has never been observed to drum a wall. Thus, the behavior is virtually limited to adult males. Metal-wall drumming is acquired at older ages because higher-ranking males tend to discourage subordinate males from doing it. No female, old or young, has been observed to drum metal walls, although juvenile females occasionally drum tree buttresses in the forest.

This remarkable difference in attitudes concerning natural and artificial substrata may be due to the outstanding quality of metal walls as sounding objects. It is plausible that young chimpanzees and females are too sensitive to overcome their fear of the very loud noise caused by such drumming as well as possible attack by older males.

Chimpanzees likely understand the identity of individuals by vocalization and the sound patterns of drumming. When the drumming sound is heard, only one or a few of the audience responds to it immediately. Thus, metal-wall drumming may also function as a way to transmit information on the performer's location.

Throwing Rocks into Water

Rock-throwing display ("Throw splash," Nishida et al. 1999) may be another example of cultural intimidation display. Rock throwing is known in many localities, but the pattern displayed at Mahale is unusual. When chimpanzees of the M group cross deep canyons, adult males and more rarely adolescent males typically throw rocks. Some adult males attain a particularly elegant style of throwing. They choose a large heavy rock, lift it with both hands, aim it carefully, and then throw it into the water. The sound of the falling rock and the sight of splashing water appear to have a dramatic effect on the other chimpanzees. I do not know when this type of display began, but as far as I know Ntologi, the alpha male who kept his top position for 13 years, was the most skillful.

It is difficult to say whether a chimpanzee is really aiming at the water while throwing a rock because any chimpanzee that throws a rock into the canyon has a good chance of hitting the water. However, older skillful males often stand on two feet close to the water and throw the rocks with both hands. Therefore, the rocks cannot go elsewhere. Consequently, my criterion of the acquisition of throw-rock-into-water display is to stand bipedally, choose a big rock, lift it with both hands, and make an aimed throw into the water (Figure 15.4). Employing this criterion, we could identify five adult males as performers of this display in 1991. However, in 1999, after many older chimpanzees had disappeared or died in 1996, we found only two adult males performing the display. Accordingly, we are concerned about whether this tradition will continue in the next generation of males.

Since most adult females do not throw rocks, young male chimpanzees cannot learn throwing behavior from their mothers. It is likely that they learn it from observing senior adult males engaged in streambed displays. Here, I suspect imitation or emulation to be at work since any alternative is hard to imagine.

Courtship Displays: Leaf Clipping and Shrub Bending

The chimpanzees of Mahale show two locality-specific courtship displays: leaf clipping (Nishida 1980a) and shrub bending (Nishida 1987, 1997). Two types of leaf clipping, which occur in similar contexts, have been observed elsewhere since the original finding at Mahale (Whiten et al. 1999). In the original type of leaf clipping with the mouth, a chimpanzee holds a leafy branch or the petiole of a leaf with one hand and pulls it repeatedly between the lips or teeth, producing a conspicuous sound that attracts attention from the prospective sex partner. In the newly recorded "leaf clipping with fingers," a chimpanzee rips a leaf or leaves with the fingers. Leaf clipping with the mouth occurs at Bossou, Taï, and Budongo. However, Bossou chimpanzees do this when they are in stressful situations, such as being surrounded by

Figure 15.4. An adult male standing on his feet throwing a large rock into water with both hands. Drawing by M. Nakamura from video footage.

observers, whereas Taï chimpanzees do so immediately before a drumming display.

In the shrub-bending display, a male approaches close to an estrous female, sits down, and begins to bend shrubs and pull overhead vines, eventually laying the vegetation on the ground. He then sits on the cushion and often stamps on the ground with one foot. He may also combine shrub bending with leaf clipping.

Cultural courtship displays develop long before sexual maturity. Ambiguous patterns of bending shrubs began to appear even among under-one-year-old males. The youngest male that showed a clear-cut pattern of shrub bending was four years old. He first showed "bipedal swagger courtship," and the estrous female assumed the presentation posture. The male mated with the female, but after the female went away, he repeated vacuum thrusting movements and then engaged in the shrub-bending display. Therefore, this was done after actual courtship, making it a behavior out of context.

Information on cultural courtship patterns cannot be transmitted from the infants' mothers because lactating females do not show courtship displays. It is likely that young male chimpanzees learn the patterns socially from observing playmates and senior males.

Grooming Behavior: Leaf Grooming, Grooming Hand-Clasp, and Social Scratch

Leaf Grooming

Leaf grooming is a behavior in which a chimpanzee takes a single leaf, brings it to his lips, transfers "an object" (a louse, according to Zamma in press) from the lip onto the leaf, folds it, and then opens it again to investigate the contents. Leaf grooming is confined to social grooming or self-grooming sessions, and it is an extremely standardized behavior pattern. A three-year-old male infant was observed picking up a leaf discarded by his mother after it was used for leaf grooming. Therefore, it is likely that infants learn leaf grooming from their mothers. The youngest infant observed to engage in leaf grooming unambiguously was a four-year-old female, but she did so when she was not engaged in either social or self-grooming. Therefore, her leaf grooming was out of context.

Grooming Hand-Clasp

Grooming hand-clasp (GHC) is the best-known cultural social behavior pattern at Mahale. According to McGrew & Tutin (1978), "two chimpanzees sitting face to face each other simultaneously extend an arm overhead; then one clasps the other's wrist or hand or both clasp each other's hand. The other's hand engages in social grooming of the underarm area revealed by the upraised limb. Either both raise their right arms and groom with their left, or vice versa." The clasping pattern has changed from palm-to-palm grasp to wrist-to-wrist contact during the decades of observation (McGrew et al.

2001). It can be seen once every five hours of observation (Nakamura in press). Alternatively, a chimpanzee may raise one of its arms to grasp a branch overhead and groom the partner with the other hand. The partner assumes a symmetrical configuration and the two chimpanzees may groom each other in the exposed armpit. This has been seen throughout all chimpanzee ranges (McGrew et al. 2001) and has been named branch clasp (BC) (Nakamura in press).

In these symmetrical grooming patterns, two chimpanzees apparently raise their arms simultaneously. However, by studying video images in detail, one can often identify the party that "proposes" BC or GHC and the partner that responds to the proposal. The party that proposes BC holds an overhead branch earlier than the counterpart. The party proposing GHC raises one of its arms to the height of its head with its elbow bent. If the partner accepts with the corresponding arm (left with left or right with right), the GHC would proceed. Thus, whether the two parties engage in GHC/BC or not appears the result of a negotiation process.

Here, discussion is limited to GHC, since it is difficult to determine whether holding an overhead branch is an invitation to BC or simply a way to stabilize the body. As a matter of fact, 123 proposals were observed in 1999. Eighty-seven proposals (71 percent) ended in GHC, while as many as 36 proposals were refused or ignored by the partner. The ignoring party often continued to groom the proposing party, but sometimes lifted the opposite arm to the one proposed (right to left or left to right).

Occasionally, apparent disagreement was noticed between parties. For example, when the adult male Dogura stopped to groom Kalunde, another adult male, and raised his left arm with the elbow bent, Kalunde raised his right arm. Dogura lowered his left arm and raised his right arm instead, thus engaging in right-arm GHC. After more than a minute, Kalunde stopped grooming. Then Dogura raised *both* arms with elbows bent! When Kalunde raised his left arm, Dogura displayed grimaces and held up his left arm. Thus, left-arm GHC was attained this time. It appeared that the older Kalunde could decide the laterality of GHC and that Dogura offered an option to Kalunde when he raised both arms.

Possibly, mothers and other older individuals teach youngsters how to groom in hand-clasp by *molding,* not unlike psychologists teaching sign language to captive apes (Nishida 1999; de Waal 2001). For example, a middle-aged mother, Xtina, lifted the right arm of her four-year-old son, Xmas, with her left hand, and then held it with her right hand while grooming the arm's outer side. This resulted in a symmetrical configuration, although GHC did not result because Xmas did not groom his mother. An old female, Gwekulo, grasped the left arm of a 10-year-old female, Pipi, with her own left hand and lifted it up immediately before they resumed GHC. Gwekulo had cared for Pipi when she was an orphaned infant, so she was like her mother. A middle-aged female, Mija, lifted the left arm of an orphaned adolescent male, Charles, with her right hand, and then re-grasped his hand with her right hand, thus taking a conventional symmetrical posture. Nakamura (unpublished) observed similar interactions between a mother and her juvenile son.

In 1999, I could confirm the proposing party on 13 occasions when GHC occurred between mothers and offspring. Mothers proposed eight times, and offspring five times. Since the offspring were older juveniles and young adolescents, they had already acquired GHC skills. The youngest individual that has been recorded performing GHC was five years old in both females (Nishida 1988) and males (Nakamura in press), and their partners were mothers. Since the GHC interaction begins between the mother and the offspring, it is likely that the former molds the latter's behavior. Moreover, the juvenile is likely motivated to respond correctly and reproduce what its mother is doing, as Matsuzawa (Chapter 14) suggested for the nut-cracking behavior. In addition, immature chimpanzees may acquire GHC by observing third parties engaging in it.

The function of GHC is not clear. The hypothesis that the proposing party intends to be groomed rather than to groom is not supported by the current evidence. The proposing party was groomed after GHC on 16 occasions, whereas the proposed party was groomed on 24 occasions. Both parties groomed each other on 25 occasions.

Social Scratch

Social scratch, a newly described pattern, is to scratch the body (in particular, the back) of another individual (Nakamura et al. 2000). This behavior is performed by almost all adult and adolescent individuals in M group, and the frequency of social scratch is extremely variable. Although simple, there are two types of social scratch: fine scratch and rough scratch. When a mother scratches her infant, she appears to try to remove detritus and ectoparasites in a meticulous way. Her fine scratch consists of repetition of short and quick scratching movements. On the other hand, when an adult male scratches the back of another adult male, his scratching stroke is usually broad and rough, and often produces a conspicuous scratching sound. The youngest chimpanzee to show this behavior was a four-year-old male, who scratched the back of his mother. It is likely that an infant chimpanzee learns to scratch socially by being scratched by its mother and by observing social scratching between third parties (Nakamura et al. 2000). At Gombe, social scratch has been observed only for a particular adult female, Madame Bee, and her infant. She "vigorously scratched the skin of her partner before intently examining the skin at that place" (Kummer & Goodall 1985). This seems an idiosyncratic rather than a cultural pattern—Madame Bee began to do so only after she lost the use of one arm during a polio endemic.

Development of Cultural Behavior Patterns

There are discrepancies in development depending on the categories of behavior. Obviously, various eating behaviors develop first, when chimpanzees are in early infancy. Then come play patterns. Next there is the development of grooming and courtship, in which leaf grooming and social scratch develop earlier than other patterns such as GHC. The latter develops around the time of weaning. Finally, intimidation displays are learned. Metal-wall drumming and throw splash are the last cultural behavior patterns that male chimpanzees of M group develop.

It appears that the critical age in the development of behavior at Mahale is four years, namely during weaning, when the majority of important cultural patterns emerge, although they still need refinement and knowledge of the appropriate contexts. In this regard, it is interesting that the most complex cultural pattern observed in chimpanzees so far—nut cracking through hammering—is said to be acquired only at the age of eight years at the site with the toughest nuts, Taï Forest (Boesch & Boesch-Achermann 2000). Hammering may be hard to acquire since it involves three objects and requires delicate adjustments of muscle power, which should be neither too strong nor too weak.

Review of Innovative Behavior

Cultural patterns begin when an individual makes an innovation, i.e., develops a solution to a novel problem or a new solution to an old problem or displays a conventional pattern in a new context (Kummer & Goodall 1985). In her pioneering study, Goodall (1973) reviewed some behavior patterns such as "wrist-shaking" initiated by a juvenile female in her study group at Gombe, which another juvenile female appeared to simulate.

New Food Repertoire

After the village people of Mahale moved out of the study area because of the government's Ujamaa policy, around 1975, the chimpanzees of M group suddenly obtained the nonrisky opportunity to experiment with farm products that had formerly been difficult to approach. Fruits such as mangos, lemons, and guavas became incorporated into the food repertoire of M group one after another (Takahata et al. 1986). Subsequently, mangos never attained an important status in the diet of M group chimpanzees because they were mostly eaten by baboons and vervet monkeys long before they were ripe. On the other hand, lemons and guavas have become important foods for them in June and August-September, respectively, in recent years. Infants learn eating these novel foods from their mothers, and adolescent female immigrants quickly learn to ingest them. Now, all M group chimpanzees eat these fruits.

Eating fresh-water algae by entering a stream was observed for the first time in M group in 1997 (Sakamaki 1998). A newly immigrated adolescent female, Sally, entered a small stream and ingested it while immersing herself up to her shoulders in the water. She ate the algae repeatedly. Since algae are distributed more abundantly in the upper streams of Mahale, it was suspected that the adolescent female had transferred from one of the groups at higher altitude. At least a few chimpanzees watched her behavior, but no other M group chimpanzee has been seen to eat this item at the time of writing (2001). Why do new immigrants ingest lemons immediately, while group members do not ingest algae? Perhaps because the first process is that of enculturation in which information moves from many to one individual, whereas in the latter information must go from a single individual to many (Nishida 1987). Moreover, in eating the algae, a chimpanzee must first enter the running water and get completely wet, which chimpanzees typically do not like to do (Nishida 1980b).

Leaf Napkin

Leaf napkin is a tool extensively developed among Gombe chimpanzees, who often wipe off feces, semen, blood, and other excreted matter with leaves (Goodall 1986). At Mahale, wiping behavior has been seen only rarely and only in a few individuals (Nishida 1994). However, around 1998 some chimpanzees suddenly began to regularly wipe their mouths with leaves and branches, without detaching them from the lemon trees (Hayaki, unpublished). There was no record of lemon wiping, even in the detailed monitoring of the entire video footage of lemon eating recorded in and before 1995 (Miho Nakamura, personal communication). Lemons became a favorite food in the 1980s (Takahata et al. 1986), and while eating them mouths get messy with lemon juice. In 1999, at least nine chimpanzees were observed wiping their mouths. The possible process of the behavioral transmission of leaf napkin will be published after compiling the data of many observers. However, the innovator was most likely Fanana, the alpha male, who has shown wiping behavior far more frequently than others, namely 75 percent of Hayaki's observations (N = 4) in 1998 and 50 percent of my own record (N = 16) in 1999.

Innovative Tool Use

An adult male was seen using a probe to open his blocked nasal passage (Nishida & Nakamura 1993). The male's behavior was repeatedly seen in subsequent years, and many chimpanzees had opportunities to observe him. However, the behavior has never been passed on to other individuals. Perhaps other chimpanzees were discouraged from simulating this behavior because they viewed it as a somewhat "bold" pattern that inserts an external object into a delicate orifice.

In the dry season, chimpanzees have occasionally been seen to dig up the wet earth of small streams in order to gain access to drinking water. Some chimpanzees have been seen to remove a heap of dead leaves or a large rock. An eight-year-old female, Serena, was once seen to use a stick to dig a water trench and drink from it. Such observations have been so rare that it cannot be determined whether digging for water is occasionally innovated individually or has been transmitted socially. However, in other dryer places such as Semliki, Uganda (Hunt 2000) or Mt. Assirik, Senegal (McGrew 2000), digging to obtain drinking water became a custom of the local populations.

Encounter with a Strange Animal

An adolescent female immigrant, Tomato, showed an extremely fearful attitude when watching a large crocodile on the sandy shore of Lake Tanganyika. The crocodile was habitually seen on the same beach from the chimpanzee path. On encountering the reptile, M group chimpanzees usually barked at it several times and soon began to ignore it. However, on one occasion Tomato climbed a tree and continued to bark at the reptile long after all the other chimpanzees, male and female and young and old, had resumed eating vine leaves. This female probably had no experience of watching a crocodile—we believe that she immigrated from a mountainous area. There is a possibility that reactions to strange animals are subject to social influence, i.e., depend on how others react.

Ground Grooming

In 1991, a young adult male, Jilba, was observed for the first time grooming the dry leaves of the ground or moss-covered rocks while he was sitting alone or being groomed by others. Later, he was seen to repeatedly perform this behavior, and I termed this "ground grooming" (see photo in Nishida 1994). This pattern may be a nonfunctional displacement activity.

Subsequently, in 1992, another young adult male, Toshibo, began to show the behavior. In 2000, a total of 10 chimpanzees, including two young adult females, were seen to engage in ground grooming. Although nothing detailed is known about the transmission of this pattern between individuals, social learning is the most likely candidate process because the pattern was unknown before 1991.

Individual Variability and Flexibility of Behavior

Drumming Display

Drumming patterns vary greatly among individuals. Some regularly stand on their feet and slap with both hands very quickly. Other males stand bipedally and slap with one hand very slowly. Other males stand on their feet and slap and kick simultaneously (Table 15.1). Variation in drumming behavior partly stems from individual preferences in the loudness and rhythm of drumming. This may be a matter of aesthetics. Laterality also differs among individuals. Thus, everybody has his own idiosyncratic drumming style.

Chimpanzees probably recognize the differences in sound production caused by differences in substrata. They differentiate kicking and slapping styles among the buttresses of trees, the ground, and the walls of metal houses. They are more likely to kick than to slap the buttresses of trees or the

Table 15.1. Male individual variation in metal-wall drumming pattern

Name	Biped	Quadruped	Biped	Quadruped	Total	Type
NT	17	8	0	16	41	Versatile slapper, quadrupedal kicker
DE	37	3	18	12	70	Quick biped slapper, versatile kicker
BE	9	0	0	3	12	Biped slapper, quadrupedal kicker
NS	36	0	2	0	38	Slow bipedal slapper
TB	17	5	0	13	35	Bipedal slapper, quadrupedal kicker
MA	12	0	6	3	21	Bipedal slapper, versatile kicker
FN	45	8	5	1	59	Slow bipedal slapper
BB	6	0	2	25	33	Quadruped kicker
DG	21	1	3	0	25	Bipedal slapper

Table 15.2. Proportion of slapping versus kicking when drumming metal walls or tree buttresses (1989–1992)

	Metal wall			Tree buttress	
Male	Percent	N		Percent	N
BA	25.0	4	NA		0
NT	58.1	43	>	11.1	18
DE	53.3	75	>	20.3	69
NS	69.2	52	>	0.0	20
AJ	100.0	6	>	30.0	20
TB	46.9	49	>	0.0	31
BE	60.0	15	>	0.0	24
JI	100.0	3	>	25.0	4
MA	57.1	21	>	0.0	2
FN	85.9	64	>	7.0	57
DG	82.1	28	>	66.7	6
AL	0.0	2		0.0	1
HB	33.3	9	>	14.3	7
NC	0.0	1		0.0	1

Note: Percentage = 100 × slapping/(slapping + kicking).

ground. However, they slap the metal walls rather than kick them. Of 14 males observed, two never slapped anything and one was not seen to drum outside of the camp. The other males on which we had information (N = 11) were more likely to slap when drumming the metal walls than when drumming tree buttresses (binomial, $p < 0.01$, Table 15.2). This is because they can easily produce sound on metal walls even by slapping.

Male chimpanzees may know the difference in sound production between one-hand and two-hand slapping. They tend to use two-hand slapping

Table 15.3. Proportion of two-hand slapping versus one-hand slapping in different substrata

	Metal wall			Tree buttress and ground	
Male	Percent	N		Percent	N
AJ	0	4	<	18.2	11
DE	52.5	40	*>	3.7	27
NT	32.0	25	>	0	3
TB	13.6	22	>	0	15
FN	7.7	52	*>	0	30
DG	42.9	21	>	25.0	8
MA	75.0	12	>	33.3	3
BB	14.2	7	>	0	1

Note: Percentage = 100 × two-hand/(two-hand + one-hand); * = significance.

against metal walls more often than against tree buttresses (30 percent versus 6 percent, Table 15.3). Seven of the eight males showed a preference for two-hand use more often when slapping the metal wall than when slapping tree buttresses or the ground (binomial, $p = 0.06$). It is difficult to produce sound on tree buttresses or the ground with two hands or two feet. Chimpanzees must hit these surfaces with one hand or one foot with a concentrated force to produce sound. Therefore, they tend to use only one limb to hit a tree buttress or the ground. Thus chimpanzees appear to understand the differences in the quality of substrata as resonant objects. Despite this, they show extreme variability in drumming patterns. This means that the lack of standardization in drumming patterns is not due to an imperfect transmission of existing habits but rather to variability in individual preferences, which has the possible advantage that individuals produce their own "signature" patterns that are potentially recognizable to the rest of the community.

Courtship Display

There exist many idiosyncratic patterns of courtship in wild chimpanzees: raking dry leaves, stripping leaves, shaking a branch with the foot, flailing a branch, and so on (Nishida et al. 1999). How can courtship display take on such diversified forms? I think this is because the penile erection is a most salient signal of sexual interest, leaving little doubt about the motivation of the male thus addressing a tumescent female. The remaining issue is only to clarify the identity of the female the male intends to mate with. Any indication of interest in her would be functional. Various patterns have the same function of producing a conspicuous sound in order to attract the attention of estrous females.

Transmission of cultural courtship gestures cannot be explained by the so-called "ontogenetic ritualization" formulated by Tomasello, in which intentional movement is ritualized through interactions between two individuals (Tomasello & Call 1997). The origin of leaf clipping is not an intentional movement, but most likely a conflict behavior. It could result from such circumstances as a willing male not being able to entice an estrous female to mate because of the presence of higher-ranking males (Nishida 1980a) or an estrous female not recognizing that he wants to mate with her. Thus, he enters a state of conflict. There are many types of conflict behavior, such as self-scratching, shrub shaking, shrub bending, branch breaking, slapping, stamping, flailing, pushing, clubbing, and throwing. Any of these could be useful to signal the intention of the frustrated suitor. In fact, all of these behaviors are occasionally employed by M group males when "herding" a disobedient estrous female.

Why have these behaviors been conventionalized as cultural signals but not others? It is likely that females are aversive to courtship gestures that have more overtly aggressive components. I dare say that such a gentle pattern as leaf clipping or shrub bending might be congruent with the aesthetics of female chimpanzees and probably have been employed by an influential male such as the alpha male faced with a hesitant or unwilling female. As already mentioned, the context of the leaf-clipping behavior is different between localities. At Bossou, chimpanzees show leaf clipping when they feel anxious in front of human observers. At Taï, chimpanzees clip leaves before a drumming display. Thus, which of the conflict behaviors is employed as a courtship signal by each local population appears to be rather arbitrary.

Discussion

In this paper, I described some innovative and cultural behavior patterns of the Mahale Mountains chimpanzees. Here I wish to discuss the functions of high-fidelity and low-fidelity information transmission. Psychologists tend to attribute a lower degree of fidelity in information transmission among chimpanzees to the cognitive inferiority of chimpanzees. There may be other possibilities, however. Only some cultural behaviors need to be copied with the greatest fidelity for them to function well. For example, to throw the largest rock possible into water with both hands may be the best way to create an intimidating effect. Only full-sized males can do so. Therefore, all of the old high-ranking males performed the same pattern. This high fidelity may have been attained by many years of trial and error with goal emulation.

However, drumming on metal walls can be performed effectively in various ways such as slapping with the hand or kicking with the foot. Slapping produces less sound, but is easier to repeat and needs less energy than kicking. Behavioral diversity should be analyzed on the basis of the cost and benefit of the patterns and the physical aptitude of individual animals. I assume that there are many optimal solutions in the case of metal-wall drumming in producing salient sounds but only one or a few solutions in the case of aimed throwing displays.

In grooming behavior, some variations can be found. The fine type of social scratch is useful for removing ectoparasites from small-bodied infants. The rough type of social scratch would be useful for adult males because this produces a conspicuous sound and as a result may broadcast the intimate social relationship between the groomer and the groomee. Consequently, I suggest that fine and rough types of social scratch have different functions.

Leaf grooming is the most stylized cultural pattern because according to the most recent observations (Zamma in press), the function of leaf grooming is to hold and kill an ectoparasite between a folded leaf.

Cultural courtship displays also have variable patterns. This is because the most salient feature of courtship by male chimpanzees is the erect penis, so the only information that needs to be transmitted is the identity of the signal sender.

Even human cultural transmission can show the same low fidelity, or as I would rather call it, high level of individual flexibility. For example, people may show different behavior patterns, which nonetheless are called by the same name. For example, the use of chopsticks is a part of Japanese culture. To confess frankly, I do not know how to use chopsticks. I do not hold chopsticks in the standard Japanese style, although I know it technically. Subsequently, I learned that I was not alone. Several of my friends also have their own unique patterns of chopstick holding. However, all of us can pick up even the smallest food items, albeit awkwardly. It follows that there are many viable solutions to one problem. The solutions are functionally of little difference in efficiency. Thus, the existence of variation in cultural behavior patterns can be interpreted as evidence of behavioral flexibility rather than as evidence of low imitative ability. Moreover, does imitation imply higher intelligence in the first place? If a person copies the same pattern demonstrated by his or her predecessor when any other pattern is also useful, does this make that person intelligent? It does not seem obvious that such is the case.

We have found a large number of differences in social behavior among chimpanzee populations. What we must do now is find the differences in the social institutions between populations, as Imanishi expected in the Japanese macaque societies. For example, the chimpanzees of Taï are said to cooperate in colobus hunting (Boesch & Boesch 1989; Boesch & Boesch-Achermann 2000), whereas there is no such evidence for the chimpanzees of Gombe (Busse 1978; Stanford 1998) or Mahale (Takahata et al. 1984; Hosaka et al 2001). The chimpanzees of Mahale seem to share meat mostly according to the sharer's selfish criteria (Nishida et al. 1992), whereas those of Taï are said to share meat according to the degree of contribution to the hunting activities (Boesch & Boesch-Achermann 2000). Are there different cultural rules? This seems a most interesting area for future investigation.

Finally, I want to stress the relevance of chimpanzee culture in the conservation of nature. Culture has been regarded as the single most important concept that differentiates humans from other animals. Thus the culture concept has provided justification for humans to monopolize natural resources and destroy environments, neglecting the survival rights of the other creatures. Therefore, consideration of cultures in the evolutionary continuum has fundamental importance for all of this planet's inhabitants.

CASE STUDY 15A

Sex Differences in Termite Fishing among Gombe Chimpanzees

STEPHANIE S. PANDOLFI, CAREL P. VAN SCHAIK, AND ANNE E. PUSEY

Social organization in primates influences not only their social behavior, but also their feeding ecology. Among our close relative the chimpanzee, for example, males are highly social and often cooperate in hunting for meat (see Chapter 4). In addition, cultural transmission allows individuals to learn complicated tool-assisted extractive foraging techniques from other members of their social group (see Chapter 16). We will show that some well-known dietary sex differences are influenced by social organization as well.

Sex differences in chimpanzee faunivory are well documented. The most striking of these differences is that females feed significantly more on insects, which they usually acquire by tool-assisted extractive foraging, than do males. These differences in insect feeding among the Gombe chimpanzees, especially on termites, have often been reported. Female chimpanzees fish for termites significantly more (4.3 percent versus 1.4 percent of total observation time) as well as for longer bouts (McGrew 1979; Goodall 1986), and fecal analysis shows that females consume more termites, as well as weaver and driver ants, than males do (McGrew 1979). Our analysis of seventeen years of the data from Gombe confirmed that females fed on insects twice as often as males—4.1 percent versus 2.0 percent of total feeding bouts, and showed that termites are, in fact, the fifth most commonly fed upon food for females (Pandolfi et al. 2001).

Only during November, the peak month when termites swarm and are relatively very abundant and easy to catch, do males spend more time than females working and feeding at termite mounds (Goodall 1986). Recalculating data reported in McGrew and Marchant (1999), we found that, on average, males are collecting 64 percent more termites per minute ($x = 6.81$) than females ($x = 4.16$). This difference is statistically significant (Wilcoxon Mann-Whitney test, $m = 7$, $n = 9$, $W_x = 36$, $p = .0058$). It is mainly due to the males making significantly more withdrawals per minute than females ($W_x = 37$, $p = .0082$), rather than getting more termites per withdrawal.

Although dietary sex differences in chimpanzee termite fishing are often cited, so far there has been no systematic attempt to analyze *why* these differences exist. One way to examine this question is to look at the effect of the presence of same-sex conspecifics on a male or female individual's likelihood to feed on termites. We examined how males and females are differentially affected by the presence of conspecifics when fishing for termites in order to test the following nonexclusive hypotheses, traditionally proposed to explain sex differences in chimpanzee faunivory.

"Care of offspring" proposes that constraints placed on females by the care of dependent offspring limits the activities available to them (McGrew 1979, 1992). Thus, we would expect females to choose foods that allow them to care for their offspring, such as termite fishing, over activities, such as hunting, that may be dangerous or require them to leave their offspring unattended. We already know that females spend more time overall fishing for termites than males, and this may be a by-product of constraints placed on the female by dependent young. However, based solely on these limitations placed on females by their offspring, we do not expect the sexes to be differentially affected by the presence of same-sex conspecifics during termite feeding.

"Nutritional constraints" is the hypothesis that gestation and lactation increase females' needs for certain nutrients, such as lipids and proteins, thus constraining their food choices (Clutton-Brock 1977; Demment 1983). If this is the case, we would expect females to specialize on high-quality foods, such as foods with high energy yields (either high lipid content or high simple carbohydrate concentration) or high protein content, or both. In fact, we do see females specializing on termites, a high-protein food (McGrew 2001).

Females may need more or different nutrients than males, and they may acquire these nutrients from insects rather than meat. Based solely on this hypothesis, we predict females will consistently feed on termites more than males, regardless of party size. In other words, while both males and females will be affected by conspecifics during termite feeding (due to limited mound space and feeding competition), females will be more likely to feed on termites in any size party than males because of their need for high-quality foods.

"Social considerations" is a hypothesis that argues that sex differences in diet exist because of differential social strategies (Boesch & Boesch 1984). Females' foremost objective is to maximize their nutritional intake. Social concerns are secondary because of their extreme sensitivity to feeding competition (Trivers 1972); thus, benefits from their social interactions cannot be expressed because of the high price of this competition. On the other hand, males' primary goal is to maximize their social position, and their feeding behavior may thus be adjusted to their social needs. Their first priority when alone is to find another male—not a desirable food source. Specifically, this hypothesis predicts that increased male sociability increases their likelihood to hunt in parties while restricting their abilities to forage with tools (because of competition for mound sites, males cannot congregate in large numbers while termite fishing). Thus, according to this hypothesis, we would expect males and females to be differentially affected by the presence of same-sex conspecifics when feeding: females will be less likely to fish for termites when other females are present, while conspecifics will have less (or no) effect on males' likelihood to fish for termites.

In order to test these hypotheses, we analyzed the effect that the presence of conspecifics has on the likelihood of either sex to fish for termites. Among independent individuals of both sexes, we determined the total number of feeding bouts spent in each of five party-size categories: alone, with one, two, and more than two same-sex conspecifics, and in bisexual parties (we are considering bisexual parties as large parties because more than 60 percent of these parties were comprised of more than ten individuals, while only 16 percent were comprised of less than four individuals). We then determined the percent of recorded feeding bouts within each of the five categories in which males and females fed on termites (see Figure 15A.1).

We ran X-squared tests using raw frequencies (each feeding bout was an independent data point as typical termite-fishing bouts on the same day were separated by at least one hour) and applied the Bonferroni correction for

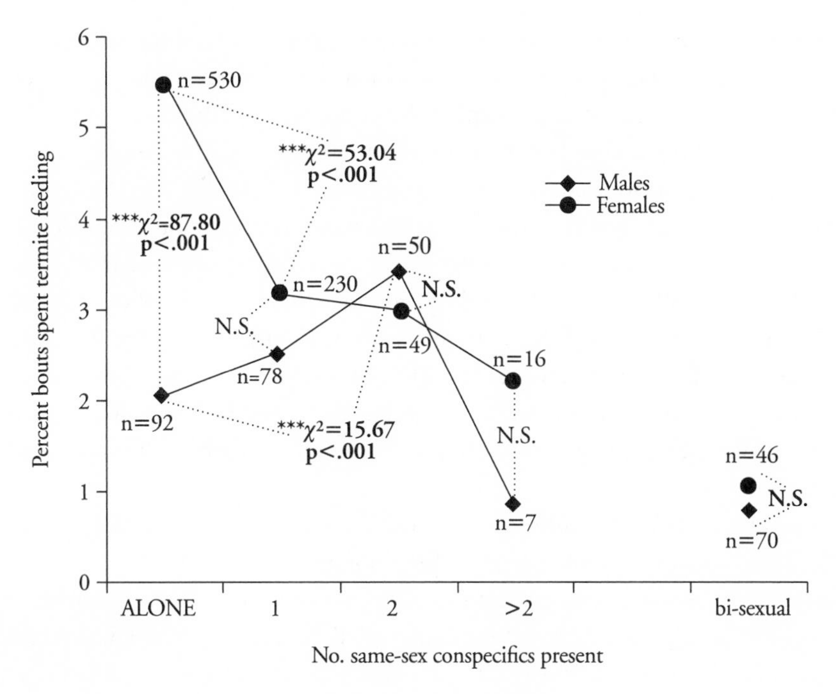

Figure 15A.1. The effect of party size during termite feeding.

multiple tests. We found that females significantly decrease termite feeding time when in the presence of same-sex conspecifics. The presence of a single same-sex conspecific significantly reduces the likelihood a female will fish for termites ($X^2 = 53.04$; $p < .001$; $\alpha = .005$). This can be explained by traditional sexual selection theory, which says that female fitness is limited by access to resources, whereas male fitness is limited by access to mates (Trivers 1972). Female apes, and female chimpanzees specifically, have a general tendency to forage alone because they are very sensitive to feeding competition (Wrangham 1996; van Schaik 1999).

Males, on the other hand, are significantly *more* likely to feed on termites with one or two other males present than when alone ($X^2 = 15.67$; $p < .001$; $\alpha = .005$). The introduction of a fourth male, however, dramatically decreases the likelihood of feeding on termites ($X^2 = 14.97$; $p < .001$). This supports the argument of Boesch and Boesch (1984) that one reason males do not participate in tool-assisted extractive foraging is that they prefer to spend time with other males. They also found that males concentrate less on the task when in the presence of same-sex conspecifics. These two arguments explain the trend in male termite fishing: males prefer feeding in the company of other males—up to a point, at which time they are probably too distracted by their conspecifics to apply the concentration necessary to fish for termites, and because in larger parties males tend to patrol or hunt.

These data show that the sex differences traditionally seen in termite fishing can be attributed to the differences when feeding alone. Females are significantly more likely to feed on termites when they are alone than males are ($X^2 = 87.80$; $p < .001$), yet there are no significant differences between the sexes' likelihood to fish for termites when in any sized party. Thus, females prefer feeding on termites alone to avoid feeding competition, whereas males prefer feeding in the presence of other males until a large enough party has formed, at which time they tend to travel, hunt, patrol, or feed on vegetable foods.

Whereas nutritional differences and child-care restraints certainly do influence chimpanzee termite-feeding behavior and the sex differences therein, they cannot explain the significant sex difference in party size effect. Much of the difference is clearly a by-product of different socioecological strategies. Males' priority is maximizing the ability to socialize with other males, whereas the main objective for females is to maximize nutritional intake. We argue that this scenario applies to other aspects of dietary sex differences in chimpanzees as well, such as hunting, frugivory, and folivory. We suspect that this trend may apply to more primate species than just chimpanzees, and is a question that should be tested on a broader basis. Perhaps even the sexual division of labor that we see in humans is primarily a product of socioecological differences between the sexes.

16

Ten Dispatches from the Chimpanzee Culture Wars

W. C. MCGREW

Culture applies equally to being yogurt or watching *kabuki.* A culture-bearer may be a petri dish or an *imam.* Herein lies the problem: The label refers to a wide range of phenomena. Somewhere in this range lie boundaries of uncertainty. The fermentation of bacteria is easy, but the fermentation of ideas raises issues. What is culture? When did it emerge? How do we know it? Who has it?

Add a heavy dollop of strong feelings to this mix, and controversy is assured. The title of this essay is taken from the current preoccupations of anthropology (and the social sciences in general), which is riven by bitter struggle. The "culture wars" in "cultural studies" are about essentials and jurisdiction, and, ultimately, about identity. The same issues affect what is called here "the chimpanzee culture wars." (For a recent and somewhat detached view of the human version, see Kuper 1999.)

Wars have battles, and from battlegrounds come dispatches, which are meant to be timely and terse. Here, these 10 dispatches are very much the latter, as depth gives way to breadth. Imagine them as being light enough to be flown by carrier pigeon from far-flung battlefields, in anthropology, biology, linguistics, psychology, and so on, amid shot and shell.

Put another way, the aim of this chapter is to look closely at cultural primatology. (Though young, the phrase already has two meanings. Human attitudes, beliefs, and treatment of apes, monkeys, and prosimians, better termed ethno-primatology, is *not* covered here. See Wheatley 1999.) Cultural primatology is analogous to cultural anthropology, as a subset of investigators interested in the culture (as opposed to the anatomy, ecology, genetics, or physiology) of nonhuman primates. However, whereas cultural primatologists assume that nonhuman primates are cultural creatures, most cultural anthropologists instead presume that only humans have culture. Both cannot be right.

More than 40 populations of wild chimpanzees across Africa make and use tools, from Tanzania in the east to Senegal in the far west. No two populations appear to have the same technological profile. Some use probes of vegetation to harvest social insects; others use stone hammers and anvils to crack open nuts. In neither case is the geography of tool use explained by absence of raw materials or targeted tasks. Even when these are constant, variations in style occur that cannot be explained by ecological factors. Such variation in humans is called material culture and ends up in museums. What to do when it occurs in apes?

Chimpanzee Culture? Absurd!

In this dispatch, from the battlefield of ethnology, culture is taken to be both universally and uniquely human, that is, all *Homo sapiens* have it and only *Homo sapiens* has it. Thus, culture is both a necessary and sufficient condition for humanity. (Modern humanity, that is. There is a problem of what to do with archaic forms like Neanderthals. Whether they were cultural or not evokes some debate.)

Some examples will show how difficult is this stand to maintain: Celestial navigation is neither universal nor unique to humans, as humans living in closed-canopy tropical forests do not show it, but migrating songbirds do. Constructed shelters may be universal to all human societies, but these are not unique, as shown by beaver lodges and orangutan beds. Writing is uniquely human, but not all humans know of it, as there are many nonliterate cultures (or were, prior to contact by outsiders). So, what is left to humanity as both universal and unique? How about the space shuttle, or for that matter, the wheel? Or, the computer, or for that matter, the abacus? Unfortunately for this line of argument, most foraging peoples, whether Inuit or San, have none of these. Either they must be excluded from cultured humanity, or we must look deeper.

A skeptical pragmatist might point to seemingly obvious truths, e.g., that all humans clearly depend on culture, while just as clearly other species do not. (This is sometimes expressed cleverly as culture being humanity's ecological niche.) To scupper this proposition, we would need to find only one human society lacking culture, but we cannot, as the ethographic record shows. However, the same logic applies to chimpanzee populations; more than 40 have been studied, and all seem to have culture. Therefore, which species is more dependent? Neither one. Moreover, as de Waal (2001) has argued, there is much evidence that chimpanzees do need culture, as unacculturated apes may be at best incompetent, and at worst, dead. Textbooks in social sciences, especially in anthropology, are full of such assertions, presented as "obvious" truths, but these need querying.

A skeptical mentalist might object that all that we know of nonhuman culture is based on behavior, which is the least notable aspect of culture. Much more interesting is the knowledge that underlies and informs behavior. Even more challenging is the meaning attributed to the knowledge that drives behavior. And finally, there are the emotions that color the meanings that pervade the knowledge that is manifest in behavior.

How can any sensible person imagine that all of this exists in animals? Surely, it is said, this is what distinguishes a mating system (pair-bonded gibbons) from the institution of marriage (wedded Mennonites). Surely this is what distinguishes an optional taboo (the English do not eat horses) from obligatory carnivory (tarsiers eat no plants). Surely, this is what distinguishes a rite of passage (Maasai initiation of young men) from puberty (adolescent male chimpanzees challenging their elders).

There are several answers to this question. One is that all that we know of knowledge, meaning, and emotion is based on behavior. We have no direct access to human or any other minds, so all is inference. Whether or not we are any better at divining human than nonhuman minds is debatable. We share the perceptual world of our fellow humans (advantage) but we are also susceptible to their mendacity (disadvantage).

It may be that the chimpanzee mind is wholly devoid of knowledge, meaning, and emotion, or it may be that some or all of these phenomena are there, in distinctively apish form. As with our fellow humans, we can only try to draw valid inferences. It would seem that cultural anthropology would be of great help to cultural primatology in this task. At the very least, the former could tell the latter what evidence would suffice in principle, so that primatologists can seek it in practice. There is a history of such aid (flirta-

Figure 16.1. Party of Gombe chimps of mixed age and sex.

tion?) in sociocultural anthropology, dating back to Kroeber (1928), but including also the works of Ruth Benedict, Marvin Harris, and G. P. Murdock.

The prospect of chimpanzee culture is absurd only if it is unimaginable.

Chimpanzee Culture? Of Course!

In this dispatch from the battlefront of ethology, the problem is not exclusion of other species from the cultured, but rather finding the limits of inclusion. In his groundbreaking book, *The Evolution of Culture in Animals,* Bonner (1980) was willing to grant candidacy to slime-molds. That is, depending on definition (see Dispatch 7, Culture Is by Definition), culture is present not only in primates, or mammals, or homiotherms, or vertebrates, or invertebrates, but also in organisms lacking a nervous system! (This is *not* just word play, nor is such inclusiveness limited to culture. See Strassmann and colleagues 2000 for the case for altruism in a protozoan.)

Textbook examples abound. In the same year as Bonner, Mundinger (1980) made the focused case for animal culture based on vocal learning in passerine birds. Invoking the research of pioneers such as Marler on white-crowned sparrows, Mundinger argued that characteristics such as plasticity, diffusion, tradition, and innovation were met. Hundreds of studies have extended these results and teased out the mechanisms of song learning and transmission (e.g., West & King 1996). In mammals, California sea otters crack mollusks on stone anvils but their Alaskan cousins do not. Among the Californian population, most crack while floating on their backs at the surface, but some take their tools underwater!

The problem with these (and others, e.g., Galapagos finch, bower-bird, black rat) marvels of natural history among nonprimates is that they seem to be mostly "one-trick ponies." That is, sea otters are great at anvil use, but do little else that is not species-typical. European blackbirds are wonderfully creative singers, but the rest of their behavioral repertory seems stereotyped by comparison. If one of the hallmarks of culture is its comprehensiveness, then single-trait candidates are bound to fall short.

None of these caveats applies to Japanese monkeys, however. Not only do these primates show an impressive array of cultural patterns, but also some of these habits have been tracked by primatologists for decades. Furthermore, their interpretation as cultural has been asserted as cultural from the outset, by Imanishi and his intellectual successors (de Waal 2001). It is no accident that sweet-potato washing, wheat sluicing, hot-spring bathing, and so on are to be found in every introductory textbook. We marvel at photographs of snow-topped monkeys immersed to their necks in hot springs and wonder at their ingenuity.

Yet there is a problem with most (but not all, see Nakamichi et al. 1998) of these classic cases. Their origins lie in human facilitation; the monkeys were lured to the beach or into the pools by artificially providing them with domesticated plant foods. This enhancement takes nothing away from what happened next, how the habits spread by horizontal or vertical transmission or continued to elaborate or became fixed. But one must always wonder: How many of these habits would have occurred without human assistance?

Within behavioral biology, crediting chimpanzees with culture or not is polarized. Ethologists (Boesch, Goodall, Nishida, Wrangham), who study chimpanzees observationally in nature, tend to say yes. Comparative psychologists (Galef, Premack, Tomasello), who study apes or rodents experimentally in captivity, tend to say no. Less of a dichotomy exists among researchers on capuchin monkeys, with field-workers (Boinski, Fedigan, Panger, Perry) largely affirmative but lab workers (Anderson, Fragaszy, Visalberghi, de Waal) of more mixed opinion. In any case, methodology and degree of direct experience are often confounded in investigators.

Notably absent (with a few exceptions, e.g., Boehm) are researchers in cultural primatology who were educated in cultural anthropology. Most primatologists are committed to the neo-Darwinian evolutionary paradigm, as natural scientists. Most anthropologists are not so committed, and think of themselves as social scientists, if they consider themselves to be scientists at all. This is a recipe for misunderstanding.

Illumination may come from a completely different mammalian order, the Cetacea. Despite the obvious logistical difficulties of studying aquatic culture, recent research on dolphins and whales is productive and provocative (Rendall & Whitehead 2001). Cetaceans do things that no ape has been seen to do: chimpanzees form small coalitions; dolphins form larger coalitions of coalitions. Cetaceans do things that we primates cannot even imagine: we can only guess at the networked communicative capacities of an echo-locating pod of orcas.

Culture Is Not Behavioral Diversity

Who can fail to be moved by the richness of human cultural diversity? Nowadays, every city has a wealth of ethnic cuisines, so that even a humble onion can turn up on a plate in a wonderful variety of forms. Satellite television brings this cross-cultural variation into our living rooms. So, it is argued that if we find behavioral diversity in apes, then they must be cultural.

True, field primatologists realized in the 1970s that we could no longer speak of The Chimpanzee. (If pressed for a milestone, one might point to Menzel's symposium on precultural behavior in primates, held at the 1972 International Congress of Primatology.) Instead, every study, whether at Bossou, Gombe, Mahale, or Taï, seemed to reveal a new twist, if not a new behavioral pattern. This eye-catching reporting continues, and the ethnography piles up, so it is hard any more to keep straight which population of apes does what. Surely, there is now a need for a Chimpanzee Relations Area File, else how can anyone keep up with the information?

However, if diversity means differences across equivalent sets, and such sets are hierarchical, there is a potential for confusion across levels. Individual variation is easy to see in primates, as Imanishi and his students showed for wild Japanese monkeys, and Köhler even earlier showed for captive chimpanzees. In chimpanzees, each alpha male has his own style. But individual differences are usually seen as something for personality psychologists to study, not as matters of culture.

At the other extreme, species differ in behavior, and sibling species like chimpanzee *(Pan troglodytes)* and bonobo *(Pan paniscus)* provide a pastiche of similarities and differences (Stanford 1999). But these are not cultural matters, any more for primates than they would be for congeneric lion and leopard, and are usually left to comparative ethology or psychology. (Such boundaries are more fuzzy for artifacts found where and when anatomically modern humans and Neanderthals coexisted. Who made what may or may not be cross-cultural diversity!)

More to the cultural point is diversity at the level of community, population, or subspecies. In chimpanzees, neighboring unit-groups at Mahale show different versions of the grooming hand-clasp (McGrew et al. 2001). The separate but nearby populations of Mahale and Gombe have contrasting cultural profiles (Nishida 1987). Far western African chimpanzees *(P.t. verus)* are nut crackers, while the other sub-species (*P.t. troglodytes* and *P.t. schweinfurthii*) show none (McGrew et al. 1997). All of this diversity could be cultural.

However, lots of species show behavioral diversity. In nonapes, hamadryas baboons on opposite sides of the Red Sea in Arabia and Ethiopia differ in group size and composition (Kummer 1995). In nonprimates, sea otter lithic technology varies along the Pacific coast. In nonmammals, scrub jay family structure differs from Florida to Arizona to California (Woolfenden & Fitzpatrick 1984). Whether any or all of this variation is cultural or merely reflects environmental constraints is a matter of investigation, but not assumption.

Kummer (1971) also pointed out a third way that behavior may vary, in addition to nature (environmental dictate) or nurture (social learning). Organisms also may learn much individually from interaction with the nonsocial environment, by trial and error. If such learning (e.g., predator avoidance) occurs in parallel, even simultaneously across individuals, it may appear social when it is not. No one needs to show us that the sun is hot; we all learn this for ourselves at a young age.

A more vexing problem in judging behavioral diversity is quality of data. Anecdotes are of little use, as a single event can be a coincidence, subject mistake, observer error, or hoax. At best, an anecdote alerts us to a possibility. Equally limited in usefulness is idiosyncrasy. This eliminates all the problems of anecdote, but an act done by only one individual, however often, can hardly be cultural, as it is asocial. Even a habit, that is, behavior done repeatedly by several group members, is but a hypothesis. Only customs, that is, acts performed normatively by appropriate subsets within a group (e.g., cooks cook, soldiers fight, elders advise), are evidence of culture. Such distinctions are often ignored by eager reporters and may make ethnology difficult (Whiten et al. 1999).

Finally, behavioral diversity is neither a necessary nor sufficient condition for culture. Kwakiutl eat elephant seal but Bantu eat elephant. This is diversity but need not be culture, as no Kwakiutl ever met an elephant, nor Bantu an elephant seal. All the world now consumes carbonated cola drinks. This is global uniformity, but it is still culture. The custom has spread, not the geographical distribution of *Cola* trees.

Thus, in seeking culture, behavioral diversity is just a possible starting point.

Culture Is beyond Social Learning

In its broadest sense, social learning occurs when information gained from others of the same species alters one's behavior, thoughts, attitudes, and feelings (although only the first of these is observable!). This contrasts with information gained from the rest of the animate or inanimate world. So, does social learning equal culture?

Social learning occurs in all classes of vertebrates and in several kinds of invertebrates, which lack true brains. In the short term, honey bees learn from their fellow workers where to find nectar. In the longer term, an octopus may learn permanently to avoid a predator, from one observation of another doing so.

Given how widespread is social learning, many investigators have focused on its mechanisms, that is, on *how* rather than *what*. The array of possible

Figure 16.2. Adult male plays with infant *(right)* while mother grooms her juvenile daughter *(left)*, Gombe.

means (and their associated jargon terms) by which information is transferred is daunting (Whiten & Ham 1993; Byrne & Russon 1998). These distinctions among mechanisms are differently emphasized, depending on the species being studied. Comparative psychologists studying nonhumans spend much time on thresholds (Is stimulus enhancement enough?) or alternatives (Is emulation *really* imitation?) or rubicons (No imitation, no culture!). This can be confusing to nonspecialists.

On the other hand, sociocultural anthropologists studying humans pay little attention to mechanism, being more interested in *what* is passed on. When information is available on processes, it turns out that most customs are transmitted by a melange of passive observational learning. For example, Aka pygmies of Congo learn most of the 50 most important activities of daily life by watching others, not by being instructed (Hewlett & Cavalli-Sforza 1986).

Of the cognitive mechanisms of social learning, teaching is deified by some comparative psychologists. It is said to be essential to culture and unique to humans, and so becomes a hallmark of "true" culture. If teaching is defined as acts by a tutor with the goal of improving the performance of a pupil, then much social learning by humans may not qualify. Formal teaching is absent in most traditional societies, except for specific contexts such as initiation rites. This makes sense: Teaching is costly, in terms of time, energy, and emotion. It is the mode of last resort for social learning, when simpler means fail. Teaching is arguably a curse, not a blessing, made necessary by a large, plastic, and expensive brain (McGrew 2001).

Riddle: When is social learning not really social learning? Answer: When human-reared apes are given "honorary" human status for the purpose of developmental cognitive studies. Thus, our closest relatives are put into experimental settings where humans are their models (and caretakers and surrogate parents and kin). Then, their ability to learn socially from human models is compared to similarly aged human children. The apes cannot win in such a setup. If they perform well, it is dismissed as "enculturation," that is, upgraded ability that is not generalizable to nonenculturated apes. If they perform badly, their inferiority is confirmed. Such an experimental design is sometimes termed "cross-fostering," but of course it is not, since no human child is ever taken from its kind and turned over to apes for rearing. It makes an interesting thought experiment: Who would show more social learning and cultural superiority, a human infant reared by an ape, or an ape infant reared by a human? Arguably, the artificiality of both conditions means that little can be learned about evolved processes of social learning or culture from them.

In summary, there are two alternatives: If culture equals social learning, then many creatures, e.g., octopus, guppy, and lizard, must be granted cultural status. If culture is more than social learning, then we must look elsewhere for essential criteria. On these grounds, it seems sensible to consider social learning as necessary but not sufficient for culture.

Tradition Is Not Enough

Tradition is continuity over time. More precisely, tradition is vertical transmission of information across generations, from old to young. Rarely, when innovation is youthful, vertical transmission may go in the reverse direction.

This occurred with the first spread of sweet-potato washing by Japanese monkeys, but it seems to be rare (Hirata et al. 2001).

Examples of traditions in animals abound. Every year, wildebeests migrate across the Serengeti, whooping cranes winter at Padre Island, monarch butterflies flit to Chihuahua, salmon surge up the Tweed. Some of these traditions have gone on since before the human species emerged. In some cases, especially with migratory birds, we can monitor how offspring retrace the routes of their parents, or even ancestors. In a few cases, we have detailed information: At Gombe, the termite fishing of wild chimpanzees has been recorded over four generations of the "F" family, starting with Goodall's observation of the matriarch Flo in the 1960s.

For human beings, the central role of tradition is clear. All human societies emphasize origin myths, however fanciful; attention to tradition, such as appeal to ancestors, is a human universal (Brown 1991). People who tell stories to sociocultural anthropologists often stress that they have *always* done things a certain way. Conversely, people who fail to keep traditions may be severely punished.

So, does culture equal tradition, and tradition equal culture? No, it is not that simple. First, some information is transmitted genetically across generations. This is deucedly difficult to establish in the wild, where variables of nature and nurture are confounded, even for behavior. Do generations of warthogs wear down a path to a water hole because it is their cultural inclination, or because it is the most energetically efficient or least predator-risky route? Even if we were lucky enough to be there to see them tread a new trail, would it be from whimsy or from changed (but unseen to us) environmental contingencies?

Second, not all transmission of information is vertical. Some is horizontal, within generations and across peers. The power of human "popular culture" is impressive—ask any teenager. But horizontal transmission of culture is more than fad. Opie and Opie (1987) showed that some aspects of children's culture, such as jumping-rope rhymes, were maintained for centuries by horizontal transmission, child to child. Thus, there is traditional culture, but no intergenerational transfer.

Finally, even if we learn from our elders, and pass on those customs to our successors, those traditions need not be cultural. Whether the models are kin, companion, or even stranger need make no difference. I learned to fish from my uncle, but I now realize that he had idiosyncratic techniques. I

Figure 16.3. Two adolescent male chimpanzees play with adolescent olive baboon, at Gombe. (This is apparently unique to Gombe.)

make pineapple upside-down cake using the recipe that is a family tradition, but it is unlike the same dish made by the other matrilines in my culture. These are surely traditions, but whether or not they are cultural, in the rich sense, depends on definition. After all, most people are not anglers, and who knows how many folks have ever made such a cake, or passed it on to their descendants? How normative does a behavioral (or cognitive or emotional) pattern have to be to be cultural?

Some particular aspects of tradition as culture have been singled out as crucial, such as the ratchet effect (Tomasello 1999). Here, information not only spreads but accumulates; thus, with each transmission, either vertically or horizontally, new "mutations" (memes?) enrich the message. We stand on the shoulders of our predecessors, making progress. This is said to be uniquely human, and so is presented as both a necessary and sufficient condition for culture. It is neither. Putting aside the problems of misreplication and maladaptation, racheting is neither unique nor universal. Since its invention in 1956, wheat sluicing or washing by Japanese monkeys has elaborated and diversified (Hirata et al. 2001). Imo's initial technique now looks crude, and successive generations have left it behind. Equally, the evidence for ratcheting in the human ethnographic literature is slim. Most ethnographers of traditional societies report stasis, not dynamic change.

So, even if tradition is a necessary condition for culture, at least in the long term, it is not a sufficient condition, at least as the term is used here.

Culture without Language

Language is everyone's favorite example of human culture. Each of us imprints upon what we hear in the cradle; some of us go on to learn more than one language. It is all a matter of exposure. No one ever suggests that the brains of infants born in Patagonia might be more genetically receptive to hearing Patagonian than Danish. Instead, we see the results of an inadvertent but global experiment in cross-fostering: A Korean newborn adopted by a Canadian grows up to speak English, or French, or both, but not Korean.

Thus, it was no surprise that the first published response to a claim of a social custom in chimpanzees (McGrew & Tutin 1978) was to deny it on the basic grounds that nonlinguistic creatures could not have culture (Washburn & Benedict 1979). This belief that language and culture are isomorphic is widespread.

For humans, the evidence is too strong to make a claim about the relationship between language and culture. All known human societies have both, so there is no informative variance. One could enable the other or vice versa, or both could be a byproduct of a third phenomenon (e.g., big-brained intelligence), or each could be independently derived (e.g., language from vocal communication and culture from extractive technology). We would need to have cultures without language or languages without culture to test the relationship. We could seek correlations between linguistic variables, such as vocabulary size, and cultural variables, such as technological complexity, but this seems not to have been done.

Or, it may be that apes are helpful models. Despite the huge and contentious published literature in "pongo-linguistics," there is no consensus (Savage-Rumbaugh 1998). There seems to be a positive correlation between time spent with chimpanzees and conviction that they are capable of language (but note the careful wording!). Thus, the most dismissive critics have spent no time with apes. On the other hand, there is another correlation among chimpologists, so that the strongest claims come from the researchers using the most artificial systems of linguistic communication, e.g., Yerkish (Savage-

Rumbaugh 1998). Until an open-minded linguist is willing to go to the field and take wild chimpanzees as they are found, we are unlikely to know more. This seems a simple request, but it is yet to be done.

One problem is definition. Clearly, full-blown human language must be both semantic and syntactic. It can be cognitive or communicative or both, but the latter is easier to measure. Only spoken language is universal across human societies, but the acoustic-auditory modality is neither necessary nor sufficient. Deaf people read lips and hearing people make signs. Both vocalize paralinguistically, as do many other organisms.

In principle, many functions of language could serve culture. Cognitively, language could have evolved as a labeling or filing system, whether for numbers, ideas, or identities. Such symbol-use need have no social function. Communicatively, language is a useful way to transmit information, especially abstract and arbitrary thoughts. Such symbol-use is necessarily social, as sender and receiver must share a common language for it to work. So far as we know, all humans normally use language both cognitively and communicatively, but these could be decoupled in apes.

In studying culture, the main strength of language is also its weakness. To get beyond behavior (which is directly observable) to knowledge and meaning (which are not), anthropologists like everyone else rely on verbal report from informants. On the other hand, much information about feelings can be inferred from "body language," especially with training. However, speech is a double-edged sword, as informants may bare their souls or lie by commission, omission, or imprecision. Deception by paralanguage seems to be harder, especially with involuntary responses, such as blushing. If your life depended on detecting deception, which would you trust, the content of a word or its spoken inflection? If you chose the latter, then you might not want to trust entirely in verbal report as the sole indicator of culture, in human or chimpanzee.

This dispatch carefully avoids passing judgment on whether or not apes have language. The aim is to show that language and culture are separable. The two are no more necessarily tied by causal co-incidence than are language and bipedality. Communicative language may be a sufficient condition for culture, but it is not a necessary one.

Culture Is by Definition

Definitions of culture are a dime a dozen, and most are of little use. Most encapsulate an idea or set of ideas, but few are heuristic for pursuing the possibility of nonhuman culture. Especially exasperating are the epigrams beloved of introductory textbook writers: "Culture is what makes us human," "Culture is the human ecological niche," or, "Culture is to human, as water is to fish." Of what empirical use are these?

Equally frustrating are the historical antecedents. Every textbook of introductory anthropology gives Tylor's (1871) seminal definition of culture as "that complex whole which includes knowledge, belief, art, law, morals, custom, and any other capabilities and habits acquired by man as a member of society." Putting aside the inherent sexism, one is left with a vague, all-embracing entity that may include all but digestion and respiration (but then think of antacid tablets and yogic breathing). Something that explains everything explains nothing.

Some have advanced checklists of features, much as Hockett (1960) did for language. Kroeber (1928) tried this for chimpanzee dancing, saying that if it showed innovation, standardization, diffusion, dissemination, durability, and tradition, it would qualify as culture. By analogy, we can recognize a luxury car by ticking off its features, and a BMW will pass and a VW will fail. Does this mean that chimpanzees are 83.3 percent of the way to being cultural if one can tick off five of the six features? Given long generation times in great apes, one would have to wait years, from innovation to tradition, assuming that the other conditions were met along the way. In studying human culture, how many ethnologists in the field have been so patient as to tick off six of six?

More productive in an operational sense may be criteria that approximate essentials, which together capture the gist of culture. Consensually, all seem to agree that culture is *learned* (rather than instinctive), *social* (rather than solitary), *normative* (rather than plastic), and *collective* (rather than idiosyncratic). This minimal combination is a starting point for necessary and sufficient conditions for attributing culture to an organism. Unfortunately for anthropocentrists, the chimpanzees' grooming hand-clasp meets all four criteria (McGrew & Tutin 1978; McGrew et al. 2001).

Another approach is to ask ordinary people what culture (in the rich sense) means to them. When anthropologists do this, and if their informants are patient enough to put up with such a simple-minded question, then the answer is usually some version of: "culture is *the way we do things.*" This elegant phrase contains at least four elements: overt action ("do things"), norms and standards ("the way"), collective consciousness ("we"), and sense of identity (as implied by the whole phrase). So, does this apply to chimpanzees?

Overt action is the easiest, as it is seen in both the behavior and the artifacts of elementary technology. Chimpanzees have both tool-kits and tool-sets (McGrew 1998). Norms and standards are revealed by behavioral diversity at the levels of group, population, and subspecies (Whiten et al. 1999). All known chimpanzee groups, populations, and subspecies scratch themselves, but only M group at Mahale in East Africa does the social scratch, which they do often and predictably (Nakamura et al. 2000). Collective consciousness is pointedly manifest in deadly xenophobia. Parties of chimpanzee males patrol boundaries and kill neighboring rivals, but usually only if three or more aggressors can catch the victim alone (Wrangham 1999). A sense of identity can be inferred when an immigrant female changes her style of doing a common behavioral pattern from that of her community of origin to that of her community of adoption. Their old way of doing things becomes her new way of doing things.

Definitions are useful only if they clarify matters. All else is pedantry. Define culture as you must to tackle the question at hand, just make it clear, fair, and most of all, productive.

Culture Is Collective

Culture is social (as opposed to solitary), but sociality is only a starting point. Collectivity implies much more. When 51 percent or more of a group behaves in concert, the act becomes a statistical norm, and therefore is typical of the group. Herring shoal, geese flock, bison stampede. Instead, collectivity entails group-oriented action, often with roles, as in an orchestra, which is more than many instruments being played at once. Empirically, roles are social traits that are not intrinsic to actors, but are transferable, able to be donned as well as shed, like clothing.

Further, to say that something is collective is not to say it is unanimous, but often just the opposite. How often do all humans in a group act, think, or feel as one? Not all humans ski, even in the Alps. Instead the collective is often a subset according to sex, age, kinship, status, and so on in relation to other subsets. In the short term, such collective action is manifest in convention: gentlemen rise, underlings bow, grandparents dote. In the long term, there are institutions: marriage, rite of passage, funeral.

The results of collectivity in culture are emergent properties. By this argument, one cannot break up culture into its components and then recombine those components to reconstitute culture. (Any more than one can reduce an animate organism to its constituent proteins, amino acids, and peptides and then bring it back to life.) Cronk (1999) made the same point about culture in another way by saying that we cannot explain behavior in terms of behavior. Thus, culture defies reductionism and is pervasive.

Finally, it is possible to be a collective creature on many levels at once. One is a European, German, Bavarian, and Münchener at the same time, and each may be indicated differently, by passport, language, patriotism, and taste in beer.

All of the above indubitably applies to human culture, whether in the New Guinea highlands or a Manila barrio. To what extent does any or all of this apply to nonhumans, such as the chimpanzee? At first glance, the task seems impossible. How can we possibly know the mind of an ape, if we have so much trouble comprehending the minds of our fellow humans?

The easiest starting point is the "quack test." The more it looks, sounds, smells, and feels like a duck, the more likely it is to be a duck. For example, grief is apparent in a chimpanzee mother after her infant dies. She is alternately agitated or subdued, distracted or focused. She may be more solicitous to her infant's body than she was to it in life. She seems to be grief-stricken, but this does not mean that she is in mourning, for that is a collective action, not individual sorrow. A sociocultural anthropologist faced with this set of circumstances would adjust her lens accordingly; cultural primatologists need to learn to do so.

One can be guided by function, for evolution ultimately boils down to outcomes in response to natural selection. Dead humans do not pass on genes any more than do dead flatworms (cryogenics apart!). Territorial aggression toward outsiders is natural; a simple rule of "Welcome familiars, but resist strangers" may be enough. Xenophobia is cultural; it is a social phenomenon based on "we" versus "they," and so comes down to collective identity. We can see this when chimpanzees immigrate or congregate. Marshall and colleagues (1999) showed that the long calls of a motley assemblage of captive chimpanzees converged to create a recognizable conventional signal. Regardless of their disparate origins, the apes created a collective dialect.

Lest we despair at the task of operationalizing beliefs and attitudes in other species, there are precedents: Tactical deception seemed intractable until Byrne and Whiten (1988) illuminated it. All is inference; the challenge is to find such ingenious ways to increase the probability of more and more accurate inference. It seems likely that social dominance in chimpanzees is a matter of personalities embedded in a collective context (de Waal 1996). Accordingly, we are likely to understand it only if we act as cultural primatologists,

Figure 16.4. Two allied males in a quiet moment of contemplation, Mahale K-group.

albeit haltingly. We will likely never interview chimpanzee informants, but we can use ethological methods to seek chimpanzee uniqueness, as well as universals.

Culture Has Escaped from Anthropology

The concept of culture emerged as the core of anthropology in the 1870s, and remained therein for more than a century. Anthropology has been termed the science of culture, and most members of the American Anthropological Association label themselves as cultural anthropologists.

Yet all along, the question of nonhuman culture has lurked in the wings. Morgan (1868), arguably the founder of American ethnology, extolled the technology of the beaver at the same time as he described Iroquois kinship. Kroeber's (1928) consideration of comparative possibilities across species was cited above. Benedict (1935) supported a graduated transition from noncultural lower animals to cultural man. The first primatologist to propose cross-cultural studies was Imanishi (1952), based on early observations of Japanese monkeys. The first explicitly titled book on the subject seems to be *Precultural Primate Behavior,* edited by Menzel (1973). The first systematic analysis across cultures of chimpanzees covered six wild populations studied for 151 years in total (Whiten et al. 1999).

Cultural primatology is not bounded by anthropology, and the culture concept has diffused to other disciplines. At least three make distinct contributions in very different ways (McGrew 1998): Anthropology asks *what* questions about the constitution of culture, whether these be artifacts in the past or rituals in the present. This is culture as phenomenology. Psychology asks *how* questions about the mechanisms and processes of culture, especially its inventions and their spread. This is culture as information transmission. Zoology asks *why* questions about the survival value and fitness of culture, using the ideas of neo-Darwinian evolutionary theory. This is culture as adaptation. Luckily, cultural primatology calls on all of these points of view.

However reluctant some anthropologists may be to give up exclusive jurisdiction over culture, or to extend the concept to nonhuman species, it is happening anyway. The best strategy for retaining culture in the field of anthropology is to set explicit standards to be met by cultural primatologists and let the chimps fall where they may. Echoing Kroeber, cultural anthropologists should operationalize their criteria and ask primatologists to meet their challenge. The latter are entitled to ask what it would take to satisfy anthropologists and then to go back to the field to seek it among the apes. It is easier to seek holy grails if you know what to look for.

Finally, there is a delicious irony. Some proportion of sociocultural anthropologists find the concept of culture to be outmoded, and even obstructive (Kuper 1999). This is hard for nonspecialists to understand, almost as if musical chairs could somehow be played in silence. How strange to think that finally when cultural primatology realizes how much it needs cultural anthropology, the latter may drop its central tenant.

Culture Is Rich and Complex (But So What?)

Paraphrasing Groucho Marx, a chimpanzee thinking of joining the Culture Club might hesitate to do so, on grounds of suspicion of any club that would have her as a member. Quoting Boy George, the androgenous pop star, if embracing cultural relativism, and so becoming, chameleon-like, "a man without conviction" is the price, it might be too high.

Culture may be a curse, as well as a blessing. Chimpanzees at Bossou have invented "pestle pounding," in which the crown of an oil palm is the mortar smashed by a detached frond as the pestle (Yamakoshi & Sugiyama 1995).

The result is a rich, pulpy soup and a good meal, but it likely kills the palm in the process. This is short-term gain but long-term loss, as the palms, like all large organisms, are slow to replace themselves.

Culture may be overrated in several ways. For example, culture is not an explanatory variable. One cannot explain behavioral diversity just by saying that culture made them do it. Cultural determinism is just as silly an idea as genetic determinism. Furthermore, there is often a misplaced value judgment: Culture does not make an organism cooperative any more than nature makes it competitive (Wrangham 1999).

Finally, culture may not be the key to understanding chimpanzee society anyway. As with humans, the more informative level may be one step down, in subculture. Just as it may be simplistic to assume that there is such a thing as (North) American culture, so it may prove for chimpanzees and other species of primates. There is plenty of evidence that shows life as a whole among lower-ranking members of the group to be very different from that of the high-rankers, from sex ratio manipulation to leisure-time pursuits. It may be that cultural primatologists will have to take account of caste or class among their subjects, and seek help from sociologists.

In any event, if nonhumans have culture in any form, then we must be concerned with cultural survival. Just as cultural anthropologists are active advocates on behalf of the traditional societies that they study, so must cultural primatologists do the same. Conservationists may seek to save the species *Pan troglodytes,* but cultural primatologists must seek to preserve cultural diversity. This means going beyond a few famous, long-term study sites like Gombe or Taï. It means safeguarding Tenkere, where the apes make cushions and sandals (Alp 1997), and Tongo, where the apes dig up tubers for moisture (Lanjouw 2002). Both of these populations are unprotected and on the verge of extinction. What a pity it would be to lose them.

Conclusions

So, are chimpanzees cultural creatures or not? We cannot yet say, to everyone's satisfaction, but the mounting evidence gives a rationale for cultural primatology. If this trend continues, then we must move on from doing beginning ethnography to doing full ethnology. Chimpanzees may use kinship terms (and a guess is that these will be found in their soft grunts, so far undeciphered) and they may have worldviews (for they seem to spend enough time musing). It may be that the ultimate function of culture for community-living apes is social identity. Just as human languages proliferated in areas where there were many distinct human groups, so may it be for our nearest living relations, where cultural identifiers tell who you are and where you come from. It is up to cultural primatologists to find ways to pursue these questions, and so to draw ever-stronger inferences.

Finally, if any of these arguments has merit, then we humans may need to re-think the boundaries of multiculturalism. We may need to be more inclusive in extending our appreciation of cultural diversity beyond anthropocentrism to admit our cousins, the great apes.

CASE STUDY 16A

Spontaneous Use of Tools by Semifree-ranging Capuchin Monkeys

EDUARDO B. OTTONI AND MASSIMO MANNU

The studies on the spontaneous use of tools and the diffusion of behavioral traditions by nonhuman primates have focused almost exclusively on Old World monkeys or hominoids. The variety and regional diversity of the tool kits used by wild chimpanzees, in particular, led to the growing use of the term "culture" applied to species other than ours. The most clear cultural difference among chimpanzees across Africa is the cracking of nuts by western populations, with the aid of stone "anvils" and "hammers" of stone (Matsuzawa 1994) or wood (Boesch & Boesch 1983). This behavioral difference cannot be explained away by simple ecological determinants. Chimpanzee nut cracking provided evidence of intergroup diffusion by migrating females (Matsuzawa 1994), and of active teaching of infants by their mothers (Boesch 1991). Though the use of stone tools by our sister species seems to anticipate the technology of early hominids, it can also be observed in a New World species, the tufted capuchin monkey *(Cebus apella)*.

Many laboratory experiments on tool use have been carried out with capuchins (see, for instance, Visalberghi 1987 and Westergaard & Suomi 1993—both involving induced nut cracking). Very few studies, though, were conducted in more naturalistic conditions, and there are only sparse reports from the wild (Fernandes 1991; Langguth & Alonso 1997). We outline here the spontaneous use of tools for cracking nuts by a semifree-ranging group (see Ottoni & Mannu 2001 for a more detailed account).

Our subjects live in a reforested area of 180,000 square meters in the Tietê Ecological Park (a restricted-access area in this public park near São Paulo, Brazil). The group was composed of 18 individuals: four adult males, three adult females, two subadult males (five to eight years old), one subadult female, four juveniles (two to five years old), and four infants. All were born in the park, with the exception of the five older adults, which were probably poached in the wild (though there are no individual records available), apprehended and released in two park islands, from where they subsequently escaped. Besides being daily provisioned by the park staff with fruits and vegetables, the group consumed a large array of naturally available plant and animal food items, whose search and consumption frequently involved complex manipulatory behaviors.

We observed the cracking of the small, hard-shelled coconuts of the palm tree *Syagrus romanzoffiana* directly ("all-occurrences" procedure) and indirectly (weekly inspecting all known nut-cracking sites). Mature nuts are cracked open with the aid of two stones, a flat "anvil" and a "hammer," in a manner very similar to what Matsuzawa (1994) described among Bossou chimpanzees. Cement pavement and partially buried stones can also be used as anvils (Figure 16A.1). Hammers are usually raised with both hands to head level, then slammed against the nut on the anvil. The tail is propped against the ground, acting as a third supporting point. Whereas hammering involves mainly strength and body equilibrium, positioning the nut requires careful manipulation.

We called nut-cracking sites all hammer-anvil sets whose use has either been directly observed or inferred from associated nutshell remnants and scratches in the stones. After one year, there were 136 mapped sites, with a monthly mean of 37.2 confirmed used in the last six months.

It was suggested that the use of stone tools by capuchins in the laboratory could be an artifact, not to be expected in the wild, owing to their arboreal lifestyle. However, in savanna-like environments, or even in the rain forest, capuchins may spend considerable time on the ground. Their nut-cracking bodily postures—as well as their dexterity when walking bipedally, sometimes carrying stones or other objects in their hands—show they are well adapted for ground activity.

We do not know whether nut cracking was brought by one of the older,

Figure 16A.1. Quinzinho (male juvenile) cracking a nut, with cement pavement as an anvil.

presumed wild-born animals, or was independently invented in the park (it cannot have been learned from humans, who do not consume these nuts), but until 1995 there were no reports of this behavior (easily detected by the noise), suggesting it was not as widespread as it has become. All monkeys were observed at least once manipulating nuts and/or stones—and all besides the three youngest infants actually cracked nuts—though there are great individual and sex-age class differences in frequency and proficiency.

Juveniles were the most active nut crackers, accounting for more than half of the 299 observed episodes, perhaps because competition restricts their access to more easily consumable items. Manipulatory activity, though, is very attractive to juvenile capuchins, even when there is no contingency between manipulation and food acquisition (Fragaszy & Adams-Curtis 1997). Besides, juveniles are bolder explorers, spending more time on the ground. Adults, though, seemed to be more efficient nut crackers, and among them, males were responsible for significantly more episodes.

Another 38 episodes of stone manipulation were classified as *inept* for involving a nonproductive procedure, resembling what Visalberghi (1987) observed in captive capuchins, and Inoue-Nakamura & Matsuzawa (1997) reported among wild infant chimpanzees. The prevalence of juveniles in the absolute frequencies of inept manipulation was to be expected, but proportional values point to a significant predominance of infants and a gradual decrease with age.

About 17 percent of the nut-cracking episodes were closely watched by other individuals, which sometimes ate leftovers or manipulated the stones —usually after the nut cracker left. Observation is mainly done by infants and juveniles, which could be due to a more intense exploratory activity, or to greater tolerance by nut crackers to youngsters. Being highly tolerated—and even carried by older individuals in their first weeks—infants can watch most nut crackers, not just their mothers.

Proficiency in nut cracking takes years to be reached, and the episodes where youngsters observed more skilled individuals are potential opportunities for some sort of observational learning, even if only as a starting point for an individual process of improvement by trial and error.

The convergence in the use of tools, and the common features between chimpanzees' and capuchin monkeys' biology (big brains and long lives) and sociality seem to favor van Schaik and colleagues' (1999) hypothesis equating the keys to tool use with the proper ecological niche (such as extractive foraging), individual cognitive abilities for acquiring complex skills, and a tolerant society.

17

Society and Culture in the Deep and Open Ocean: The Sperm Whale and Other Cetaceans

HAL WHITEHEAD

In this chapter I move out to sea and encounter a species and a habitat that are less accessible than the subjects of the other chapters of this volume. We cannot keep adult sperm whales (*Physeter macrocephalus;* Figure 17.1) in captivity or live-capture them, even temporarily; we see very little of their behavior; individuals cannot usually be identified in real-time, but their ranges exceed one kilometer vertically and 1,000 kilometers horizontally; and the only experiments we even contemplate are the crudest of playbacks. Despite these drawbacks, the deep and open ocean may hold important clues as to the evolution of social complexity and culture. There also may be sophisticated intelligence out there, although any substantive discussion of it goes beyond current information.

In 1976, Nicholas Humphrey published a remarkable essay, which in some ways sparked the Machiavellian intelligence hypothesis: that there is an evolutionary link between social complexity and cognitive evolution (Byrne & Whiten 1988). Here is part of what he said: "the open sea is probably an environment where technical knowledge can bring little benefit and thus complex societies—and high intelligence—are contraindicated (dolphins and whales provide, maybe, a remarkable and unexplained exception)"

Figure 17.1. A group of female sperm whales. Illustration by E. Paul Oberlander, Woods Hole Oceanographic Institution, from a photograph by Flip Nicklin.

(Humphrey 1976, p. 311). In this chapter, I will try to go a little way toward explaining Humphrey's conundrum.

The social structure of a population is to a large extent an adaptation to the environment, although there are feedbacks. For instance, the social structure of males is often considered to be consequent on that of the females (Wrangham & Rubenstein 1986). Culture, which I will define as "information or behavior shared by a population or subpopulation which is acquired from conspecifics through some form of social learning" (Rendell & Whitehead 2001b), is highly related to both environment and social structure, although once again there are feedbacks, as when a cultural badge of group membership can provoke cultural group conformity (Richerson & Boyd 1998). Culture can itself have a large bearing on social structure—most elements of human social structures are culturally determined.

In this chapter, I will discuss society and culture in the context of the deep and open ocean and one of its most emblematic inhabitants, the sperm whale. The ocean far from shore has characteristics radically different from the habitats of the terrestrial mammals and birds considered in other chapters of this book, but there are also contrasts with the near-shore habitats of the more easily studied cetaceans (see Chapters 2 and 13 and Case Study 4B). How have these attributes affected the social systems of the animals that live out there? Of the animals that spend their entire lives in deep waters, the society of the sperm whale has been best studied, and will be the focus of much of the chapter. There seem to be some elements of sperm whale behavior that are culturally determined, and I will move on to a consideration of culture in whales and dolphins and a discussion of gene-culture coevolution in these animals.

The Deep and Open Ocean as a Habitat for Large Animals

Some of the characteristics of the sperm whale's habitat are obvious, but bear repeating because they contrast with the terrestrial habitats of humans and most other species being considered in this volume. The deep and open ocean is buoyant, three-dimensional, and highly connected, with no barriers (except continents) and few immovable features (the occasional seamount, perhaps).

Other characteristics are less obvious but probably no less important. One of these is variability. Compared with terrestrial habitats, the *physical* ocean is remarkably stable over small spatial and temporal scales: move a few meters, or kilometers (horizontally), wait an hour or even a few days, and the temperature, salinity and other physical characteristics of the ocean will likely have changed very little (Steele 1985). However, over larger scales of hundreds of kilometers, months, and more, variability becomes more substantial (Steele 1985): an "El Niño" may have struck, or a "cold-core ring" spun off the Gulf Stream bringing very different physical characteristics to large ocean areas. The smaller short-lived oceanic organisms near the base of the food web seen have adapted to this pattern of variability, often using specialist-plus-dispersal strategies. A species will thrive in its preferred conditions, which usually persist throughout a lifetime, but then, because its descendants may have to deal with a very different environment, an organism spreads its offspring widely, using the spatial variability and connectedness of the habitat to compensate for its long-term temporal variability. Thus, in the ocean, there are blooms and patches and turnovers of plankton. And so, for larger organisms that depend directly or indirectly on plankton for sustenance, the ocean is a particularly unstable *biological* environment over medium to large scales (days to decades, and hundreds of meters to thousands of kilometers), in fact less predictable than the biotic environment on land (Steele 1985).

Thus, in any place the feeding success of the sperm whale, and probably most other deep ocean vertebrates, varies enormously over time scales of months and more (Whitehead 1996b). How do the larger animals of the ocean deal with such variability? Like the smaller organisms, they use spatial variability to average out temporal variability, but must do this within their own lifetimes, in fact within the interval within which they can store sufficient food to survive (perhaps three months for female sperm whales, Whitehead 1996b), so effective movements are crucial. Luckily, other characteristics of the ocean, particularly its buoyancy and connectedness, make such long-distance movements relatively cheap, fast, and easy (Williams et al. 1992). Any traits that improve the efficiency of such movements, such as learning good movement strategies from more experienced individuals, or directing the movement of inexperienced but related animals, will also be advantageous.

Over medium scales (kilometers and hours to days), the clumping of prey also has potential social consequences. Social behavior may help animals find and efficiently exploit large but rather unpredictable patches of prey through cooperative searching and foraging (Norris & Schilt 1988; Barrett-Lennard et al. 2001). Such resources are often large enough for many animals to feed plentifully, and can rarely be monopolized in any effective manner.

Living in three dimensions makes defending anything, be it a food patch or female, considerably more difficult than for species of two-dimensional habitats. There is no evidence of territoriality among the animals of the pelagic (three-dimensional) ocean, and we might, in such a situation, expect low levels of conflict within or between groups.

Most animals are both predators *and* prey. Even the very large are vulnerable; for the sperm whale, killer whales *(Orcinus orca)* are a real threat (Jefferson et al. 1991; Pitman et al. 2001). The structure of the oceanic habitat has consequences for confronting such threats. There is nowhere to hide, and the three-dimensional habitat structure makes defense against predators, and perhaps also vigilance, more difficult than on land. However, social structures can help, and communal vigilance for predators (Norris & Schilt 1988) and communal defense against them (Arnbom et al. 1987) have both been considered as important functions of the social structures of pelagic animals.

Social Structure of Sperm Whales

Sperm Whale: An Animal of Extremes

For more information on sperm whales, see Rice (1989) and Whitehead & Weilgart (2000). All information on sperm whales given in this section can be found in these reviews, except where a specific reference is cited.

The sperm whale (Figure 17.1), the largest of the toothed whales, may reach 18 meters and 57 metric tons. It has the largest brain on Earth (new evidence is suggesting, at least within the primates, that absolute brain size is a better predictor of some cognitive abilities than relative brain size; Beran et al. 1999). With the possible exception of the bottlenose whales (*Hyperoodon* spp.; Hooker & Baird 1999), sperm whales are the longest and deepest mammalian divers (more than 1,000 meters for more than one hour, although dives for 35 minutes to about 400 meters separated by about eight minutes of breathing at the surface are more normal; Papastavrou et al. 1989). They possess in the spermaceti organ (a huge nose highly modified for sound production) what is almost certainly the most powerful sonar system in the natural world (see Møhl et al. 2000). The sperm whale is also extreme among the cetaceans in sexual dimorphism. Adult males (at about 16 meters in length and 45 metric tons) are roughly one and a half times as long and three times the mass of adult females (11 meters and 15 metric tons). The sexes differ in other respects. Whereas males may be found over any deep water between the Arctic and Antarctic ice packs, females are generally restricted to latitudes less than 40°S and 45°N. The sexes also possess conspicuously different social structures, as I will discuss in the next sections. These and other characteristics give sperm whales a remarkable ability to extract resources from their habitat, particularly midwater and bottom-dwelling squid of sizes ranging from a few hundred grams to a few hundred kilograms. Although no reliable estimates of sperm whale populations exist, a very rough estimate is that the sperm whales currently extract about 100 million metric tons of biomass from the ocean each year (Clarke 1977). This is similar to the total from all human marine fisheries, and perhaps greater than that of any other vertebrate species.

Male Sociality

The social structure of many mammal species is built around females and their dependent young, but as in so many other biological characteristics, sperm whales take this to an extreme. However, the males are interesting, both as a contrast to females, and for their own role in sperm whale society.

A male spends the first few years of his life with his mother in her social unit. The age of dispersal from the maternal unit is very variable but seems to average about six years (Best 1979; Richard et al. 1996). He becomes gradually sexually mature in his teens but continues growing until he is about 35–60 years old (Rice 1989). During this period, there is a general displacement to higher latitudes, so that the largest males can be seen just off the pack-ice in both hemispheres. These mature and maturing males form seemingly loose "bachelor" groups, within which some elements of social structure, such as preferential affiliation, appear absent (e.g., Whitehead et al. 1992; but see Christal 1998). However, groups of males can mass strand, fatally, on beaches (Rice 1989), and sometimes they purposefully swim ashore beside doomed companions in the absence of any common environmental trigger (Lucas & Hooker 2000). What is more "social" than choosing to die together?

Beginning at about age 27, males return to lower latitudes to breed (Best 1979). It seems that they do not necessarily breed in or near the home range of their mothers (Lyrholm et al. 1999), and that they often spend at least two months on a breeding ground (Whitehead 1993), but otherwise we know very little of the frequency, duration, seasonality, or consistency of the males' breeding migrations. While on the breeding grounds, males rove singly among groups of females, usually spending just a few hours with each, but sometimes revisiting a particular group several times over a period of days (Whitehead 1993). While roving, they make distinctive and very loud "slow clicks," which may be a signal to females or other males (Weilgart & Whitehead 1988). There are occasional fights between males (Kato 1984), so that male-male competition is likely an important part of the mating process. However, there are also some indications that female choice may be significant (Whitehead & Weilgart 2000).

Female Social Structure

Most results on the social structure of female sperm whales come from my work and that of my colleagues in the eastern tropical Pacific (but see Gordon 1987; Kahn 1991; Gordon et al. 1998), so the following characterization of the social milieu of a female sperm whale will be based on these studies (Whitehead & Weilgart 2000). However, sperm whale social structures vary with place, probably with time, and have likely been affected by whaling (Whitehead & Kahn 1992). There is also much variation among groups in any place at a particular time (Christal et al. 1998).

Female sperm whales are generally nomadic animals, moving a straight-line distance of between 10–90 kilometers in 24 hours, depending on feeding conditions (Whitehead 1996b). In the eastern tropical Pacific, a female has a range spanning of the order of 1,000 kilometers, which overlaps with those of very roughly 15,000 others. At any time, she is aggregated with, and in acoustic range of, about 20–200 animals, but moves as part of a cohesive group of about 20 animals (Whitehead et al. 1991). However, groups themselves often contain more than one long-term social unit which move together for periods of days (Whitehead et al. 1991).

The social unit consists of about 11 females and immatures (Whitehead et al. 1991). Although matrilineal structures exist within social units, the expected matriline size (numbers of animals with a common oldest living female ancestor) given what is known of sperm whale demography is about two to four (H. Whitehead, unpublished), and not all members of a unit are necessarily matrilineally related (Christal 1998; Mesnick 2001; Case Study 6A; Figure 17.2). Thus, social units contain both matrilineally related and unrelated animals. Although most relationships among females within social units last years, and probably decades, some splits, merges, and transfers between social units have been recorded (Christal et al. 1998), but it is not known whether the splits and transfers divided matrilines.

A foraging group of sperm whales often spreads out into a rank about one kilometer long aligned perpendicular to the direction of movement, and traveling through the ocean at about four kilometers per hour (Whitehead 1989). Young calves, who do not dive for as long as their mothers, move between those adult or sub-adult members of the group (or perhaps unit—it is very difficult to distinguish units within groups at sea) that are at the surface between dives (Gordon 1987). When an adult companion dives, the calf often swims over toward another. Groups containing young calves have generally less synchronous dives than those that do not, suggesting active babysitting (Whitehead 1996a). Groups defend themselves communally against predators; when faced with attack by killer whales or other potential predators, members of a group/unit may adopt either "heads-out" or "marguerite" (tails out) formations (Arnbom et al. 1987; Pitman et al. 2001). There is also good, but not conclusive, evidence for allosuckling within sperm whale units/groups (Best et al. 1984; Gordon 1987).

These structures and behavior, together with the more solitary nature of the much larger mature males, suggest that combating predation, especially on calves, is a major function of the group/unit social structure of female sperm whales (see Best 1979; Gordon 1987; Whitehead & Weilgart 2000). However, this social arrangement may also bring foraging benefits both in the short term, if animals gain information about prey distributions from other members of the feeding rank or can usefully herd prey (Whitehead 1989), and in the long term, if there is communal memory of ephemeral food sources (e.g., those that are available during El Niño conditions; Whitehead 1996b).

Within a cohesive group, animals are more likely to associate with members of their own social unit—i.e., long-term companions—than with members of other social units—i.e., short-term acquaintances (Christal & Whitehead 2001). At least sometimes, this seems to result from the units that constitute the group maintaining, for periods of hours, distinctive positions in the foraging rank, with unit A on the left and unit B on the right, for instance (Christal & Whitehead 2001). In contrast, even in rather well-studied units we find little evidence of social substructure at the sub-unit level (Christal & Whitehead 2001; Figure 17.2), or any clear correlation between genetic relatedness and level of association (Christal 1998; Figure 17.2). Members of units have rather homogeneous social relationships, seemingly treating other unit members as "equivalents" (see Chapter 7). This homogeneity of social relationships within a long-term vertebrate social structure is unusual, and may perhaps be related to the low potential for conflict within units resulting from the structure of the sperm whale's oceanic habitat. A consequence is that we need to look at sperm whale societies at both the level of the individual, and at that of the long-term social unit.

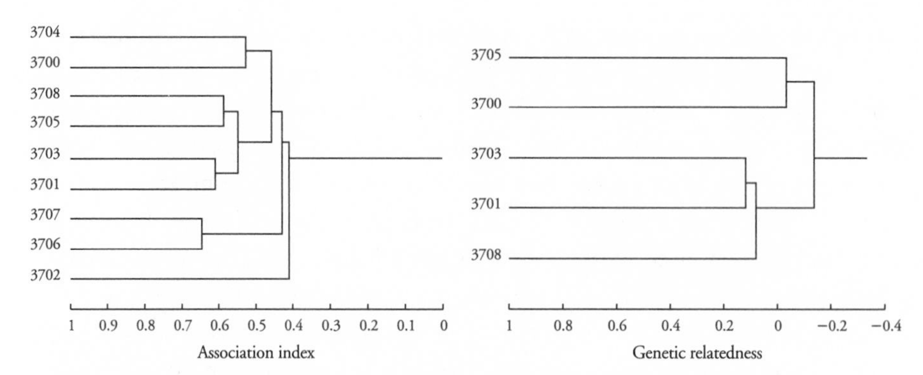

Figure 17.2. Average linkage cluster analyses showing social relationships *(left)* and genetic relatedness *(right)* among a long-term social unit of nine female and immature sperm whales studied off the Galápagos Islands for 18 days in 1999, who were also identified together in 1998. Noteworthy are the homogeneity of social relationships (Christal & Whitehead in press), the low level of genetic relatedness (see Case Study 6A), and the lack of any obvious correlation between genetic relatedness and association (matrix correlation $r = 0.02$, $P = 0.52$, Mantel test). The x-axis of the left-hand diagram is an estimate of the proportion of days on which pairs were photographed diving within 10 minutes of each other, a "simple ratio" association index. The x-axis of the right-hand diagram is an estimate of the genetic relatedness among clusters of animals calculated using nine microsatellite loci for those five of the nine animals for which we have samples. Genetic analysis courtesy of S. Mesnick; see Case Study 6A.

As social units are not readily distinguishable at sea, the latter is not straightforward. However, by comparing occasions separated by weeks or more when photographic individual identifications indicate that basically the same set of whales was present, we have been able to look for characteristics of sperm whale social units. Not surprisingly, there are genetic similarities within groups and units (Lyrholm & Gyllensten 1998), but members also share patterns of markings on their flukes (tails), perhaps because of similar experiences of predation or harassment by killer whales, sharks, or other threats (Dufault & Whitehead 1998). Units have characteristic repertoires of "codas"—patterns of clicks (e.g., "Click-Click-pause-Click") used for communication (Watkins & Schevill 1977)—overlaid on broad geographical dialects (Weilgart & Whitehead 1997). There is also some, not conclusive, evidence that units have characteristic ways of moving through the habitat (Whitehead 1999b). Most surprisingly, we found correlations between the predominant mitochondrial DNA (mtDNA) haplotype of a unit and both its fluke markings and its coda repertoire, although in the latter case the sample size was small—six units—(Whitehead et al. 1998). The most plausible explanation for these results is that the units possess stable cultures that are being transmitted matrilineally in parallel with mtDNA. However, this seems to conflict with results on the multimatrilineality of sperm whale units (Christal 1998; Mesnick et al. 1999; Mesnick 2001; Case Study 6A; see above). Some potential resolutions to this paradox that consider the effects of whaling, currently unrecognized population or social structures, as well as other effects have been suggested (Whitehead 1999a; Mesnick 2001).

Culture in Whales and Dolphins

Sperm whales are not the only cetaceans for which culture has been suggested to be an important determinant of behavior. Luke Rendell and I have recently reviewed the case for culture in whales and dolphins (Rendell & Whitehead 2001a). We conclude that, although the quality of the evidence is very variable, taken as a whole it strongly suggests that culture is an important determinant of many aspects of the behavior of these animals. However, echoing the "chimpanzee culture wars" (see Chapter 16), our conclusions are controversial to some scientists (Rendell & Whitehead 2001b). They question cetacean culture on the same two principal grounds used to attack other nonhuman cultures: by restricting culture only to behavior or information that is transmitted by imitation or teaching and not other forms of social learning (see Whiten & Ham 1992 for a taxonomy of social learning), and by suggesting ways that patterns of behavior, which we ascribe to culture, could have arisen without social learning (Rendell & Whitehead 2001b). Here, I will first justify the broad definition of culture given in the introduction and then summarize the evidence for it in whales and dolphins (for more details, see Rendell & Whitehead 2001a, b and Chapter 2, and, for a primate perspective, Chapters 15 and 16 or de Waal 2000).

Defining "Culture"

Imitation and teaching are important forms of social learning because they can more easily give rise to stable cultures than other types of social learning (Boyd & Richerson 1996), and stability may be a prerequisite for culture to have much effect on genetic evolution (Laland 1992). However, restricting culture only to patterns transmitted by teaching or imitation is unhelpful because:

1. The general theory of cultural transmission and evolution (see Cavalli-Sforza & Feldman 1981; Boyd & Richerson 1985) does not depend directly on transmission mechanisms;

2. It is not clear that much of what we call culture in humans is transmitted by these mechanisms (Boesch 2001; Hewlett and Cavalli-Sforza 1986);

3. Imitation and teaching are not well defined (McGrew 1998; Whiten 2000a);

4. Imitation and teaching are not necessarily more cognitively advanced mechanisms than other forms of social learning such as "goal emulation" (Whiten 2000a);

5. It is usually impossible (especially with cetaceans) to prove, or disprove, that a pattern of behavior common to a population or subpopulation of animals in the wild is transmitted by imitation or teaching, rather than some other type of social learning. Consequently, in many important cases, the presence of culture under such a definition cannot be determined one way or the other (McGrew 1998).

Thus, in common with other potential restrictions to the broad definition that we have adopted (see the beginning of this chapter), the imitation/teaching qualification suffers from both denying important parts of human culture and being unworkable for many nonhuman cultures (Rendell & Whitehead 2001b). However, not all culture is equal, and within the broad definition it is important to consider cultural characteristics such as transmission mechanisms (where they can be determined), type (vocal, material, social), patterns of transmission (e.g., from elder relatives (vertical), or within generations (horizontal); Cavalli-Sforza et al. 1982), stability, evolution, adaptiveness, the level of cultural similarities (group, subpopulation, geographical area, and so on), and the amount of conformity (Rendell & Whitehead 2001b).

The Evidence for Cetacean Culture

Before summarizing the patterns of behavior that suggest culture in whales and dolphins, I first wish to note that there is independent evidence for sophisticated social learning abilities in a few cetacean species (see Boran & Heimlich 1999 for a review of social learning in whales and dolphins). Bottlenose dolphins (*Tursiops* spp.), which have been studied extensively in captivity and the wild for many years, show sophisticated vocal and motor imitation at both the "action" (or "simple imitation") and "program" (or "goal emulation") levels, and can comprehend the concept of "imitation" (Herman in press). There is some evidence for imitation in other dolphins or whales, but it is generally based on either anecdotes or deductions from "ethnographic" field data (see below), as the bottlenose dolphin is the only cetacean species to have been extensively studied in controlled conditions (Rendell & Whitehead 2001a). Killer whales appear to fulfill most of the conditions for teaching (as defined by Caro & Hauser 1992) when they accompany their young during "practice" in the demanding and dangerous behavior of stranding on beaches to catch seals (Guinet & Bouvier 1995; Boran & Heimlich 1999; Baird 2000), although not all accept this as good evidence for teaching (e.g., Maestripieri & Whitham 2001).

However, there is compelling evidence from the wild that at least some elements of whale and dolphin behavior are culturally determined. To establish this, it is necessary to rule out that the observed conformity and variation in behavior are either produced by genetic variation, or by individual learning in different ecological conditions. Three general patterns of behavioral conformity have been used to indicate culture in whales and dolphins: mother-offspring similarity, rapid-spread of new behavioral variants, and group conformity (Rendell & Whitehead 2001a).

Probably the best documented, and most well-known, case of cetacean culture is the song of the humpback whale *(Megaptera novaeangliae)*. The song, a structured series of vocalizations cycling with a period of about 20 minutes, is principally sung by male humpback whales on the warm-water breeding grounds (K. Payne 1999). In any location, all males sing more or less the same song at any time, but the song evolves over periods of months and years (K. Payne & Payne 1985). There is remarkable conformity in the song on different breeding grounds within an ocean basin at any time, but the songs in different oceans usually evolve independently (R. Payne & Guinee 1983). However, in 1997, the singers off the east coast of Australia in the South Pacific basin unanimously abandoned their own song type and adopted an Indian Ocean type originally sung by a few "minstrel" whales from Australia's west coast, the only documented case of a "revolution" in a nonhuman culture (Noad et al. 2000). These patterns are not consistent with either genetic determination or individual learning in different ecological conditions, and indicate sophisticated vocal learning, as well as the importance of sexual selection in the ocean (K. Payne 1999). Other vocal and behavioral patterns of the mysticetes (baleen whales) also suggest important and sometimes quite rapidly evolving cultures, but are not as well documented as is humpback song. Examples include the traditional migratory paths of several baleen species, the songs of the bowhead whales *(Balaena mysticetus),* and the spread of new methods of feeding in humpback whales (Rendell & Whitehead 2001a).

In contrast, one of the most interesting characteristics of what we know of odontocete (toothed whale and dolphin) cultures is their stability. The best example of multifaceted stable cetacean cultures are the dialects, foraging specializations, and other group-specific behavioral characteristics of killer whales (see Case Study 17A). Sperm whale cultures (see above) seem to possess many of the same elements—group-specific, stable, sympatric, multifaceted—as those of killer whales, but we know fewer of the details. In addition to killer and sperm whales, a third odontocete species has been extensively studied in the wild, the bottlenose dolphin. Bottlenose dolphins show an impressive array of foraging specializations both between and within research sites (Connor 2001; Rendell & Whitehead 2001a). In many of these cases it is impossible to conclusively rule out individual learning in varying ecological environments as a cause for these patterns of behavioral variation. However, in some parts of the world, bottlenose dolphins, as well as other small cetaceans (such as the Irrawaddy dolphin, *Orcaella brevirostris,* in Burma, Rendell & Whitehead 2001a), have complex, ritualized, and generations-old fishing cooperatives with humans. In the best-studied cases, off Brazil, the dolphin behavior is actively coordinated group activity so some form of social learning, and thus culture, is present (Pryor et al. 1990; Simöes-Lopes et al. 1998).

The Question of Cetacean Culture

Despite the strong evidence for imitative abilities in the bottlenose dolphin, in none of the cases discussed above has the transmission mechanism for an ethnographic pattern been proven experimentally, and given the difficulties of studying cetaceans in their natural habitat, and the lack of success in this endeavor with much more tractable ape species, it is likely that they never will be (Rendell & Whitehead 2001a). Thus, adoption of a definition of culture that includes experimental demonstration of imitation or teaching will not be fruitful. In contrast, the ethnographic evidence is remarkably strong given the difficulty of study. Cetaceans possess some cultural attributes that have no known parallels outside humans (Rendell & Whitehead 2001a, b). Particularly unusual are the stable, sympatric, and multifaceted cultures of killer whales (see Case Study 17A) and the humpback whale songs evolving in parallel over huge geographic areas (K. Payne, 1999). Many of us who spend substantial portions of our time with these animals at sea are seriously considering the hypothesis that much of their behavior is culturally determined.

Evolutionary and Ecological Effects of Cetacean Culture

If this hypothesis is correct, then perhaps some of the other strange attributes of cetacean biology become more explicable. There have been a number of suggestions that culture has driven evolution and affected ecology in whales and dolphins.

For instance, it is notable that species for which we have the best evidence for cultural transmission of behavior patterns—humans, killer whales, sperm whales, bottlenose dolphins—have some of the widest geographical ranges and variety of foraging strategies of all mammals (Rendell & Whitehead 2001a). Culture theoretically allows animals to take better advantage of new opportunities, as a beneficial behavior can spread through a population after it has been discovered by only one innovator (Boyd & Richerson 1985). There is some evidence for this in the cetaceans, particularly in situations in which human activity has altered the marine environment (Rendell & Whitehead 2001a). But culture can also have negative impacts on individual fitness (Boyd & Richerson 1985), for instance when "doing the done thing" has adverse consequences. Thus culture may have a role in explaining the strange phenomenon of mass stranding (Rendell & Whitehead 2001a), when groups of cetaceans run up onto shorelines and often die, although they seem to be healthy (Simmonds 1997).

In a cultural environment and society, certain traits become more useful and may be selected for. Barrett-Lennard and colleagues (2001) have proposed that culture preceded, and selected for, vocal variation and vocal copying in cetaceans. In stable, group-specific, kin-based societies and cultures—such as those that seem to exist in killer and sperm whales—animals that possess the most useful information will be particularly able to increase their inclusive fitness. With the high variability of the ocean over long temporal scales, it is easy to imagine selective pressure for longevity (Barrett-Lennard et al. 2001) and even menopause (Whitehead & Mann 2000) in such situations. It is notable that the only two species, other than humans, for which there is good evidence that females routinely live decades after their last birth are matrilineal group–living whales: the killer whale and the short-finned pilot whale, *Globicephala macrorhynchus* (Marsh & Kasuya 1986; Olesiuk et al. 1990).

Culture can also affect genetic evolution in ways that have little consequence for fitness. For instance, it has been suggested that the two sympatric forms of killer whales, "residents," which eat fish, and "transients," which eat marine mammals, and are distinguishable morphologically and genetically in ways approaching sub-species level, may have initially diverged because of cultural differences between groups (see Boran & Heimlich 1999; Baird 2000). I am going to consider a further type of gene-culture coevolution in more detail in the following subsection.

Mitochondrial Genetic Diversity and Cultural Hitchhiking in the Matrilineal Whales

Genetic diversity is generally expected to increase with the range of a species for at least three reasons: a more widely ranging population will generally have a larger population size, a longer evolutionary history, and a greater diversity of habitats, all of which promote genetic diversity (see Avise et al. 1984). The "control region" of the maternally-inherited mitochondrial DNA of cetacean species shows this general pattern (Figure 17.3). For instance, the Gulf of California harbor porpoise or vaquita, *Phocoena sinus,* which is found only in the northern part of the Gulf of California, is genetically monomorphic, whereas the wide-ranging fin *(Balaenoptera physalus)* and humpback whales have, by cetacean standards, substantial diversity (Figure 17.3). However, there are four apparent outlying species with substantial ranges but low mtDNA diversity: the killer whale, sperm whale, long-finned pilot whale *(Globicephala melas),* and short-finned pilot whale. These are the only four species for which there is good evidence for matrilineal societies (i.e., females remaining grouped with their mothers while they are both alive).

Traditional explanations for low genetic diversity—population bottlenecks or selection—have been suggested in these cases (e.g., Lyrholm et al. 1996), but do not explain why these effects are only present in the species with matrilineal social structures (Whitehead 1999a). More convincing are demographic explanations incorporating the animals' social structure (Siemann 1994; Amos 1999; Tiedemann & Milinkovitch 1999). However, the models used by these authors make either assumptions or predictions that are not consistent with what is known of the species' biology (Whitehead 1998, 1999a).

As discussed above, two of these species (the killer and sperm whale) show

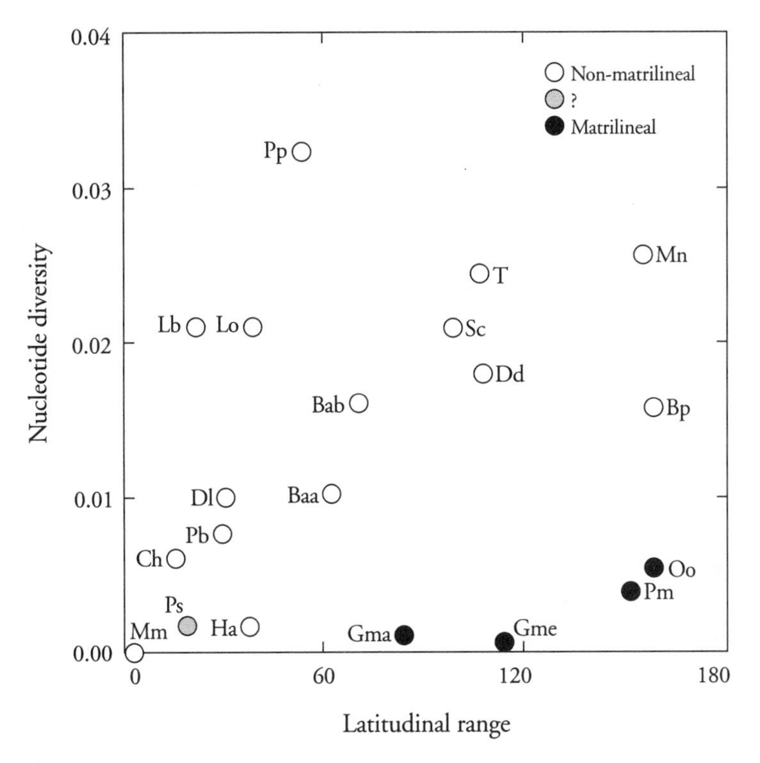

Figure 17.3. Nucleotide diversity in the control region of mtDNA (as estimated from sequencing studies with sample sizes of at least 10 animals from Bérubé et al. 1998; Secchi et al. 1998; Whitehead 1998; Rosel & Rojas-Bracho 1999; Dalebout et al. 2001) plotted against latitudinal range (calculated from maps in Jefferson et al. 1993) for species and subspecies of cetaceans: *Balaenoptera acutorostrata acutorostrata* (Baa), *B. acutorostrata bonaerensis* (Bab), *Megaptera novaeangliae* (Mn), *Delphinapterus leucas* (Dl), *Tursiops* spp. (T), *Lissodelphis borealis* (Lb), *Delphinus delphis* (Dd), *Phocoena phocoena* (Pp), *Cephalorhynchus hectori* (Ch), *Lagenorhynchus obliquidens* (Lo), *Stenella coeruleoalba* (Sc), *Monodon monoceros* (Mm), *Physeter macrocephalus* (Pm), *Orcinus orca* (Oo), *Globicephala macrorhynchus* (Gma), *Globicephala melas* (Gme), *Phocoena sinus* (Ps), *Balaenoptera physalus* (Bp), *Pontoporia blainvillei* (Pb), and *Hyperoodon ampullatus* (Ha). The four species known to have at least partially matrilineal social systems are marked, as is the narwhal *(Monodon monoceros),* for which matrilineality has been suggested.

evidence for stable, group-specific, and, to some extent, matrilineal cultures. Thus I have made the suggestion (Whitehead 1998) that selection on these cultures has reduced the diversity of similarly transmitted mtDNA through a process that I call cultural hitchhiking. Cultural hitchhiking can be thought of as analogous to genetic hitchhiking, in which the diversity of a neutral gene is reduced by selection on a linked gene (Maynard Smith & Haigh

1974). In the original formulation of cultural hitchhiking, a female in a geographically unstructured population has a cultural innovation, which gives her a fitness advantage. She passes both the advantage and her mtDNA genome to her descendants. As long as the innovation is reliably passed from mother to daughter, and not outside the matrilineal line, mtDNA diversity can be decimated as the haplotype of the innovator takes over the population, hitchhiking on the culturally advantageous fitness trait (Whitehead 1998). However, if more than a few females do not exactly copy their mother's cultural phenotype, then the process breaks down (Whitehead 1998). Consequently, results indicating imperfect matrilineality in sperm whales (see above; Case Study 6A; Mesnick et al. 1999) and the evolution of killer whale cultures (Deecke et al. 2000) have been interpreted as counter-indicating cultural hitchhiking as a cause of the low mtDNA diversity in these species.

Probably more realistic than the single, stable, advantageous innovation used in the original formulation is a model of cultural evolution in which multiple innovations build upon one another. Modeling cultural hitchhiking in early human populations in such a way showed that genetic diversity can be reduced in a population of territorial tribes even with quite substantial gene and culture flows between tribes (Whitehead et al. in press). Taking this approach with the nonterritorial whales also shows that mtDNA diversity can be substantially reduced even when social structures are not particularly stable and cultures evolve (Figure 17.4).

These new models increase the range of circumstances in which cultural hitchhiking reduces genetic diversity, and thus boost its plausibility as an explanation for the low mtDNA diversity in the matrilineal whales. However, we are a long way from proving that cultural hitchhiking was the cause for the outliers of Figure 17.3. The hypothesis is being tested further by examining patterns of genetic variation at other loci with different transmission patterns, as well as the social structures of other cetacean species. But the biggest gap in our knowledge is in the cultures of the whales and dolphins, which need to be studied directly using new and innovative techniques.

Society, Culture, Cognition, and Survival in the Open Ocean

Out at sea there are: the largest brains on Earth; sophisticated abilities in communication, cognition, and social learning (Boran & Heimlich 1999; Herman in press); complex social structures (Connor et al. 1998); and cultures possessing attributes that have otherwise, so far, been found only

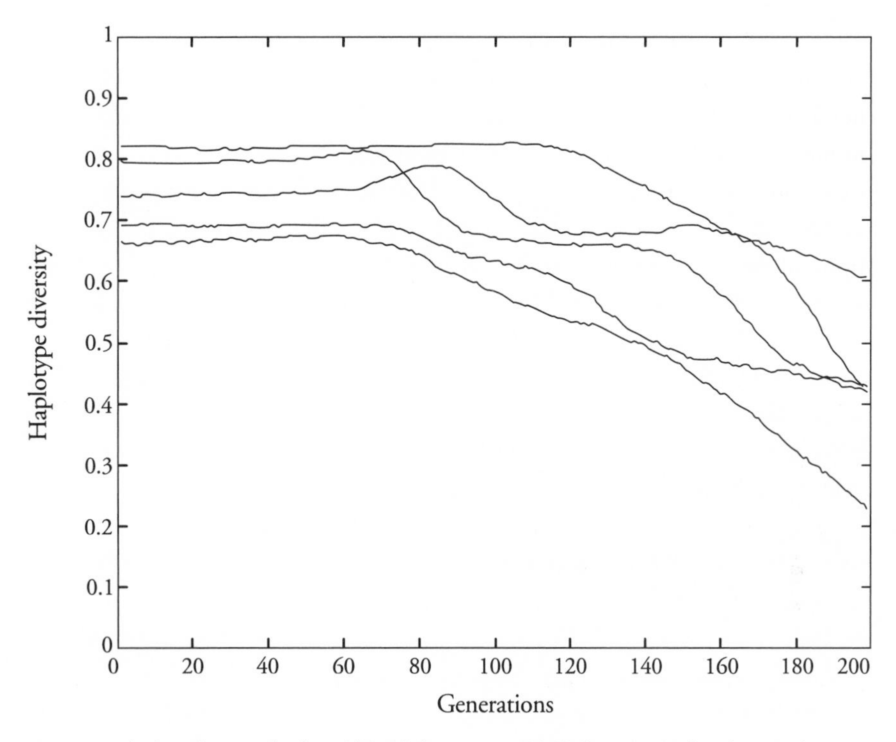

Figure 17.4. Simulation of cultural hitchhiking on mtDNA diversity in five density-dependent populations of 100,000 female whales. Each female has a culturally determined fitness that she usually passes to her daughters along with her mtDNA haplotype. New haplotypes are formed through mutation at a rate of 10^{-5}/animal/generation. At generation 0 all females have the same culturally determined fitness. Innovations occur at a rate of 10^{-4}/animal/generation, with each innovation changing the fitness of the innovator and her descendants by a factor of $(1+ \mid N(0,0.05) \mid)$, where N is a normal random variable. One percent of the females adopt the culturally determined fitness of a randomly chosen member of the population rather than their mother, and pass this on to their daughters. In any generation, the number of daughters of a female is determined by a Poisson random variable with mean = fitness of female × carrying capacity of population / (mean fitness of population × population size). All the populations show reduced haplotype diversity as their cultures evolve.

among humans (Rendell & Whitehead 2001a). Thus there has been a parallel evolution of the social-cognitive-cultural complex in a radically different environment from that of the great apes or the other terrestrial animals being considered in this volume. The cetaceans and perhaps other marine animals made this passage and evolved these abilities in a habitat with little potential for technology, no refuges against predators, few barriers and cheap movement, excellent acoustic propagation, and high biological variability over medium to large temporal and spatial scales. There are a range of implications for our consideration of social, cognitive, and cultural evolution.

The first is that these are very likely to be linked. That cetaceans evolved large brains and substantial cognitive abilities in a marine habitat without much potential for tools or a stable fine-scale resource structure, but with complex social structures, argues for the Machiavellian intelligence hypothesis and against its competitors. Similarly, as has been suggested in the case of primates (see for instance Whiten 2000a, b), cognitive abilities and particular types of social structure seem to at least favor and perhaps be necessary for cultural transmission to become an important determinant of behavior.

Similar elements of social structure and culture are found both in the ocean and on land. For instance, the African elephant *(Loxodonta africana)* has a remarkably congruent social organization to that of the sperm whale, although one is a terrestrial herbivore and the other a marine carnivore (Weilgart et al. 1996). The two animals also share life-history parameter values, large size, large brain size, substantial sexual dimorphism, wide movements, acoustic communication over large ranges, and specialized noses (the spermaceti organ and trunk). To achieve these convergences, some common evolutionary mechanism seems to have been operating—a mechanism disconnected from the underlying habitat. We have suggested that a primary feature may have been a feedback between "K-selected" life history characteristics and sociality (Weilgart et al. 1996). Although this likely operated in other socially complex, long-lived animals such as chimpanzees and humans, in elephants and sperm whales it may have been given an extra twist by the ecological success of the species resulting from their highly modified noses. With the trunk and spermaceti organ producing feeding efficiencies not enjoyed by competing members of their ecological guilds, sperm whales and elephants found themselves largely constrained by scramble competition with conspecifics for food, i.e., "K-selection." Increased feeding efficiency also allows larger sizes, larger brains, and so greater safety from predators. Thus, in these species, anatomical innovations giving foraging efficiencies over a large dietary range may have allowed a parallel suite of life history, social, and cognitive developments. Although elephant cultures have been studied rather differently from those of sperm whales (see McComb et al. 2001), I suspect that similarities will emerge.

Also instructive when examining the evolution of social-cognitive-cultural complexes are differences between the marine and terrestrial species (Connor et al. 1998). For instance, both sexes of resident killer whales remain grouped with their mothers while she is alive, a pattern unknown among terrestrial mammals, which may be partially a consequence of the reduced travel costs of the marine environment (Baird 2000; Connor 2000). As another example, the apparently homogenous relationships within long-term social units of sperm whales (Christal & Whitehead 2001) are unusual on land, perhaps reflecting differences between the spatial and temporal distributions of resources in the two environments, and thus the potential for competition and cooperation.

There is a growing emphasis on studying animal societies at the level of the individual (reflected in the subtitle of this book), but the results on sperm whales suggest that this should be tempered. The homogeneity of relationships within sperm whale units, and the potentially important characteristics of the social units in this species, as well as killer whales, indicate that the unit is, in some respects, as important a structural element of the population as the individual. Thus research should proceed at both the level of the individual, and, where it exists, the long-term social unit.

Cultural comparisons between land and sea are also interesting, but here there is a major confounding factor. Although the terrestrial species for which we have most information on culture are almost all territorial (chimpanzees, humans, rodents, songbirds), no cetacean is known to defend geographical territory. Thus, among primates, cultural comparisons are usually made between noninteracting groups (e.g., Whiten et al. 1999), whereas killer and sperm whale groups with different cultures frequently intermingle. The closest nonhuman parallels on land we have to the sympatric, multifaceted, group-specific cultures of killer, and perhaps sperm, whales, may be the case of the spear-nosed bats *(Phyllostomus hastatus),* in which long-term groups that share roosting caves have specific calls (Boughman & Wilkinson 1998; Chapter 12). The evolving songs of the humpback whale share some similarities with birdsong (e.g., Boran & Heimlich 1999), but the geographic scale (at the level of oceanic basins) of the parallel progression is extreme. Conversely, there are elements of nonhuman terrestrial cultures that seem absent or severely reduced in the ocean. The most obvious of these are tool and material use. Tools are prominent elements of chimpanzee culture (Whiten et al. 1999), but have only been suggested in one limited case for cetaceans: the use of sponges by a few female bottlenose dolphins in the shallow waters of Shark Bay, Australia (Smolker et al. 1997).

As I have alluded to in this chapter, the populations, societies, and perhaps cultures of sperm whales have been heavily affected by an element of human culture—whaling. Sperm whales were killed in two massive worldwide hunts, first by the "Moby Dick" whalers of the eighteenth and nineteenth centuries, and then, from about 1945–1980, by more lethal mechanized whaling. Although each of these hunts each made substantial reductions to sperm whale populations, we do not know the extent, nor how sperm whale society was affected, although there are strong indications that both in terms of numbers and social structure the effects on population size and growth rate were important and long-lasting (Whitehead 1995; Whitehead et al. 1997). There is little current whaling for sperm whales, although it is recommencing in Japan. The major threats to the species and most other cetaceans are unintentional. Despite their deep ocean habitat, and thus distance from the most prominent sources of human impact on the ocean, sperm whales are accumulating chemical pollutants (Law et al. 1996), being disturbed by noise (Watkins et al. 1985), caught in fishing gear (Haase & Félix 1994), and hit by ships (André & Potter 2000). The huge ranges of sperm whales and the connectedness of their ocean habitat through which both noise and chemical pollutants spread with ease mean that habitat protection has to be practiced on a grand scale. To preserve the sperm whale and its society and culture, we must protect the integrity of virtually the entire deep waters of our planet.

Summary

The sperm whale is extreme or unusual in many areas of its biology, including size, brain size, the spermaceti organ, ecological success, and sexual dimorphism (in size, distribution, and social structure). Female sperm whales have complex societies similar to those of elephants, which include stable units containing matrilines, and within which there is communal care for the young. However, in contrast to the general terrestrial picture, relationships among unit members seem remarkably homogeneous. Sympatric units also appear to have distinctive cultures, a pattern better known for killer whales. Cetacean cultures seem to have some characteristics rare among terrestrial nonhumans such as conformist evolution and revolution over very large geographical scales (humpback whale songs), and multifaceted distinctive cultures of sympatric groups (killer and perhaps sperm whales). These societies and cultures evolved in a habitat with little potential for technology, no refuges against predators, few barriers and cheap movement, excellent acoustic propagation, and high biological variability over large temporal and spatial scales.

CASE STUDY 17A

Do Killer Whales Have Culture?

HARALD YURK

In the surf of the Southern Indian Ocean around the Crozet Archipelago, a killer whale rides an incoming wave that completely covers its body. The whale causes no disturbance in the wave but uses its movement swiftly to maneuver itself directly onto the beach to the place where an unsuspecting elephant seal is getting ready to enter the water.

Guinet (1991) described this intentional stranding behavior as a hunting tool used by killer whales at the Crozet Archipelago. Intentional stranding has also been observed in killer whales at beaches of Punta Norte, Argentina (Figure 17A.1), where southern sea lions are the primary target of this hunting strategy (Lopez & Lopez 1985). However, mammal-eating killer whales in the northeastern Pacific, which predominantly prey on harbor seals and to a lesser degree on California and Steller sea lions, show very different hunting tactics (Morton 1990; Baird & Dill 1995; Ford & Ellis 1999). Whales of this area pursue their prey mainly in the water. It is possible that the existence of particular beach types in Argentina and on the Crozet Islands might have driven the development of the intentional stranding behavior. The northeastern Pacific is in contrast mostly characterized by rocky shorelines, where seals and sea lions appear to prefer rocks to the few existing beaches to haul out. This environmental difference might explain why different hunting strategies exist, but they do not necessarily explain why some individuals or groups in the Crozet Islands use the technique while others do not. Interesting questions that arise from this behavioral diversity between individuals

Figure 17A.1. A killer whale catches a young southern sea lion off a beach at Punta Norte, Argentina by stranding itself. Photograph courtesy of John Ford.

and groups are whether or not behaviors, such as intentional stranding, are learned and if so, whether or not they are passed on from generation to generation by teaching or copying and therefore can be called cultural? In the next paragraphs, I will examine some of the behaviors shown by killer whales to see if they could be considered cultural.

Let us start with the seal-snapping whales of the Crozet Islands. Guinet and Bouvier (1995) described a number of incidents in which two young killer whales practiced the intentional stranding behavior together with their mother or another adult female of the same group. During these incidents, Guinet and Bouvier (1995) also observed how one female killer whale pushed her young onto the beach, presumably to "teach" it to catch seals. These accounts are very impressive examples of potential social learning in these killer whales, but do they provide enough evidence to call the behavior cultural? So far there is no evidence that the intentional stranding behavior is spreading to other whales of the same group, although one female of a different group also shows the behavior (Guinet & Bouvier 1995). However, this does not rule out that the behavior is transmitted through copying. After all, it is obvious that intentional stranding is quite dangerous for a five-tonne whale—if it gets stuck on the beach it will most certainly die. Some animals might just not take the risk, and perhaps it is also helpful when one learns such a behavior at a younger age. What needs to be seen, though, is if the juveniles that have been "taught" to hunt this way picked up the teaching habit and will pass it on to their offspring. Intentional stranding is a good candidate behavior for culture, but still further evidence is needed, particularly on the transmission process of this behavior, to consider it unequivocally as such.

Other examples of candidate behaviors for culture are seen in a fish-eating killer whale population, called "residents," that inhabits the northeastern Pacific. This killer whale population is extremely well studied. Beginning in the early 1970s, field studies were initiated to identify individuals and to determine population size, life history traits, and social organization. These ongoing studies have revealed that residents are exclusive fish-eaters in contrast to the aforementioned mammal-eaters that live sympatrically in the same area (Ford et al. 1998; Saulitis et al. 2000) and have a highly unusual social system (Bigg 1982; Bigg et al. 1990; Matkin et al. 1999; Ford et al. 2000). They live their entire life in stable groups, called matrilines, which consist of a female and all her living offspring, male and female (up to four generations) (Matkin et al. 1999; Ford et al. 2000). Matrilines appear to use discrete group-specific call dialects to stay in contact (Ford 1989; Deecke et al. 2000; Miller & Bain 2000; Ford 2002), and call sharing between matrilines reflects maternal relationships among them (Ford 1991; Yurk et al. 2002). Although the function of these discrete calls is still under investigation, the question of whether the call-sharing pattern is the result of social learning of calls or whether it is only the passive result of the group-splitting process will affect whether or not it may be considered culture.

Matrilines that spend the majority of observed time travelling together, and which presumably are closely related (Bigg et al. 1990; Barrett-Lennard 2000) are called pods. Pods in British Columbia and southern Alaska have distinct repertoires of 7–17 discrete call types (Ford 1989; Yurk et al. 2002), typically produced when the whales are spread out or when whales from different pods meet (Ford 1989). Residents produce a number of nondiscrete vocalizations that most often are heard during social play when the whales are in close proximity (Ford 1989). Ford (1991) noted that discrete call repertoires of some pods had remained relatively constant for more than 25 years. Some pods share parts of their repertoires, forming acoustic clusters based on vocal similarity. These clusters are called vocal clans or simply clans.

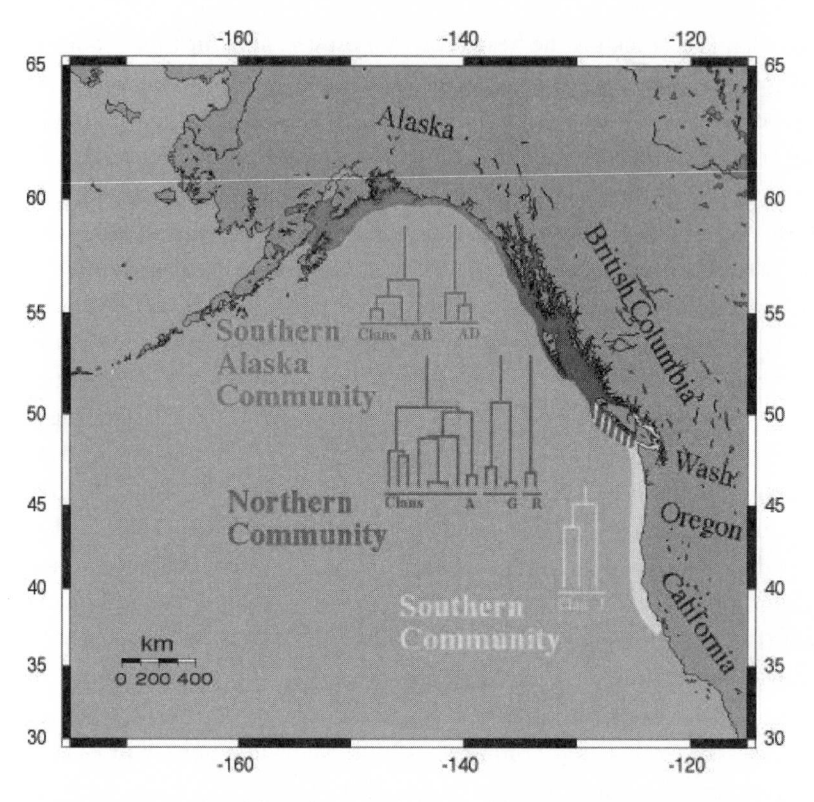

Figure 17A.2. Resident killer whale communities of the northeastern Pacific and vocal lineages (clans) that exist in each community. The endpoints of each line at the bottom of a dendrogram represent one pod. The horizontal lines within each dendrogram represent the acoustic or vocal similarity between respective pods based on call sharing. Pods from different clans do not share discrete calls.

Clans appear to represent lineages of vocally and matrilineally related pods (Figure 17A.2).

The maintenance of a vocal clan could be considered a vocal tradition, if it were shown that the calls were indeed learned within the matriline. Thus far conclusive proof for call learning, e.g., through playback experiments that show vocal mimicry, which exist for songbirds (Marler & Tamura 1962) and some mammals, e.g., bottlenose dolphins (Caldwell & Caldwell 1972; Janik & Slater 1997), has not been produced. However, a number of anecdotal observations (Bain 1988; Ford 1991) support the learning notion, and inferences made from parallel maternal relatedness and vocal similarity (Barrett-Lennard 2000) make call learning within the matriline the best explanation for observed call-sharing patterns. Observations of captive killer whales with different regional ancestry (Bain 1988; Ford 1991) have shown that whales regularly imitate calls of their tank mates. Furthermore, Ford (1991) observed true vocal mimicry in the wild, and Deecke and colleagues (2000) documented horizontal or oblique transmission of structural features of calls between closely related matrilines. If calls are not learned, but instead genetically inherited, the following findings are hard to explain. All members of a matriline use the same set of calls, and mating usually takes place between pods (Barrett-Lennard 2000). Therefore, paternal genetic inheritance is unlikely, because it would over time produce different repertoires for individuals within the matriline. Maternal genetic inheritance of call types either through the influence of mitochondrial DNA, maternal sex chromosomes, or through genomic imprinting, is also very unlikely, because many call types are highly complex (Ford 1989). From this body of evidence, it seems most likely that calls are learned within the matriline and that vocal clans represent true vocal traditions that could be defined as culture.

Residents appear to have other candidates for traditions, such as beach rubbing on gravel at particular beaches, a behavior that is exhibited by members of the northern resident community (Figure 17A.2) but not by the southern residents (Figure 17A.2), and only by one pod of the Alaskan residents (Bigg et al. 1990; Matkin & Saulitis 1997). Another candidate could be the so-called "greeting ceremony" that Osborne and colleagues (Osborne 1986) observed in the southern residents (Figure 17A.2). There are more potentially learned traditions in other killer whale populations than those mentioned here, for example the so-called "carousel feeding" technique used by whales off northern Norway to herd and prey on herring schools (Simila & Ugarte 1993), but this case study is only meant to solicit interest in the question of whether culture exists in killer whales, not to answer it.

18

Discovering Culture in Birds: The Role of Learning and Development

MEREDITH J. WEST, ANDREW P. KING, AND
DAVID J. WHITE

There are 9,000 species of birds in the world; this is a report on two of them, cowbirds *(Molothrus ater)* and starlings *(Sturnus vulgaris)* (Bird 1999). We note the discrepancy in numbers to put into perspective the task of characterizing culture and social learning in birds as a group. But as daunting as the task might sound, the diversity of behaviors and ecologies in birds offers a window on social learning, culture, and evolution that is hard to match. Bird species afford diverse opportunities for comparative analysis at the level of peas in a pod; a taxonomic point at which natural and sexual selection operate (King & West 1990).

Our interest in cultural processes in birds grew out of an interest in birds' songs: how songs develop and how they are used. Vocal learning and the transmission of song information across generations have been documented in many species (Freeberg 2000; Kroodsma & Miller 1996). Song is used in territorial defense and in the attraction of mates. Moreover, there is no question that birds make use of social and auditory learning to craft their repertoires. It is the nature of the crafting process that appeals to us, especially the need for social interactions to build an inventory of vocalizations and behaviors important for reproductive success.

In our minds, culture is a verb; it is the "growing of biological material in a nutrient substance," according to *Webster's Collegiate Dictionary.* The "material" is the organism along with its behaviors-to-be and the "nutrient substance" is the organism's developmental ecology. The cultural transmission of behaviors across generations is therefore the emergent product of an active organism and the nutrient substance. Such a metaphor may seem innocuous, but at its core is an assumption we believe is fundamental to theories of culture or social learning. In the laboratory analogue, we assume different properties of the biological material or organism and of the medium or nutrients in which it is grown. Scientists focus most on changes in the former, not the latter. But what if the material affects the medium, the presumed passive source of growth? In a completely fictional example, what if bacteria being cultured produced heat that then affected the chemical properties of the medium, which then affected the nutrient value of the medium? Thus, the nature of the outcome is neither "in" the organism nor "in" the medium but in the synergy between the two. How could we know this to be the case? We would have to begin by measuring the medium before, during, and after the experiment, just as we would the properties of the organism.

Scientists of social behavior generally accept the fact of synergy as a given, but that does not make it easier to study (Cairns 1979). In this chapter, we are concerned with social dynamics and understanding culture in animals. By culture, we mean the social transmission of specific behaviors from one generation to another within a species. The study of culture in animals is a growing field, with much effort directed toward documenting the comparative existence of and behavioral transmission of cultural information by mechanisms such as imitation and teaching (Whiten et al. 1999). Although these mechanisms are important, we are most interested in the processes responsible for the emergence of cultural information, i.e., how do young learners come to detect and choose to adopt traditional behaviors? Said another way, what processes lead to selectivity in the specificity of imitation or the learning of the informational properties of peers or adults?

Our studies of only two avian species lead us to believe that "cultural information" is not simply there in an environment to be copied or observed, but that it becomes available only by virtue of social interactions by which the needed information emerges. In other words, cultural information, e.g., a certain song variant or a certain way of eating, is not bio-available in all social ecologies because the information is neither in the animal nor the environment. Thus, to study culture, one must first study the social discovery processes by which organisms make the needed information accessible in a dynamic environment. As will be seen, such an approach requires study of many more aspects of the organism's behavior than the trait of interest, i.e., birdsong. One must learn how animals negotiate their everyday social world to uncover and learn the meaning of specific behaviors.

Our plan in this chapter is to describe this social discovery process in cowbirds and starlings with reference to song learning, including 1) the acoustic nature of the songs, 2) the function of songs, and 3) nature of the behaviors used by the birds to elicit information from their surroundings. We follow with some thoughts about the role of imitation and culture, whether song is a sufficiently self-contained behavior to support cultural arguments, and whether culture of the whole organism can be studied.

The two species we discuss are ones not often associated with culture, but crime. Cowbirds are disliked because they are brood parasites, laying their eggs in the nests of over 200 different species and subspecies in North America (Friedmann, Kiff, & Rothstein 1977). This habit elicits quite uncharacteristic attributions from usually staid writers of field guides: "the cowbird is an acknowledged villain, and has no standing in the bird world . . . No self-respecting American bird should be found in his company" (Chapman 1934, p. 359). European starlings earned a nasty reputation after being imported to North America, where their numbers swelled from 200 to 200 million in less than a century. According to Chapman (1934), the species "is a distinctly foreign element in our bird-life, and seems out of place among the species with which we share the bond of birthplace in common" (p. 355). Once in America, starlings had to carve out a niche and carve they did, finding nesting and roosting sites in nooks and crannies in cities and countrysides. To them, a parking garage is the concrete equivalent of evenly spaced trees and perching posts. Starlings are now found in every state and province in North America.

Another way to look at these species is that they are extremely successful. Some have attributed that success to their adaptability with respect to food habits: both are omnivores and both show considerable local variation in mating systems. Our focus, however, is on their social flexibility and how it intersects with the transmission of vocal traditions. In ontogenetic terms, the species represent opposite developmental endpoints, with cowbirds expected to have closed systems for species and mate recognition as they are not raised by conspecifics, and with starlings expected to be exemplars of open systems as they show life-long vocal learning. But, as we hope to show, when it comes to discovering and attending to the social affordances of their environments, these two gregarious species have much in common.

The Cowbird: Vocal Athlete at Work

The song of the cowbird is a bubbly mixture of notes followed by a modulated whistle, from 300 to 13,000 kHz. Crawford Greenewalt, who produced much of the definitive work on how birds sing, pronounced, "This undistinguished bird . . . is the undisputed winner in the decathlon of avian vocalizations" (1968, p. 119). Among the cowbird's achievements is use of a wider frequency margin, nearly four octaves, than in any other bird's song; a maximum frequency higher than that found for any other bird; a frequency spread between the two voices of a full two octaves; a first introductory note with the most rapid glissando—5–8 kHz in four milliseconds; and the highest modulating frequency in the high voice. Greenewalt had no explanation for the convergence of all of the vocal accomplishments in cowbirds—"only madam cowbird will know," he speculated (p. 120).

Greenewalt captures the vocal challenges for the young cowbird. He also alludes to a social dimension, female cowbirds. We have explored these vocal challenges using many approaches, from aviary studies of mating success to acoustic playback of song segments to female cowbirds. Here we will focus on three studies chosen to illustrate the proximate or social dynamics of song learning. First, we give an example of a female cowbird's potential influence on males during song ontogeny. We follow that with an example of young males learning to sing when exposed to different contexts. Finally, we show how those learning processes play out in richer social contexts.

Female Nourishment of Song

Some years ago, we studied a cowbird flock overwintering in North Carolina. During the time of year when juveniles' songs develop most rapidly, i.e., late winter and early spring, we found that juvenile males affiliated with females in small groups of half a dozen birds or so (King & West 1988). These observations were of great interest because we had accumulated several lines of evidence implicating female cowbirds in the development of song. Her role did not seem to be one of simply providing motivation for males to sing, but rather the role appeared to be one of motivating a male to change what he was singing (King & West 1983a; West & King 1985). How females could do this was puzzling, as they themselves do not sing.

We had uncovered her role while trying to change her preferences for song (in a perfect example of the effects of synergy discussed earlier). Female cowbirds, when tested by acoustic playback, show reliable preferences for natal song as measured by frequency of copulatory responses given within one second of the song's onset. We attempted to modify female preferences for song by housing a local female with a "distant" adult male, for example, females from Texas (TX) with male cowbirds from North Carolina (NC) (King & West 1983b).

These studies showed that it was the males, not the females, who changed. NC males began to sound more like TX males, including different acoustic structures in their songs, structures simply not found in NC males (King & West 1987). Birdsong learning is usually described as vocal imitation: a male hears songs of his species, memorizes them, and later practices them, honing their final songs by counter-singing with other males. Thus, males certainly need access to male stimulation, but how could females serve a meaningful role?

To find out what female cowbirds were doing, we videotaped pairs of young males and adult females housed together. We chose the time of year (March) when the discovery process for the young males is in full gear. At this time in the wild, males and females are on prospective breeding grounds and younger males now have a chance to see what song is all about. The clue to what was happening came from watching the reactions of the males to the females: every once in a while when a male sang, he would suddenly change the pace of his singing, as well as sometimes move abruptly toward the female. We found that such changes in the males' behavior were often preceded by wing actions by females, movements we called wing strokes. These data suggested to us that female cowbirds used a gestural signal system to communicate about an acoustic signal system and that males had to know how to read these gestures to use the information available. Thus, a male's discovery process was a multimodal one; he had to look at what happened when he sang a certain song and remember what he was singing when reactions occurred. The males often repeated the song that had elicited the wing stroke, suggesting a means of memory rehearsal (West & King 1988).

To find the functional value of wing strokes, we employed a playback test during the breeding season to see if females would adopt copulatory postures to songs that had elicited wing strokes as opposed to ones that had not. The answer was clear: wing-stroke song elicited significantly more copulatory postures from a different set of females who knew no more about the singers than their songs (West & King 1988). These data led us to view song learning as a process of social shaping. Wing stroking was a form of contingent learning, produced on a partial schedule, serving to bias the male's singing toward more female-preferred signals. But the males had to perform, to sing time and time again, in order for female reactions to emerge. These data on social shaping complement those of Marler and Nelson's process of selective attrition and action-based learning in recognition of the role of male counter-singing in the sparrows they studied (Marler & Nelson 1993; Nelson & Marler 1994).

Recently, we looked again at male-female interactions throughout the year, studying young males individually housed with two adult females. We wanted to find out just what parts of male vocal ontogeny females could modulate. Some males were housed with adult females from their local region to maximize social and acoustic synchrony between the two sexes with respect to local song preferences and timing of breeding. Some males were housed with adult females from a distant population, females with different song preferences and a different time course for breeding readiness (Smith et al. 2000). We found that specific actions of adult females predicted different rates of vocal progress throughout the stages of development. Young males housed with local adult females produced stereotyped song earlier, ceased motor practice earlier, and produced more effective playback songs than did males with females from the distant population. The males with the "distant" females showed slower growth and more variable song well into the spring. The major difference between the groups was that the local females stayed instead of flying away more often when males sang, whereas distant females left, sometimes even before the song was finished. Proximity between male and females gave females the ability to hear more details in the song and gave males an opportunity to inspect female movements more closely (Smith et al. 2000).

From In Vitro to In Situ

In the work just described the males' songs were the focus of the culturing process; we did not look at the birds themselves and how they behaved. Although we had looked at courtship in wild-caught males, neither we nor other labs had observed courtship in the laboratory specimens of birds housed under restrictive conditions (but see Williams et al. 1993). We chose to take a look at what happens when males face a more competitive and responsive environment. We asked two questions: did they show species recognition at the time of mating and were they successful at obtaining mates? The first study included wild-caught juvenile male cowbirds from South Dakota (SD; *M. a. artemisiae*), the same population studied in Smith and colleagues (2000). We housed young SD juvenile males with either SD females (the female-housed or FH males) or canaries (the canary-housed or CH males) in the laboratory in individual contexts, one male with two females or canaries. Thus, we were using housing conditions we had used many times before to look at species recognition. Moreover, instead of concluding the study when the males' songs were mature, we continued on, asking how males would use their learned vocalizations (Freeberg et al. 1995). By use, we meant whether or not they would direct their songs repeatedly to females and whether or not they would sing to adult males, as these two skills are necessary to obtain a mate.

First, we observed all of the males when individually placed in a new flight cage with new SD females and new canaries. In the test flight cage, we watched in astonishment as the CH males chased and sang to new canaries. The CH males' persistence was striking in that the canaries offered no incentives and simply evaded their odd consort-to-be. The FH males vocalized to female conspecifics in the flight cages and ignored the canaries, behaviors that made sense to us. And so, we moved the birds to the next setting, outdoor aviaries with multiple conspecifics and heterospecifics, but no adult male cowbirds. We had thought that once the CH males saw FH males court female cowbirds, some sort of light bulb would turn on and they would adopt typical cowbird courtship. It did not happen. Their behavior remained significantly oriented toward the canaries, not toward the female cowbirds.

But their behavior did not represent a case of ignoring available information, i.e., observing competent courters, because the FH males behaved differently in the aviary than in the flight cage. The FH males rarely sang to females and had little success courting female cowbirds. The major failure was in song use; they did not direct their vocalizations to females, instead singing to each other. After documenting the behavior of the CH and FH males with several sets of females, we introduced adult males to the aviary with FH and CH males to give them competent models to observe. The introduction of the adults had an immediate and positive impact on the females. The day the adults were introduced we saw more social activity by the female cowbirds than at any time prior. The adults successfully courted and copulated with females. But their success did not change the behavior of the FH or CH males, except perhaps to suppress what little courtship we had seen. Thus, learning by observing did not appear to have any effect at this point in development.

By the end of the breeding season, we also knew something about the attractiveness of the songs of the FH and the CH males to playback females. The females had responded significantly more often to the songs of the FH males and the adult males compared to the CH males. Thus, the FH males had learned something from their winter with their female cowbird companions, but it was not enough to make them competitive in the setting we chose because they did not vocalize in the right context.

In past studies, we had assumed that playback females could recognize effective songs, but we had also assumed that males could recognize when to sing such songs. It had simply not occurred to us that song use was a skill in the same way that learning to sing was a skill. Clearly, we were wrong.

In the following two years, we did more work that exposed the multiple layers of learning needed to connect social behaviors, singing, and courting, connections that we had not known could be dissociated by social manipulation. Several lessons were primary. First, the CH males' orientation toward canaries made clear that species recognition is not a sure thing in cowbirds. Neither innate mechanisms nor early learning guarantee discrimination of potential mates. Second, the cage and aviary tests revealed the need to consider social proximity as a possible scaffolding mechanism to facilitate learning. The size of the flight cage represented a supportive mechanism for the males as the females could not get very far away and the males did not need to go very far to find a female. We would not have recognized such support had we stopped our studies after observing the CH and FH males in the confines of a flight cage. Without the aviary tests, we would have concluded that FH males showed appropriate recognition and attention to potential mates and we would have been wrong.

Finally, the studies showed that new tools were needed to confront the multidimensional nature of social influences in complex settings. Songs can be recorded and analyzed and matched on objective acoustic criteria. In contrast, the quality and timing of song overtures, a male's ability to maintain proximity to a female, his reaction to being approached by an older male, a

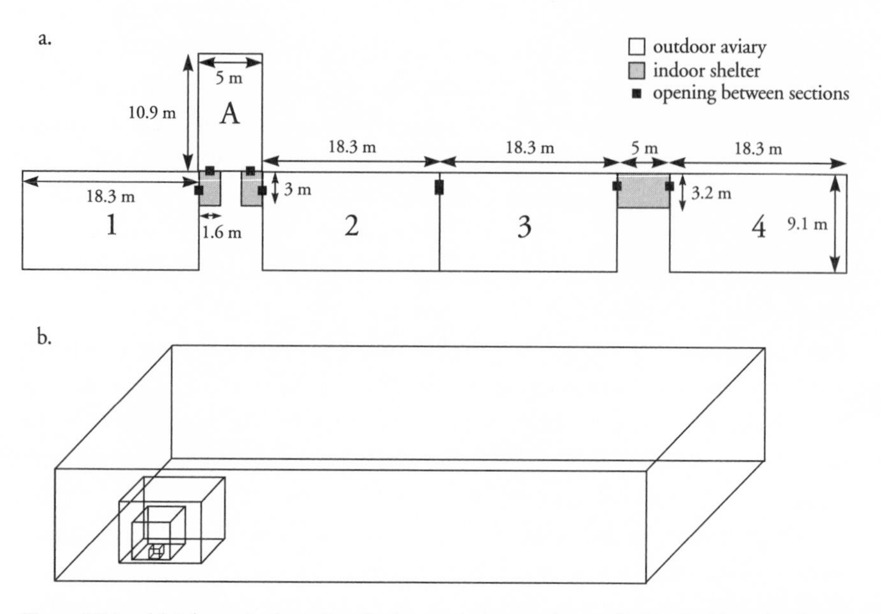

Figure 18.1. *(a)* Schematic depicting the large aviary complex studied in Smith and colleagues (in press). *(b)* Scale diagram of housing conditions for the various studies mentioned. The smallest box is the size of sound attenuating chambers (one cubic meter; e.g., West & King 1988). The next larger box is the size of the flight cages (e.g., Freeberg et al. 1995). The next larger box is the size of our typical aviaries (e.g., Smith et al. in press) and the largest box is the size of the aviary complex studied in Smith et al. (in press).

female's willingness to be close to him, and a male's persistence in following and guarding her from other males would require new metrics to find the social pathways to competence.

Breaking Away: Cowbirds in Flocks

The realization that such practical factors were so important to an individual's reproductive success encouraged us to shift the way we studied social learning. We moved out of the laboratory and studied groups of birds in large outdoor aviaries where we could provide them with social stimulation similar to their natural environment. To begin, we placed a group of 74 cowbirds, differing in sex and age class (young and adult) into a football field–sized aviary and observed them for a year (Smith et al. in press; Figure 18.1). We then looked to see how the birds would organize their environment.

The birds quickly self-selected their own social affiliation patterns, organizing themselves by age and by sex. Assortment based on similarity in age and sex persisted throughout the entire year, with separation by sex being stronger than separation by age. For example, some of the young males associated more with adult males than other juvenile males: these males turned out to be more successful in courting females than were young males that associated less with adult males. Also, young males that sang more to females stopped singing earlier; a pattern we have seen in the past that suggests these males progressed to crystallized song faster. The correlations between patterns of social assortment prior to the breeding season and courtship competence in the breeding season show that self-selection affected vocal and social development and reinforced the point that young learners have to act on their social environment to develop social skills—simply "being there" is not enough (Smith et al. in press).

The large aviary experiment showed us that social learning was open to study in complex social environments and differences among individuals could be measured. Moreover, we now had correlations suggesting that self-selected social environments influenced developmental trajectories of functionally important behaviors. The next step was to test these correlations. To do this meant manipulating social composition to obtain a more rigorous assessment of social influences.

We created groups with different social compositions, manipulating the types of social experience that were available (White et al. in press a). Thus, in four groups we varied the age class of males present. Each group contained young and adult females, one group contained both adult and young males (AY group), one contained only young males without adults (Y group), one group contained only adult males without young males (A group), and one group did not contain any males at all. Groups had approximately 24 birds in each. We watched the groups daily from September through the breeding season in late spring. The results showed that for any class of birds—young males, young females, adult females, and even adult males—the social environment in which they were housed influenced their development (Figure 18.2).

Let's begin with the young males. As the year progressed, the young males in the two different groups developed reliably different courtship and communicative skills. Between August and April, the young males in the AY group assorted by sex and age and frequently sang to other males and to females. In May, during the breeding season, the AY males were aggressive with other males, and courted and copulated with females in patterns typical for

Figure 18.2. A male *(right)* and female *(left)* cowbird. Photograph by Andrew King.

cowbirds in the Midwest. In stark contrast, throughout the year, the Y males showed little social structure, avoiding other males and females, paid little attention to females, and sang in long bouts of undirected soliloquies. In the breeding season, while males in both groups got copulations, the Y males showed different patterns of courtship and copulation and displayed no aggression with other males. The two groups of young males also developed structurally different songs at different rates, with Y males advancing sooner to stereotyped song and developing song more effective at eliciting females' copulatory responses than AY males. But, in the aviaries, the Y males did not show much interest in following females and directing their songs to them. Without adult males, either to challenge the young males or energize the females or both, the Y males settled into a highly stable state of atypical social patterns.

Differences in male behavior across the conditions also related to differences in female behavior across conditions. The large aviaries gave females the space to meter their social proximity with males and with other females. They could engage in or avoid social interactions by choice. In the A male condition, females assorted with males far more than in the other two conditions that contained young males. In the Y male condition, females rarely came near males, affiliating mainly with other females. Females assorted by age class in the three conditions where they were housed with males, but not in the one condition where they were housed without males. In the breeding season, the females housed with Y males produced fewer eggs and destroyed more of them than AY and A females (West et al. in review). Why females react so differently in response to the different males is unclear. It seems that females are stimulated by male-male competitive interactions and may use them for mate assessment.

Finally, there were differences in adult male behavior. It is generally thought that there is little modifiability in adulthood, rendering development and learning of little importance. For most of the year, this conceptualization held true for the A males in our study; there were few differences between the two groups of adult males (housed with or without young males). Both groups sang approximately the same amount. The clearest difference was that A males sang more songs to females over the year, probably a result of the females remaining closer to these males. In the breeding season, we brought the two groups of adult males together in a new aviary, allowing them to compete directly against each other for access to a new group of females. Here a difference between adult males from the two groups became stunningly apparent. The AY males vastly out-competed the A males. Every single AY male got copulations faster than the A males. In fact, only one of the A males ever managed to copulate. Young males, constantly exploring their physical and social environments while not knowing the rules of social interaction, provided the AY males with constant challenges. As a result, living and interacting with young males kept the competitive abilities of AY males honed (White et al. in press b).

We have now studied the origins of the effect of adults on juvenile male development (White et al. in press c) and have also followed the Y males into their second breeding season. The Y males' patterns of atypical behavior persisted into adulthood and were resistant to our making changes to their social environment by exchanging the females in their group. The next step will be to use the Y males as adults for a new generation of young males and chart the development of this F1 culture of males exposed to adults displaying atypical behavior.

These new experiments in socially complex aviaries have revealed that some of the emergent information necessary for development may not be available in simpler social contexts. Moreover, the different cultures that emerged from differences in composition of the groups demonstrate how different opportunities for learning are catalyzed.

One result that impressed us most from watching these different groups was how social behavior could serve as a variation generator. Social groups rapidly diverged from one another as different behaviors became common in the different groups. In some groups, males focused on singing only to themselves, in others, to other males and to females. In some groups males adopted fighting as a response to confrontation, others did not. In some groups males guarded females, in others, males simply moved on to other females. And there were also reliable differences in song potency and song structure across groups. The differences we saw throughout the year translated into differences in mating success with direct implication for survival of phenotypes.

These data support the idea that cowbirds have a facultative system of development, with variation in behavioral outcomes reflecting differential phenotypic responsiveness to local ecologies. Such variation brings to mind the concept of developmental reaction norms as defined by Schlichting and Pigliucci (1998): "the complete set of multivariate ontogenetic trajectories that can be produced by a single genotype exposed to all biologically relevant environments" (p. 22). The distinction to be made in our work is that the environments differed based on the presence or absence of other individuals in the group. Therefore, just as there is an individual's norm of reaction that reflects how the individual differs in response to the environment, so too there is an environmental norm of reaction that reflects how the environment differs in response to the individuals within it. This interconnectedness between the environment's effects on development and development's effects on the environment results in the merging of developmental and cultural trajectories.

Part of the above-mentioned definition of reaction norms stresses the importance of the biological relevance of the environment. Cowbirds provide an excellent species to explore different social ecologies because as a species they are second to none in being successful in just about every type of ecology available. When cowbirds overwinter in flocks, they sometimes find themselves with other species and in classes of cowbirds of mixed age and sex, but field data also report all female flocks or flocks of juvenile males (Friedmann 1929). It is at this level, within the range of normal contexts, that phenotypic plasticity evolved and yet it is at this level that we still know the least. In cowbirds and other songbirds, for example, some young males reliably hear song in their first summer and others do not (Kroodsma & Pickert 1980; Nelson et al. 1995; O'Loghlen & Rothstein 1995). Such environmental variation may have led to the evolution of multiple modes of information transfer and differential dependence on mechanisms such as imitation or trial-and-error learning.

Cowbirds are brood parasites and have often been expected to be exemplars of closed programs of development, i.e., they were not supposed to have open programs highly dependent on social learning, nor would theorists have predicted the amount of geographic variation found in modes of learning across cowbird populations (Lehrman 1970, 1974; Mayr 1974). But, as several decades of research have shown, the conceptual pendulum has swung. Ironically, in recent reviews of our work, the idea has been put forth by reviewers that the cowbird probably is an exemplar of an open program of development because they have to rely more on learning to make up for their odd early environment! And so, we turn to a more traditional parental species, the starling, and to quite different methods to expose the role of the kinds of social learning seen in cowbirds.

Starlings: Composers at Work

Let us begin by providing some perspective. An adult male cowbird with a large repertoire has seven seconds of sound to call his own, give or take a few seconds for whistles. This output would produce a paper portfolio of seven pages of 1.2-second sonograms. A starling, in contrast, has a basic song consuming up to one minute, 60 times longer than the average cowbird song. And starlings have many songs, requiring not a slim file folder or sonograms but an entire book, complexly indexed to capture their ability to repeat, improvise, and acquire new material. And if the male cowbird's ability to traverse several musical octaves is impressive, so is the starling's ability to sustain and organize singing for periods over minutes. Adult male starlings have repertoires of 60–80 song units called motifs, a subset of which appear in individual songs. Motifs are most comparable to note clusters in cowbird song; they are arranged and rearranged into predictable sequences to configure songs, often repeating motifs at least once. The mimicry of an entire cowbird song qualifies as a single motif. Motifs depend on the learning of many sounds including conspecifics, heterospecifics, and inanimate and nonavian sounds (cell phones are fashionable right now). An entire repertoire may consume 10 or more continuous minutes of song, a 600-fold increase over the average cowbird song.

Like cowbird song, the major proposals for the function of starling song include mate attraction, in addition to individual recognition and male-male competition. There is a growing consensus that the size of the repertoires and perhaps aspects of song delivery (rate of vocalizing) are products of sexual selection, with female preferences driving male song learning toward larger repertoires (Ball & Hulse 1998; Mountjoy & Lemon 1991).

With respect to song development, studies using restrictive housing, usually birds in individual isolation, have shown that young starlings housed without adults produce smaller and atypical repertoires. Young starlings housed with peers do better but still show smaller repertoires than do wild starlings. Starlings copy tutors beyond their first year, with a sensitive period occurring between six and 15 months (Chaiken et al. 1993; Chaiken et al. 1997). Starlings keep most of their repertoire from year to year and simply add to it. There is growing evidence suggesting that a starling's social ecology affects its vocal repertoire throughout its life (Hausberger 1997; Figure 18.3).

Life with Starlings: An Insider's View

But how does the social environment guide the repertoire building process? We were fortunate to stumble on evidence suggesting strong links between repertoire content and social histories between individuals, and to find a method to watch how sounds morphed from one acoustic state into another as starlings went about the "processing" of sounds. To accomplish both goals, we used ourselves and other humans as potential sources of sounds and placed the birds in different human contexts to see how social interactions affected vocal composition.

Starlings appear to have a long history of association with humans; the Latin scholar Pliny reported that he taught starlings to repeat Latin and Greek phrases. Mozart owned a starling that produced notes from one of his concertos. He had the bird for three years and accorded him a funeral, complete with a poem written by the composer (West & King 1990).

Our first experience with starlings came about by accident; we were actually studying his companion, a male cowbird. Until the bird uttered something that sounded like "Good Morning" we were unaware how much the starling had observed us: we knew nothing of their mimicking abilities. This starling muscled his way into our psyches before we knew what was happening. Rex (so named because as a baby he looked like a dinosaur) simply insisted on being part of our affairs: he shared our morning coffee, vocalized to music and songs whistled to him, chattered at lab meetings, and perched on the shower curtain watching events inside. But it was when he uttered the phrase, "Does Hammacher Schlemmer have a toll free number?" that we paid attention.

Figure 18.3. A male starling. Photograph by Andrew King.

We knew he only heard this phrase once and we knew the context. From this unlikely start, we thought we had a way to trace the origins of mimicked phrases, as utterances such as Hammacher Schlemmer have a high signal to noise ratio for humans compared to the sounds of other birds or even the sounds such as squeaking doors, microwave ovens, coughing and sneezing, and other sounds so familiar that we filter them out. Thus, we saw Rex's assimilation of sounds with a known history as a way to trace the selectivity and flexibility of vocal mimicry much as one might use a radioactive isotope to watch its course through some physiological system (West et al. 1983).

Rex's mimicry and that of the other birds we went on to study suggested to us that the motivating conditions for vocal learning were not simple: sounds heard most frequently by the birds were not present in their mimicry (such as "no" or "here's your food") and sounds only rarely heard seemed over-represented (such as "defense, defense" chanted after observing a televised basketball game).

Most important, we found that social interaction with humans was necessary for human mimicry of words or whistles or any other human-derived sounds (e.g., clearing of the throat). We also housed some starlings in homes in cages with another bird but instructed the caregivers to provide good care but no extra interaction. Thus, the birds heard human sounds as often as the birds that also experienced human social interaction (West et al. 1983). Although these birds mimicked extensively, none of them copied human sounds.

Quantifying Sociality

To add to our observations, Marianne Engle (2001) undertook a longitudinal study of starlings raised under several conditions, varying the amount of human interactions as we had. She also recorded local, wild birds to gather baseline information. Two of her findings are relevant here. First, she operationalized the social relationship between the human caregivers and the birds by using a behavioral protocol to gauge degree of social interaction. The major variable was degree and duration of social proximity, the same variables we had found to be important in cowbirds. To be scored as interactive, the bird had to 1) attend to the caregiver, 2) approach, 3) create contexts for interaction by 4) getting near the caregiver, or 5) attend to a novel object to provide a topic. The more of these acts completed, as well as some others, the higher the bird's score. Engle found significant correlation between protocol scores and human-derived mimicry for both males and females.

Here was another social discovery–based system. Starlings appeared to throw out a sound and follow what happened: we dubbed the system "social sonar" (West & King 1990). The repertoires of the birds contained greetings, farewells, conversational fragments, and time-locked sequences (from the starling's point of view), such as 1) squeaking door, 2) dogs barking, 3) dishes clinking, 4) mimicry of "good morning." Events 1–4 represent the morning sequence for that household. All of these mimicries stress social attention and emphasize events where humans and starlings were apt to interact on a routine basis. The rules for social learning by starlings seem to be compounded from degree and duration of proximity and what happens during proximate interactions.

Engle's work also provides an instructive perspective on the developmental data from other laboratories cited earlier. Chaiken and colleagues (1993) found that birds deprived of adult male companions or deprived of any social companions developed smaller repertoires than wild males. Although Engle's starlings were equally deprived of starling input, as were Chaiken and colleagues' (1993), their social surroundings were far more complex, albeit atypical. Engle's birds developed repertoires overlapping in size with wild males from her sample and from Chaiken and colleagues' (1993). Engle also found, as had we, that noninteractive exposure to humans did not lead to the use of human sounds for mimicry (West et al. 1983). Engle's home-based, interactive birds did show some similarities to isolated birds in lab studies, though, in their failure to produce certain wild-type sounds such as the distinctive "click" trains. The home-based starling also differed in the preferred duration of their song production (generally shorter segments than wild birds), perhaps being influenced by the average length of speech utterances heard in their homes.

Starlings are vocal mirrors, reflecting back exactly what they hear, with no filtering. Their mimicry of human sounds included the sounds of human speech production: our lips smacking and swallowing. Despite the accuracy of the imitation, the mimicry lacked contextual elements that are usually associated with social transmission. First, no vocal contingencies developed. Even if a human immediately repeated the sound mimicked by the starling, the bird was unlikely to respond in kind. Second, the mimicked sounds were used in many contexts not associated with their original context. Among our home birds, one screeched, "I have a question" when held to have her toenails clipped. During an agonistic encounter over suet, all sorts of mimicries seemed literally squeezed out of the very excited birds, as is often seen in the wild at a feeder. If we were to sample the behavior of our starling only at breakfast and suppertime, we would be led to conclude that they use sounds referentially, as 15 minutes or so before those times, the starling increases his mimicry of feeding-related utterances, such as "breakfast" or "hi guys" or imitations of the microwave. But, if one were to visit the same birds at other times, the chances are high of hearing the very same sounds mixed into the potpourri. It may be the case that what distinguishes the feeding times is the possible greater focus on specific mimicries, but that important distinction remains to be tested. Thus, in our view, vocal imitation is a cheap way to acquire potential song material, with the particular selection of sounds to be imitated influenced by social learning.

Whiten (2000) lists "social knowledge" as one of the components of social intelligence in primates, i.e., "a store of social information about their companion's characteristics and past behavior, which they can apply to novel situations" (p. 193). The selectivity of the starlings' mimicry to those humans who interact with them suggests an analogous ability, but one humans can learn about if they know the animal's social history. Imitation is a way of storing and expressing social knowledge. Thus, the transmission of culture in starlings involves imitation, but imitation guided by social learning.

What Is the Nature of Avian Culture?

What we have outlined is a different way of thinking about culture, not limited to transmission of compartmentalized innovations in a population (potato washing or termite fishing, for example) or isolated species-typical traits such as birdsong, but an approach that encompasses the development of the whole animal (becoming a cowbird). If, as we argued at the outset, culture is a process, then it may be too narrow a view to look only for domain-specific instances. We originally compared the process of culture as the growing of biological material in a nutrient medium under controlled conditions. As we review what we have written here, we make one amendment: the biological material is the whole organism—it is not the song or any behavior separated from the bird. In study after study, for example, we have shown that cowbirds must learn how, when, and to whom to sing.

In a review of culture in nonhuman primates, Whiten and colleagues (1999) document innovations among different groups of primates to reveal the diversity of cultural innovations found. They distinguish primates from other taxa based on the number of cultural processes reported, suggesting that cultural processes typically exist for only single behavior patterns in other mammals and birds. We would suggest this might be an artifact of the compartmentalized way in which culture has been studied in other taxa. Birds raised in socially barren conditions cannot show any other potential forms of culture. With respect to birdsong, it is not only the song that is affected but also other behaviors of the singer and the perceiver. Let's consider the starling first. In order for a novel form of behavioral transmission to occur, incorporation of human sounds, the starlings a) were raised from an early age with heterospecifics (humans); b) lived in the heterospecifics' environment, meaning that even basic resources were delivered in new ways (food from a human's dish, water from a faucet); c) absorbed an ambient visual and auditory world biased toward human activity; d) physically interacted with humans, allowing extreme proximity; e) became attuned to a temporal schedule different from that of wild conspecifics; and f) were deprived of aggressive feedback for almost any action (not including someone throwing a sock at the cage to quiet them down). In Hausberger's (1997) work with wild starlings, the control parameter predicting song and whistle sharing is close contact. But in order to achieve close contact, all of the processes we saw in the home that lead up to directing song to a neighbor must also fall into place.

Thus, the object/subject of the culturing is the animal, not simply one of its behaviors. In cowbirds, we have shown that the following behaviors, which are obligatory for song to be used as a mating signal, can be altered by postnatal experience: species and mate recognition, that is, to whom to affiliate and sing; male-male relationships including counter-singing and dominance; vigilance to potential mates, especially after copulation; and populational selectivity in mate choice. When viewed in this context, the male cowbird's total seven seconds of song is embedded in a far more complex series of events that must occur and must occur in a certain order and in a certain time frame. Lehrman (1970) noted, "nature selects for outcomes"; the same is true for culture, as outcomes determine what is transmitted.

We have focused here on the processes that can give rise to cultural transmission. The orthodox cultural researcher would correctly point out that the true test in thinking about culture and whole organisms comes, however, by asking if we can actually affect generational transfer of behavioral traditions experimentally by following the methods and assumptions outlined here. We have data for cowbirds, but not yet for starlings. Todd Freeberg (1996, 1997, 1998) provided the data for cowbirds. Freeberg put together knowledge from previous work to propose a strict test of cultural transmission: could he alter mating preferences and would his subjects carry on the tradition for him? He collected cowbirds from South Dakota. All the birds lived in large aviaries where they could interact with other juvenile birds from their original population and adults of one of two experimental populations. The first experimental adult population was birds from the same geographic area as the juveniles; the other population was adults from Indiana, a population 1,500 kilometers away. The young birds, male and female pupils, were randomly assigned to these conditions. Thus, from the young bird's point of view, there was no designated model: all of the adults were potential models and from the young bird's point of view, so were the other juveniles. The test worked: young males and females preferred as mates birds from their postnatal experimental context. Moreover, in the next year, the birds, in the absence of any "true" models, transmitted their culture to a new group of South Dakota juveniles.

The change in female preference was the most striking from the point of view of the history of our lab's work. With one exception, we had not found a way to modify female cowbirds' natal song preferences even if they were hand-reared and exposed only to males from a distant population. And so why did females from South Dakota develop a preference for Indiana males and their songs when they could also hear and interact with other South Dakota males, albeit juveniles? We believe the answer lies in the social context provided: we gave females other females, as well as males. Although we have not tested the idea, we now wonder if housing females alone with males is stressful for them, as they seem to always be on guard, aware of the male's proximity to them. Nadler and colleagues (1994) showed striking differences in chimpanzees *(Pan troglodytes)* if they could regulate access to males. Perhaps female cowbirds need the sense of self-regulation to allow them to absorb information from adults or maybe they need to see both male and female adults. Female starlings are also less likely to sing if males are present, suggesting social inhibition of learned behaviors (Hausberger 1997). As a field, we may have to approach questions of male and female modifiability from different perspectives. Recent work on stress in humans documents different stress regulatory styles, with females showing a distinctly more social approach, labeled "tend and befriend" as opposed to the male pattern of "flight-or-fight" (Taylor et al. 2000).

Summary: A Test of Culture as a Social Enterprise

We have provided abbreviated tales of song learning in birds as examples of how cultural transmission may come about. We have focused on the role of social interaction and feedback as a guiding mechanism. Michael Goldstein (2001) completed studies of human infants in our lab where he tested some of our ideas about cultural transmission as an emergent property of social interactions. Imitation and teaching play an obvious role in the learning of speech sounds, but many theorists have denied any role for social learning, proposing that development advances automatically via biologically privileged mental modules (Pinker 1994; Wexler 1990). We began our study by borrowing two methods from our avian studies. First, we carried out playback experiments with mothers, recording how they categorized the infant sounds of babbling, which occur over many months in an infant's first year. Mothers reliably identified specific information in these quite variable sounds, even when listening and looking at very short videotape clips of unfamiliar infants (Goldstein & West 1999). Moreover, as infants made more phonologically advanced sounds, mothers relied more on sounds than facial expressions to determine meaning. Next, we asked if infants' babbling was sensitive to social stimulation from mothers, much as we had asked how young male cowbirds respond to stimulation from adult females. In an ABA design, after a baseline play session, mothers provided nonvocal stimulation to infant sounds when directed to do so by an experimenter. An extinction period followed in which mothers were told to behave as they had in the first trial. Compared to baseline, the seven- or eight-month-old infants used more mature forms of infant speech during the interactive session and retained many of these sounds in the extinction phase. A control group of infants, given the same stimulation but not linked to their vocalizations, did not show the developmental changes in their babbling, although they vocalized as much as the experimental group. The infants' interest in social contingencies thus led to the emergence of the most advanced sounds in their repertoires, sounds then bio-available for further shaping in the course of the mother's and infant's play. These results are the first to show the sensitivity of the phonetic trajectory of babbling to social stimulation; previous work found no such relationship, but previous research did not employ a design allowing for social interactions to occur. Much like the starlings' use of social sonar, the babies threw out sounds, acted on their immediate consequences, and produced a new sound. The tendency to use developmentally more advanced sounds may have been due to arousal and the infant's association of more advanced sounds with more maternal attention. Only more research can distinguish these possibilities. But these data are an example of how the social environment can influence an important developmental skill through social learning that is based neither on imitation or teaching, but based on social discovery of contingencies. Thus, the emergence of cultural information through social interaction begins much earlier in our own species than is assumed by many linguists (Goldstein 2000; Goldstein & West in prep.).

In summary, the study of even two species of birds reveals a potentially important mechanism for many taxa: the role of social interaction to elicit feedback that helps the young learners to identify cultural information. Culture or traditions are not simply passed on as one might pass on ownership of a home to an heir. Social knowledge is earned by organisms acting on their particular social surroundings to probe its properties. Although skills can come from imitation and teaching, these mechanisms ultimately depend on successful social performance and the feedback it engenders.

Thus, studying the behaviors by which animals uncover culture is a critical next step. The uncovering of culture involves attention to social detail, knowledge of the social properties of conspecifics and of one's own behavior, and pragmatic skills to manipulate another's behavior within a group. For those accustomed to thinking of these characteristics as existing only in big-brained mammals, think again. The next time you see a flock of blackbirds projected against the sky, we hope you can see beyond the grace of their flight to their social intelligence writ large, a collective brain, as it were, over our heads, but not beyond our grasp.

REFERENCES

ACKNOWLEDGMENTS

CONTRIBUTORS

INDEX

References

1. Life History and Cognitive Evolution in Primates

Aiello, L. C. & Wheeler, P. W. 1995. The expensive-tissue hypothesis. *Current Anthropology,* 36(2):199–221.

Allman, J. M. 1999. *Evolving Brains.* New York: Scientific American Library.

Austad, S. N. & Fischer, K. E. 1992. Primate longevity: its place in the mammalian scheme. *American Journal of Primatology,* 28:251–261.

Balda, R. P., Kamil, A. C. & Bednekoff, P. A. 1996. Predicting cognitive capacity from natural history: examples from four species of corvids. In *Current Ornithology,* ed. E. D. Ketterson, pp. 33–36. New York: Plenum Press.

Balda, R. P., Pepperberg, I. M. & Kamil, A. C. 1998. *Animal Cognition in Nature: The Convergence of Psychology and Biology in Laboratory and Field.* San Diego: Academic Press.

Barton, R. A. 1996. Neocortex size and behavioral ecology in primates. *Proceedings of the Royal Society of London,* ser. B, 263:173–177.

——— 1998. Visual specialization and brain evolution in primates. *Proceedings of the Royal Society of London,* ser. B, 265:1933–1937.

——— 1999. The evolutionary ecology of the primate brain. In *Comparative Primate Socioecology,* ed. P. C. Lee, pp. 167–194. New York: Cambridge University Press.

Barton, R. & Dunbar, R. 1997. Evolution of the social brain. In *Machiavellian Intelligence II: Extensions and Evaluations,* 2d ed., ed. A. Whiten & R. W. Byrne, pp. 240–263. Cambridge: Cambridge University Press.

Barton, R. A. & Purvis, A. 1994. Primate brains and ecology: looking beneath the surface. In *Current Primatology: Proceedings of the XIVth Congress International Primatological Society,* ed. J. R. Anderson, B. Thierry, & N. Herrenschmidt, pp. 1–11. Strasbourg: University of Strasbourg.

Basil, J. A., Kamil, A. C., Balda, R. P. & Fite, K. V. 1996. Differences in hippocampal volume among food storing corvids. *Brain, Behaviour and Evolution,* 47:156–164.

Bennett, P. M. & Harvey, P. H. 1985. Relative brain size and ecology in birds. *Journal of Zoology,* ser. A, 207(2):151–169.

Bernard, R. T. F. & Nurton, J. 1993. Ecological correlates of relative brain size in some South African rodents. *South African Journal of Zoology,* 28(2):95–98.

Bradbury, J. W. 1986. Social complexity and cooperative behavior in delphinids. In *Dolphin Cognition and Behavior: A Comparative Approach,* ed. R. J. Schusterman, J. A. Thomas, & F. G. Wood, pp. 361–372. Hillsdale, NJ: Lawrence Erlbaum Associates.

Brooks, D. R. & McLennan, D. A. 1991. *Phylogeny, Ecology, and Behavior: A Research Program in Comparative Biology.* Chicago: University of Chicago Press.

Byrne, R. W. 1995. *The Thinking Ape.* Oxford: Oxford University Press.

——— 1997. The technical intelligence hypothesis: an additional evolutionary stimulus to intelligence? In *Machiavellian Intelligence II: Extensions and Evaluations,* ed. A. Whiten & R. W. Byrne, pp. 289–311. Cambridge: Cambridge University Press.

Byrne, R. W. & Whiten, A. 1988. *Machiavellian Intelligence: Social Expertise and the Evolution of Intellect in Monkeys, Apes, and Humans.* Oxford: Clarendon Press.

——— 1997. Machiavellian intelligence. In *Machiavellian Intelligence II: Extensions and Evaluations,* ed. A. Whiten & R. W. Byrne, pp. 1–23. Cambridge: Cambridge University Press.

Charnov, E. L. 1993. *Life History Invariants: Some Explorations of Symmetry in Evolutionary Ecology.* Oxford: Oxford University Press.

Cheney, D. L., Seyfarth, R. M. & Smuts B. B. 1986. Social relationships and social cognition in non-human primates. *Science,* 234:1361–1366.

Clutton-Brock, T. H. & Harvey, P. 1980. Primates, brains, and ecology. *Journal of Zoology* (London), 190:309–323.

Coddington, J. A. 1988. Cladistic tests of adaptational hypotheses. *Cladistics,* 4:3–22.

Cole, L. C. 1954. The population consequences of life history phenomena. *Quarterly Review of Biology,* 29:103–137.

Cummins, D. D. 1998. Social norms and other minds: the evolutionary roots of higher cognition. In *The evolution of Mind,* ed. D. D. Cummins & C. Allen, pp. 30–50. Oxford: Oxford University Press.

Deacon, T. W. 1997. *The Symbolic Species: The Co-evolution of Language and the Brain.* New York: W. W. Norton and Company.

Deaner, R. O., Barton, R. A. & van Schaik, C. P. In press. Primate brains and life histories: renewing the connection. In *Primate Life Histories and Socioecology,* ed. P. M. Kappeler & M. E. Pereira. Chicago: Chicago University Press.

Deaner, R. O., Nunn, C. L. & van Schaik, C. P. 2000. Comparative tests of primate cognition: different scaling methods produce different results. *Brain, Behavior and Evolution,* 55:44–52.

Dennett, D. C. 1996. *Kinds of Minds: Toward an Understanding of Consciousness.* New York: Basic Books.

Dukas, R. 1998. Evolutionary ecology of learning. In *Cognitive Ecology: The Evolutionary Ecology of Information Processing and Decision Making,* ed. R. Dukas, pp. 129–174. Chicago: University of Chicago Press.

Dunbar, R. I. M. 1992. Neocortex size as a constraint on group size in primates. *Journal of Human Evolution,* 20:469–493.

——— 1995. Neocortex size and group size in primates: a test of the hypothesis. *Journal of Human Evolution,* 28:287–296.

——— 1998. The social brain hypothesis. *Evolutionary Anthropology,* 6:178–190.

Dunbar, R. I. M. & Bever, J. 1998. Neocortex size predicts group size in carnivores and some insectivores. *Ethology,* 104:695–708.

Economos, A. C. 1980. Brain–life span conjecture: a re-evaluation of the evidence. *Gerontology,* 26:82–89.

Eisenberg, J. F. 1981. *The Mammalian Radiations.* Chicago: University of Chicago Press.

Eisenberg, J. F. & Wilson, D. E. 1981. Relative brain size and demographic strategies in didelphid marsupials. *American Naturalist,* 118(1):1–15.

Felsenstein, J. 1985. Phylogenies and the comparative method. *American Naturalist,* 125(1):1–15.

Gibson, K. R. 1999. Social transmission of facts and skills in the human species: neural mechanisms. In *Mammalian Social Learning: Comparative and Ecological Perspectives,* ed. H. O. Box & K. R. Gibson, pp. 351–366. Cambridge: Cambridge University Press.

Gittleman, J. L. 1986. Carnivore brain size, behavioral ecology, and phylogeny. *Journal of Mammalogy,* 67:23–36.

——— 1994. Female brain size and parental care in carnivores. *Proceedings of the National Academy of Sciences USA,* 91:5495–5497.

Glander, K. E. 1981. Feeding patterns in mantled howling monkeys. In *Foraging Behavior: Ecological, Ethological and Psychological Perspectives,* ed. A. Kamil & T. D. Sargent. New York: Garland Press.

Harvey, P. H. & Krebs, J. R. 1990. Comparing brains. *Science,* 249:140–146.

Harvey, P. H. & Pagel, M. D. 1991. *The Comparative Method in Evolutionary Biology.* Oxford: Oxford University Press.

Harvey, P. H., Martin, R. D. & Clutton-Brock, T. H. 1987. Life histories in comparative perspective. In *Primate Societies,* ed. B. B. Smuts, D. L. Cheney, R. M. Seyfarth, R. W. Wrangham, & T. T. Struhsaker, pp. 181–196. Chicago: Chicago University Press.

Harvey, P. H., Read, A. F. & Promislow, D. E. L. 1989. Life history variation in placental mammals: unifying the data with theory. *Oxford Surveys in Evolutionary Biology,* 6:13–31.

Jerison, H. J. 1973. *Evolution of the Brain and Intelligence.* New York: Academic Press.

——— 1991. *Brain Size and the Evolution of Mind.* New York: American Museum of Natural History.

Joffe, T. H. & Dunbar, R. I. M. 1997. Visual and socio-cognitive information processing in primate brain evolution. *Proceedings of the Royal Society of London,* ser. B, 264:1303–1307.

Johnson, V. E., Deaner, R. O. & van Schaik, C. P. In press. Bayesian analysis of multi-study rank data with application to primate intelligence ratings. *Journal of the American Statistical Association.*

Kamil, A. C. 1988. A synthetic approach to the study of animal intelligence. In *Comparative Perspectives in Modern Psychology: Nebraska Symposium on Motivation,* ed. D. W. Leger, pp. 258–308. Lincoln: University of Nebraska Press.

Lefebvre, L. & Giraldeau, L. A. 1996. Is social learning an adaptive specialization? In *Social Learning in Animals,* ed. C. M. Heyes, & B. G. Galef, Jr., pp. 107–128. New York: Academic Press.

Lemen, C. 1980. Relationship between relative brain size and climbing ability in *Peromyscus. Journal of Mammalogy,* 61:360–364.

Mace, G. A. & Eisenberg, J. F. 1982. Competition, niche specialization and the evolution of brain size in the genus *Peromyscus. Biological Journal of the Linnean Society,* 17:243–257.

Macphail, E. M. 1982. *Brain and Intelligence in Vertebrates.* Oxford: Clarendon Press.

Marino, L. 1998. A comparison of encephalization between odontocete cetaceans and anthropoid primates. *Brain, Behavior, and Evolution,* 51:230–238.

Martin, R. D. 1990. *Primate Origins and Evolution: A Phylogenetic Reconstruction.* London: Chapman & Hall.

Martins, E. P. & Hansen, T. F. 1996. The statistical analysis of interspecific data: a review and evaluation of phylogenetic comparative methods. In *Phylogenies and the Comparative Method in Animal Behavior,* ed. E. P. Martin, pp. 22–75. New York: Oxford University Press.

Meier, P. T. 1983. Relative brain size within the North American Sciuridae. *Journal of Mammalogy,* 64(4):642–647.

Middleton, F. & Strick, P. 1994. Anatomical evidence for cerebellar and basal ganglia involvement in higher cognitive functions. *Science,* 266:458–461.

Milton, K. 1988. Foraging behaviour and the evolution of primate intelligence. In *Machiavellian Intelligence II: Extensions and Evaluations,* ed. A. Whiten & R. W. Byrne, pp. 285–305. Oxford: Clarendon Press.

Nunn, C. L. & Barton, R. A. 2001. Comparative methods for studying primate adaptation and allometry. *Evolutionary Anthropology,* 10:81–98.

Pagel, M. D. & Harvey, P. H. 1993. Evolution of the juvenile period in mammals. In *Juvenile Primates: Life History, Development, and Behavior,* ed. M. E. Pereira & L. A. Fairbanks, pp. 28–37. New York: Oxford University Press.

Paradiso, S., Andreasen, N. C., O'Leary, D. S., Arndt, S. & Robinson, R. G. 1997. Cerebellar size and cognition: correlations with IQ, verbal memory and motor dexterity. *Neuropsychiatry, Neuropsychology and Behavioral Neurology,* 10:1–8.

Parker, S. T. & Gibson, K. R. 1977. Object manipulation, tool use and sensorimotor intelligence as feeding adaptations in cebus monkeys and great apes. *Journal of Human Evolution,* 6:623–641.

——— 1979. A developmental model of the evolution of language and intelligence in early hominids. *Behavioral and Brain Sciences,* 2:367–407.

Passingham, R. E. 1975. The brain and intelligence. *Brain, Behavior, and Evolution,* 11:1–15.

Pomeroy, D. 1990. Why fly? The possible benefits for lower mortality. *Biological Journal of the Linnean Society,* 40:53–65.

Povinelli, D. J. & Cant, J. G. H. 1995. Arboreal clambering and the evolution of self-conception. *Quarterly Review of Biology,* 70(4):393–421.

Promislow, D. E. L. 1991. Senescence in natural populations of mammals: a comparative study. *Evolution,* 45:1869–1887.

Promislow, D. E. L. & Harvey, P. H. 1990. Living fast and dying young: a comparative analysis of life-history variation among mammals. *Journal of Zoology* (London), 220:417–437.

Prothero, J. & Jürgens, K. D. 1987. Scaling of maximal lifespan in mammals: a review. In *Evolution of Longevity in Animals: A Comparative Approach,* ed. A. D. Woodhead & K. H. Thompson. New York: Plenum Press.

Purvis, A. & Rambaut, A. 1995. Comparative analysis by independent contrasts (CAIC): an Apple Macintosh application for analysing comparative data. *Computer Applications in the Biosciences,* 11:247–251.

Purvis, A. & Webster, A. J. 1999. Phylogenetically independent comparisons and primate phylogeny. In *Comparative Primate Socioecology,* ed. P. C. Lee, pp. 44–70. New York: Cambridge University Press.

Read, A. F. & Harvey, P. H. 1989. Life history differences among the eutherian radiations. *Journal of Zoology* (London), 219:329–353.

Ross, C. 1988. The intrinsic rate of natural increase and reproductive effort in primates. *Journal of Zoology* (London), 214:199–219.

Rumbaugh, D. M. 1997. Competence, cortex, and primate models: a comparative primate perspective. In *Development of the Prefrontal Cortex,* ed. N. A. Krasnegor, G. R. Lyon & P. S. Goldman-Rakic, pp. 117–139. Baltimore, MD: Paul H. Brookes.

Sacher, G. A. 1959. Relation of lifespan to brain weight and body weight in mammals. In *CIBA Foundation Symposium on the Lifespan of Animals,* ed. G. E. W. Wolstenholme & M. O'Connor, pp. 115–133. Boston: Little, Brown, and Company.

Sacher, G. A. & Staffeldt, E. F. 1974. Relation of gestation time to brain weight for placental mammals: implications for the theory of vertebrate growth. *American Naturalist,* 108:593–615.

Savage-Rumbaugh, E. S. & Lewin, R. 1994. *Kanzi: The Ape at the Brink of the Human Mind.* New York: Wiley & Sons.

Sawaguchi, T. & Kudo, H. 1990. Neocortical development and social structure in primates. *Primates,* 31(2):283–289.

Shea, B. T. 1987. Reproductive strategies, body size, and encephalization in primate evolution. *International Journal of Primatology,* 8:139–156.

Shettleworth, S. J. 1998. *Cognition, Evolution, and Behavior.* Oxford: Oxford University Press.

Stearns, S. C. 1992. *The Evolution of Life Histories.* Oxford: Oxford University Press.

——— 2000. Life history evolution: successes, limitations, and prospects. *Naturwissenschaften,* 87:476–486.

Strum, S. C., Forster, D. & Hutchins, E. 1997. Why Machiavellian intelligence may not be Machiavellian. In *Machiavellian Intelligence II: Extensions and Evaluations,* ed. A. Whiten & R. W. Byrne, pp. 50–85. Cambridge: Cambridge University Press.

Tomasello, M. 1999. *The Cultural Origins of Human Cognition.* Cambridge, MA: Harvard University Press.

Tomasello, M. & Call, J. 1994. Social cognition of monkeys and apes. *Yearbook of Physical Anthropology,* 37:273–305.

Tyack, P. L. 1999. Communication and cognition. In *Biology of Marine Mammals,* vol. 1, ed. J. E. Reynolds III & J. R. Twiss Jr., pp. 287–323. Washington, DC: Smithsonian Institution Press.

van Schaik, C. P. 2000. Vulnerability to infanticide: patterns among mammals. In *Infanticide by Males and Its Implications,* ed. C. P. van Schaik & C. H. Janson, pp. 61–71. Cambridge: Cambridge University Press.

Whiten, A. 2000. Social complexity and social intelligence. In *The Nature of Intelligence,* pp. 185–196. Novartis Foundation Symposium 233. Chichester: Wiley.

Case Study 1A. *Sociality and Disease Risk*

Barratt, C. L. R., Bolton, A. E. & Cooke, I. D. 1990. Functional significance of white blood-cells in the male and female reproductive tract. *Human Reproduction,* 5:639–648.

Côté, I. M. & Poulin, R. 1995. Parasitism and group size in social animals: a meta-analysis. *Behavioral Ecology,* 6:159–165.

Daszak, P., Cunningham, A. A. & Hyatt, A. D. 2000. Wildlife ecology: emerging infectious diseases of wildlife. *Science,* 287:443–449.

Dudley, R. & Milton, K. 1990. Parasite deterrence and the energetic costs of slapping in howler monkeys, *Alouatta palliata. Journal of Mammalogy,* 71:463–465.

Felsenstein, J. 1985. Phylogenies and the comparative method. *American Naturalist,* 125:1–15.

Freeland, W. J. 1976. Pathogens and the evolution of primate sociality. *Biotropica,* 8:12–24.

Grenfell, B. T. & Dobson, A. P., eds. 1995. *Ecology of Infectious Diseases in Natural Populations.* Cambridge: Cambridge University Press.

Harcourt, A. H., Harvey, P. H., Larson, S. G. & Short, R. V. 1981. Testis weight, body weight and breeding system in primates. *Nature,* 293:55–57.

Harcourt, A. H., Purvis, A. & Liles, L. 1995. Sperm competition: mating system, not breeding system, affects testes size of primates. *Functional Ecology,* 9:468–476.

Hart, B. L. 1990. Behavioral adaptations to pathogens and parasites: five strategies. *Neuroscience and Biobehavioral Reviews,* 14:273–294.

Hart, B. J., Korinek, E. & Brennan, P. 1987. Postcopulatory genital grooming in male rats: prevention of sexually transmitted infections. *Physiology and Behavior,* 41:321–325.

Harvey, P. H. & Pagel, M. D. 1991. *The Comparative Method in Evolutionary Biology.* Oxford: Oxford University Press.

Hausfater, G. & Meade, B. J. 1982. Alternation of sleeping groves by yellow baboons *(Papio cynocephalus)* as a strategy for parasite avoidance. *Primates,* 23:287–297.

Holmes, K. K., Sparling, P. F., Mardh, P. A., Lemon, S. M., Stamm, W. E., Piot, P. & Wasserheit, J. N., eds. 1994. *Sexually Transmitted Diseases.* New York: McGraw-Hill.

Huffman, M. A. & Caton, J. M. 2001. Self-induced increase of gut motility and the control of parasitic infections in wild chimpanzees. *International Journal of Primatology,* 22:329–346.

Jolly, A. 1966. *Lemur Behavior.* Chicago: University of Chicago Press.

Kiesecker, J. M., Skelly, D. K., Beard, K. H. & Preisser, E. 1999. Behavioral reduction of infection risk. *Proceedings of the National Academy of Sciences USA,* 96:9165–9168.

Knell, R. J. 1999. Sexually transmitted disease and parasite-mediated sexual selection. *Evolution,* 53:957–961.

Lockhart, A. B., Thrall, P. H. & Antonovics, J. 1996. Sexually transmitted diseases in animals: ecological and evolutionary implications. *Biological Reviews,* 71:415–471.

Møller, A. P., Dufva, R. & Allander, K. 1993. Parasites and the evolution of host social behavior. *Advances in the Study of Behavior,* 22:65–102.

Moret, Y. & Schmid-Hempel, P. 2000. Survival for immunity: the price of immune system activation for bumblebee workers. *Science,* 290:1166–1168.

Nunn, C. L. In revision. A comparative study of leukocyte counts and disease risk in primates. *Evolution.*

Nunn, C. L. & Barton, R. A. 2001. Comparative methods for studying primate adaptation and allometry. *Evolutionary Anthropology,* 10:81–98.

Nunn, C. L. & van Schaik, C. P. 2001. Reconstructing the behavioral ecology of extinct primates. In *Reconstructing Behavior in the Fossil Record,* ed. J. M. Plavcan, R. F. Kay, W. L. Jungers, & C. P. van Schaik, pp. 159–216. New York: Plenum.

Nunn, C. L., Gittleman, J. L. & Antonovics, J. 2000. Promiscuity and the primate immune system. *Science,* 290:1168–1170.

Pandya, I. J. & Cohen, J. 1985. The leukocytic reaction of the human uterine cervix to spermatozoa. *Fertility and Sterility,* 43:417–421.

Purvis, A. 1995. A composite estimate of primate phylogeny. *Philosophical Transactions of the Royal Society of London,* ser. B, 348:405–421.

Purvis, A. & Rambaut, A. 1995. Comparative analysis by independent contrasts (CAIC): an Apple Macintosh application for analysing comparative data. *Computer Applications in the Biosciences,* 11:247–251.

Roitt, I. M., Brostoff, J. & Male, D. K. 1998. *Immunology.* London: Gower Medical Publishing.

Rubenstein, D. I. & Hohmann, M. E. 1989. Parasites and social behavior of island feral horses. *Oikos,* 55:312–320.

Sheldon, B. C. & Verhulst, S. 1996. Ecological immunology: costly parasite defences and trade-offs in evolutionary ecology. *Trends in Ecology and Evolution,* 11:317–321.

Smuts, B. B. & Smuts, R. W. 1993. Male aggression and sexual coercion of females in nonhuman primates and other mammals: evidence and theoretical implications. *Advances in the Study of Behavior,* 22:1–63.

van Schaik, C. P. & Janson, C., eds. 2000. *Infanticide by Males and Its Implications.* Cambridge: Cambridge University Press.

van Schaik, C. P., van Noordwijk, M. A. & Nunn, C. L. 1999. Sex and social evolution in primates. In *Comparative Primate Socioecology,* ed. P. C. Lee, pp. 204–240. Cambridge: Cambridge University Press.

2. Dolphin Social Complexity

Altmann, J. 1974. Observational study of behavior: sampling methods. *Behaviour,* 49:227–267.

Barros, N. B. & Wells, R. S. 1998. Prey and feeding patterns of resident bottlenose dolphins *(Tursiops truncatus)* in Sarasota Bay, Florida. *Journal of Mammalogy,* 79:1045–1059.

Busnel, R. G. 1973. Symbiotic relationship between man and dolphins. *Transactions of the New York Academy of Sciences,* ser. 2, 35:112–131.

Cairns, S. J. & Schwager, S. 1987. A comparison of association indices. *Animal Behaviour,* 35:1455–1469.

Caldwell, D. K. & Caldwell, M. C. 1972. *The World of the Bottlenosed Dolphin.* New York: Lippincott.

Connor, R. C., Smolker, R. A. & Richards, A. F. 1992a. Dolphin alliances and coalitions. In *Coalitions and Alliances in Humans and Other Animals,* ed. A. H. Harcourt & F. B. M. de Waal, pp. 415–444. Oxford: Oxford University Press.

——— 1992b. Two levels of alliance formation among bottlenose dolphins (*Tursiops* sp.). *Proceedings of the National Academy of Sciences USA,* 89:987–990.

Connor, R. C., Heithaus, M. R. & Barre, L. M. 1999. Superalliance of bottlenose dolphins. *Nature,* 397:571–572.

Connor, R. C., Wells, R. S., Mann, J. & Read, A. J. 2000a. The bottlenose dolphin, *Tursiops* spp.: social relationships in a fission-fusion society. In *Cetacean Societies: Field Studies of Dolphins and Whales,* ed. J. Mann, R. C. Connor, P. L. Tyack, & H. Whitehead, pp. 91–126. Chicago: University of Chicago Press.

Connor, R. C., Heithaus, M. R., Berggren, P. & Miksis, J. L. 2000b. "Kerplunking": surface fluke-splashes during shallow water bottom foraging by bottlenose dolphins. *Marine Mammal Science,* 16:646–653.

Corkeron, P. J., Bryden, M. M. & Hedstrom, K. E. 1990. Feeding by bottlenose dolphins in association with trawling operations in Moreton Bay, Australia. In *The Bottlenose Dolphin,* ed. S. Leatherwood & R. R. Reeves, pp. 329–336. San Diego: Academic Press.

Duffield, D. A. & Wells, R. S. 1991. The combined application of chromosome, protein and molecular data for the investigation of social unit structure and dynamics in *Tursiops truncatus.* In *Genetic Ecology of Whales and Dolphins,* ed. A. R. Hoelzel, pp. 155–169. Cambridge: Report of the International Whaling Commission, Special Issue 13.

——— In press. The molecular profile of a resident community of bottlenose dolphins, *Tursiops truncatus.* In *Molecular and Cell Biology of Marine Mammals,* ed. C. J. Pfeiffer. Melbourne: Krieger Publishing Company.

Herzing, D. L. 1996. Vocalizations and associated underwater behavior of free-ranging Atlantic spotted dolphins, *Stenella frontalis* and bottlenose dolphins, *Tursiops truncatus. Aquatic Mammals,* 22:61–79.

——— 1997. The life history of free-ranging Atlantic spotted dolphins *(Stenella frontalis):* age classes, color phases, and female reproduction. *Marine Mammal Science,* 13:576–595.

Irvine, B. & Wells, R. S. 1972. Results of attempts to tag Atlantic bottlenose dolphins (*Tursiops truncatus*). *Cetology,* 13:1–5.

Irvine, A. B., Scott, M. D., Wells, R. S. & Kaufmann, J. H. 1981. Movements and activities of the Atlantic bottlenose dolphin, *Tursiops truncatus,* near Sarasota, Florida. *Fishery Bulletin US,* 79:671–688.

Küss, K. M. 1998. The occurrence of PCBs and chlorinated pesticide contaminants in bottlenose dolphins in a resident community: comparison with age, gender and birth order. M.Sc. thesis, Nova Southeastern University.

Lahvis, G. P., Wells, R. S., Kuehl, D. W., Stewart, J. L., Rhinehart, H. L. & Via, C. S. 1995. Decreased lymphocyte responses in free-ranging bottlenose dolphins (*Tursiops truncatus*) are associated with increased concentrations of PCB's and DDT in peripheral blood. *Environmental Health Perspectives,* 103:67–72.

Moors, T. L. 1997. Is 'menage a trois' important in dolphin mating systems? Behavioral patterns of breeding female bottlenose dolphins. M.Sc. thesis, University of California, Santa Cruz.

Nowacek, D. P. 1999. Sound use, sequential behavior and foraging ecology of foraging bottlenose dolphins, *Tursiops truncatus.* Doctoral diss., Massachusetts Institute of Technology and Woods Hole Oceanographic Institution. MIT/WHOI 99–16.

Nowacek, D. P., Tyack, P. L., Wells, R. S. & Johnson, M. P. 1998. An onboard acoustic data logger to record biosonar of free-ranging bottlenose dolphins. *Journal of the Acoustical Society of America,* 103:1409–1410.

Nowacek, D. P., Tyack, P. L. & Wells, R. S. 2001a. A platform for continuous behavioral and acoustic observation of free-ranging marine mammals: overhead video combined with underwater audio. *Marine Mammal Science,* 17:191–199.

Nowacek, S. M., Wells, R. S. & Solow, A. R. 2001b. The effects of boat traffic on bottlenose dolphins, *Tursiops truncatus,* in Sarasota Bay, Florida. *Marine Mammal Science* 17:673–688.

Owen, C. F. W. 2001. A comparison of maternal care by primiparous and multiparous bottlenose dolphins *(Tursiops truncatus):* does parenting improve with experience? M.Sc. thesis, University of California, Santa Cruz.

Pryor, K., Lindbergh, J., Lindbergh, S. & Milano, R. 1990. A dolphin-human fishing cooperative in Brazil. *Marine Mammal Science,* 6:77–82.

Read, A. J., Wells, R. S., Hohn, A. A. & Scott, M. D. 1993. Patterns of growth in wild bottlenose dolphins, *Tursiops truncatus. Journal of Zoology* (London), 231:107–123.

Ridgway, S. H., Kamolnick, T., Reddy, M., Curry, C. & Tarpley, R. J. 1995. Orphan-induced lactation in *Tursiops* and analysis of collected milk. *Marine Mammal Science,* 11:172–182.

Rossbach, K. A. & Herzing, D. L. 1997. Underwater observations of benthic-feeding bottlenose dolphins (*Tursiops truncatus*) near Grand Bahama Island, Bahamas. *Marine Mammal Science,* 13:498–504.

Sayigh, L. S., Wells, R. S. & Tyack, P. L. 1993. Recording underwater sounds of free-ranging dolphins while underway in a small boat. *Marine Mammal Science,* 9:209–213.

Sayigh, L. S., Tyack, P. L., Wells, R. S., Scott, M. D. & Irvine, A. B. 1995. Sex difference in whistle production of free-ranging bottlenose dolphins, *Tursiops truncatus. Behavioral Ecology and Sociobiology,* 36:171–177.

Sayigh, L. S., Tyack, P. L., Wells, R. S., Solow, A. R., Scott, M. D. & Irvine, A. B. 1999. Individual recognition in wild bottlenose dolphins: a field test using playback experiments. *Animal Behaviour,* 57:41–50.

Scott, M. D., Wells, R. S. & Irvine, A. B. 1990a. A long-term study of bottlenose dolphins on the west coast of Florida. In *The Bottlenose Dolphin,* ed. S. Leatherwood & R. R. Reeves, pp. 235–244. San Diego: Academic Press.

Scott, M. D., Wells, R. S., Irvine, A. B. & Mate, B. R. 1990b. Tagging and marking studies on small cetaceans. In *The Bottlenose Dolphin,* ed. S. Leatherwood & R. R. Reeves, pp. 489–514. San Diego: Academic Press.

Scott, M. D., Westgate, A. J., Wells, R. S., Pabst, D. A., McLellan, W. A., Read, A. J., Townsend, F. I., Rhinehart, H. L., Hanson, M. B. & Rommel, S. A. In press. Dorsal fin morphology and attachment of radiotags to small cetaceans. *Marine Mammal Science.*

Sellas, A. B. 2002. Population structure and group relatedness of bottlenose dolphins *(Tursiops truncatus)* in the coastal Gulf of Mexico using mitochondrial DNA and nuclear microsatellite markers. M.Sc. thesis, University of California, Santa Cruz.

Shane, S. H. 1990a. Behavior and ecology of the bottlenose dolphin at Sanibel Island, Florida. In *The Bottlenose Dolphin,* ed. S. Leatherwood & R. R. Reeves, pp. 245–265. San Diego: Academic Press.

——— 1990b. Comparison of bottlenose dolphin behavior in Texas and Florida, with a critique of methods for studying dolphin behavior. In *The Bottlenose Dolphin,* ed. S. Leatherwood & R. R. Reeves, pp. 541–558. San Diego: Academic Press.

Shane, S. H., Wells, R. S. & Würsig, B. 1986. Ecology, behavior, and social organization of the bottlenose dolphin: a review. *Marine Mammal Science,* 2:34–63.

Smolker, R., Richards, A. F., Connor, R. C., Mann, J. & Berrgren, P. 1997. Sponge carrying by Indian Ocean bottlenose dolphins: possible tool use by a delphinid. *Ethology,* 103:454–465.

Tolley, K. A., Read, A. J., Wells, R. S., Urian, K. W., Scott, M. D., Irvine, A. B. & Hohn, A. A. 1995. Sexual dimorphism in wild bottlenose dolphins (*Tursiops truncatus*) from Sarasota, Florida. *Journal of Mammalogy,* 76:1190–1198.

Tyack, P. L. 1997. Development and social functions of signature whistles in bottlenose dolphins, *Tursiops truncatus. Bioacoustics,* 8:21–46.

Tyack, P. L. & Sayigh, L. S. 1997. Vocal learning in cetaceans. In *Social Influences on Vocal Development,* ed. C. Snowdon & M. Hausberger, pp. 208–233. Cambridge: Cambridge University Press.

Urian, K. W., Duffield, D. A., Read, A. J., Wells, R. S. & Shell, D. D. 1996. Seasonality of reproduction in bottlenose dolphins, *Tursiops truncatus. Journal of Mammalogy,* 77:394–403.

Vedder, J. 1996. Levels of organochlorine contaminants in milk relative to health of bottlenose dolphins *(Tursiops truncatus)* from Sarasota, Florida. M.Sc. thesis, University of California, Santa Cruz.

Wells, R. S. 1991. The role of long-term study in understanding the social structure of a bottlenose dolphin community. In *Dolphin Societies: Discoveries and Puzzles,* ed. K. Pryor & K. S. Norris, pp. 199–225. Berkeley: University of California Press.

——— 2000. Reproduction in wild bottlenose dolphins: overview of patterns observed during a long-term study. In *Bottlenose Dolphin Reproduction Workshop Report,* ed. D. Duffield & T. Robeck, pp. 57–74. Silver Springs, MD: AZA Marine Mammal Taxon Advisory Group.

Wells, R. S. & Duffield, D. A. In preparation. Patterns of paternity in bottlenose dolphins.

Wells, R. S. & Scott, M. D. 1990. Estimating bottlenose dolphin population parameters from individual identification and capture-release techniques. In *Individual Recognition of Cetaceans: Use of Photo-Identification and Other Techniques to Estimate Population Parameters,* ed. P. S. Hammond, S. A. Mizroch, & G. P. Donovan, pp. 407–415. Cambridge: Report of the International Whaling Commission, Special Issue 12.

——— 1994. Incidence of gear entanglement for resident inshore bottlenose dolphins near Sarasota, Florida. In *Gillnets and Cetaceans,* ed. W. F. Perrin, G. P. Donovan, & J. Barlow, p. 629. Cambridge: Report of the International Whaling Commission, Special Issue 15.

——— 1997. Seasonal incidence of boat strikes on bottlenose dolphins near Sarasota, Florida. *Marine Mammal Science,* 13:475–480.

——— 1999. Bottlenose dolphin *Tursiops truncatus* (Montagu, 1821). In *Handbook of Marine Mammals,* vol. 6, *The Second Book of Dolphins and Porpoises,* ed. S. H. Ridgway & R. Harrison, pp. 137–182. San Diego: Academic Press.

Wells, R. S., Hofmann, S. & Moors, T. L. 1998. Entanglement and mortality of bottlenose dolphins (*Tursiops truncatus*) in recreational fishing gear in Florida. *Fishery Bulletin,* 96:647–650.

Wells, R. S., Irvine, A. B. & Scott, M. D. 1980. The social ecology of inshore odontocetes. In *Cetacean Behavior: Mechanisms and Functions,* ed. L. M. Herman, pp. 263–317. New York: J. Wiley & Sons.

Wells, R. S., Scott, M. D. & Irvine, A. B. 1987. The social structure of free-ranging bottlenose dolphins. In *Current Mammalogy,* vol. 1, ed. H. Genoways, pp. 247–305. New York: Plenum Press.

Wells, R. S., Boness, D. J. & Rathbun, G. B. 1999. Behavior. In *Biology of Marine Mammals,* ed. J. E. Reynolds III & S. A. Rommel, pp. 324–422. Washington, DC: Smithsonian Institution Press.

Whiten, A., Goodall, J., McGrew, W. C., Nishida, T., Reynolds, V., Sugiyama, Y., Tutin, C. E. G., Wrangham, R. W. & Boesch, C. 1999. Cultures in chimpanzees. *Nature,* 399:682–685.

Wilson, D. R. B. 1995. The ecology of bottlenose dolphins in the Moray Firth, Scotland: a population at the northern extreme of the species' range. Ph.D. diss., University of Aberdeen, Scotland.

Würsig, B. & Jefferson, T. A. 1990. Methods of photo-identification for small cetaceans. In *Individual Recognition of Cetaceans: Use of Photo-Identification and Other Techniques to Estimate Population Parameters,* ed. P. S. Hammond, S. A. Mizroch, & G. P. Donovan, pp. 43–52. Cambridge: Report of the International Whaling Commission, Special Issue 12.

3. Sources of Social Complexity in the Three Elephant Species

Charif, R., Payne, K., Ramey, R., Langbauer, Jr., W. R. & Brown, L. In review. Home ranges, movement patterns & genetic relatedness in African elephants. *Behavioral Ecology and Sociobiology.*

Douglas-Hamilton, I. 1972. On the ecology and behaviour of the African elephant: the elephants of Lake Manyara. Ph.D. diss., Oxford University.

Douglas-Hamilton, I. & Douglas-Hamilton, O. 1975. *Among the Elephants.* New York: Viking Press.

Dublin, H. T. 1983. Cooperation & competition among female African elephants. In *Social Behavior of Female Vertebrates,* ed. S. K. Wasser, pp. 291–313. New York: Academic Press.

Dunbar, R. I. M. 1992. Neocortex size as a constraint on group size in primates: a test of the hypothesis. *Journal of Human Evolution,* 28:287–296.

Eltringham, S. K. 1982. *Elephants.* Dorset: Blandford Press.

Estes, R. D. 1991. *The Behavior Guide to African Mammals.* Berkeley: University of California Press.

Fernando, P. & Lande, R. 2000. Molecular genetic & behavioral analysis of social organization in the Asian elephant (*Elephas maximus*). *Behavioral Ecology and Sociobiology,* 48:84–91.

Heffner, R. & Heffner, H. 1980. Hearing in the elephant (*Elephas maximus*). *Science,* 208:518–520.

Kummer, H. 1968. *Social Organisation of Hamadryas Baboons.* Chicago: University of Chicago Press.

Langbauer, Jr., W. R., Payne, K., Charif, R., Rappaport, E. & Osborn, F. 1991. African elephants respond to distant playbacks of low-frequency conspecific calls. *Journal of Experimental Biology,* 157:35–46.

Larom, D., Garstang, M., Payne, K., Raspet, R. & Lindeque, M. 1997. The influence of surface atmospheric conditions on the range & area reached by animal vocalizations. *Journal of Experimental Biology,* 200:421–431.

Laws, R. M. 1969. Aspects of reproduction in the African elephant, *Loxodonta africana. Journal of Reproduction and Fertility,* Suppl. 6:193–217.

Lee, P. C. 1987. Allomothering among African elephants. *Animal Behaviour,* 35:278–291.

——— 1989. Family structure, communal care, and female reproductive effort. In *Comparative Socioecology,* ed. V. Standen & R. A. Foley, pp. 323–340. Oxford: Blackwell Scientific Publications.

——— 1991. Social life. In *The Illustrated Encyclopedia of Elephants,* ed. S. K. Eltringham, pp. 48–63. London: Salamander.

Lee, P. C. & Moss, C. J. 1986. Early maternal investment in male and female African elephant calves. *Behavioral Ecology and Sociobiology,* 18:352–361.

Martin, R. M. 1978. Aspects of elephant social organization. *Rhodesia Science News,* 12:184–187.

McComb, K., Moss, C. J., Sayialel, S. & Baker, L. 2000. Unusually extensive networks of vocal recognition in African elephants. *Animal Behaviour,* 59:1103–1109.

McComb, K., Moss, C. J., Durant, S. M., Baker, L., & Sayialel, S. 2001. Matriarchs as repositories of social knowledge in African elephants. *Science,* 292:491–494.

Moss, C. J. 1982. *Portraits in the Wild: Animal Behavior in East Africa.* Chicago: University of Chicago Press.

——— 1988. *Elephant Memories.* New York: Fawcett Columbine.

——— 1992. Elephant calves: the story of two sexes. In *Elephants,* ed. J. Shoshoni & J. W. Owen, pp. 106–113. San Francisco: Weldon Owen.

Moss, C. J. & Poole, J. 1983. Relationships and social structure of African elephants. In *Primate Social Relationships: An Integrated Approach,* ed. R. Hinde, pp. 315–325. Oxford: Blackwell Scientific Publications.

Osborn, F. V. 1998. The ecology of crop-raiding elephants in Zimbabwe. Ph.D. diss., University of Cambridge.

Payne, K. 1998. *Silent Thunder: In the Presence of Elephants.* New York: Simon and Schuster.

Payne, K., Langbauer, Jr., W. R., & Thomas, E. 1986. Infrasonic calls of the Asian elephant (*Elephas maximus*). *Behavioral Ecology and Sociobiology,* 18:297–301.

Payne, K., Thompson, M., & Kramer, L. In review. Elephant calling patterns as indicators of group size and composition: the basis for an acoustic monitoring system. *African Journal of Ecology.*

Poole, J. H. 1982. Musth and male-male competition in the African elephant. Ph.D. diss., University of Cambridge.

——— 1987. Rutting behaviour in African elephants: the phenomenon of musth. *Behaviour,* 102:283–316.

——— 1989. Announcing intent: the aggressive state of musth in African elephants. *Animal Behaviour,* 37:140–152.

——— 1999. Signals and assessment in African elephants: evidence from playback experiments. *Animal Behaviour,* 58:185–193.

Poole, J. H. & Moss, C. J. 1989. Elephant mate searching: group dynamics and vocal and olfactory communication. In *The Biology of Large African Mammals in Their Environment,* ed. P. A. Jewell & G. M. O. Maloiy. *Symposia of the Zoological Society of London,* 61:111–125.

Poole, J. H., Payne, K. B., Langbauer, Jr., W. R., & Moss, C. J. 1988. The social contexts of some very low frequency calls of African elephants. *Behavioral Ecology and Sociobiology,* 22:385–392.

Rasmussen, L. E. L. & Krishnamurthy, V. 2000. How chemical signals integrate Asian elephant society: the known and the unknown. *Zoo Biology,* 19(5):405–423.

Rasmussen, L. E. L., Hallmartin, A. J., & Hess, D. L. 1996. Chemical profiles of male African elephants, *Loxodonta africana:* physiological and ecological implications. *Journal of Mammalogy,* 77:422–439.

Roca, A. L., Goergiadis, N., Pecon-Slattery, J., & O'Brien, S. J. 2001. Genetic evidence for two species of elephant in Africa. *Science,* 293:1473–1477.

Seeley, T. D. & Buhrman, S. C. 1999. Group decision making in swarms of honey bees. *Behavioral Ecology and Sociobiology,* 45:19–31.

Slotow, R., van Dyk, G., Poole, J., Page, B. & Klocke, A. 2000. Older bull elephants control young males. *Nature,* 408:426–426.

Sukumar, R. 1989. *The Asian Elephant: Ecology and Management.* Cambridge: Cambridge University Press.

——— 1994. *Elephant Days and Nights: Ten Years with the Indian Elephant.* New Delhi: Oxford University Press.

Turkalo, A. 1996. Studying forest elephants by direct observation in the Dzanga clearing: an update. *Pachyderm,* 22:59–60.

Viljoen, P. J. 1982. Western Kaokoland, Damaraland and the Skeleton Coast Park aerial game census. Namibia Wildlife Trust Publication. 30 pp.

——— 1989. Spatial distribution and movements of elephants (*Loxodonta africana*) in the northern Namib region of Kaokoveld, South West Africa (Namibia). *Journal of Zoology* (London), 219:1–19.

——— 1990. Daily movements of desert-dwelling elephants in the northern Namib Desert. *South African Wildlife Research,* 20(2):69–72.

4. Complex Cooperation among Taï Chimpanzees

Bekkering, H., Wohlschläger, A. & Gattis, M. 2000. Imitation of gestures in children is goal-directed. *Quarterly Journal of Experimental Psychology,* 53A(1):153–164.

Boesch, C. 1994. Cooperative hunting in wild chimpanzees. *Animal Behaviour,* 48:653–667.

Boesch, C. & Boesch, H. 1989. Hunting behavior of wild chimpanzees in the Taï National Park. *American Journal of Physical Anthropology,* 78:547–573.

Boesch, C. & Boesch-Achermann, H. 2000. *The Chimpanzees of the Taï Forest: Behavioural Ecology and Evolution.* Oxford: Oxford University Press.

Byrne, R. 1995. *The Thinking Ape.* Oxford: Oxford University Press.

Carruthers, P. & Smith, P. 1996. *Theories of Theories of Mind.* Cambridge: Cambridge University Press.

Chalmeau, R. 1994. Do chimpanzees cooperate in a learning task? *Primates,* 35:385–392.

Chalmeau, R., Visalberghi, E. & Gallo, A. 1997. Capuchin monkeys, *Cebus apella,* fail to understand a cooperative task. *Animal Behaviour,* 54:1215–1225.

Cooper, S. M. 1990. The hunting behaviour of spotted hyenas (*Crocuta crocuta*) in a region containing both sedentary and migratory populations of herbivores. *African Journal of Ecology,* 28:131–141.

Creel, S. 1997. Cooperative hunting and group size: assumptions and currencies. *Animal Behaviour,* 54:1319–1324.

Creel, S. & Creel, N. M. 1995. Communal hunting and pack size in African wild dogs, *Lycaon pictus. Animal Behaviour,* 50:1325–1339.

de Waal, F. & Berger, M. 2000. Payment for labour in monkeys. *Nature,* 404:563.

Dunbar, R. 1992. Neocortex size as a constraint on group size in primates. *Journal of Human Evolution,* 20:469–493.

Foley, R. 1995. *Humans before Humanity: An Evolutionary Perspective.* Oxford: Blackwell Publishers.

Goodall, J. 1986. *The Chimpanzees of Gombe: Patterns of Behavior.* Cambridge: The Belknap Press of Havard University Press.

Grinnell, J., Packer, C. & Pusey, A. E. 1995. Cooperation in male lions, kinship, reciprocity or mutualism? *Animal Behaviour,* 49:95–105.

Gurven, M., Alen-Arave, W., Hill, K. & Hurtado, M. 2000. "It's a wonderful life": signaling generosity among the Achè of Paraguay. *Evolution and Human Behavior,* 21:263–282.

Harcourt, A. H. & de Waal, F., eds. 1992. *Cooperation in Competition in Animals and Humans.* Oxford: Oxford University Press.

Hawkes, K. 1991. Showing off: tests of an hypothesis about men's foraging goals. *Ethology and Sociobiology,* 12:19–54.

Hawkes, K., O'Connell, J. & Blurton-Jones, N. 1997. Hadza women's time allocation, offspring provisioning, and the evolution of long postmenopausal life spans. *Current Anthropology,* 38(4):551–577.

Heyes, C. M. 1998. Theory of mind in nonhuman primates. *Behavioral and Brain Sciences,* 21(1):101–134.

Humphrey, N. K. 1976. The social function of intellect. In *Growing Points in Ethology,* ed. P. P. Bateson & R. Hinde, pp. 303–317. Cambridge: Cambridge University Press.

Isaac, G. 1978. The food sharing behavior of protohuman hominids. *Scientific American,* 238:90–108.

Kaplan, H. & Hill, K. 1985. Hunting ability and reproductive success among male Achè foragers: preliminary results. *Nature,* 26(1):131–133.

Kaplan, H., Hill, K., Lancaster, J. & Hurtado, A. 2000. A theory of human life history evolution: diet, intelligence and longevity. *Evolutionary Anthropology,* 9(4):156–185.

Manson, J. & Wrangham, R. 1991. Intergroup aggression in chimpanzees and humans. *Current Anthropology,* 32(4):369–390.

Marlowe, F. 2000. Male care and mating effort among Hadza foragers. *Behavioral Ecology and Sociobiology,* 46:57–64.

Mendres, K. & de Waal, F. 2000. Capuchins do cooperate: the advantage of an intuitive task. *Animal Behaviour,* 60:523–529.

Menzel, E. W. 1974. A group of young chimpanzees in a one-acre field. In *Behavior of Nonhuman Primates,* ed. A. M. Schrier & F. Stollnitz, pp. 83–153. New York: Academic Press.

Mitani, J. & Watts, D. 1999. Demographic influences on the hunting behavior of chimpanzees. *American Journal of Physical Anthropology,* 109:439–454.

Mithen, S. 1996. *The Prehistory of Mind: The Cognitive Origin of Art and Science.* London: Thames and Hudson.

Nishida, T. & Turner, L. 1996. Food transfer between mother and infant chimpanzees of the Mahale Moutains National Park, Tanzania. *International Journal of Primatology,* 17(6):947–968.

Nishida, T., Uehara, S. & Nyondo, R. 1983. Predatory behavior among wild chimpanzees of the Mahale Mountains. *Primates,* 20:1–20.

Packer, C., Scheel, D. & Pusey, A. E. 1990. Why lions form groups: food is not enough. *American Naturalist,* 136:1–19.

Savage-Rumbaugh, E. S., Rumbaugh, D. M. & Boysen, S. 1978. Linguistically mediated tool use and exchange by chimpanzees (*Pan troglodytes*). *Behavioral and Brain Sciences,* 201:641–644.

Scheel, D. & Packer, C. 1991. Group hunting behaviour of lions: a search for cooperation. *Animal Behaviour,* 41:697–709.

Silk, J. B. 1978. Patterns of food sharing among mother and infant chimpanzees at Gombe National Park, Tanzania. *Folia primatologica,* 29:129–141.

Stander, P. E. 1992. Cooperative hunting in lions: the role of the individual. *Behavioral Ecology and Sociobiology,* 29:445–454.

Stanford, C. 1998. *Chimpanzee and Red Colobus: The Ecology of Predator and Prey.* Cambridge: Harvard University Press.

Stanford, C., Wallis, J., Mpongo, E. & Goodall, J. 1994. Hunting decisions in wild chimpanzees. *Behaviour,* 131:1–20.

Strum, S. C. 1981. Processes and products of change: baboon predatory behavior at Gilgil, Kenya. In *Omnivorous Primates: Gathering and Hunting in Human Evolution,* ed. R. S. O. Harding & G. Teleki, pp. 255–302. New York: Columbia University Press.

Takahata, Y., Hasegawa, T. & Nishida, T. 1984. Chimpanzee predation in the Mahale Mountains from August 1979 to May 1982. *International Journal of Primatology,* 5:213–233.

Teleki, G. 1973. *The Predatory Behavior of Wild Chimpanzees.* Brunswick: Bucknell University Press.

——— 1975. Primate subsistence patterns: collector-predators and gatherer-hunters. *Journal of Human Evolution,* 4:125–184.

Tomasello, M. & Call, J. 1997. *Primate Cognition.* Oxford: Oxford University Press.

Trivers, R. L. 1985. *Social Evolution.* Menlo Park, CA: Benjamin/Cummings Publishing Company.

Uehara, S., Nishida, T., Hamai, M., Hasegawa, T., Hayaki, H., Huffman, M., Kawanaka, K., Kobayashi, S., Mitani, J., Takahata, Y., Takasaki, H. & Tsukahara, T. 1992. Characteristics of predation by the chimpanzees in the Mahale Mountains National Park, Tanzania. In *Topics in Primatology,* vol. 1, *Human Origins,* ed. T. Nishida, W. C. McGrew, P. Marler, M. Pickford & F. de Waal, pp. 143–158. Tokyo: University of Tokyo Press.

van der Dennen, J. 1995. *The Origin of War.* Groningen: Origin Press.

Whiten, A. & Byrne, R. W. 1988. Tactical deception in primates. *Behavioral and Brain Sciences,* 11:233–273.

Wilson, E. O. 1975. *Sociobiology.* Cambridge: Harvard University Press.

Wrangham, R. & Peterson, D. 1996. *Demonic Males: Apes and the Origins of Human Violence.* Boston: Houghton Mufflin Co.

Zemel, A. & Lubin, Y. 1995. Inter-group competition and stable group sizes. *Animal Behaviour,* 50:485–488.

Case Study 4A. *Coalitionary Aggression in White-Faced Capuchins*

de Waal, F. B. M. 1978. Exploitative and familiarity-dependent support strategies in a colony of semi–free living chimpanzees. *Behaviour,* 66:268–312.

de Waal, F. B. M. & Harcourt, A. H. 1992. Coalitions and alliances: a history of ethological research. In *Coalitions and Alliances in Humans and Other Animals,* ed. A. H. Harcourt & F. B. M. de Waal, pp. 1–19. Oxford: Oxford University Press.

Fedigan, L. M. 1993. Sex differences and intersexual relations in adult white-faced capuchins (*Cebus capucinus*). *International Journal of Primatology,* 14:853–877.

Fedigan, L. M., Rose, L. M. & Avila, R. M. 1996. See how they grow: tracking capuchin monkey populations in a regenerating Costa Rican dry forest. In *Adaptive Radiations of Neotropical Primates,* ed. M. Norconck, A. Rosenberger, & P. Garber, pp. 289–307. New York: Plenum Press.

Harvey, P., Martin, R. D. & Clutton-Brock, T. H. 1987. Life histories in comparative perspective. In *Primate Societies,* ed. B. B. Smuts, D. L. Cheney, R. M. Seyfarth, R. W. Wrangham, & T. T. Struhsaker, pp. 181–196. Chicago: Chicago University Press.

Manson, J. H., Rose, L. M., Perry, S. & Gros-Louis, J. 1999. Dynamics of female-female relationships in wild *Cebus capucinus:* Data from two Costa Rican sites. *International Journal of Primatology,* 20:679–706.

Nishida, T. & Hosaka, K. 1996. Coalition strategies among adult male chimpanzees of the Mahale Mountains, Tanzania. In *Great Ape Societies,* ed. W. C. McGrew, L. F. Marchant, & T. Nishida, pp. 114–134. Cambridge: Cambridge University Press.

Perry, S. 1995. Social relationships in wild white-faced capuchin monkeys, *Cebus capucinus.* Ph.D. diss., University of Michigan, Ann Arbor.

——— 1996a. Female-female relationships in wild white-faced capuchins (*Cebus capucinus*). *International Journal of Primatology,* 40:167–182.

——— 1996b. Intergroup encounters in wild white-faced capuchins (*Cebus capucinus*). *American Journal of Primatology,* 17:309–330.

——— 1997. Male-female social relationships in wild white-faced capuchin monkeys (*Cebus capucinus*). *Behaviour,* 134:477–510.

——— 1998a. Male-male relationships in wild white-faced capuchins (*Cebus capucinus*). *Behaviour,* 135:139–172.

——— 1998b. A case report of a male rank reversal in a group of wild white-faced capuchins (*Cebus capucinus*). *Primates,* 39:51–69.

Rose, L. M. 1994. Benefits and costs of resident males to females in white-faced capuchins, *Cebus capucinus. American Journal of Primatology,* 32:235–248.

——— 1998. Behavioral ecology of white-faced capuchins *(Cebus capucinus)* in Costa Rica. Ph.D. diss., Washington University, St. Louis.

CASE STUDY 4B. *Levels and Patterns in Dolphin Alliance Formation*

Cairns, S. J. & Schwager, S. 1987. A comparison of association indices. *Animal Behaviour,* 3:1454–1469.

Chapais, B. 1992. The role of alliances in social inheritance of rank among female primates. In *Coalitions and Alliances in Animals and Humans,* ed. A. H. Harcourt & F. B. M. de Waal, pp. 29–59. Oxford: Oxford University Press.

Connor, R. C., Smolker, R. A. & Richards, A. F. 1992a. Two levels of alliance formation among male bottlenose dolphins (*Tursiops* sp.). *Proceedings of the National Academy of Sciences USA,* 89:987–990.

——— 1992b. Dolphin alliances and coalitions. In *Coalitions and Alliances in Animals and Humans,* ed. A. H. Harcourt & F. B. M. de Waal, pp. 415–444. Oxford: Oxford University Press.

Connor, R. C., Richards, A. F., Smolker, R. A. & Mann, J. 1996. Patterns of female attractiveness in Indian Ocean bottlenose dolphins. *Behaviour,* 133:37–69.

Connor, R. C., Heithaus, R. M. & Barre, L. M. 1999. Superalliance of bottlenose dolphins. *Nature,* 371:571–572.

——— 2001. Complex social structure, alliance stability and mating access in a bottlenose dolphin 'super-alliance.' *Proceedings of the Royal Socety of London,* ser. B, 268:263–267.

de Waal, F. B. M. 1982. *Chimpanzee Politics.* London: Jonathan Cape.

Harcourt, A. H. 1992. Coalitions and alliances: are primates more complex than nonprimates? In *Coalitions and Alliances in Humans and Other Animals,* ed. A. H. Harcourt & F. B. M. de Waal, pp. 445–472. Oxford: Oxford University Press.

Keller, L. & Reeve, H. K. 1994. Partitioning of reproduction in animal societies. *Trends in Ecology and Evolution,* 9:98–102.

Nishida, T. 1983. Alpha status and agonistic alliance in wild chimpanzees *(Pan troglodytes schweinfurthii). Primates,* 24:318–336.

Noe, R. 1990. A veto game played by baboons: shopping for profitable partners. In *Coalitions and Alliances in Animals and Humans,* ed. A. H. Harcourt & F. B. M. de Waal. Oxford: Oxford University Press.

Schusterman, R. J., Reichmuch, C. J. & Kastak, D. 2000. How animals classify friends and foes. *Current Directions in Psychological Science,* 9:1–6.

5. The Social Complexity of Spotted Hyenas

Allman, J. 1999. *Evolving Brains.* New York: Scientific American Library.

Altmann, J. & Alberts, S. 1987. Body mass and growth rates in a wild primate population. *Oecologia,* 72:15–20.

Altmann, J., Alberts, S. C., Haines, S. A., Dubach, J., Muruthi, P., Coote, T., Gefen, E., Cheesman, D. J., Mututua, R. S., Saiyalel, S. N., Wayne, R. K., Lacy, R. C. & Bruford, M. W. 1996. Behavior predicts genetic structure in a wild primate group. *Proceedings of the National Academy of Sciences USA,* 93:5797–6801.

Apps, P. J., Viljoen, H. W., Richardson, P. R. K. & Pretorius, V. 1989. Volatile components of anal gland secretion of aardwolf (*Proteles cristatus*). *Journal of Chemical Ecology,* 15:1681–1688.

Asa, C. S. & Valdespino, C. 1998. Canid reproductive biology: an integration of proximate mechanisms and ultimate causes. *American Zoologist,* 38:251–259.

Aureli, F. & de Waal, F. B. M. 2000. *Natural Conflict Resolution.* Berkeley: University of California Press.

Barrett, P. & Bateson, P. 1978. The development of play in cats. *Behaviour,* 66:106–120.

Bearder, S. K. 1977. Feeding habits of spotted hyaenas in a woodland habitat. *East African Wildlife Journal,* 15:236–280.

Bearder, S. K. & Randall, R. M. 1978. The use of fecal marking sites by spotted hyenas and civets. *Carnivore,* 2:32–48.

Beck, B. B. 1973. Cooperative tool use by captive hamadryas baboons. *Science,* 182:594–597.

Bekoff, M. 1972. The development of social interaction, play, and metacommunication in mammals: an ethological perspective. *Quarterly Review of Biology,* 47:412–434.

——— 1974. Social play and play-soliciting by infant canids. *American Zoologist,* 14:323–340.

——— 2001. Social play behaviour: cooperation, fairness, trust, and the evolution of morality. *Journal of Consciousness Studies,* 8:81–90.

Bekoff, M., Daniels, T. J. & Gittleman, J. L. 1984. Life history patterns and the comparative social ecology of carnivores. *Annual Review of Ecology and Systematics,* 15:191–232.

Bernstein, I. S. & Ehardt, C. L. 1985. Agonistic aiding: kinship, rank, age, and sex influences. *American Journal of Primatology,* 8:37–52.

Biben, M. 1983. Comparative ontogeny of social behaviour in three South American canids, the maned wolf, crab-eating fox and bush dog: implications for sociality. *Animal Behaviour,* 31:814–826.

Boesch, C. & Boesch, H. 1989. Hunting behavior of wild chimpanzees in the Taï National Park. *American Journal of Physical Anthropology,* 78:547–573.

Bourne, G. H. 1975. *The Rhesus Monkey,* vol. 1, *Anatomy and Physiology.* New York: Academic Press.

Bowlby, J. 1969. *Attachment and Loss.* New York: Basic Books.

Boydston, E. E., Morelli, T. L. & Holekamp, K. E. 2001. Sex differences in territorial behavior exhibited by the spotted hyena (Hyaenidae, *Crocuta crocuta*). *Ethology,* 107:369–385.

Brown, R. E. & Macdonald, D. W. 1985. *Social Odours in Mammals,* vol. 2. Oxford: Clarendon Press.

Buglass, A. J., Darling, F. M. C. & Waterhouse, J. S. 1990. Analysis of the anal sac secretion of the Hyaenidae. In *Chemical Signals in Vertebrates 5,* ed. D. W. Macdonald, D. Müller-Schwarze, & S. E. Natynczuk, pp. 65–69. New York: Oxford University Press.

Burton, J. J. 1977. Absence de comportement coopératif spontané dans une troupe de *Macaca fuscata* en présence de pierres appâtées. *Primates,* 18:359–366.

Butynski, T. M. 1982. Vertebrate predation by primates: a review of hunting patterns and prey. *Journal of Human Evolution,* 11:421–430.

Byrne, R. W. & Whiten, A. 1988. *Machiavellian Intelligence: Social Expertise and the Evolution of Intellect in Monkeys, Apes, and Humans.* Oxford: Clarendon Press.

Calderone, J. B., Reese, B. E. & Jacobs, G. H. 1995. Retinal ganglion cell distribution in the spotted hyena, *Crocuta crocuta. Society for Neuroscience Abstracts,* 21:1418.

Chalmeau, R. 1994. Do chimpanzees cooperate in a learning task? *Primates,* 35:385–392.

Chalmeau, R. & Gallo, A. 1993. Social constraints determine what is learned in the chimpanzee. *Behavioural Processes,* 28:173–180.

——— 1995. La coopération chez les primates. *L'Année Psychologique,* 95:119–130.

——— 1996. Cooperation in primates: critical analysis of behavoural criteria. *Behavioural Processes,* 35:101–111.

Chalmeau, R., Visalberghi, E. & Gallo, A. 1997. Capuchin monkeys, *Cebus apella,* fail to understand a cooperative task. *Animal Behaviour,* 54:1215–1225.

Cheney, D. L. & Seyfarth, R. M. 1990. *How Monkeys See the World: Inside the Mind of Another Species.* Chicago: University of Chicago Press.

Clutton-Brock, T. H. & Harvey, P. H. 1980. Primates, brains and ecology. *Journal of Zoology* (London), 190:309–323.

Cooper, S. M. 1990. The hunting behaviour of spotted hyaenas (*Crocuta crocuta*) in a region containing both sedentary and migratory populations of herbivores. *African Journal of Ecology,* 28:131–141.

——— 1991. Optimal hunting group size: the need for lions to defend their kills against loss to spotted hyaenas. *African Journal of Ecology,* 29:130–136.

——— 1993. Denning behavior of spotted hyaenas (*Crocuta crocuta*) in Botswana. *African Journal of Ecology,* 31:178–180.

Cooper, S. M., Holekamp, K. E. & Smale, L. 1999. A seasonal feast: long-term analysis of feeding behaviour in the spotted hyaena (*Crocuta crocuta*). *African Journal of Ecology,* 37:149–160.

Coussi-Korbel, S. 1994. Learning to outwit a competitor in mangabeys. *Journal of Comparative Psychology,* 108:164–171.

Crawford, M. P. 1937. The coöperative solving of problems by young chimpanzees. *Comparative Psychology Monographs,* 14:1–88.

Creel, S., Creel, N. M., Wildt, D. E. & Monfort, S. L. 1991. Behavioural and endocrine mechanisms of reproduction suppression in Serengeti dwarf mongooses. *Animal Behaviour,* 43:231–245.

Creel, S., Creel, N. M., Mills, M. G. L. & Monfort, S. L. 1997. Rank and reproduction in cooperatively breeding African wild dogs: Behavioral and endocrine correlates. *Behavioral Ecology,* 8:298–306.

D'Amato, M. R. & Colombo, M. 1989. Serial learning with wild card items by monkeys *(Cebus apella):* implications for knowledge of ordinal position. *Journal of Comparative Psychology,* 103:252–261.

de Waal, F. 1982. *Chimpanzee Politics: Power and Sex among Apes.* Baltimore: The Johns Hopkins University Press.

Drea, C. M. 1998. Status, age, and sex effects on performance of discrimination tasks in group-tested rhesus monkeys. *Journal of Comparative Psychology,* 112:170–182.

Drea, C. M. & Wallen, K. 1999. Low-status monkeys "play dumb" when learning in mixed social groups. *Proceedings of the National Academy of Sciences USA,* 96:12965–12969.

——— In press. Female sexuality and the myth of male control. In *Evolution, Gender, and Rape,* ed. C. B. Travis. Cambridge: MIT Press.

Drea, C. M., Hawk, J. E. & Glickman, S. E. 1996a. Aggression decreases as play emerges in infant spotted hyaenas: preparation for joining the clan. *Animal Behaviour,* 51:1323–1336.

Drea, C., Neves, A., Lopez, V. & Glickman, S. 1996b. Cooperation in captive spotted hyenas (*Crocuta crocuta*). Paper presented at the annual meeting of the Animal Behavior Society, Flagstaff, AZ.

Drea, C. M., Weldele, M. L., Forger, N. G., Coscia, E. M., Frank, L. G., Licht, P. & Glickman, S. E. 1998. Androgens and masculinization of genitalia in the spotted hyaena (*Crocuta crocuta*). 2. Effects of prenatal anti-androgens. *Journal of Reproduction and Fertility,* 113:117–127.

Drea, C. M., Coscia, E. M. & Glickman, S. E. 1999. Hyenas. In *Encyclopedia of Reproduction,* vol. 2, ed. E. Knobil, J. Neill, & P. Licht, pp. 718–725. San Diego: Academic Press.

Drea, C. M., Vignieri, S. N., Cunningham, S. B. & Glickman, S. E. In press a. Responses to olfactory stimuli in spotted hyenas (*Crocuta crocuta*): I. Investigation of environmental odors and the function of rolling. *Journal of Comparative Psychology.*

Drea, C. M., Vignieri, S. N., Kim, H.S., Weldele, M. L. & Glickman, S. In press b. Responses to olfactory stimuli in spotted hyenas (*Crocuta crocuta*): II. Discrimination of conspecific scent. *Journal of Comparative Psychology.*

East, M. L. & Hofer, H. 1991a. Loud calling in a female-dominated mammalian society: I. Structure and composition of whooping bouts of spotted hyaenas, *Crocuta crocuta. Animal Behaviour,* 42:637–649.

——— 1991b. Loud calling in a female-dominated mammalian society: II. Behavioural contexts and functions of whooping of spotted hyaenas, *Crocuta crocuta. Animal Behaviour,* 42:651–669.

——— 2001. Male spotted hyenas (*Crocuta crocuta*) queue for status in social groups dominated by females. *Behavioral Ecology,* 12:558–568.

East, M., Hofer, H. & Turk, A. 1989. Functions of birth dens in spotted hyaenas (*Crocuta crocuta*). *Journal of Zoology* (London), 219:690–697.

East, M. L., Hofer, H. & Wickler, W. 1993. The erect 'penis' is a flag of submission in a female-dominated society: greetings in Serengeti spotted hyenas. *Behavioral Ecology and Sociobiology,* 33:355–370.

Eisenberg, J. F. 1973. Mammalian social systems: Are primate social systems unique? In *Symposia of the Fourth International Congress of Primatology,* vol. 1, *Precultural Primate Behavior,* ed. W. Montagna & E. W. Menzel, Jr., pp. 232–249. Basel: Karger.

Eloff, F. C. 1964. On the predatory habits of lions and hyaenas. *Koedoe,* 7:105–112.

Engh, A. L., Esch, K., Smale, L. & Holekamp, K. E. 2000. Mechanisms of maternal rank 'inheritance' in the spotted hyaena, *Crocuta crocuta. Animal Behaviour,* 60:323–332.

Ewer, R. F. 1973. *The Carnivores.* Ithaca, NY: Cornell University Press.

Fady, J. C. 1972. Absence de coopération de type instrumental en milieu naturel chez *Papio papio. Behaviour,* 43:157–164.

Fagen, R. 1994. Primate juveniles and primate play. In *Juvenile Primates: Life History, Development and Behavior,* ed. M. E. Periera & L. A. Fairbanks, pp. 182–196. New York: Oxford University Press.

Frank, L. G. 1986a. Social organization of the spotted hyaena (*Crocuta crocuta*). I. Demography. *Animal Behaviour,* 34:1500–1509.

——— 1986b. Social organization of the spotted hyaena *Crocuta crocuta.* II. Dominance and reproduction. *Animal Behaviour,* 34:1510–1527.

——— 1996. Female masculinization in the spotted hyena: endocrinology, behavioral ecology and evolution. In *Carnivore Behavior, Ecology, and Evolution,* vol. 2, ed. J. L. Gittleman, pp. 78–131. Ithaca, NY: Comstock Publishing Associates.

——— 1997. Evolution of genital masculinization: why do female hyaenas have such a large 'penis.' *Trends in Ecology and Evolution,* 12:58–62.

Frank, L. G. & Glickman, S. E. 1994. Giving birth through a penile clitoris: parturition and dystocia in the spotted hyaena (*Crocuta crocuta*). *Journal of Zoology* (London), 234:659–665.

Frank, L. G., Glickman, S. E. & Zabel, C. J. 1989. Ontogeny of female dominance in the spotted hyaena: perspectives from nature and captivity. *Symposia of the Zoological Society of London,* 61:127–146.

Frank, L. G., Glickman, S. E. & Powch, I. 1990. Sexual dimorphism in the spotted hyaena (*Crocuta crocuta*). *Journal of Zoology* (London), 221:308–313.

Frank, L. G., Glickman, S. E. & Licht, P. 1991. Fatal sibling aggression, precocial development, and androgens in neonatal spotted hyenas. *Science,* 252:702–704.

Frank, L. G., Holekamp, K. E. & Smale, L. 1995a. Dominance, demography, and reproductive success of female spotted hyaenas. In *Serengeti II: Dynamics, Management, and Conservation of an Ecosystem,* ed. A. R. E. Sinclair & P. Arcese, pp. 364–384. Chicago: University of Chicago Press.

Frank, L. G., Weldele, M. & Glickman, S. E. 1995b. Masculinization costs in hyaenas. *Nature,* 377:584–585.

Gasaway, W. C., Mossestad, K. T. & Stander, P. E. 1991. Food acquisition by spotted hyaenas in Etosha National Park, Namibia: predation versus scavenging. *African Journal of Ecology,* 29:64–75.

Gibson, K. R. 1986. Cognition, brain size and the extraction of embedded food resources. In *Ontogeny, Cognition and Social Behaviour,* ed. J. G. Else & P. C. Lee, pp. 93–104. Cambridge: Cambridge University Press.

Gittleman, J. L. 1989. *Carnivore Behaviour, Ecology, and Evolution.* Ithaca: Comstock Publishing Associates.

Glickman, S. E. 1995. The spotted hyena from Aristotle to the Lion King: reputation is everything. *Social Research,* 62:501–537.

——— 2000. Culture, disciplinary tradition, and the study of behavior: sex, rats, and spotted hyenas. In *Primate Encounters: Models of Science, Gender, and Society,* ed. S. C. Strum & L. M. Fedigan, pp. 275–295. Chicago: University of Chicago Press.

Glickman, S. E., Frank, L. G., Davidson, J. M., Smith, E. R. & Siiteri, P. K. 1987. Androstenedione may organize or activate sex-reversed traits in female spotted hyenas. *Proceedings of the National Academy of Sciences USA,* 84:3444–3447.

Glickman, S. E., Frank, L. G., Licht, P., Yalcinkaya, T. M., Siiteri, P. K. & Davidson, J. M. 1992a. Sexual differentiation of the female spotted hyena: one of nature's experiments. *Annals of the New York Academy of Sciences,* 662:135–159.

Glickman, S. E., Frank, L. G., Pavgi, S. & Licht, P. 1992b. Hormonal correlates of 'masculinization' in female spotted hyaenas (*Crocuta crocuta*). 1. Infancy to sexual maturity. *Journal of Reproduction and Fertility,* 95:451–462.

Glickman, S. E., Frank, L. G., Holekamp, K. E., Smale, L. & Licht, P. 1993. Costs and benefits of "androgenization" in the female spotted hyena: the natural selection of physiological mechanisms. In *Perspectives in Ethology,* vol. 10, *Behavior and Evolution,* ed. P. P. G. Bateson et al., pp. 87–134. New York: Plenum Press.

Glickman, S. E., Zabel, C. J., Yoerg, S. I., Weldele, M., Drea, C. M. & Frank, L. G. 1997. Social facilitation, affiliation, and dominance in the social life of spotted hyenas. In *Annals of the New York Academy of Sciences,* vol. 807, *The Integrative Neurobiology of Affiliation,* ed., C. S. Carter, I. I. Lederhendler, & B. Kirkpatrick, pp. 175–184. New York: New York Academy of Sciences.

Glickman, S. E., Coscia, E. M., Frank, L. G., Licht, P., Weldele, M. L. & Drea, C. M. 1998. Androgens and masculinization of genitalia in the spotted hyaena (*Crocuta crocuta*). 3. Effects of juvenile gonadectomy. *Journal of Reproduction and Fertility,* 113:129–135.

Golla, W., Hofer, H. & East, M. L. 1999. Within-litter sibling aggression in spotted hyaenas: effect of maternal nursing, sex and age. *Animal Behaviour,* 58:715–726.

Gorman, M. L. & Mills, M. G. L. 1984. Scent marking strategies in hyaenas (Mammalia). *Journal of Zoology* (London), 202:535–547.

Gosling, L. M. 1982. A reassessment of the function of scent marking in territories. *Zeitschrift für Tierpsychologie,* 60:89–118.

Gosling, S. D. 1998. Personality dimensions in spotted hyenas (*Crocuta crocuta*). *Journal of Comparative Psychology,* 112:107–118.

Haber, G. C. 1996. Biological, conservation, and ethical implications of exploiting and controlling wolves. *Conservation Biology,* 10:1068–1081.

Harcourt, A. H. & de Waal, F. B. M. 1992. *Coalitions and Alliances in Humans and Other Animals.* New York: Oxford University Press.

Harvey, P. H., Martin, R. D. & Clutton-Brock, T. H. 1987. Life histories in comparative perspective. In *Primate Societies,* ed. B. B. Smuts, D. L. Cheney, R. M. Seyfarth, R. W. Wrangham, & T. T. Struhsaker, pp. 181–196. Chicago: University of Chicago Press.

Hemelrijk, C. K. & Ek, A. 1991. Reciprocity and interchange of grooming and 'support' in captive chimpanzees. *Animal Behaviour,* 41:923–935.

Hendrickx, A. G. 1971. *Embryology of the Baboon.* Chicago: University of Chicago Press.

Henschel, J. R. & Skinner, J. D. 1987. Social relationships and dispersal patterns in a clan of spotted hyaenas *Crocuta crocuta* in the Kruger National Park. *South African Journal of Zoology,* 22:18–24.

——— 1990a. Parturition and early maternal care of spotted hyaenas *Crocuta crocuta:* a case report. *Journal of Zoology* (London), 222:702–704.

——— 1990b. The diet of the spotted hyaenas *Crocuta crocuta* in Kruger National Park. *African Journal of Ecology,* 28:69–82.

——— 1991. Territorial behaviour by a clan of spotted hyaenas *Crocuta crocuta. Ethology,* 88:223–235.

Henschel, J. R. & Tilson, R. L. 1988. How much does a spotted hyaena eat? Perspective from the Namib Desert. *African Journal of Ecology,* 26:247–255.

Heyes, C. M. 1994. Social cognition in primates. In *Animal Learning and Cognition,* ed. N. J. Mackintosh, pp. 281–305. New York: Academic Press.

Hill, K. 1982. Hunting and human evolution. *Journal of Human Evolution,* 11:521–544.

Hinde, R. A. & White, L. E. 1974. Dynamics of a relationship: rhesus mother-infant ventro-ventral contact. *Journal of Comparative and Physiological Psychology,* 86:8–23.

Hofer, H. & East, M. L. 1993a. The commuting system of Serengeti spotted hyaenas: how a predator copes with migratory prey: I. Social organization. *Animal Behaviour,* 46:547–557.

——— 1993b. The commuting system of Serengeti spotted hyaenas: how a predator copes with migratory prey: II. Intrusion pressure and commuters' space use. *Animal Behaviour,* 46:559–574.

——— 1993c. The commuting system of Serengeti spotted hyaenas: how a predator copes with migratory prey: III. Attendance and maternal care. *Animal Behaviour,* 46:575–589.

——— 1995. Population dynamics, population size, and the commuting system of Serengeti spotted hyenas. In *Serengeti II: Dynamics, Mangagement, and Conservation of an Ecosystem,* ed. A. R. E. Sinclair & P. Arcese, pp. 332–363. Chicago: University of Chicago Press.

——— 2000. Conflict management in female-dominated spotted hyenas. In *Natural Conflict Resolution,* ed. F. Aureli & F. B. M. de Waal, pp. 232–234. Berkeley: University of California Press.

Holekamp, K. E. & Smale, L. 1990. Provisioning and food sharing by lactating spotted hyaenas, *Crocuta crocuta* (Mammalia: Hyaenidae). *Ethology,* 86:191–202.

——— 1991. Dominance acquisition during mammalian social development: the "inheritance" of maternal rank. *American Zoologist,* 31:306–307.

——— 1993. Ontogeny of dominance in free-living spotted hyaenas: juvenile rank relations with other immature individuals. *Animal Behaviour,* 46:451–466.

——— 1998a. Behavioral development in the spotted hyena. *Bioscience,* 48:997–1005.

——— 1998b. Dispersal status influences hormones and behavior in the male spotted hyena. *Hormones and Behavior,* 33:205–216.

Holekamp, K. E., Smale, L. & Szykman, M. 1996. Rank and reproduction in the female spotted hyaena. *Journal of Reproduction and Fertility,* 108:229–237.

Holekamp, K. E., Smale, L., Berg, R. & Cooper, S. M. 1997. Hunting rates and hunting success in the spotted hyena (*Crocuta crocuta*). *Journal of Zoology* (London), 242:1–15.

Holekamp, K. E., Boydston, E. E., Szykman, M., Graham, I., Nutt, K. J., Birch, S., Piskiel, A., & Singh, M. 1999. Vocal recognition in the spotted hyaena and its possible implications regarding the evolution of intelligence. *Animal Behaviour,* 58(2):383–395.

Holekamp, K. E., Boydston, E. E. & Smale, L. 2000. Group travel in social carnivores. In *On the Move: How and Why Animals Travel in Groups,* ed. S. Boinski & P. A. Garber, pp. 587–627. Chicago: University of Chicago Press.

Humphrey, N. K. 1976. The social function of intellect. In *Growing Points in Ethology,* ed. P. P. G. Bateson & R. A. Hinde, pp. 303–317. Cambridge: Cambridge University Press.

Jenks, S. M. & Werdelin, L. 1998. Taxonomy and systematics of living hyaenas (Family Hyaenidae). In *Hyaenas: Status Survey and Conservation Action Plan,* ed. G. Mills & H. Hofer, pp. 8–17. Gland, Switzerland: IUCN.

Jenks, S. M., Weldele, M. L., Frank, L. G. & Glickman, S. E. 1995. Acquisition of matrilineal rank in captive spotted hyenas: emergence of a natural social system in peer-reared animals and their offspring. *Animal Behaviour,* 50:893–904.

Jerison, H. J. 1973. *Evolution of the Brain and Intelligence.* New York: Academic Press.

Jolly, A. 1966a. *Lemur Behavior.* Chicago: University of Chicago Press.

——— 1966b. Lemur social behaviour and primate intelligence. *Science,* 153:501–506.

Jones, M. L. 1982. Longevity of captive mammals. *Der Zoologische Garten,* 52:113–128.

Kaplan, H., Hill, K., Lancaster, J. & Hurtado, A. M. 2000. A theory of human life history evolution: diet, intelligence, and longevity. *Evolutionary Anthropology,* 9:156–185.

Kappeler, P. 1990. Female dominance in *Lemur catta:* more than just female feeding priority? *Folia primatologica,* 55:92–95.

King, G. E. 1976. Society and territory in human evolution. *Journal of Human Evolution,* 5:323–332.

——— 1980. Alternative uses of primates and carnivores in the reconstruction of early hominid behavior. *Ethology and Sociobiology,* 1:99–109.

Knight, M. H., Van Jaarsveld, A. S. & Mills, M. G. L. 1992. Allo-suckling in spotted hyaenas (*Crocuta crocuta*): an example of behavioural flexibility in carnivores. *African Journal of Ecology,* 30:245–251.

Krusko, N., Weldele, M. & Glickman, S. E. 1988. Meeting ceremonies in spotted hyenas (*Cocuta rocuta*). Poster presented at the annual meeting of the Animal Behavior Society, Missoula, MT.

Kruuk, H. 1972. *The Spotted Hyena: A Study of Predation and Social Behavior.* Chicago: University of Chicago Press.

——— 1975. Functional aspects of social hunting by carnivores. In *Function and Evolution in Behaviour: Essays in Honour of Professor Niko Tinbergen, F.R.S.,* ed. G. Baerends, C. Beer, & A. Manning, pp. 119–141. Oxford: Clarendon Press.

——— 1976. Feeding and social behaviour of the striped hyaena (*Hyaena vulgaris* Desmarest). *East African Wildlife Journal,* 14:91–111.

Kruuk, H. & Sands, W. A. 1972. The aardwolf (*Proteles cristatus* Sparrman) 1783 as a predator of termites. *East African Wildlife Journal,* 10:211–227.

Kuhme, W. 1965. Communal food distribution and division of labor in hunting dogs. *Nature,* 205:443–444.

Kummer, H. 1971. *Primate Societies.* Chicago: Aldine.

Lee, P. C. 1983. Play as a means for developing relationships. In *Primate Social Relationships,* ed. R. A. Hinde, pp. 82–89. Oxford: Blackwell Scientific Publications.

——— 1987. Sibships: cooperation and competition among immature vervet monkeys. *Primates,* 28:47–59.

Licht, P., Frank, L. G., Pavgi, S., Yalcinkaya, T. M., Siiteri, P. K. & Glickman, S. E. 1992. Hormonal correlates of 'masculinization' in female spotted hyaenas (*Crocuta crocuta*). 2. Maternal and fetal steroids. *Journal of Reproduction and Fertility,* 95:463–474.

Licht, P., Hayes, T., Tsai, P. S., Cunha, G. R., Hayward, S., Martin, M., Jaffe, R., Golbus, M., Kim, H. S. & Glickman, S. E. 1998. Androgens and masculinization of genitalia in the spotted hyaena (*Crocuta crocuta*). 1. Urogenital morphology and placental androgen production during fetal life. *Journal of Reproduction and Fertility,* 113:105–116.

Loeven, J. C. 1994. The ontogeny of social play in timber wolves, *Canis lupus.* M.S. thesis, Dalhousie University, Halifax, Canada.

Maestripieri, D. & Wallen, K. 1997. Affiliative and submissive communication in rhesus macaques. *Primates,* 38:127–138.

Malcolm, J. R. & van Lawick, H. 1975. Notes on wild dogs (*Lycaon pictus*) hunting zebras. *Mammalia,* 39:231–240.

Mason, W. A. & Hollis, J. H. 1962. Communication between young rhesus monkeys. *Animal Behaviour,* 10:211–221.

Matthews, L. H. 1939. Reproduction in the spotted hyaena, *Crocuta crocuta,* (Erxleben). *Philosophical Transactions of the Royal Society of London,* ser. B, 230:1–78.

McNab, B. K. 1963. Bioenergetics and the determination of home range size. *American Naturalist,* 97:133–140.

Mech, L. D. 1970. *The Wolf: The Ecology and Behavior of an Endangered Species.* Garden City, NY: The Natural History Press.

Mendres, K. A. & de Waal, F. B. M. 2000. Capuchins do cooperate: the advantage of an intuitive task. *Animal Behaviour,* 60:523–529.

Menzel, E. W. 1974. A group of chimpanzees in a 1-acre field: leadership and communication. In *Behavior of Nonhuman Primates,* ed. A. M. Schrier & F. Stollnitz, pp. 83–153. New York: Academic Press.

Mills, M. G. L. 1978. The comparative socio-ecology of the Hyaenidae. *Carnivore,* 1:1–7.

——— 1985. Related spotted hyaenas forage together but do not cooperate in rearing young. *Nature,* 316:61–62.

——— 1990. *Kalahari Hyaenas: Comparative Behavioural Ecology of Two Species.* London: Unwin-Hyman.

Mills, M. G. L. & Gorman, M. L. 1987. The scent-marking behaviour of the spotted hyaena *Crocuta crocuta* in the southern Kalahari. *Journal of Zoology* (London), 212:483–497.

Mills, M. G. L., Gorman, M. L. & Mills, M. E. J. 1980. The scent marking behaviour of the brown hyaena *Hyaena brunnea. South African Journal of Zoology,* 15:240–248.

Milton, K. 1988. Foraging behaviour and the evolution of primate intelligence. In *Machiavellian Intelligence: Social Expertise and the Evolution of Intellect in Monkeys, Apes, and Humans,* ed. R. W. Byrne & A. Whiten, pp. 285–305. Oxford: Clarendon Press.

Neaves, W. B., Griffin, J. E. & Wilson, J. D. 1980. Sexual dimorphism of the phallus in spotted hyaena (*Crocuta crocuta*). *Journal of Reproduction and Fertility,* 59:509–513.

Oftedal, O. T. & Gittleman, J. L. 1989. Patterns of energy output during reproduction in carnivores. In *Carnivore Behavior, Ecology, and Evolution,* ed. J. L. Gittleman, pp. 355–378. Ithaca, NY: Comstock Publishing Associates.

Owens, D. D. & Owens, M. J. 1979. Communal denning and clan associations in brown hyenas (*Hyaena brunnea,* Thunberg) of the central Kalahari Desert. *African Journal of Ecology,* 17:35–44.

——— 1984. Helping behaviour in brown hyenas. *Nature,* 308:843–845.

——— 1996. Social dominance and reproductive patterns in brown hyaenas, *Hyaena brunnea,* of the central Kalahari desert. *Animal Behaviour,* 51:535–551.

Owens, M. J. & Owens, D. D. 1978. Feeding ecology and its influence on social organization in brown hyenas (*Hyaena brunnea,* Thunberg) of the central Kalahari Desert. *East African Wildlife Journal,* 16:113–135.

Packard, J. M., Seal, U. S., Mech, L. D. & Plotka, E. D. 1985. Causes of reproductive failure in two family groups of wolves (*Canis lupus*). *Zeitschrift für Tierpsychologie,* 68:24–50.

Packer, C. & Ruttan, L. 1988. The evolution of cooperative hunting. *American Naturalist,* 132:159–198.

Packer, C., Lewis, S. & Pusey, A. 1992. A comparative-analysis of non-offspring nursing. *Animal Behaviour,* 43:265–281.

Parker, S. T. 1990. Why big brains are so rare: energy costs of intelligence and brain size in anthropoid primates. In *"Language" and Intelligence in Monkeys and Apes,* ed. S. T. Parker & K. R. Gibson, pp. 129–154. Cambridge: Cambridge University Press.

Parker, S. T. & Gibson, K. R. 1977. Object manipulation, tool use and sensorimotor intelligence as feeding adaptations in cebus monkeys and great apes. *Journal of Human Evolution,* 6:623–641.

Passingham, R. E. 1978. Brain size and intelligence in primates. In *Recent Advances in Primatology*, vol. 3, *Evolution*, ed. D. J. Chivers & K. A. Joysey, pp. 85–86. New York: Academic Press.

Pedersen, J. M., Glickman, S. E., Frank, L. G. & Beach, F. A. 1990. Sex differences in the play behavior of immature spotted hyenas, *Crocuta crocuta*. *Hormones and Behavior*, 24:403–420.

Petit, O., Desportes, C. & Thierry, B. 1992. Differential probability of "coproduction" in two species of macaque (*Macaca tonkeana, M. mulatta*). *Ethology*, 90:107–120.

Pournelle, G. H. 1965. Observations on birth and early development of the spotted hyena. *Journal of Mammalogy*, 46:503.

Pusey, A. E. & Packer, C. 1987. Dispersal and philopatry. In *Primate Societies*, ed. B. B. Smuts, D. L. Cheney, R. M. Seyfarth, R. W. Wrangham, & T. T. Struhsaker, pp. 250–266. Chicago: University of Chicago Press.

Racey, P. A. & Skinner, J. D. 1979. Endocrine aspects of sexual mimicry in spotted hyaenas *Crocuta crocuta*. *Journal of Zoology* (London), 187:315–326.

Ralls, K. 1976. Mammals in which females are larger than males. *Quarterly Review of Biology*, 51:245–276.

Rasa, O. A. E. 1987. The dwarf mongoose: a study of behavior and social structure in relation to ecology in a small, social carnivore. *Advances in the Study of Behavior*, 17:121–163.

Richard, A. F. 1987. Malagasy prosimians: female dominance. In *Primate Societies*, ed. B. B. Smuts, D. L. Cheney, R. M. Seyfarth, R. W. Wrangham, & T. T. Struhsaker, pp. 25–33. Chicago: University of Chicago Press.

Richardson, P. R. K. 1987. Aardwolf mating system: overt cuckoldry in an apparently monogamous mammal. *South African Journal of Science*, 83:405–410.

——— 1990. Scent marking and territoriality in the aardwolf. In *Chemical Signals in Vertebrates 5*, ed. D. W. Macdonald, D. Müller-Schwarze, & S. E. Natynczuk, pp. 378–387. New York: Oxford University Press.

Rieger, I. 1979. Scent rubbing in carnivores. *Carnivore*, 2:17–25.

——— 1981. Hyaena hyaena. *Mammalian Species*, 150:1–5.

Rifkin, A. 2000. Attachment and separation in the spotted hyena (*Crocuta crocuta*). M.A. thesis, University of California, Berkeley.

Rood, J. P. 1980. Mating relationships and breeding suppression in the dwarf mongooses. *Animal Behaviour*, 28:143–150.

Schaller, G. B. & Lowther, G. R. 1969. The relevance of carnivore behavior to the study of early hominids. *Southwestern Journal of Anthropology*, 25:307–341.

Schusterman, R. J. & Kastak, D. 1998. Functional equivalence in a California sea lion: relevance to animal social and communicative interactions. *Animal Behaviour*, 55:1087–1095.

Seyfarth, R. M. & Cheney, D. L. 1984. Grooming, alliances, and reciprocal altruism in vervet monkeys. *Nature*, 308:541–543.

Silk, J. B. 1997. The function of peaceful post-conflict contacts among primates. *Primates,* 38:265–279.

Sillero-Zubiri, C. & MacDonald, D. W. 1998. Scent-marking and territorial behaviour of Ethiopian wolves. *Journal of Zoology* (London), 245:351–361.

Skinner, J. D., Henschel, J. R. & Van Jaarsveld, A. S. 1986. Bone-collecting habits of spotted hyaenas *Crocuta crocuta* in the Kruger National Park. *South African Journal of Zoology,* 21:303–308.

Sliwa, A. & Richardson, P. R. K. 1998. Responses of aardwolves, *Proteles cristatus,* Sparrman 1783, to translocated scent marks. *Animal Behaviour,* 56:137–146.

Smale, L., Frank, L. G., Holekamp, K. Nishida, T., Reynolds, V., Sugiyama, Y., Tutin, C. E. G., Wrangham, R. W. & Boesch, C. 1999. Cultures in chimpanzees. *Nature,* 399:682–685.

Smale, L., Holekamp, K. E., Weldele, M., Frank, L. G., & Glickman, S. E. 1995. Competition and cooperation between littermates in the spotted hyaena, *Crocuta crocuta. Animal Behaviour,* 50:671–682.

Stander, P. E. 1992. Cooperative hunting in lions: the role of the individual. *Behavioral Ecology & Sociobiology,* 29:445–454.

Suomi, S. J. 1991. Early stress and adult emotional reactivity in rhesus monkeys. *Ciba Foundation Symposium,* 156:171–183.

Suomi, S. 1995. Influence of attachment theory on ethological studies of biobehavioral development in nonhuman primates. In *Attachment Theory: Social, Developmental, and Clinical Perspectives* (S. Goldberg & R. Muir, eds.), pp. 185–201. Hillsdale, NJ: Analytic Press.

Suomi, S. 1999. Attachment in rhesus monkeys. In *Handbook of Attachment: Theory, Research, and Clinical Applications* (J. Cassidy & P. Shaver, eds.), pp. 181–197. New York: The Gulford Press.

Sutcliffe, A. J. 1970. Spotted hyaena: crusher, gnawer, digester and collector of bones. *Nature (Lond.),* 227:1110–1113.

Szykman, M., Engh, A. L., Van Horn, R. C., Funk, S. M., Scribner, K. T., & Holekamp, K. E. 2001. Association patterns among male and female spotted hyenas (*Crocuta crocuta*) reflect male mate choice. *Behavioral Ecology & Sociobiology,* 50:231–238.

Teleki, G. 1973. *The Predatory Behavior of Wild Chimpanzees.* Lewisburg, PA: Bucknell University Press.

Thompson, P. R. 1975. A cross-species analysis of carnivore, primate, and hominid behaviour. *Journal of Human Evolution,* 4:113–124.

Tilson, R. L. & Hamilton, W. J., III. 1984. Social dominance and feeding patterns of spotted hyaenas. *Animal Behaviour,* 32:715–724.

Tilson, R. L. & Henschel, J. R. 1986. Spatial arrangement of spotted hyaena groups in a desert environment, Namibia. *African Journal of Ecology,* 24:173–180.

Tilson, R. L., Von Blottnitz, F., & Henschel, J. R. 1980. Prey selection by spotted hyaenas (*Crocuta crocuta*) in the Namib Desert. *Madoqua,* 12:41–49.

Van Jaarsveld, A. S. 1993. A comparative investigation of hyaenid and aardwolf life-histories, with notes on spotted hyaena mortality patterns. *Transactions of the Royal Society of South Africa,* 48:219–232.

Van Jaarsveld, A. S., Skinner, J. D., & Lindeque, M. 1988. Growth, development and parental investment in the spotted hyaena, *Crocuta crocuta. Journal of Zoology, London,* 216:45–53.

Van Jaarsveld, A. S., McKenzie, A. A., & Skinner, J. D. 1992. Changes in concentration of serum prolactin during social and reproductive development of the spotted hyaena (*Crocuta crocuta*). *Journal of Reproduction & Fertility,* 95:765–773.

Walters, J. R. & Seyfarth, R. M. 1987. Conflict and cooperation. In *Primate Societies* (B. B. Smuts, D. L. Cheney, R. M. Seyfarth, R. W. Wrangham, & T. T. Struhsaker, eds.), pp. 306–317. Chicago: University of Chicago Press.

Watson, M. 1877. On the female generative organs of *Hyaena crocuta. Proceedings of the Zoological Society, London,* 24:369–379.

Whiten, A., Goodall, J., McGrew, W. C., Nishida, T., Reynolds, V., Sugiyama, Y., Tutin, C. E. G., Wrangham, R. W., & Boesch, C. 1999. Cultures in chimpanzees. *Nature,* 399:682–685.

Woodmansee, K. B., Zabel, C. J., Glickman, S. E., Frank, L. G. & Keppel, G. 1991. Scent marking (pasting) in a colony of immature spotted hyenas (*Crocuta crocuta*): a developmental study. *Journal of Comparative Psychology,* 105:10–14.

Wrangham, R. W. & Peterson, D. 1996. *Demonic Males: Apes and the Origins of Human Violence.* Boston: Houghton Mifflin.

Yalcinkaya, T. M., Siiteri, P. K., Vigne, J. L., Licht, P., Pavgi, S., Frank, L. G. & Glickman, S. E. 1993. A mechanism for virilization of female spotted hyenas *in utero. Science,* 260:1929–1931.

Yoerg, S. I. 1991. Social feeding reverses learned flavor aversions in spotted hyenas (*Crocuta crocuta*). *Journal of Comparative Psychology,* 105:185–189.

Zabel, C. J., Glickman, S. E., Frank, L. G., Woodmansee, K. B. & Keppel, G. 1992. Coalition formation in a colony of prepubertal spotted hyenas. In *Coalitions and Alliances in Humans and Other Animals,* ed. A. H. Harcourt & F. B. M. de Waal, pp. 113–135. New York: Oxford University Press.

Case Study 5A. *Maternal Rank "Inheritance" in the Spotted Hyena*

Byrne, R. W. 1994. The evolution of intelligence. In *Behaviour and Evolution,* ed. P. J. B. Slater & T. R. Halliday, pp. 223–265. Cambridge: Cambridge University Press.

Chapais, B. 1992. The role of alliances in social inheritance of rank among female primates. In *Coalitions and Alliances in Humans and Other Animals,* ed. A. H. Harcourt and F. B. M. de Waal, pp. 29–59. Oxford: Oxford Scientific.

Cheney, D. L. & Seyfarth, R. M. 1980. Vocal recognition in free-ranging vervet monkeys. *Animal Behaviour,* 28:362–367.

East, M. L., Hofer, H. & Wickler, W. 1993. The erect 'penis' is a flag of submission in a female-dominated society: greetings in Serengeti spotted hyenas. *Behavioral Ecology and Sociobiology,* 33:355–370.

Engh, A. L., Esch, K., Smale, L. & Holekamp, K. E. 2000. Mechanisms of maternal rank 'inheritance' in the spotted hyaena (*Crocuta crocuta*). *Animal Behaviour,* 60:323–332.

Frank, L. G. 1986. Social organisation of the spotted hyaena (*Crocuta crocuta*). II. Dominance and reproduction. *Animal Behaviour,* 35:1510–1527.

Holekamp, K. E. & Smale, L. 1991. Dominance acquisition during mammalian social development: the 'inheritance' of maternal rank. *American Zoologist,* 31:306–317.

——— 1993. Ontogeny of dominance in free-living spotted hyaenas: juvenile rank relations with other immature individuals. *Animal Behaviour,* 46:451–466.

Holekamp, K. E., Cooper, S. M., Katona, C. I., Berry, N. A., Frank, L. G. & Smale, L. 1997. Patterns of association among female spotted hyenas (*Crocuta crocuta*). *Journal of Mammalogy,* 78:55–64.

Holekamp, K. E., Boydston, E. E., Szykman, M., Graham, I., Nutt, K., Piskiel, A. & Singh, M. 1999. Vocal recognition in the spotted hyaena and its possible implications regarding the evolution of intelligence. *Animal Behaviour,* 58:383–395.

Holekamp, K. E., Boydston, E. E. & Smale, L. 2000. Group travel in social carnivores. In *On the Move: How and Why Animals Travel in Groups,* ed. S. Boinski & P. A. Garber, pp. 587–627. Chicago: University of Chicago Press.

Horrocks, J. & Hunte, W. 1983. Maternal rank and offspring rank in vervet monkeys: an appraisal of the mechanisms of rank acquisition. *Animal Behaviour,* 31:772–782.

Jenks, S., Weldele, M., Frank, L. & Glickman, S. E. 1995. Acquisition of matrilineal rank in captive spotted hyaenas: emergence of a natural social system in peer-reared animals and their offspring. *Animal Behaviour,* 50:893–904.

Melnick, D. J. & Pearl, M. C. 1987. Cercopithecines in multimale groups: genetic diversity and population structure. In *Primate Societies,* ed. B. Smuts, D. L. Cheney, R. M. Seyfarth, R. W. Wrangham, & T. T. Struhsaker, pp. 121–134. Chicago: University of Chicago Press.

Seyfarth, R. M. 1980. The distribution of grooming and related behaviours among adult female vervet monkeys. *Animal Behaviour,* 28:798–813.

Seyfarth, R. M. & Cheney, D. L. 1984. Grooming, alliances, and reciprocal altruism in vervet monkeys. *Nature,* 308:541–543.

Smale, L., Frank, L. G. & Holekamp, K. E. 1993. Ontogeny of dominance in free-living spotted hyaenas: juvenile relations with adult females and immigrant males. *Animal Behaviour,* 46:467–477.

Tilson, R. T. & Hamilton, W. J. 1984. Social dominance and feeding patterns of spotted hyaenas. *Animal Behaviour,* 32:715–724.

Tomasello, M. & Call, J. 1997. *Primate Cognition.* Oxford: Oxford University Press.

Wrangham, R. W. & Waterman, P. G. 1981. Feeding behavior of vervet monkeys on *A. tortilis* and *A. xanthophloea:* with special reference to reproductive strategies and tannin production. *Journal of Animal Ecology,* 50:715–731.

Zabel, C. J., Glickman, S. E., Frank, L. G., Woodmansee, K. B. & Keppel, G. 1992. Coalition formation in a colony of prepubertal spotted hyenas. In *Coalitions and Alliances in Humans and Other Animals,* ed. A. H. Harcourt & F. B. M. de Waal, pp. 113–135. New York: Oxford University Press.

6. Is Social Stress a Consequence of Subordination or a Cost of Dominance?

Arnold, W. & Dittami, J. 1997. Reproductive suppression in male alpine marmots. *Animal Behaviour,* 53:53–66.

Bercovitch, F. B. 1993. Dominance rank and reproductive maturation in male rhesus macaques (*Macaca mulatta*). *Journal of Reproduction and Fertility,* 99:113–120.

Blanchard, D. C., Spencer, R. L., Weiss, S. M., Blanchard, R. J., McEwen, B. & Sakai, R. 1995. Visible burrow system as a model of chronic social stress: behavioral and neuroendocrine correlates. *Psychoneuroendocrinology,* 20:117–134.

Bronson, F. H. & Eleftheriou, B. E. 1964. Chronic physiological effects of fighting in mice. *General and Comparative Endocrinology,* 4:9–14.

Burrows, R. 1995. Demographic changes and social consequences in wild dogs, 1964–1992. In *Serengeti II: Dynamics, Management and Conservation of an Ecosystem,* ed. A. R. E. Sinclair & P. Arcese, pp. 400–420. Chicago: University of Chicago Press.

Cavigelli, S. 1999. Behavioural patterns associated with faecal cortisol levels in free-ranging female ring-tailed lemurs, *Lemur catta. Animal Behaviour,* 57:935–944.

Chrousos, G. P. & Gold, P. W. 1992. The concepts of stress and stress system disorders: overview of physical and behavioral homeostasis. *Journal of the American Medical Association,* 267:1244–1252.

Creel, S. 2001. Social stress and dominance. *Trends in Ecology and Evolution,* 16:491–497.

Creel, S. & Creel, N. M. In press. *The African Wild Dog: Behavior, Ecology and Evolution.* Princeton: Princeton University Press.

Creel, S. & Waser, P. M. 1994. Inclusive fitness and reproductive strategies in dwarf mongooses. *Behavioral Ecology,* 5:339–348.

——— 1996. Variation in reproductive suppression among dwarf mongooses: interplay between mechanisms and evolution. In *Cooperative Breeding in Mammals,* ed. N. Solomon & J. French, pp. 150–170. Cambridge: Cambridge University Press.

Creel, S., Creel, N. M., Wildt, D. E. & Monfort, S. L. 1992. Behavioral and endocrine mechanisms of reproductive suppression in Serengeti dwarf mongooses. *Animal Behaviour,* 43:231–245.

Creel, S., Wildt, D. E. & Monfort, S. L. 1993. Aggression, reproduction and androgens in wild dwarf mongooses: a test of the challenge hypothesis. *American Naturalist,* 141:816–825.

Creel, S., Creel, N. M. & Monfort, S. L. 1996. Social stress and dominance. *Nature,* 379:212.

Creel, S., Creel, N. M., Mills, M. G. L. & Monfort, S. L. 1997. Rank and reproduction on cooperatively breeding African wild dogs: behavioral and endocrine correlates. *Behavioral Ecology,* 8:298–306.

Creel, S., Fox, J. E., Hardy, A., Sands, J. L., Garrott, R. & Peterson, R. O. In press. Snowmobile activity and glucocorticoid stress responses in wolves and elk. *Conservation Biology.*

Faulkes, C. G. & Abbott, D. H. 1997. The physiology of a reproductive dictatorship: regulation of male and female reproduction by a single breeding female in colonies of naked mole-rats. In *Cooperative Breeding in Mammals,* ed. N. G. Solomon & J. A. French, pp. 302–334. Cambridge: Cambridge University Press.

Frame, L. H., Malcolm, J. R., Frame, G. W. & van Lawick, H. 1979. Social organization of African wild dogs *Lycaon pictus* on the Serengeti plains, Tanzania, 1967–1978. *Zeitschrift für Tierpsychologie,* 50:225–249.

Friend, T. H., Polan, C. E. & McGilliard, M. C. 1977. Free stall and feed bunk requirements relative to behavior, production and individual intake in dairy cows. *Journal of Dairy Science,* 60:108–115.

Fuller, T. K., Kat, P. W., Bulger, J. B., Maddock, A. H., Ginsberg, J. R., Burrows, R., McNutt, J. W. & Mills, M. G. L. 1992. Population dynamics of African wild dogs. In *Wildlife 2001: Populations,* ed. D. R. McCullough & R. H. Barret, pp. 1125–1139. London: Elsevier Applied Science.

Ginther, A. J., Ziegler, T. E. & Snowdon, C. T. 2001. Reproductive biology of captive male cottontop tamarin monkeys as a function of social environment. *Animal Behaviour,* 61:65–78.

Girman, D. J., Mills, M. G. L., Geffen, E. & Wayne, R. K. 1997. A genetic analysis of social structure and dispersal in African wild dogs (*Lycaon pictus*). *Behavioral Ecology and Sociobiology,* 40:187–198.

Jameson, K. A., Appleby, M. C. & Freeman, L. C. 1999. Finding an appropriate order for a hierarchy based on probabilistic dominance. *Animal Behaviour,* 57:991–998.

Keane, B., Waser, P., Creel, S., Creel, N. M., Elliott, L. F. & Minchella, D. J. 1994. Subordinate reproduction in dwarf mongooses. *Animal Behaviour,* 47:65–75.

Louch, C. D. & Higginbotham, M. 1967. The relation between social rank and plasma corticosterone levels in mice. *General and Comparative Endocrinology,* 8:441–444.

Malcolm, J. R. & Marten, K. 1982. Natural selection and the communal rearing of pups in African wild dogs (*Lycaon pictus*). *Behavioral Ecology and Sociobiology,* 10:1–13.

Manogue, K. R. 1975. Dominance status and adrenocortical reactivity to stress in squirrel monkeys (*Saimiri sciureus*). *Primates,* 14:457–463.

Mays, N. A., Vleck, C. M. & Dawson, J. 1991. Plasma luteinizing hormone, steroid hormones, behavioral role, and nest stage in cooperatively breeding Harris' hawks (*Parabuteo unicinctus*). *Auk,* 108:619–637.

McLeod, P. J., Moger, W. H., Ryon, J., Gadbois, S. & Fentress, J. C. 1996. The relation between urinary cortisol levels and social behavior in captive timber wolves. *Canadian Journal of Zoology,* 74:209–216.

Munck, A., Guyre, P. M. & Holbrook, N. J. 1984. Physiological functions of glucocorticoids in stress and their relation to pharmacological actions. *Endocrine Reviews,* 5:25–44.

Packer, C., Collins, D. A., Sindimwo, A. & Goodall, J. 1995. Reproductive constraints on aggressive competition in female baboons. *Nature,* 373:60–63.

Pottinger, T. G. 1999. The impact of stress on animal reproductive activities. In *Stress Physiology in Animals,* ed. P. H. M. Baum, pp. 130–177. Boca Raton: CRC Press.

Rood, J. P. 1990. Group size, survival, reproduction and routes to breeding in the dwarf mongoose. *Animal Behaviour,* 39:566–572.

Saltzmann, W., Schultz-Darken, N. J., Wegner, F. H., Wittwer, D. J. & Abbott, D. H. 1998. Suppression of cortisol levels in subordinate female marmosets: reproductive and social contributions. *Hormones and Behavior,* 33:58–74.

Sapolsky, R. M. 1982. The endocrine stress response and social status in the wild baboon. *Hormones and Behavior,* 15:27–284.

——— 1983. Endocrine aspects of social instability in the olive baboon. *American Journal of Primatology,* 5:365–372.

——— 1990. Adrenocortical function, social rank, and personality among wild baboons. *Biological Psychiatry,* 28:862–878.

——— 1992. Neuroendocrinology of the stress response. In *Behavioral Endocrinology,* ed. J. B. Becker et al., pp. 287–324. Cambridge: MIT Press.

Sapolsky, R. M. & Ray, J. 1989. Styles of dominance and their physiological correlates among wild baboons. *American Journal of Primatology,* 18:1–9.

Schoech, S., Mumme, R. L. & Moore, M. C. 1991. Reproductive endocrinology and mechanisms of breeding inhibition in cooperatively breeding Florida scrub jays (*Aphelocoma c. coerulescens*). *Condor,* 93:354–364.

Smith, T. E. & French, J. A. 1997. Social and reproductive conditions modulate urinary cortisol excretion in black tufted-ear marmosets (*Callithrix kuhli*). *American Journal of Primatology,* 42:253–267.

van Lawick, H. 1973. *Solo: The Story of an African Wild Dog.* London: Collins.

Virgin, C. E. & Sapolsky, R. M. 1997. Styles of male social behavior and their endocrine correlates among low-ranking baboons. *American Journal of Primatology,* 42:25–39.

Wasser, S. K. 1995. Costs of conception in baboons. *Nature,* 376:219–220.

Wingfield, J. C. 1994. Modulation of the adrenocortical response to stress in birds. In *Perspectives in Comparative Endocrinology,* ed. K. G. Davey et al., pp. 520–528. Ottawa: National Research Council.

Wingfield, J. C. & Ramenofsky, M. 1999. Hormones and the behavioral ecology of stress. In *Stress Physiology in Animals,* ed. P. H. M. Baum, pp. 1–51. Boca Raton: CRC Press.

Wingfield, J. C., Hegner, R. E. & Lewis, D. M. 1991. Circulating levels of luteinizing hormone and steroid hormones in relation to social status in the cooperatively breeding white-browed sparrow weaver, *Plocepasser mahali. Journal of Zoology,* 225:43–58.

Ziegler, T. E., Scheffler, G. & Snowdon, C. T. 1995. The relationship of cortisol levels to social environment and reproductive functioning in female cotton-top tamarins, *Sanguinus oedipus. Hormones and Behavior,* 29:407–424.

Case Study 6A. *Sperm Whale Social Structure*

Best, P. B. 1979. Social organization in sperm whales, *Physeter macrocephalus.* In *Behavior of Marine Mammals,* ed. H. E. Winn & B. L. Olla, pp. 227–289. New York: Plenum Press.

Bond, J. 1999. Genetical analysis of the sperm whale using microsatellites. Ph.D. diss., Cambridge University.

Caldwell, M. C. & Caldwell, D. K. 1966. Epimeletic (care-giving) behavior in cetacea. In *Whales, Dolphins, and Porpoises,* ed. K. S. Norris, pp. 755–789. Berkeley: University of California Press.

Caldwell, D. K., Caldwell, M. C. & Rice, D. W. 1966. Behavior of the sperm whale. In *Whales, Dolphins, and Porpoises,* ed. K. S. Norris, pp. 677–717. Berkeley: University of California Press.

Christal, J. 1998. An analysis of sperm whale social structure: patterns of association and genetic relatedness. Ph.D. diss., Dalhousie University, Halifax, Canada.

Christal, J., Whitehead, H. & Lettevall, E. 1998. Sperm whale social units: variation and change. *Canadian Journal of Zoology,* 76:1431–1440.

Connor, R. C. 1995. Pseudo-reciprocity: investing in mutualism. *Animal Behaviour,* 34:1562–1566.

Connor, R. C. & Norris, K. S. 1982. Are dolphins reciprocal altruists? *American Naturalist,* 119:358–374.

Dillon, M. C. 1996. Genetic structure of sperm whale populations assessed by mitochondrial DNA sequence variation. Ph.D. diss., Dalhousie University, Halifax, Canada.

Dugatkin, L. A. 1997. The evolution of cooperation. *BioScience,* 47:355–362.

Hamilton, W. D. 1964. The genetical evolution of social behaviour, II. *Journal of Theoretical Biology,* 7:17–52.

——— 1971. Geometry for the selfish herd. *Journal of Theoretical Biology,* 31:295–311.

Kumar, S., Tamura, K. & Nei, M. 1993. *Molecular Evolutionary Genetics Analysis,* version 1.01. State College: Pennsylvania State University.

Lima, S. 1989. Iterated prisoner's dilemma: an approach to evolutionarily stable cooperation. *American Naturalist,* 134:828–834.

Mesnick, S. L. 2001. Genetic relatedness in sperm whales: evidence and cultural implications. *Behavior and Brain Science,* 24:346–347.

Nicolson, N. 1987. Infants, Mothers and Other Females. In *Primate Societies,* ed. B. B. Smuts, D. L. Cheney, R. M. Seyfarth, R. W. Wrangham, & T. T. Struhsaker, pp. 330–342. Chicago: University of Chicago Press.

Norris, K. S. & Dohl, T. P. 1980. The structure and function of cetacean schools. In *Cetacean Behavior: Mechanisms and Functions,* ed. L. M. Herman, pp. 211–262. New York: John Wiley and Sons.

Ohsumi, S. 1971. Some investigations on the school structure of sperm whale. *Scientific Reports Whales Research Institute,* 23:1–25.

Pitman, R. L., Ballance, L. T., Mesnick, S. I., and Chivers, S. J. 2001. Killer whale predation on sperm whales: observations and implications. *Marine Mammal Science,* 17:494–507.

Queller, D. C. & Goodnight, K. F. 1989. Estimating relatedness using genetic markers. *Evolution,* 43:258–275.

Richard, K. R., Dillon, M. C., Whitehead, H. & Wright, J. M. 1996. Pattterns of kinship in groups of free-living sperm whales (*Physeter macrocephalus*) revealed by multiple molecular genetic analyses. *Proceedings of the National Academy of Sciences,* 93:8792–8795.

Scammon, C. M. 1874 [1968]. *The Marine Mammals of the Northwestern Coast of North America.* New York: Dover Publications.

Trivers, R. L. 1971. The evolution of reciprocal altruism. *Quarterly Review of Biology,* 46:35–57.

Whitehead, H. 1996. Babysitting, dive synchrony, and indications of alloparental care in sperm whales. *Behavioral Ecology and Sociobiology,* 38:237–244.

Whitehead, H. & Kahn, B. 1992. Temporal and geographic variation in the social structure of female sperm whales. *Canadian Journal of Zoology,* 70:2145–2149.

Whitehead, H., Waters, S. & Lyrholm, T. 1991. Social organization of female sperm whales and their constant companions and casual acquaintances. *Behavioral Ecology and Sociobiology,* 29:385–389.

7. Equivalence Classification as an Approach to Social Knowledge

Bachman, C. & Kummer, H. 1980. Male assessment of female choice in hamadryas baboons. *Behavioural Ecology and Sociobiology,* 6:315–321.

Campagña, C., Le Boeuf, B. J. & Cappozzo, H. L. 1988. Group raids: a mating strategy of male southern sea lions. *Behaviour,* 105:224–249.

Carr, D., Wilkinson, K. M., Blackman, D. & McIlvane, W. J. 2000. Equivalence classes in individuals with minimal verbal repertoires. *Journal of the Experimental Analysis of Behavior,* 74:101–114.

Carter, D. E. & Werner, T. J. 1978. Complex learning and information processing by pigeons: a critical analysis. *Journal of the Experimental Analysis of Behavior,* 29:565–601.

Chen, S. F., Swartz, K. B., & Terrace, H. S. 1997. Knowledge of the ordinal position of list items in rhesus monkeys. *Psychological Science,* 8:80–86.

Cheney, D. L., & Seyfarth, R. M. 1980. Vocal recognition in free-ranging vervet monkeys. *Animal Behaviour,* 28:362–367.

——— 1986. The recognition of social alliances among vervet monkeys. *Animal Behaviour,* 34:1722–1731.

——— 1989. Reconciliation and redirected aggression in vervet monkeys, *Cercopithecus aethiops. Behaviour,* 110:258–275.

——— 1990. *How Monkeys See the World: Inside the Mind of Another Species.* Chicago: University of Chicago Press.

——— 1999. Recognition of other individuals' social relationships by female baboons. *Animal Behaviour,* 58:67–75.

Connor, R. C., Heithaus, R. M. & Barre, L. M. 1999. Superalliance of bottlenose dolphins. *Nature,* 371:571–572.

——— 2001. Complex social structure, alliance stability and mating access in a bottlenose dolphin 'super-alliance.' *Proceedings of the Royal Society of London,* ser. B, 268:263–267.

D'Amato, M., Salman, D. P., Loukas, E. & Tomie, A. 1985. Symmetry and transitivity of conditional relations in monkeys (*Cebus apella*) and pigeons (*Columba livia*). *Journal of the Experimental Analysis of Behavior,* 44:35–47.

de Waal, F. B. M. 1982. *Chimpanzee Politics: Power and Sex among Apes.* London: Jonathan Cape.

——— 1992. Coalitions as part of reciprocal relations in the Arnhem chimpanzee colony. In *Coalitions and Alliances in Humans and Other Animals,* ed. A. H. Harcourt & F. B. M. de Waal, pp. 233–257. Albany, NY: SUNY Press.

Dube, W. V., McIlvane, W. J., Callahan, T. D. & Stoddard, L. T. 1993. The search for stimulus equivalence in nonverbal organisms. *Psychological Record,* 43:761–778.

Eibl-Eibesfeldt, I. 1955. Ethologische studien am Galapagos-seelowen, *Zalophus wollebaeki* Sivertson. *Zeitshrift für Tierpsychologie,* 12:286–303. (English translation on file at Marine Mammal Biological Laboratory, Seattle.)

Galef, B. G., Jr. 1996. Social enhancement of food preferences in Norway rats: a brief review. In *Social Learning in Animals: The Roots of Culture,* ed. C. M. Heyes & B. G. Galef Jr., eds., pp. 49–64. New York: Academic Press.

Gentry, R. L. 1970. Social behavior of the Steller sea lion. Ph.D. thesis, University of California, Santa Cruz.

Gisiner, R. C. 1985. Male territoriality and reproductive behavior in the Steller sea lion, *Eumetopias jubatus.* Unpublished Ph.D. thesis, University of California, Santa Cruz.

Goodall, J. 1986. *The Chimpanzees of Gombe: Patterns of Behavior.* Cambridge, MA: Belknap Press of Harvard University Press.

Hanggi, E. B. & Schusterman, R. J. 1990. Kin recognition in captive California sea lions (*Zalophus californianus*). *Journal of Comparative Psychology,* 104:368–372.

Hauser, M. D. 1998. Functional referents and acoustic similarity: field playback experiments with rhesus monkeys. *Animal Behaviour,* 55:1647–1658.

Heath, C. B. 1989. The behavioral ecology of the California sea lion, *Zalophus californianus.* Unpublished Ph.D. thesis, University of California, Santa Cruz.

Herman, L. M., Richards, D. G. & Wolz, J. P. 1984. Comprehension of sentences by bottlenosed dolphins. *Cognition,* 16:129–219.

Herman, L. M., Hovancik, J. R., Gory, J. D. & Bradshaw, G. L. 1989. Generalization of visual matching by a bottlenose dolphin (*Tursiops truncatus*): evidence for invariance of cognitive performance with visual and auditory materials. *Journal of Experimental Psychology: Animal Behavior Processes,* 15:124–136.

Holekamp, K. E., Boydston, E. E., Szykman, M., Graham, I., Nutt, K. J., Birch, S., Piskiel, A., Singh, M. 1999. Vocal recognition in the spotted hyaena and its possible implications regarding the evolution of intelligence. *Animal Behaviour,* 58:383–395.

Horne, P. J. & Lowe, C. F. 1996. On the origins of naming and other symbolic behavior. *Journal of the Experimental Analysis of Behavior,* 65:185–241.

——— 1997. Toward a theory of verbal behavior. *Journal of the Experimental Analysis of Behavior,* 68:271–296.

Insley, S. J. 2001. Mother-offspring vocal recognition in northern fur seals is mutual but asymmetrical. *Animal Behaviour,* 61:129–137.

——— 2000. Long-term vocal recognition in the northern fur seal. *Nature,* 406:404–405.

Kastak, D. & Schusterman, R. J. 1994. Transfer of visual identity matching-to-sample in two California sea lions (*Zalophus californianus*). *Animal Learning and Behavior,* 22:427–435.

Kawai, M. 1958. On the system of social ranks in a natural group of Japanese monkeys: basic rank and dependent rank. *Primates,* 1:111–148. (In Japanese with English summary.)

Kummer, H. 1967. Tripartite relations in hamadryas baboons. In *Social Communication among Primates,* ed. S. A. Altmann, pp. 63–71. Chicago: University of Chicago Press.

——— 1995. *In Quest of the Sacred Baboon.* Princeton, NJ: Princeton University Press.

Oden, D. L., Thompson, R. K. R. & Premack, D. 1988. Spontaneous transfer of matching by infant chimpanzees (*Pan troglodytes*). *Animal Behavior Processes,* 14:140–145.

Pearce, J. M. 1994. Discrimination and categorization. In *Animal Learning and Cognition,* ed. N. J. Mackintosh, pp. 109–134. San Diego, CA: Academic Press.

Pepperberg, I. M. 1986. Acquisition of anomalous communicatory systems: implications for studies on interspecies communication. In *Dolphin Cognition and Behavior: A Comparative Approach,* ed. R. J. Schusterman, J. A. Thomas, & F. G. Wood, pp. 289–302. Hillsdale, NJ: Erlbaum.

Peterson, R. S. 1968. Social behavior in pinnipeds. In *The Behavior and Physiology of Pinnipeds,* ed. R. J. Harrison, R. C. Hubbard, R. S. Peterson, C. E. Rice, & R. J. Schusterman, pp. 3–53. New York: Appleton-Century-Crofts.

Peterson, R. S. & Bartholomew, G. A. 1967. The natural history and behavior of the California sea lion. *American Society of Mammalogists,* Special publication no. 1.

Porter, F. L. 1979. Social behavior in the leaf-nosed bat *Carollia perspicillata* II: Social communication. *Zeitshrift für Tierpsychologie,* 50:1–8.

Reichmuth Kastak, C., Schusterman, R. J. & Kastak, D. 2001. Equivalence classification by California sea lions using class-specific reinforcers. *Journal of the Experimental Analysis of Behavior,* 76:131–158.

Reidman, M. 1990. *The Pinnipeds: Seals, Sea Lions, and Walruses.* Berkeley, CA: University of California Press.

Riess, B. F. 1940. Semantic conditioning involving the galvanic skin reflex. *Journal of Experimental Psychology,* 36:143–152.

Sandegren, F. E. 1970. Breeding and maternal behavior of the Steller sea lion *(Eumetopias jubata).* Unpublished M.Sc. thesis, University of Alaska, Fairbanks.

——— 1976. Courtship display, agonistic behavior and social dynamics in the Steller sea lion (*Eumetopias jubatus*). *Behaviour,* 55:159–172.

Savage-Rumbaugh, E. S., Rumbaugh, D. M., Smith, S. T. & Lawson, J. 1980. Reference: the linguistic essential. *Science,* 210:922–924.

Schusterman, R. J. & Dawson, R. G. 1968. Barking, dominance, and territoriality in male sea lions. *Science,* 160:434–436.

Schusterman, R. J. & Gisiner, R. C. 1997. Pinnipeds, porpoises and parsimony: animal language research viewed from a bottom-up perspective. In *Anthropomorphism, Anecdotes and Animals: The Emperor's New Clothes?,* ed. R. W. Mitchell, N. S. Thompson, & H. L. Miles, pp. 370–382. Albany, NY: SUNY Press.

Schusterman, R. J. & Kastak, D. 1993. A California sea lion (*Zalophus californianus*) is capable of forming equivalence relations. *Psychological Record,* 43:823–839.

——— 1998. Functional equivalence in a California sea lion: relevance to social and communicative interactions. *Animal Behaviour,* 55:1087–1095.

——— In press. Problem solving and memory. In *Marine Mammal Biology: An Evolutionary Approach,* ed. A. R. Hoelzel. London: Blackwells.

Schusterman, R. J., Gisiner, R. & Hanggi, E. 1992a. Imprinting and other aspects of pinniped/human interactions. In *The Inevitable Bond,* ed. H. Davis & D. Balfour, pp. 334–356. New York: Cambridge University Press.

Schusterman, R. J., Hanggi, E. & Gisiner, R. 1992b. Acoustic signaling in mother-pup reunions, interspecies bonding, and affiliation by kinship in California sea lions (*Zalophus californianus*). In *Marine Mammal Sensory Systems,* ed. J. A. Thomas, R. A. Kastelein, & Y. A. Supin, pp. 533–551. New York: Plenum Press.

Schusterman, R. J., Reichmuth, C. J. & Kastak, D. 2000. How animals classify friends and foes. *Current Directions in Psychological Science,* 9:1–6.

Seyfarth, R. M. & Cheney, D. L. In press. The structure of social knowledge in monkeys. In *The Cognitive Animal,* ed. M. Bekoff, C. Allen, & G. Burghardt. Cambridge, MA: MIT Press.

Sidman, M. 1971. Reading and auditory-visual equivalences. *Journal of Speech and Hearing Research,* 14:5–13.

——— 1994. *Equivalence Relations and Behavior: A Research Story.* Boston: Author's Cooperative.

——— 2000. Equivalence relations and the reinforcement contingency. *Journal of the Experimental Analysis of Behavior,* 74:127–146.

Sidman, M. & Tailby, W. 1982. Conditional discrimination vs. matching to sample: an expansion of the testing paradigm. *Journal of the Experimental Analysis of Behavior,* 37:5–22.

Sidman, M., Wynne, C. K., Maguire, R. W. & Barnes, T. 1989. Functional classes and equivalence relations. *Journal of the Experimental Analysis of Behavior,* 52:261–274.

Tomasello, M. 2000. Two hypotheses about primate cognition. In *The Evolution of Cognition,* ed. C. Heyes & L. Huber, pp. 165–183. Cambridge, MA: MIT Press.

Tomasello, M. & Call, J. 1997. *Primate Cognition.* Oxford: Oxford University Press.

Trillmich, F. 1996. Parental investment in pinnipeds. In *Advances in the Study of Behavior,* pp. 533–577. Academic Press.

Vaughan, W., Jr. 1988. Formation of equivalence sets in pigeons. *Journal of Experimental Psychology: Animal Behavior Processes,* 14:36–42.

Wasserman, E. A., DeVolder, C. L. & Coppage, D. J. 1992. Non-similarity based conceptualization in pigeons via secondary or mediated generalization. *Psychological Science,* 3:374–379.

Zajonc, R. B. 1972. *Animal Social Behavior.* Morristown, NJ: General Learning Press.

Zentall, T. R. 1998. Symbolic representation in animals: emergent stimulus relations in conditional discrimination learning. *Animal Learning and Behavior,* 26:363–377.

Zuberbühler, K. 2000a. Causal cognition in a non-human primate: field playback experiments with Diana monkeys. *Cognition,* 76:195–207.

——— 2000b. Causal knowledge of predators' behavior in wild Diana monkeys. *Animal Behaviour,* 59:209–220.

8. The Structure of Social Knowledge in Monkeys

Altmann, J. 1980. *Baboon Mothers and Infants.* Cambridge, MA: Harvard University Press.

Altmann, J., Alberts, S. C., Haines, S. A., Dubach, J., Muruthi, P., Coote, T., Geffen, E., Cheesman, D. J., Mututua, R. S., Saiyalel, S. N., Wayne, R. K., Lacy, R. C. & Bruford, M. W. 1996. Behavior predicts genetic structure in a wild primate group. *Proceedings of the National Academy of Sciences USA,* 93:5797–5801.

Aureli, F., Cozzolino, R., Cordischi, C. & Scucchi, S. 1992. Kin-oriented redirection among Japanese macaques: an expression of a revenge system? *Animal Behaviour,* 44:283–291.

Bousfield, W. A. 1953. The occurrence of clustering in the recall of randomly arranged associates. *Journal of General Psychology,* 49:229–240.

Brannon, E. M. & Terrace, H. S. 1998. Ordering of the numerosities 1 to 9 by monkeys. *Science,* 282:746–749.

Bulger, J. & Hamilton, W. J. 1988. Inbreeding and reproductive success in a natural chacma baboon, *Papio cynocephalus ursinus,* population. *Animal Behaviour,* 36:574–578.

Carpenter, M., Tomasello, M. & Savage-Rumbaugh, E. S. 1995. Joint attention and imitative learning in children, chimpanzees, and enculturated chimpanzees. *Social Development,* 4:217–237.

Chapais, B. 1988. Experimental matrilineal inheritance of rank in female Japanese macaques. *Animal Behaviour,* 36:1025–1037.

Cheney, D. L. & Seyfarth, R. M. 1980. Vocal recognition in free-ranging vervet monkeys. *Animal Behaviour,* 28:362–367.

——— 1982. Recognition of individuals within and between groups of free-ranging vervet monkeys. *American Zoologist,* 22:519–529.

——— 1986. The recognition of social alliances among vervet monkeys. *Animal Behaviour,* 34:1722–1731.

——— 1989. Reconciliation and redirected aggression in vervet monkeys. *Behaviour,* 110:258–275.

——— 1990. *How Monkeys See the World: Inside the Mind of Another Species.* Chicago: University of Chicago Press.

——— 1999. Recognition of other individuals' social relationships by female baboons. *Animal Behaviour,* 58:67–75.

Cheney, D. L., Seyfarth, R. M. & Silk, J. B. 1995a. The responses of female baboons (*Papio cynocephalus ursinus*) to anomalous social interactions: evidence for causal reasoning? *Journal of Comparative Psychology,* 109:134–141.

——— 1995b. The role of grunts in reconciling opponents and facilitating interactions among adult female baboons. *Animal Behaviour,* 50:249–257.

Dallal, N. & Meck, W. 1990. Hierarchical structures: chunking by food type facilitates spatial memory. *Journal of Experimental Psychology: Animal Behavior Processes,* 16:69–84.

D'Amato, M. & Colombo, M. 1988. Representation of serial order in monkeys (*Cebus apella*). *Journal of Experimental Psychology: Animal Behavior Processes,* 14:131–139.

Dasser, V. 1988. A social concept in Java monkeys. *Animal Behaviour,* 36:225–230.

Dube, W. V., McIlvaine, W. J., Callahan, T. D. & Stoddard, L. T. 1993. The search for stimulus equivalence in nonverbal organisms. *Psychological Record,* 43:761–778.

Fields, L. 1993. Foreword: special issue on stimulus equivalence. *Psychological Record,* 43:543–546.

Fountain, S. B. & Annau, Z. 1984. Chunking, sorting, and rule learning from serial patterns of brain-stimulation reward by rats. *Animal Learning and Behavior,* 12:265–274.

Fountain, S. B., Henne, D. R. & Hulse, S. H. 1984. Phrasing cues and hierarchical organization in serial pattern learning by rats. *Journal of Experimental Psychology: Animal Behavior Processes,* 10:30–45.

Gillan, D. 1981. Reasoning in the chimpanzee. II. Transitive inference. *Journal of Experimental Psychology: Animal Behavior Processes,* 7:150–164.

Gouzoules, S., Gouzoules, H. & Marler, P. 1984. Rhesus monkey (*Macaca mulatta*) screams: representational signaling in the recruitment of agonistic aid. *Animal Behaviour,* 32:182–193.

Gygax, L., Harley, N. & Kummer, H. 1997. A matrilineal overthrow with destructive aggression in *Macaca fascicularis. Primates,* 38:149–158.

Hamilton, W. J., Buskirk, R. E. & Buskirk, W. H. 1976. Defense of space and resources by chacma (*Papio ursinus*) baboon troops in an African desert and swamp. *Ecology,* 57:1264–1272.

Hansen, E. W. 1976. Selective responding by recently separated juvenile rhesus monkeys to the calls of their mothers. *Developmental Psychobiology,* 9:83–88.

Harcourt, A. H. 1988. Alliances in contests and social intelligence. In *Machiavellian Intelligence: Social Expertise and the Evolution of Intellect in Monkeys, Apes, and Humans,* ed. R. W. Byrne & A. Whiten, pp. 132–152. Oxford: Oxford University Press.

Hauser, M. D., MacNeilage, P. & Ware, M. 1996. Numerical representations in primates. *Proceedings of the National Academy of Sciences USA,* 93:1514–1517.

Heyes, C. M. 1994. Social cognition in primates. In *Animal Learning and Cognition,* ed. N. J. Mackintosh, pp. 281–305. New York: Academic Press.

Judge, P. 1982. Redirection of aggression based on kinship in a captive group of pigtail macaques. *International Journal of Primatology,* 3:301.

Kanwisher, N., McDermott, J. & Chun, M. M. 1997. The fusiform face area: a module in human extrastriate cortex specialized for face perception. *Journal of Neuroscience,* 17:4302–4311.

Macuda, T. & Roberts, W. A. 1995. Further evidence for hierarchical chunking in rat spatial memory. *Journal of Experimental Psychology: Animal Behavior Processes,* 21:20–32.

Miller, G. A. 1956. The magical number seven, plus or minus two: some limits on our capacity for processing information. *Psychological Review,* 63:81–97.

Palombit, R. A., Seyfarth, R. M. & Cheney, D. L. 1997. The adaptive value of "friendships" to female baboons: experimental and observational evidence. *Animal Behaviour,* 54:599–614.

Palombit, R., Cheney, D., Seyfarth, R., Rendall, D., Silk, J., Johnson, S. & Fischer, J. 2000. Male infanticide and defense of infants in chacma baboons. In *Infanticide by Males and Its Implications,* ed. C. van Schaik & C. Janson, pp. 123–153. Cambridge: Cambridge University Press.

Parr, L. A. & de Waal, F. B. M. 1999. Visual kin recognition in chimpanzees. *Nature,* 399:647–648.

Premack, D. 1983. The codes of man and beast. *Behavioral Brain Science,* 6:125–167.

Rendall, D., Rodman, P. S. & Emond, R. E. 1996. Vocal recognition of individuals and kin in free-ranging rhesus monkeys. *Animal Behaviour,* 51:1007–1015.

Restle, F. 1972. Serial patterns: The role of phrasing. *Journal of Experimental Psychology,* 92:385–390.

Samuels, A., Silk, J. B. & Altmann, J. 1987. Continuity and change in dominance relations among female baboons. *Animal Behaviour,* 35:785–793.

Schusterman, R. J. & Kastak, D. A. 1993. A California sea lion (*Zalophus californianus*) is capable of forming equivalence relations. *Psychological Record,* 43:823–839.

——— 1998. Functional equivalence in a California sea lion: relevance to animal social and communicative interactions. *Animal Behaviour,* 55:1087–1095.

Seeley, T. D. 1995. *The Wisdom of the Hive: The Social Physiology of Honey Bee Colonies.* Cambridge, MA: Harvard University Press.

Seyfarth, R. M. 1976. Social relationships among adult female baboons. *Animal Behaviour,* 24:917–938.

——— 1977. A model of social grooming among adult female monkeys. *Journal of Theoretical Biology,* 65:671–698.

——— 1978. Social relations among adult male and female baboons. II. Behavior throughout the female reproductive cycle. *Behaviour,* 64:227–247.

——— 1980. The distribution of grooming and related behaviors among adult female vervet monkeys. *Animal Behaviour,* 28:798–813.

Seyfarth, R. M. & Cheney, D. L. 1984. Grooming, alliances, and reciprocal altruism in vervet monkeys. *Nature,* 308:541–543.

Shettleworth, S. 1998. *Cognition, Evolution, and Behaviour.* Oxford: Oxford University Press.

Sidman, M. 1994. *Equivalence Relations and Behavior: A Research Story.* Boston: Authors' Cooperative.

Silk, J. B. 1993. Does participation in coalitions influence dominance relationships among male bonnet macaques? *Behaviour,* 126:171–189.

——— 1999. Male bonnet macaques use information about third-party rank relationships to recruit allies. *Animal Behaviour,* 58:45–51.

Silk, J. B., Seyfarth, R. M. & Cheney, D. L. 1999. The structure of social relationships among female savanna baboons. *Behaviour,* 136:679–703.

Simon, H. 1974. How big is a chunk? *Science,* 183:482–488.

Smuts, B. 1985. *Sex and Friendship in Baboons.* Chicago: Aldine.

Swartz, K. B., Chen, S. & Terrace, H. S. 1991. Serial learning by rhesus monkeys: I. Acquisition and retention of multiple four-item lists. *Journal of Experimental Psychology: Animal Behavior Processes,* 17:396–410.

Terrace, H. S. 1987. Chunking by a pigeon in a serial learning task. *Nature,* 325:149–151.

Thompson, R. K. R. 1995. Natural and relational concepts in animals. In *Comparative Approaches to Cognitive Science,* ed. H. Roitblat & J. A. Meyer, pp. 175–224. Cambridge, MA: MIT Press.

Thompson, R. K. & Oden, D. L. 1995. A profound disparity revisited: perception and judgment of abstract identity relations by chimpanzees, human infants, and monkeys. *Behavioural Processes,* 35:149–161.

Tomasello, M., Savage-Rumbaugh, E. S. & Kruger, A. C. 1993. Imitative learning of actions on objects by children, chimpanzees, and enculturated chimpanzees. *Child Development,* 64:1688–1705.

Tulving, E. 1962. Subjective organization in the free recall of "unrelated" words. *Psychological Review,* 69:344–354.

Walters, J. R. & Seyfarth, R. M. 1987. Conflict and cooperation. In *Primate Societies,* ed. B. Smuts, D. L. Cheney, R. M. Seyfarth, R. W. Wrangham & T. Struhsaker, pp. 306–317. Chicago: University of Chicago Press.

Wasserman, E. A. & Astley, S. L. 1994. A behavioral analysis of concepts: application to pigeons and children. In *Psychology of Learning and Motivation,* vol. 31, ed. D. L. Medin, pp. 73–132. New York: Academic Press.

Wasserman, E. A., Hugart, J. A. & Kirkpatrick-Steger, K. 1995. Pigeons show same-different conceptualization after training with complex visual stimuli. *Journal of Experimental Psychology: Animal Behavior Processes,* 21:248–252.

Whiten, P. L. 1983. Diet and dominance among female vervet monkeys. *American Journal of Primatology,* 5:139–159.

Wright, A., Cook, R. & Rivera, J. 1988. Concept learning by pigeons: matching to sample with trial-unique video picture stimuli. *Animal Learning and Behavior,* 16:436–444.

9. Social Syntax

Aureli, F. & de Waal, F. B. M. 2000. *Natural Conflict Resolution.* Berkeley: University of California Press.

Aureli, F. & van Schaik, C. P. 1991. Post-conflict behaviour in long-tailed macaques (*Macaca fascicularis*): I. The social events. II. Coping with the uncertainty. *Ethology,* 89:89–114.

Bickerton, D. 2000. Foraging versus social intelligence in the evolution of protolanguage. In *The Evolution of Language,* ed. J. L. Dessalles & L. Ghadakpour, p. 20. Paris: Ecole Nationale Supérieure des Télécommunications.

Byrne, R. W. 1996. The misunderstood ape: Cognitive skills of the gorilla. In *Reaching into Thought: The Minds of the Great Apes,* ed. A. E. Russon, K. A. Bard, & S. T. Parker, pp. 111–130. Cambridge: Cambridge University Press.

Byrne, R. W. & Whiten, A. 1997. Machiavellian intelligence. In *Machiavellian Intelligence II: Extensions and Evaluations,* ed. A. Whiten & R. W. Byrne, pp. 1–23. Cambridge: Cambridge University Press.

Calvin, W. H. & Bickerton, D. 2000. *Lingua ex Machina: Reconciling Darwin and Chomsky with the Human Brain.* Cambridge, MA: MIT Press.

Chance, M., R. A. Emory, G. & Payne, R. 1977. Status referents in long-tailed macaques (*Macaca fascicularis*): precursors and effects of a female rebellion. *Primates,* 18:611–632.

Cheney, D. L. & Seyfarth, R. M. 1990. *How Monkeys See the World: Inside the Mind of Another Species.* Chicago: University of Chicago Press.

Connor, R. C., Smolker, R. A. & Richards, A. F. 1992. Dolphin alliances and coalitions. In *Coalitions and Alliances in Humans and Other Animals,* ed. A. H. Harcourt & F. B. M. de Waal, pp. 415–444. Oxford: Oxford University Press.

Cords, M. 1992. Post-conflict reunions and reconciliation in long-tailed macaques. *Animal Behaviour,* 44:57–61.

Cords, M. & Aureli, F. 1996. Reasons for reconciling. *Evolutionary Anthropology,* 5:42–45.

——— 2000. Reconciliation and relationship qualities. In *Natural Conflict Resolution,* ed. F. Aureli & F. B. M. de Waal, pp. 177–198. Berkeley: University of California Press.

Cords, M. & Thurnheer, S. 1993. Reconciliation with valuable partners by long-tailed macaques. *Ethology,* 93:315–325.

Crawford, M. P. 1937. The cooperative solving of problems by young chimpanzees. *Comparative Psychology Monographs,* 14:1–88.

de Waal, F. B. M. 1978. Exploitative and familiarity-dependent support strategies in a colony of semi-free-living chimpanzees. *Behaviour,* 66:268–312.

——— 1982. *Chimpanzee Politics: Power and Sex among Apes.* London: Jonathan Cape.

——— 1986. Integration of dominance and social bonding in primates. *Quarterly Review of Biology,* 61:459–479.

——— 1989. Food-sharing and reciprocal obligations in chimpanzees. *Journal of Human Evolution,* 18:433–459.

——— 1997a. Food-transfers through mesh in brown capuchins. *Journal of Comparative Psychology,* 111:370–378.

——— 1997b. The chimpanzee's service economy: food for grooming. *Evolution and Human Behavior,* 18:375–386.

——— 2000a. The first kiss: foundations of conflict resolution research in animals. In *Natural Conflict Resolution,* ed. F. Aureli & F. B. M. de Waal, pp. 15–33. Berkeley: University of California Press.

——— 2000b. Attitudinal reciprocity in food sharing among brown capuchins. *Animal Behaviour,* 60:253–261.

——— 2001. *The Ape and the Sushi Master: Cultural Reflections by a Primatologist.* New York: Basic Books.

de Waal, F. B. M. & Berger, M. L. 2000. Payment for labour in monkeys. *Nature,* 404:563.

de Waal, F. B. M. & Luttrell, L. M. 1988. Mechanisms of social reciprocity in three primate species: symmetrical relationship characteristics or cognition? *Ethology and Sociobiology,* 9:101–118.

de Waal, F. B. M. & van Roosmalen, A. 1979. Reconciliation and consolation among chimpanzees. *Behavioral Ecology and Sociobiology,* 5:55–66.

de Waal, F. B. M. & Yoshihara, D. 1983. Reconciliation and redirected affection in rhesus monkeys. *Behaviour,* 85:224–241.

Dugatkin, L. A. 1997. *Cooperation among Animals: An Evolutionary Perspective.* New York: Oxford University Press.

Gallup, G. 1970. Chimpanzees: self-recognition. *Science,* 167:86–87.

Goodall, J. 1971. *In the Shadow of Man.* Boston: Houghton Mifflin.

——— 1990. *Through a Window.* Boston: Houghton Mifflin.

Greenfield, P. 1991. Language, tools, and brains: the ontogeny and phylogeny of hierarchically organized sequential behavior. *Behavioral and Brain Sciences,* 14:531–551.

Hall, K. R. L. & DeVore, I. 1965. Baboon social behavior. In *Primate Behavior,* ed. I. DeVore, pp. 53–110. New York: Holt, Rinehart and Winston.

Harcourt, A. H. 1992. Coalitions and alliances: are primates more complex than nonprimates? In *Coalitions and Alliances in Humans and Other Animals,* ed. A. H. Harcourt & F. B. M. de Waal, pp. 445–472. Oxford: Oxford University Press.

Harcourt, A. H. & de Waal, F. B. M., eds. 1992. *Coalitions and Alliances in Humans and Other Animals.* Oxford: Oxford University Press.

Hinde, R. A. 1976. Interactions, relationships and social structure. *Man,* 11:1–17.

Hofer, H. & East, M. L. 2000. Conflict management in female-dominated spotted hyenas. In *Natural Conflict Resolution,* ed. F. Aureli & F. B. M. de Waal, pp. 232–234. Berkeley: University of California Press.

Humphrey, N. K. 1976. The social function of intellect. In *Growing Points in Ethology,* ed. P. Bateson & R. Hinde, pp. 303–321. Cambridge: Cambridge University Press.

Judge, P. G. 1991. Dyadic and triadic reconciliation in pigtail macaques (*Macaca nemestrina*). *American Journal of Primatology,* 23:225–237.

Judge, P. G. & de Waal, F. B. M. 1994. Intergroup grooming relations between alpha females in a population of free-ranging rhesus macaques. *Folia primatologica,* 63:63–70.

Kappeler, P. M. & van Schaik, C. P. 1992. Methodological and evolutionary aspects of reconciliation among primates. *Ethology,* 92:51–69.

Kawai, M. 1958. On the system of social ranks in a natural troop of Japanese monkey. I: Basic rank and dependent rank. II: Ranking order as observed among the monkeys on and near the test box. *Primates,* 1:111–148. (In Japanese.)

Kawamura, S. 1958. Matriarchal social ranks in the Minoo-B troop: a study of the rank system of Japanese monkeys. *Primates,* 1:148–156. (In Japanese.)

Köhler, W. 1925. *The Mentality of Apes.* New York: Vintage Books.

Kummer, H. 1971. *Primate Societies: Group Techniques of Ecological Adaptation.* Chicago: Aldine.

——— 1978. On the value of social relationships to nonhuman primates: a heuristic scheme. *Social Science Information,* 17:687–705.

Lasswell, H. 1936. *Who Gets What, When and How.* New York: McGraw-Hill.

Matsuzawa, T. 1996. Chimpanzee intelligence in nature and captivity: isomorphism of symbol use and tool use. In *Great Ape Societies,* ed. W. C. McGrew, L. F. Marchant, & T. Nishida, pp. 196–209. Cambridge: Cambridge University Press.

Menzel, E. W. 1974. A group of young chimpanzees in a one-acre field. In *Behavior of Non-human Primates,* vol. 5, ed. A. M. Schrier & F. Stollnitz, pp. 83–153. New York: Academic Press.

Perry, S. & Rose, L. 1994. Begging and transfer of coati meat by white-faced capuchin monkeys, *Cebus capucinus. Primates,* 35:409–415.

Premack, D. & Woodruff, G. 1978. Does the chimpanzee have a theory of mind? *Behavioral and Brain Sciences,* 1:515–526.

Riss, D. & Goodall, J. 1977. The recent rise to the alpha-rank in a population of free-living chimpanzees. *Folia primatologica,* 27:134–151.

Rose, L. 1997. Vertebrate predation and food-sharing in *Cebus* and *Pan. International Journal of Primatology,* 18:727–765.

Rothstein, S. I. & Pierotti, R. 1988. Distinctions among reciprocal altruism, kin selection, and cooperation and a model for the initial evolution of beneficent behavior. *Ethology and Sociobiology,* 9:189–209.

Rumbaugh, D. M., Washburn, D. A. & Hillix, W. A. 1996. Respondents, operants, and emergents: toward an integrated perspective on behavior. In *Learning as a Self-Organizing Process,* ed. K. Pribram & J. King, pp. 57–73. Hillsdale, NJ: Lawrence Erlbaum.

Sade, D. S. 1967. Determinants of dominance in a group of free-ranging rhesus monkeys. In *Social Communication among Primates,* ed. S. Altmann, pp. 99–114. Chicago: University of Chicago Press.

Samuels, A. & Flaherty, C. 2000. Peaceful conflict resolution in the sea? In *Natural Conflict Resolution,* ed. F. Aureli & F. B. M. de Waal, pp. 229–231. Berkeley: University of California Press.

Schino, G. 1998. Reconciliation in domestic goats. *Behaviour,* 135:343–356.

Silk, J. B. 1996. Why do primates reconcile? *Evolutionary Anthropology,* 5:39–42.

Trivers, R. L. 1971. The evolution of reciprocal altruism. *Quarterly Review of Biology,* 46:35–57.

van Schaik, C. P. & van Hooff, J. A. R. A. M. 1983. On the ultimate causes of primate social systems. *Behaviour,* 85:91–117.

Veenema, H. C., Das, M. & Aureli, F. 1994. Methodological improvements for the study of reconciliation. *Behavioural Processes,* 31:29–38.

Walker Leonard, J. 1979. A strategy approach to the study of primate dominance behaviour. *Behavioral Processes,* 4:155–172.

Westergaard, G. C. 1999. Structural analysis of tool-use by tufted capuchins (*Cebus apella*) and chimpanzees (*Pan troglodytes*). *Animal Cognition,* 2:141–145.

Yerkes, R. M. 1943. *Chimpanzees: A Laboratory Colony.* New Haven: Yale University Press.

Case Study 9A. *Conflict Resolution in the Spotted Hyena*

Andelman, S. J. 1985. *Ecology and Reproductive Strategies of Vervet Monkeys* (Cercopithecus aethiops) *in Amboseli National Park, Kenya.* University of Washington.

Aureli, F. & de Waal, F. B. M., eds. 2000. *Natural Conflict Resolution.* Berkeley: University of California Press.

Chaffin, C. L., Friedlen, K. & de Waal, F. B. M. 1995. Dominance style of Japanese macaques compared with rhesus and stumptail macaques. *American Journal of Primatology,* 35:103–116.

Cheney, D. L. & Seyfarth, R. M. 1989. Redirected aggression and reconciliation among vervet monkeys, *Cercopithecus aethiops. Behaviour,* 110:258–275.

Cords, M. 1988. Resolution of aggressive conflicts by immature long-tailed macaques. *Animal Behaviour,* 36:1124–1135.

Cords, M. & Aureli, F. 2000. Reconciliation and relationship qualities. In *Natural Conflict Resolution,* ed. F. Aureli & F. B. M. de Waal, pp. 177–198. Berkeley: University of California Press.

de Waal, F. B. M. 1986. Conflict resolution in monkeys and apes. In *Primates: The Road to Self-sustaining Populations,* ed. K. Benirsche, pp. 341–350. Berlin: Springer Verlag.

de Waal, F. B. M. & van Roosmalen, A. 1979. Reconciliation and consolation among chimpanzees. *Behavioral Ecology and Sociobiology,* 5:55–66.

de Waal, F. B. M. & Yoshihara, D. 1983. Reconciliation and redirected affection in rhesus monkeys. *Behaviour,* 85:224–241.

Engh, A., Esch, K., Smale, L. & Holekamp, K. E. 2000. Mechanisms of maternal rank 'inheritance' in the spotted hyaena. *Animal Behaviour,* 60:323–332.

Flynn, J. J. 1996. Carnivorian phylogeny and rates of evolution: morphological, taxic, and molecular. In *Carnivore Behavior, Ecology, and Evolution,* ed. J. L. Gittleman, pp. 542–581. Ithaca, NY: Cornell University Press.

Frank, L. G. 1986. Social organization of the spotted hyaena (*Crocuta crocuta*). I. Demography. *Animal Behaviour,* 34:1500–1509.

Harcourt, A. H. 1992. Coalitions and alliances: are primates more complex than non-primates? In *Coalitions and Alliances in Humans and Other Animals,* ed. A. H. Harcourt & F. B. M. de Waal, pp. 445–471. Oxford: Oxford Science Publications.

Henschel, J. R. & Skinner, J. D. 1991. Territorial behaviour by a clan of spotted hyaenas *Crocuta crocuta. Ethology,* 88:223–235.

Hofer, H. & East, M. 2000. Conflict management in female-dominated spotted hyenas. In *Natural Conflict Resolution,* ed. F. Aureli & F. B. M. de Waal, pp. 232–234. Berkeley: University of California Press.

Holekamp, K. E., Smale, L., Berg, R. & Cooper, S. M. 1997. Hunting rates and hunting success in the spotted hyena (*Crocuta crocuta*). *Journal of Zoology,* 242:1–15.

Holekamp, K. E., Boydston, E. E., Szykman, M., Graham, I., Nutt, K. J., Birch, S., Piskiel, A. & Singh, M. 1999. Vocal recognition in the spotted hyena and its possible implications regarding the evolution of intelligence. *Animal Behaviour,* 58:383–395.

Kruuk, H. 1972. *The Spotted Hyaena: A Study of Predation and Social Behavior.* Chicago: University of Chicago Press.

Kummer, H. 1978. On the value of social relationships to non-human primates: a heuristic scheme. *Social Science Information,* 17:687–705.

Petit, O., Abegg, C. & Thierry, B. 1997. A comparative study of aggression and conciliation in three cercopithecine monkeys (*Macaca fuscata, Macaca nigra, Papio papio*). *Behaviour,* 134:415–432.

Samuels, A. & Flaherty, C. 2000. Peaceful conflict resolution in the sea? In *Natural Conflict Resolution,* ed. F. Aureli & F. B. M. de Waal, pp. 229–231. Berkeley: University of California Press.

Schino, G. 2000. Beyond the primates: expanding the reconciliation horizon. In *Natural Conflict Resolution,* ed. F. Aureli & F. B. M. de Waal, pp. 225–242. Berkeley: University of California Press.

Smale, L., Holekamp, K. E., Weldele, M., Frank, L. G. & Glickman, S. E. 1995. Competition and cooperation between littermates in the spotted hyaena, *Crocuta crocuta. Animal Behaviour,* 50:671–682.

van den Bos, R. 1997. Conflict regulation in groups of domestic cats (*Felis silvestris catus*) living in confinement. *Advances in Ethology,* 32:149.

Veenema, H. C., Das, M. & Aureli, F. 1994. Methodological improvements for the study of reconciliation. *Behavioural Processes,* 31:29–38.

Wahaj, S. A., Guse, K. R. & Holekamp, K. E. 2001. Reconciliation in the spotted hyena (*Crocuta crocuta*). *Ethology,* 107:1057–1074.

Zabel, C. J., Glickman, S. E., Frank, L. G., Woodmansee, K. B. & Keppel, G. 1992. Coalition formation in a colony of prepubertal spotted hyenas. In *Coalitions and Alliances in Humans and Other Animals,* ed. A. H. Harcourt & F. B. M. de Waal, pp. 112–135. Oxford: Oxford University Press.

10. Laughter and Smiling

Andrew, R. J. 1963. The origin and evolution of the calls and facial expressions of the primates. *Behaviour,* 20:1–109.

Bachorowski, J. A. & Owren, M. J. 2001. Not all laughs are alike: voiced but not unvoiced laughter elicits positive affect in listeners. *Psychological Science,* 12:252–257.

Bachorowski, J. A., Smoski, M. J. & Owren, M. J. In press. The acoustic features of human laughter. *Journal of the Acoustical Society of America.*

Bachorowski, J. A., Smoski, M. J. & Owren, M. J. Submitted. Laugh rate and acoustics are associated with context: I. Empirical outcomes.

Bekoff, M. & Allen, C. 1998. Intentional communication and social play: how and why animals negotiate and agree to play. In *Animal Play: Evolutionary, Comparative and Ecological Perspectives,* ed. M. Bekoff & J. A. Byers, pp. 97–114. Cambridge: Cambridge University Press.

Bekoff, M. & Byers, J. A. 1985. The development of behavior from evolutionary and ecological perspectives in mammals and birds. *Evolutionary Biology,* 19:215–286.

Berlyne, D. E. 1969. Laughter, humour and play. In *The Handbook of Social Psychology,* ed. G. Lindzey & E. Aronson, vol. 3, pp. 795–852. Reading: Addison-Wesley.

Bradbury, J. W. & Vehrencamp, S. L. 1998. *Principles of Animal Communication.* Sunderland, MA: Sinauer Associates.

Castles, D. L., Aureli, F. & de Waal, F. B. M. 1996. Variation in conciliatory tendency and relationship quality across groups of pigtail macaques. *Animal Behaviour,* 52:389–403.

Chevalier-Skolnikoff, S. 1982. A cognitive analysis of facial behavior in Old World monkeys, apes, and human beings. In *Primate Communication,* ed. C. T. Snowdon, H. Brown, & M. R. Peterson, pp. 303–368. Cambridge: Cambridge University Press.

Darwin, C. 1872. *The Expression of the Emotions in Man and Animals.* London: Murray.

Duecker, S. 1996. Soziale Funktionen der affiliativen Gesichtsausdruecke bei *Theropithecus gelada.* Diss., Ruhr-Universitaet Bochum.

Eibl-Eibesfeldt, I. 1973. The expressive behavior of the deaf-and-blind born. In *Social Communication and Movement,* ed. M. von Cranach & I. Vine, pp. 163–194. London: Academic Press.

——— 1997. *Die Biologie des menschlichen Verhaltens: Grundriß der Humanethologie.* Weyarn: Seehammer.

Ekman, P. 1973. Cross-cultural studies of facial expression. In *Darwin and Facial Expression,* ed. P. Ekman, pp. 169–222. New York: Academic Press.

——— 1989. The argument and evidence about universals in facial expressions of emotion. In *Handbook of Social Psychophysiology,* ed. H. Wagner & A. Manstead, pp. 143–164. Chichester: John Wiley.

——— 1994. Strong evidence for universals in facial expressions: a reply to Russell's mistaken critique. *Psychological Bulletin,* 115:268–287.

Ekman, P. & Rosenberg, E. L. 1997. *What the Face Reveals.* New York: Oxford University Press.

Fa, J. E. 1989. The genus *Macaca:* a review of taxonomy and evolution. *Mammal Review,* 19(2):45–81.

Fagen, R. 1981. *Animal Play Behavior.* New York: Oxford University Press.

Fernández-Dols, J. M. & Ruiz-Belda, M. A. 1995. Are smiles a sign of happiness? Gold medal winners at the Olympic Games. *Journal of Personality and Social Psychology,* 69:1113–1119.

Ficken, M. S. 1977. Avian play. *Auk,* 94:573–582.

Fox, M. W. 1970. A comparative study of the development of facial expressions in canids; wolf, coyote and foxes. *Behaviour,* 36:4–73.

Frijda, N. 1986. *The Emotions.* Cambridge: Cambridge University Press.

Goldenthal, P., Johnston, R. E. & Kraut, R. E. 1981. Smiling, appeasement and the silent, bared teeth display. *Ethology and Sociobiology,* 2:127–133.

Gouzoules, H., Gouzoules, S. & Tomaszycki, M. 1998. Agonistic screams and the classification of dominance relationships: are monkeys fuzzy logicians? *Animal Behaviour,* 55:51–60.

Green, S. 1975. Variation of vocal pattern with social situation in the Japanese monkey *(Macaca fuscata):* a field study. In *Primate Behavior, Development and Laboratory Research,* vol. 4, ed. L. A. Rosenblum, pp. 1–102. New York, Academic Press.

Haanstra, B., Adang, O. J. M. & van Hooff, J. A. R. A. M. 1982. *The Family of Chimps.* Laren: Haanstra Film Productions.

Hand, J. L. 1986. Resolution of social conflicts: dominance, egalitarism, spheres of dominance, and game theory. *Quarterly Review of Biology,* 61:201–220.

Hayworth, D. 1928. The social origin and function of laughter. *Psychological Review,* 35:367–384.

Hennig, W. 1979. *Phylogenetic Systematics.* Urbana: University of Illinois Press.

Hinde, R. A. 1985. Expression and negotiation. In *The Development of Expressive Behavior,* ed. G. Zivin, pp. 103–116. New York: Academic Press.

Jeanotte, L. A. & de Waal, F. B. M. 1996. *Play Signaling in Juvenile Chimpanzees: Play Intensity and Social Context.* Poster presented at the ISP Congress, Madison, WI.

Kaufmann, C. & Rosenblum, L. A. 1966. A behavioral taxonomy for *Macaca nemestrina* and *Macaca radiata:* based on longitudinal observations of family groups in the laboratory. *Primates,* 7:205–258.

Koestler, A. 1949. *Insight and Outlook.* London: MacMillan.

Kohts, N. 1937. La conduite du petit chimpanzé et de l'enfant de l'homme. *Journal de Psychologie Normale et Pathologique,* 34:494–531.

Kraut, R. E. & Johnston, R. E. 1979. Social and emotional messages of smiling: an ethological approach. *Journal of Personality and Social Psychology,* 37:1539–1553.

Lakoff, G. 1987. *Women, Fire and Dangerous Things: What Categories Reveal about the Mind.* Chicago: University of Chicago Press.

Lorenz, K. 1953. Verstandigung unter Tieren. Zurich: Fontana, 1953.

Manning, A. & Stamp Dawkins, M. 1992. *An Introduction to Animal Behaviour,* 4th ed. Cambridge: Cambridge University Press.

Martin, P. & Caro, T. M. 1985. On the functions of play and its role in behavioral development. *Advances in the Study of Behavior,* 15:59–103.

Martin, R. D. 1990. *Primate Origins and Evolution.* London: Chapman and Hall.

Mason, W. A. 1965. Determinants of social behavior in young chimpanzees. In *Behavior of Nonhuman Primates,* vol. 2, ed. A. M. Schrier, H. F. Harlow, & F. Stollnitz, pp. 335–364. New York: Academic Press.

Melnick, D. J. & Hoelzer, G. A. 1996. The population genetic consequences of macaque social organisation and behaviour. In *Evolution and Ecology of Macaque Societies,* ed. J. E. Fa & D. G. Lindberg, pp. 413–443. Cambridge: Cambridge University Press.

Miller, R. E., Caul, W. F. & Mirsky, J. A. 1967. Communication of affect between feral and socially isolated monkeys. *Journal of Personality and Social Psychology,* 7:231–239.

Morris, D. 1957. "Typical intensity" and its relation to the problem of ritualisation. *Behaviour,* 11:1–12.

Ortega, J. C. & Bekoff, M. 1987. Avian play: comparative evolutionary and developmental trends. *Auk,* 104:338–341.

Pellis, S. M. & Iwaniuk, A. N. 2000. Adult-adult play in primates: comparative analyses of its origin, distribution and evolution. *Ethology,* 106:1083–1104.

Preuschoft, S. 1992. 'Laughter' and 'smile' in Barbary macaques *(Macaca sylvanus). Ethology,* 91:220–236.

——— 1995. *'Laughter' and 'Smiling' in Macaques: An Evolutionary Approach.* Utrecht: Universiteit Utrecht.

——— 1999. Are primates behaviorists? Formal dominance, cognition, and free floating rationales. *Journal of Comparative Psychology,* 113:91–95.

——— 2000. Primate faces and facial expressions. *Social Research,* 67:245–271.

Preuschoft, S. & Preuschoft, H. 1994. Primate nonverbal communication: our communicatory heritage. In *Origins of Semiosis,* ed. W. Nöth, pp. 61–100. Berlin: Mouton de Gruyter.

Preuschoft, S. & van Hooff, J. A. R. A. M. 1995. Homologizing primate facial displays: a critical review of methods. *Folia primatologica,* 65:121–137.

——— 1997. The social function of 'smile' and 'laughter': variations across primate species and societies. In *Nonverbal Communication: Where Nature Meets Culture,* ed. U. Segerstråle & P. Molnàr, pp. 171–189. Mahwah, NJ: Erlbaum.

——— 1999. *'Laughter' and 'Smiling' in Evolutionary Perspective* (film). Institut für den wissenschaftlichen Film, Göttingen.

Provine, R. R. 1993. Laughter punctuates speech: linguistic, social and gender contexts of laughter. *Ethology,* 95:291–298.

——— 1996. Laughter. *American Scientist,* 84:38–45.

Redican, W. K. 1975. Facial expressions in non-human primates. In *Primate Behavior: Developments in Field and Laboratory Research,* vol. 4, ed. L. A. Rosenblum, pp. 103–194. London: Academic Press.

Ruch, W. 1997. The FACS in humor research. In *What the Face Reveals,* ed. P. Ekman & E. L. Rosenberg, pp. 109–111. New York: Oxford University Press.

Schenkel, R. 1947. Ausdrucksstudien an Wölfen. *Behaviour,* 1:81–129.

——— 1964. Zur Ontogenese des Verhaltens bei Gorilla und Mensch. *Zeitschrift Morphologie und Anthropologie,* 54:233–259.

Schilder, M. B. H., van Hooff, J. A. R. A. M., van Geer-Plesman, C. J. & Wensing, J. B. 1984. A quantitative analysis of facial expressions in the plains zebra. *Zeitschrift für Tierpsychologie,* 66:11–32.

Smith, W. J. 1965. Message, meaning and context in ethology. *American Naturalist,* 908:405–409.

Smoski, M. J. & Bacharowski, J. A. In press. Antiphonal laughter between friends and strangers. *Cognition and Emotion.*

Thierry, B. 2000. Covariation of conflict management patterns across macaque species. In *Natural Conflict Resolution,* ed. F. Aureli & F. B. M. de Waal. Berkeley: University of California Press.

Thierry, B., Demaria, C., Preuschoft, S. & Desportes, C. 1989. Structural convergence between silent bared-teeth display and relaxed open-mouth display in the Tonkean macaque (*Macaca tonkeana*). *Folia primatologica,* 52:178–185.

Thompson, K. V. 1998. Self assessment in juvenile play. In *Animal Play: Evolutionary, Comparative and Ecological Perspectives,* ed. M. Bekoff & J. A. Byers, pp. 183–204. Cambridge: Cambridge University Press.

Tinbergen, N. 1952. "Derived" activities: their causation, biological significance, origin, and emancipation during evolution. *Quarterly Review of Biology,* 27:1–32.

Trumler, E. 1959. Das Rossigkeitsgesicht und ähliches Ausdrucksverhalten bei Einhufern. *Zeitschrift für Tierpsychologie,* 16:478–488.

van Hooff, J. A. R. A. M. 1967. The facial displays of catarrhine monkeys and apes. In *Primate Ethology,* ed. D. Morris, pp. 7–68. London: Weidenfeld and Nikolson.

——— 1972. A comparative approach to the phylogeny of laughter and smiling. In *Non-verbal Communication,* ed. R. A. Hinde, pp. 209–241. London: Cambridge University Press.

——— 1973. The Arnhem Zoo chimpanzee consortium: An attempt to create an ecologically and socially acceptable habitat. *International Zoological Yearbook,* 13:195–205.

——— 1976. The comparison of facial expressions in man and higher primates. In

Methods of Inference from Animal to Human Behavior, ed. M. von Cranach, pp. 165–196. The Hague: Mouton-Aldine.
——— 1989. Laughter and humour, and the "duo-in-uno" of nature and culture. In *The Nature of Culture,* ed. W. A. Koch, pp. 120–149. Bochum: Brockmeyer.
Vehrencamp, S. 1983. A model for the evolution of despotic versus egalitarian societies. *Animal Behaviour,* 31:667–682.

CASE STUDY 10A. *Emotional Recognition by Chimpanzees*

Brothers, L. 1989. A biological perspective on empathy. *American Journal of Psychiatry,* 146:10–19.
Davis, M. H. 1994. *Empathy: A Social Psychological Approach.* Madison, WI: Brown and Benchmark.
de Waal, F. B. M. 1996. *Good Natured: The Origin of Right and Wrong in Humans and Other Animals.* Cambridge, MA: Harvard University Press.
de Waal, F. B. M. & van Roosmalen, A. 1979. Reconciliation and consolation among chimpanzees. *Behavioral Ecology and Sociobiology,* 5:55–66.
Ekman, P. 1973. Cross-cultural studies of facial expressions. In *Darwin and Facial Expressions,* ed. P. Ekman, pp. 169–222. New York: Academic Press.
Ekman, P., Friesen, W. V. & Ellsworth, P. 1972. *Emotion in the Human Face: Guidelines for Research and an Integration of Findings.* New York: Pergamon Press.
Ekman, P., Levenson, R. W. & Friesen, W. V. 1983. Autonomic nervous system activity distinguishes among emotions. *Science,* 221:1208–1210.
Flack, J. & de Waal, F. B. M. 2000. 'Any animal whatever': Darwinian building blocks of morality in monkeys and apes. *Journal of Consciousness Studies,* 7.
Goodall, J. van Lawick. 1968. The behavior of free-living chimpanzees in the Gombe Stream Reserve. *Animal Behavior Monographs,* 1:165–311.
Hatfield, E., Cacioppo, J. T. & Rapson, R. L. 1994. *Emotional Contagion.* Paris: Cambridge University Press.
Izard, C. E. 1971. *The Face of Emotion.* New York: Appleton-Century-Crofts.
Kappas, A., Hess, U. & Banse, R. 1993. Empathy: The Role of Physiological Synchronization. Paper presented at the Convention of the American Psychological Association, Toronto, Canada, 20–24 August.
Levenson, R. W. & Ruef, A. M. 1992. Empathy: a physiological substrate. *Journal of Personality and Social Psychology,* 63:234–246.
Ortony, A., Clore, G. L. & Collins, A. 1988. *The Cognitive Structure of Emotions.* Cambridge, MA: Cambridge University Press.
Parr, L. A. 2001. Cognitive and physiological markers of emotional awareness in chimpanzees, *Pan troglodytes. Animal Cognition,* 4:223–229.
Parr, L. A. & de Waal, F. B. M. 1999. Visual kin recognition in chimpanzees. *Nature,* 399:647–648.
Parr, L. A., Hopkins, W. D. & de Waal, F. B. M. 1998. The perception of facial ex-

pressions in chimpanzees (*Pan troglodytes*). *Evolution of Communication,* 2:1–23.

Parr, L. A., Winslow, J. T., Hopkins, W. D. & de Waal, F. B. M. 2000. Recognizing facial cues: individual recognition in chimpanzees (*Pan troglodytes*) and rhesus monkeys (*Macaca mulatta*). *Journal of Comparative Psychology,* 114:47–60.

Preuschoft, S. & van Hooff, J. A. R. A. M. 1995. Homologizing primate facial displays: a critical review of methods. *Folia primatologica,* 65:121–137.

11. Vocal Communication in Wild Parrots

Acedo, V. 1992. Ecology of the yellow-naped amazon in Guatemala. *AFA Watchbird,* 19:31–34.

Ali, N. J., Farabaugh, S. M. & Dooling, R. J. 1993. Recognition of contact calls by the budgerigar (*Melopsittacus undulatus*). *Bulletin of the Psychonomic Society,* 31:468–470.

Arrowood, P. C. 1988. Duetting, pair bonding, and agonistic display in parakeet pairs. *Behaviour,* 106:129–157.

Ballance, L. T. 1990. Residence patterns, group organization, and surfacing associations of bottlenose dolphins in Kino Bay, Gulf of California, Mexico. In *The Bottlenose Dolphin,* ed. S. Leatherwood & R. R. Reeves, pp. 267–283. San Diego: Academic Press.

Bartlett, P. & Slater, P. J. B. 1999. The effect of new recruits on the flock specific call of budgerigars (*Melopsittacus undulatus*). *Ethology, Ecology and Evolution,* 11:139–147.

Bradbury, J. W. & Vehrencamp, S. L. 1998. *Principles of Animal Communication.* Sunderland, MA: Sinauer Associates.

Bradbury, J. W., Cortopassi, K. A. & Clemmons, J. R. In press. Geographical variation in the contact calls of orange-fronted conures. *Auk.*

Brereton, J. L. G. & Pidgeon, R. W. 1966. The language of the eastern rosella. *Australian Natural History,* 15:225–229.

Brittan-Powell, E. F., Dooling, R. J. & Farabaugh, S. M. 1997. Vocal development in budgerigars (*Melopsittacus undulatus*): contact calls. *Journal of Comparative Psychology,* 111:226–241.

Brockway, B. F. 1962. Ethological studies of the budgerigar: reproductive behavior. *Behaviour,* 23:294–324.

——— 1965. Stimulation of ovarian development and egg laying by male courtship vocalization in budgerigars (*Melopsittacus undulatus*). *Animal Behaviour,* 12:493–501.

——— 1969. Roles of budgerigar vocalization in the integration of breeding behaviour. In *Bird Vocalizations,* ed. R. A. Hinde, pp. 131–158. Cambridge: Cambridge University Press.

Brown, E. D. & Farabaugh, S. M. 1991. Song sharing in a group-living songbird, the Australian magpie *Gymnorhina tibicen.* Part III. Sex specificity and individual specificity of vocal pats in communal chorus and duet songs. *Behaviour,* 118:244–274.

Brown, S. D., Dooling, R. J. & O'Grady, K. J. 1988. Perceptual organization of acoustic stimuli by budgerigars (*Melopsittacus undulatus*): III. Contact calls. *Comparative Psychology,* 102:236–247.

Catchpole, C. K. & Slater, P. J. B. 1995. *Bird Song: Biological Themes and Variations.* Cambridge: Cambridge University Press.

Chapman, C. A. 1990. Association patterns of spider monkeys: the influence of ecology and sex on social organization. *Behavioral Ecology and Sociobiology,* 26:409–414.

Chapman, C. A. & Weary, D. M. 1990. Variability in spider monkey vocalizations may provide basis for individual recognition. *American Journal of Primatology,* 22:279–284.

Clout, M. N. 1989. Foraging behaviour of glossy black cockatoos. *Australian Wildlife Research,* 16:467–473.

Cortopassi, K. A. & J. W. Bradbury. 2000. The comparison of harmonically rich sounds using spectrographic cross-correlation and principal coordinates analysis. *Bioacoustics,* 11:89–127.

Cruickshank, A. J., Gautier, J. P. & Chappuis, C. 1993. Vocal mimicry in wild African grey parrots *Psittacus erithacus. Ibis,* 135:293–299.

Dooling, R. J. 1986. Perception of vocal signals by the budgerigar (*Melopsittacus undulatus*). *Journal of Experimental Biology,* 45:195–218.

Durand, S. E., Heaton, J. T., Amateau, S. K. & Brauth, S. E. 1997. Vocal control pathways through the anterior forebrain of a parrot *(Melopsittacus undulatus*). *Journal of Comparative Neurology,* 377:179–206.

Farabaugh, S. M. 1982. The ecological and social significance of duetting. In *Acoustic Communication of Birds* (D. E. Kroodsma & E. H. Miller, eds.), pp. 85–124. New York: Academic Press

Farabaugh, S. M. & Dooling, R. J. 1996. Acoustic communication in parrots: laboratory and field studies of budgerigars, *Melopsittacus undulatus.* In *Ecology and Evolution of Acoustic Communication in Birds,* ed. D. E. Kroodsma & E. H. Miller, pp. 97–117. Ithaca, NY: Comstock Publishing Associates.

Farabaugh, S. M., Linzenbold, A. & Dooling, R. J. 1994. Vocal plasticity in budgerigars (*Melopsittacus undulatus*): evidence for social factors in the learning of contact calls. *Journal of Comparative Psychology,* 108:81–92.

Feekes, F. 1977. Colony-specific song in *Cacicus cela* (Icteridae, Aves): the password hypothesis. *Ardea,* 3:197–202.

Ficken, M. S., Ficken, R. W. & Witkin, S. R. 1978. Vocal repertoire of the black-capped chickadee. *Auk,* 95:34–48.

Forshaw, J. M. 1989. *Parrots of the World.* London: Blandford.

Galetti, M. 1993. Diet of the scaly-headed parrot (*Pionus maximiliani*) in a semi-deciduous forest in southeastern Brazil. *Biotropica,* 25:419–425.

Garnetzke-Stollman, K. & Franck, D. 1991. Socialization tactics of the spectacled parrotlet (*Forpus conspicillatus*). *Behaviour,* 119:1–29.

Gaunt, A. S. & Gaunt, S. L. L. 1985. Electromyographic studies of the syrinx in parrots (Aves, Psittacidae). *Zoomorphology,* 105:1–11.

Gilardi, J. D. 1996. Ecology of parrots in the Peruvian Amazon: habitat use, nutrition, and geophagy. Ph.D. diss., University of California, Davis.

Gilardi, J. D., Duffey, S. S., Munn, C. A. & Tell, L. A. 1999. Biochemical functions of geophagy in parrots: detoxification of dietary toxins and cytoprotective effects. *Journal of Chemical Ecology,* 25:897–922.

Greene, T. C. 1998. Foraging ecology of the red-crowned parakeet (*Cyanoramphus novaezelandiae novaezelandiae*) and yellow-crowned parakeet (*C-auriceps auriceps*) on Little Barrier Island, Hauraki Gulf, New Zealand. *New Zealand Journal of Ecology,* 22:161–171.

Hall, M. L. 2000. The function of duetting in magpie-larks: conflict, cooperation, or commitment? *Animal Behaviour,* 60:667–677.

Hardy, J. W. 1963. Epigamic and reproductive behavior of the orange-fronted parakeet. *Condor,* 65:169–199.

——— 1965. Flock social behavior of the orange-fronted parakeet. *Condor,* 67:140–156.

Hausberger, M., Jenkins, P. F. & Keene, J. 1991. Species-specificity and mimicry in bird song: are they paradoxes? *Behaviour,* 117:53–81.

Hile, A. G., Plummer, T. K. & Striedter, G. F. 2000. Male vocal imitation produces call convergence during pair bonding in budgerigars, *Melopsittacus undulatus. Animal Behaviour,* 59:1209–1218.

Hope, S. 1980. Call form in relation to function in the Steller's jay. *American Naturalist,* 116:788–820.

Janzen, D. H. 1981. *Ficus ovalis* seed predation by an orange-chinned parakeet (*Brotogeris jugularis*) in Costa Rica. *Auk,* 98:841–844.

Jarvis, E. D. & Mello, C. V. 2000. Molecular mapping of brain areas involved in parrot vocal communication. *Journal of Comparative Neurology,* 419:1–31.

Juniper, T. & Parr, M. 1998. *Parrots: A Guide to Parrots of the World.* New Haven, CT: Yale University Press.

Kroodsma, D. E. & Miller, E. H. 1982. *Acoustic Communication in Birds.* New York: Academic Press.

——— 1996. *Ecology and Evolution of Acoustic Communication in Birds.* Ithaca, NY: Comstock Publishing Associates.

Langmore, N. E. 1998. Functions of duet and solo songs of female birds. *Trends in Ecology and Evolution,* 13:136–140.

Levin, R. N. 1996a. Song behaviour and reproductive strategies in a duetting wren, *Thryothorus nigricapillus:* I. Removal experiments. *Animal Behaviour,* 52:1093–1106.

——— 1996b. Song behaviour and reproductive strategies in a duetting wren, *Thryothorus nigricapillus:* II. Playback experiments. *Animal Behaviour,* 52:1107–1117.

Löhrl, H. 1958. Das Verhalten des Kleibers. *Zeitschrift für Tierpsychologie,* 15:191–252.

Mammen, D. L. & Nowicki, S. 1981. Individual differences and within-flock convergence in chickadee calls. *Behavioral Ecology and Sociobiology,* 9:179–186.

Martella, M. B. & Bucher, E. H. 1990. Vocalizations of the monk parakeet. *Bird Behavior,* 8:101–110.

Martuscelli, P. 1995. Ecology and conservation of the red-tailed amazon, *Amazona brasiliensis,* in south-eastern Brazil. *Bird Conservation International,* 5:405–420.

Mebes, H. D. 1978. Pair-specific duetting in the peach-faced lovebird *Agapornis roseicollis. Naturwissenschaften,* 65:66–67.

Merton, D. V., Morris, R. B. & Atkinson, I. E. 1984. Lek behavior in a parrot: the kakapo, *Strigops habroptilus,* of New Zealand. *Ibis,* 126:277–283.

Morton, E. S. 1996. A comparison of vocal behavior among tropical and temperate passerine birds. In *Ecology and Evolution of Acoustic Communication in Birds,* ed. D. E. Kroodsma & E. H. Miller, pp. 258–268. Ithaca, NY: Comstock Publishing Associates.

Moss, C. J. 1988. *Elephant Memories.* Boston: Houghton Mifflin.

Mundinger, P. C. 1970. Vocal imitation and individual recognition of finch calls. *Science,* 168:480–482.

——— 1979. Call learning in the Carduelinae: ethological and systematic considerations. *Systematic Zoology,* 28:270–283.

Nespor, A. A. 2000. Comparative neuroendocrine mechanisms mediating sex differences in reproductive and vocal behavior and the related brain regions in songbirds, budgerigars and quail. *Avian and Poultry Biology Reviews,* 11:45–62.

Norconk, M. A. & Kinzey, W. G. 1994. Challenge of neotropical frugivory: travel patterns of spider monkeys and bearded sakis. *American Journal of Primatology,* 34:171–183.

Norconk, M. A., Wertis, C. & Kinzey, W. G. 1997. Seed predation by monkeys and macaws in eastern Venezuela: preliminary findings. *Primates,* 38:177–184.

Noske, S. 1980. Aspects of the behaviour and ecology of the white cockatoo *(Cacatua galerita)* and the galah *(C. roseicapilla)* in croplands in north-east New South Wales. Master's thesis, University of New England, Armidale, Australia.

Nottebohm, F. 1972. The origins of vocal learning. *American Naturalist,* 106:116–140.

——— 1976. Phonation in the orange-winged Amazon parrot, *Amazona amazonica. Journal of Comparative Physiology,* 108:157–170.

Nowicki, S. 1983. Flock-specific recognition of chickadee calls. *Behavioral Ecology and Sociobiology,* 12:317–320.

Pepperberg, I. M. 1987. Acquisition of the same/different concept by an African grey parrot *(Psittacus erithacus)*: learning with respect to color, shape, and material. *Animal Learning Behavior,* 15:423–432.

——— 1988. Comprehension of "absence" by an African grey parrot: learning with respect to questions of same/different. *Journal of Experimental Analytical Behavior,* 50:553–564.

——— 1990. Some cognitive abilities of an African grey parrot. In *Advances in the Study of Behavior,* ed. P. J. B. Slater, J. S. Rosenblatt & C. Beer, pp. 357–409. New York: Academic Press.

——— 1994. Evidence for numerical competence in an African grey parrot (*Psittacus erithacus*). *Journal of Comparative Psychology,* 108:36–44.

——— 1999. *The Alex Studies: Cognitive and Communicative Abilities of Grey Parrots.* Cambridge, MA: Harvard University Press.

Pidgeon, R. 1970. The individual and social behaviour of the galah. Master's thesis, University of New England, Armidale, Australia.

——— 1981. Calls of the galah, *Cacatua roseicapilla,* and some comparisons with four other species of Australian parrot. *Emu,* 81:158–168.

Power, D. M. 1966a. Agonistic behavior and vocalizations of orange-chinned parakeets in captivity. *Condor,* 8:562–581.

——— 1966b. Antiphonal duetting and evidence for the auditory reaction time in the orange-chinned parakeet. *Auk,* 83:314–319.

Robbins, D., Chapman, C. A. & Wrangham, R. W. 1991. Group size and stability: why do gibbons and spider monkeys differ? *Primates,* 32:301–305.

Rowley, I. 1980. Parent-offspring recognition in a cockatoo, the galah, *Cacatua roseicapillus. Australian Journal of Zoology,* 28:445–456.

Rowley, I. & Chapman, G. 1986. Cross-fostering, imprinting, and learning in two sympatric species of cockatoo. *Behaviour,* 96:1–16.

Saunders, D. A. 1980. Food and movements of the short-billed form of the white-tailed black cockatoo. *Australian Wildlife Research,* 7:257–269.

——— 1982. The breeding behaviour and biology of the short-billed form of the white-tailed black cockatoo (*Calyptorhynchus funereus*). *Ibis,* 124:422–455.

——— 1983. Vocal repertoire and individual vocal recognition in the short-billed white-tailed black cockatoo, *Calyptorhynchus funereus latirostris* Carnaby. *Australian Wildlife Research,* 10:527–536.

Scott, M. D., Wells, R. S. & Blair Irvine, A. 1990. A long-term study of bottlenose dolphins on the west coast of Florida. In *The Bottlenose Dolphin,* ed. S. Leatherwood & R. R. Reeves, pp. 235–244. San Diego, CA: Academic Press.

Seibt, U. & Wickler, W. 1977. Duettieren als Revier-Anzeife bei Völgeln. *Zeitschrift für Tierpsychologie,* 43:80–87.

Serpell, J. 1981. Duets, greetings, and triumph ceremonies: analogous displays in the parrot genus *Trichoglossus. Zeitschrift für Tierpsychologie,* 55:268–283.

Skeate, S. T. 1984. Courtship and reproductive behavior of captive white-fronted Amazon parrots *Amazona albifrons. Bird Behavior,* 5:103–109.

Snyder, N. F. R., Wiley, J. W. & Kepler, C. B. 1987. *The Parrots of Luquillo: Natural History and Conservation of the Puerto Rican Parrot.* Los Angeles: Western Foundation of Vertebrate Zoology.

Sonnenschein, E. & Reyer, H. E. 1983. Mate-guarding and other functions of antiphonal duets in the slate-coloured boubou (*Laniarius funebris*). *Zeitschrift für Tierpsychologie,* 63:112–140.

Stiles, F. G. & Skutch, A. 1989. *A Guide to the Birds of Costa Rica.* Ithaca, NY: Comstock Publishing Associates.

Streidter, G. F. 1994. The vocal control pathways in budgerigars differ from those in songbirds. *Comparative Neurology,* 343:35–36.

Teixidor, P. & Byrne, R. W. 1997. Can spider monkeys (*Ateles geoffroyi*) discriminate vocalizations of familiar individuals and strangers? *Folia primatologica,* 68:254–264.

——— 1999. The 'whinny' of spider monkeys: individual recognition before situational meaning. *Behaviour,* 136:279–308.

Trainer, J. M. 1989. Cultural evolution in song dialects of yellow-rumped caciques in Panama. *Ethology,* 80:190–204.

Vehrencamp, S. L. 2000. Handicap, index, and conventional signal elements of bird song. In *Animal Signals: Signalling and Signal Design in Animal Communication,* ed. Y. Espmark, T. Amundsen & G. Rosenqvist, pp. 277–300. Trondheim: Tapir Publishers.

Vehrencamp, S. L., Stile, F. G. & Bradbury, J. W. 1977. Observations on the foraging behavior and avian prey of the neotropical carnivorous bat, *Vampyrum spectrum. Journal of Mammalogy,* 58:469–478.

Waltman, J. R. & Beissinger, S. R. 1992. Breeding behavior of the green-rumped parrotlet. *Wilson Bulletin,* 104:62–84.

Wanker, R., Bernate, N. C. & Franck, D. 1996. Socialization of spectacled parrotlets *Forpus conspicillatus:* the role of parents, creches, and sibling groups in nature. *Journal für Ornithologie,* 137:447–461.

Wanker, R., Apcin, J., Jennerjahn, B. & Waibel, B. 1998. Discrimination of different social companions in spectacled parrotlets *(Forpus conspicillatus*): evidence for individual vocal recognition. *Behavioral Ecology and Sociobiology,* 43:197–202.

Wermundsen, T. 1997. Seasonal change in the diet of the Pacific parakeet *Aratinga strenua* in Nicaragua. *Ibis,* 139:566–568.

Wickler, W. 1980. Vocal duetting and the pair bond: I. Coyness and partner commitment. A hypothesis. *Zeitschrift für Tierpsychologie,* 52:201–209.

Wiley, R. H. & Wiley, M. S. 1977. Recognition of neighbor's duets by stripe-backed wrens, *Campylorhynchus nuchalis. Behaviour,* 62:10–34.

Wright, T. F. 1996. Regional dialects in the contact call of a parrot. *Proceedings of the Royal Society of London,* ser. B, 263:867–872.

——— 1997. Vocal communication in the yellow-naped amazon (*Amazona auropalliata*). Ph.D. diss., University of California, San Diego.

Wright, T. F. & Dorin, M. 2001. Pair duets in the yellow-naped amazon (*Amazona auropalliata*): responses to playbacks of different dialects. *Ethology,* 107:111–124.

Wright, T. F. & Wilkinson, G. S. 2001. Population genetic structure and vocal dialects in an Amazon parrot. *Proceedings of the Royal Society of London,* ser. B, 268:609–616.

Wyndham, E. 1980. Diurnal cycle, behaviour, and social organization of the budgerigar (*Melopsittacus undulatus*). *Emu,* 80:25–33.

——— 1983. Movements and breeding seasons of the budgerigar. *Emu,* 82:276–282.

Case Study 11A. *Representational Vocal Signaling in the Chimpanzee*

Goodall, J. 1986. *The Chimpanzees of Gombe: Patterns of Behavior.* Cambridge, MA: The Belknap Press of Harvard University Press.

Marler, P. 1985. Representational vocal signals of primates. *Fortschritte der Zoologie,* 31:211–221.

Marler, P. & Tenaza, R. 1977. Signaling behavior of apes with special reference to vocalization. In *How Animals Communicate,* ed. T. A. Sebeok, pp. 965–1033. Bloomington, IN: Indiana University Press.

12. Social and Vocal Complexity in Bats

Aiello, L. C. & Dunbar, R. I. M. 1993. Neocortex size, group size and the evolution of language. *Behavioral and Brain Sciences,* 34:184–193.

August, P. V. 1979. Distress calls in *Artibeus jamaicensis:* ecology and evolutionary implications. In *Vertebrate Ecology in the Northern Neotropics,* ed. J. F. Eisenberg, pp. 151–159. Washington: Smithsonian Institution Press.

Balcombe, J. P. 1990. Vocal recognition of pups by mother Mexican free-tailed bats, *Tadarida brasiliensis mexicana. Animal Behaviour,* 39:960–966.

Balcombe, J. P. & McCracken, G. F. 1992. Vocal recognition in Mexican free-tailed bats: do pups recognize mothers? *Animal Behaviour,* 43:79–88.

Barak, Y. & Yom-Tov, Y. 1991. The mating system of *Pipistrellus kuhli* (Microchiroptera) in Israel. *Mammalia,* 55:285–292.

Barbour, R. W. & Davis, W. H. 1969. *Bats of America.* Lexington: University Press of Kentucky.

Barclay, R. M. R., Fenton, M. B. & Thomas, D. 1979. Social behavior of the little brown bat, *Myotis lucifugus.* II. Vocal communication. *Behavioral Ecology and Sociobiology,* 6:137–146.

Barlow, K. E. & Jones, G. 1997. Differences in songflight calls and social calls between two phonic types of the vespertilionid bat *Pipistrellus pipistrellus. Journal of Zoology,* 241:315–324.

Baron, G., Stephan, H. & Frahm, H. E. 1996. *Comparative Neurobiology in Chiroptera: Macromorphology, Brain Structures, Tables and Atlases.* Boston: Birkhäuser Verlag.

Barton, R. A. & Dunbar, R. I. M. 1997. Evolution of the social brain. In *Machiavellian Intelligence II: Extensions and Evaluations,* ed. A. Whiten & R. W. Byrne, pp. 240–263. Cambridge: Cambridge University Press.

Bearzi, G., Notarbartolo di Sciara, G. & Politi, E. 1997. Social ecology of bottlenose dolphins in the Kvarneric (northern Adriatic Sea). *Marine Mammal Science,* 13:650–668.

Beecher, M. D. 1989. Signalling systems for individual recognition: an information theory approach. *Animal Behaviour,* 38:248–261.

Bogdanowicz, W., Fenton, M. B. & Daleszczyk, K. 1999. The relationships between echolocation calls, morphology and diet in insectivorous bats. *Journal of Zoology,* 247:381–393.

Boughman, J. W. 1997. Greater spear-nosed bats give group-distinctive calls. *Behavioral Ecology and Sociobiology,* 40:61–70.

——— 1998. Vocal learning in greater spear-nosed bats. *Proceedings of the Royal Society of London,* ser. B, 265:227–233.

Boughman, J. W. & Wilkinson, G. S. 1998. Bats distinguish group members by vocalizations. *Animal Behaivour,* 55:1717–1732.

Bradbury, J. W. 1977a. Lek mating behavior in the hammer-headed bat. *Zeitschrift für Tierpsychologie,* 45:225–255.

——— 1977b. Social organization and communication. In *The Biology of Bats,* ed. W. A. Wimsatt, pp. 1–72. New York: Academic Press.

——— 1981. The evolution of leks. In *Natural Selection and Social Behavior,* ed. R. D. Alexander & D. W. Tinkle, pp. 138–169. New York: Chiron.

Bradbury, J. W. & Emmons, L. H. 1974. Social organization of some Trinidad bats. I. Emballonuridae. *Zeitschrift für Tierpsychologie,* 36:137–183.

Bradbury, J. W. & Vehrencamp, S. L. 1976a. Social organization and foraging in emballonurid bats. I. Field studies. *Behavioral Ecology and Sociobiology,* 1:337–381.

——— 1976b. Social organization and foraging in emballonurid bats. II. A model for the determination of group size. *Behavioral Ecology and Sociobiology,* 1:383–404.

Brooke, A. P. 1997. Organization and foraging behaviour of the fishing bat, *Noctilio leporinus* (Chiroptera:Noctilionidae). *Ethology,* 103:421–436.

Brosset, A. 1976. Social organization in the African bat, *Myotis bocagei. Zeitschrift für Tierpsychologie,* 42:50–56.

Brown, P. 1976. Vocal communication in the pallid bat, *Antrozous pallidus. Zeitschrift für Tierpsychologie,* 41:34–54.

Brown, P., Brown, T. & Grinnell, A. 1983. Echolocation, development, and vocal communication in the lesser bulldog bat, *Noctilio albiventris. Behavioral Ecology and Sociobiology,* 13:287–298.

Burland, T. M., Barratt, E. M., Nichols, R. A. & Racey, P. A. 2001. Mating patterns, relatedness and the basis of natal philopatry in the brown long-eared bat, *Plecotus auritus. Molecular Ecology.*

Catchpole, C. K. 1980. Sexual selection and the evolution of complex song among warblers of the genus *Acrocephalus. Behaviour,* 74:149–166.

Davidson, S. M. & Wilkinson, G. S. 2001. Function of male song in the greater white-lined bat, *Saccopteryx bilineata. Animal Behaviour.*

——— 2002. Geographic and individual variation in vocalizations by male *Saccopteryx bilineata. Journal of Mammalogy* 83:526–535.

DeVoogd, T. J., Krebs, J. R., Healy, S. D. & Purvis, A. 1993. Relations between song repertoire size and the volume of brain nuclei related to song: comparative evolutionary analyses amongst oscine birds. *Proceedings of the Royal Society of London, ser. B,* 254:75–82.

Dunbar, R. I. M. 1991. Functional significance of social grooming in primates. *Folia primatologica,* 57:121–131.

——— 1992. Neocortex size as a constraint on group size in primates. *Journal of Human Evolution,* 20:469–493.

——— 1993. Coevolution of neocortical size, group-size and language in humans. *Behavioral and Brain Sciences,* 16:681–694.

——— 1995. Neocortex size and group-size in primates: a test of the hypothesis. *Journal of Human Evolution,* 28:287–296.

Dunbar, R. I. M. & Bever, J. 1998. Neocortex size predicts group size in carnivores and some insectivores. *Ethology,* 104:695–708.

Entwhistle, A. C., Racey, P. A. & Speakman, J. R. 2000. Social and population structure of a gleaning bat, *Plecotus auritus. Journal of Zoology,* 252:11–17.

Esser, K. H. 1994. Audio-vocal learning in a non-human mammal: the lesser spear-nosed bat *Phyllostomus discolor. Neuroreport,* 5:1718–1720.

Esser, K. H. & Schmidt, U. 1989. Mother-infant communication in the lesser spear-nosed bat *Phyllostomus discolor* (Chiroptera, Phyllostomidae): evidence for acoustic learning. *Ethology,* 82:156–168.

Esser, K. H. & Schubert, J. 1998. Vocal dialects in the lesser spear-nosed bat *Phyllostomus discolor. Naturwissenschaften,* 85:347–349.

Esser, K. H., Condon, C. J., Suga, N. & Kanwal, J. S. 1997. Syntax processing by auditory cortical neurons in the FM-FM area of the mustached bat *Pteronotus parnellii. Proceedings of the National Academy of Sciences USA,* 94:14019–14024.

Fayenuwo, J. O. & Halstead, L. B. 1974. Breeding cycle of straw-colored fruit bat, *Eidolon helvum,* Ile-Ife, Nigeria. *Journal of Mammalogy,* 55:453–454.

Fenton, M. B. 1985. *Communication in the Chiroptera.* Bloomington: Indiana University Press.

——— 1995. Natural history and biosonar signals. In *Hearing by Bats* (A. N. Popper & R. R. Fay, eds.), pp. 37–86. New York: Springer-Verlag.

Fenton, M. B., Belwood, J. J. & Fullard, J. H. 1976. Responses of *Myotis lucifugus* (Chiroptera: Vespertilionidae) to calls of conspecifics and to other sounds. *Canadian Journal of Zoology,* 54:1443–1448.

Fenton, M. B., Boyle, N. G. H., Harrison, T. M. & Oxley, D. J. 1977. Activity patterns, habitat use and prey selection by some African insectivorous bats. *Biotropica,* 9:73–85.

Fenton, M. B., Portfors, C. V., Rautenbach, I. L. & Waterman, J. M. 1998. Compromises: sound frequencies used in echolocation by aerial-feeding bats. *Canadian Journal of Zoology,* 76:1174–1182.

Findley, J. S. 1993. *Bats: A Community Perspective.* Cambridge: Cambridge University Press.

Fleming, T. H. 1988. *The Short-Tailed Fruit Bat.* Chicago: Chicago University Press.

Gardner, A. L., Handley, C. O., Jr. & Wilson, D. E. 1991. Survival and relative abundance. In *Demography and Natural History of the Common Fruit Bat,* Artibeus jamaicensis, *on Barro Colorado Island, Panama,* ed. C. O. Handley, Jr., D. E. Wilson, & A. L. Gardner, pp. 53–75. Washington, DC: Smithsonian Institution Press.

Gelfand, D. L. & McCracken, G. F. 1986. Individual variation in the isolation calls of Mexican free-tailed bat pups (*Tadarida brasiliensis mexicana*). *Animal Behaviour,* 34:1078–1086.

Gerell-Lundberg, K. & Gerell, R. 1994. The mating behaviour of the pipistrelle and the Nathusius' pipistrelle (Chiroptera): a comparison. *Folia zoologica,* 43:315–324.

Goodall, J. 1986. *The Chimpanzees of Gombe: Patterns of Behavior.* Cambridge, MA: Belknap Press of Harvard University Press.

Gould, E. 1975. Neonatal vocalizations in bats of eight genera. *Journal of Mammalogy,* 56:15–29.

——— 1977. Echolocation and communication. In *Biology of Bats of the New World Family Phyllostomatidae,* Part II, ed. R. J. Baker, J. K. Jones, Jr. & D. C. Carter, pp. 247–280. Lubbock: Texas Tech Press.

Gould, E., Woolf, N. & Turner, D. C. 1973. Double-note communication calls in bats: occurrence in three families. *Journal of Mammalogy,* 54:998–1001.

Grafen, A. 1990. Biological signals as handicaps. *Journal of Theoretical Biology,* 144:517–546.

Guillen, A., Juste, J. & Ibanez, C. 2000. Variation in the frequency of the echolocation calls of *Hipposideros ruber* in the Gulf of Guinea: an exploration of the adaptive meaning of the constant frequency value in rhinolophoid CF bats. *Journal of Evolutionary Biology,* 13:70–80.

Guinee, L., Chu, K. & Dorsey, E. M. 1983. Changes over time in the songs of known individual humpback whales *(Megaptera novaeangliae).* In *Communication and Behavior of Whales,* ed. R. Payne. Boulder: Westview Press.

Habersetzer, J. 1981. Adaptive echolocation sounds in the bat *Rhinopoma hardwickei. Journal of Comparative Physiology,* ser. A, 144:559–566.

Hauser, M. D. & Fowler, C. 1991. Declination in fundamental frequency is not unique to human speech: evidence from nonhuman primates. *Journal of the Acoustical Society of America,* 91:363–369.

Heckel, G., Voigt, C. C., Mayer, F. & Von Helversen, O. 1999. Extra-harem paternity in the white-lined bat *Saccopteryx bilineata* (Emballonuridae). *Behaviour,* 136:1173–1185.

Heibert, S. M., Stoddard, P. K. & Arcese, P. 1989. Repertoire size, territory acquisition and reproductive success in song sparrows. *Animal Behaviour,* 37:266–277.

Heller, K. G. & Helverson, O. V. 1989. Resource partitioning of sonar frequency bands by rhinolophoid bats. *Oecologia,* 80:178–186.

Howell, D. J. 1979. Flock foraging in nectar-feeding bats: advantages to the bats and to the host plants. *American Naturalist,* 114:23–49.

Janik, V. M. & Slater, P. J. B. 1997. Vocal learning in mammals. *Advances in the Study of Behaviour,* 26:59–99.

Jones, G. 1999. Scaling of echolocation call parameters in bats. *Journal of Experimental Biology,* 202:3359–3367.

Jones, G. & Ransome, R. D. 1993. Echolocation calls of bats are influenced by maternal effects and change over a lifetime. *Proceedings of the Royal Society of London,* ser. B, 252:125–128.

Jones, G., Gordon, T. & Nightingale, J. 1992. Sex and age differences in the echolocation calls of the lesser horseshoe bat, *Rhinolophus hipposideros. Mammalia,* 56:189–193.

Jones, G., Hughes, P. M. & Rayner, J. M. V. 1991. The development of vocalizations in *Pipistrellus pipistrellus* (Chiroptera: Vespertilionidae) during post-natal growth and the maintenance of vocal signatures. *Journal of Zoology* (London), 225:71–84.

Kanwal, J., Suga, N. & Matsumura, Y. 1994. The vocal repertoire of the moustached bat, *Pteronotus parnelli. Journal of the Acoustical Society of America,* 96:1229–1254.

Keen, R. & Hitchcock, H. B. 1980. Survival and longevity of the little brown bat (*Myotis lucifugus*) in southeastern Ontario. *Journal of Mammalogy,* 61:1–7.

Kerth, G. & Konig, B. 1999. Fission, fusion and nonrandom associations in female Bechstein's bats (*Myotis bechsteinii*). *Behaviour,* 136:1187–1202.

Kerth, G., Mayer, F. & König, B. 2000. Mitochondrial DNA (mtDNA) reveals that female Bechstein's bats live in closed societies. *Molecular Ecology,* 9:481–492.

Kuhl, P. K. & Miller, J. D. 1975. Speech perception by the chinchilla: voiced-voiceless distinction in alveolar plosive consonants. *Science,* 190:69–72.

Kunz, T. H. & Stern, A. A. 1995. Maternal investment and post-natal growth in bats. *Symposium of the Zoological Society of London,* 67:123–138.

Lehmann, J., Jenni, L. & Maumary, L. 1992. A new longevity record for the long-eared bat (*Plecotus auritus,* Chiroptera). *Mammalia,* 56:316–318.

Lewis, S. E. 1992. Behavior of Peter's tent-making bat, *Uroderma bilobatum,* at maternity roosts in Costa Rica. *Journal of Mammalogy,* 73:541–546.

——— 1995. Roost fidelity of bats: a review. *Journal of Mammalogy,* 76:481–496.

——— 1996. Low roost-site fidelity in pallid bats: associated factors and effect on group stability. *Behavioral Ecology and Sociobiology,* 39:335–344.

Masters, W. M., Raver, K. A. S. & Kazial, K. A. 1995. Sonar signals of big brown bats (*Eptesicus fuscus*) contain information about individual identity and family affiliation. *Animal Behaviour,* 50:1243–1260.

McCracken, G. F. 1984. Communal nursing in Mexican free-tailed bat maternity colonies. *Science,* 233:1090–1091.

McCracken, G. F. & Bradbury, J. W. 1981. Social organization and kinship in the polygynous bat *Phyllostomus hastatus. Behavioral Ecology and Sociobiology,* 8:11–34.

McCracken, G. F. & Wilkinson, G. S. 2000. Bat mating systems. In *Reproductive Biology of Bats,* ed. P. H. Krutszch & E. G. Crichton. New York: Academic Press.

McWilliam, A. N. 1987. The reproductive and social biology of *Coleura afra* in a seasonal environment. In *Recent Advances in the Study of Bats,* ed. M. B. Fenton, P. A. Racey & J. M. V. Rayner, pp. 324–350. Cambridge: Cambridge University Press.

——— 1988. Social organisation of the bat *Tadarida (Chaerephon) pumila* (Chiroptera: Molossidae) in Ghana, West Africa. *Ethology,* 77:115–124.

Merwe, M. v. d. & Rautenbach, I. L. 1987. Reproduction in Schlieffen's bat, *Nycticeius schlieffenii,* in the eastern Transvaal lowvel, South Africa. *Journal of Reproduction and Fertility,* 81:41–50.

Miller, L. A. & Degn, H. J. 1981. The acoustic behavior of four species of vespertilionid bats studied in the field. *Journal of Comparative Physiology,* ser. A, 142:67–74.

Mitani, J. C. & Gros-Louis, J. 1998. Chorusing and call convergence in chimpanzees: tests of three hypotheses. *Behaviour,* 135:1041–1064.

Morrison, D. W. 1980. Foraging and day roosting dynamics of canopy fruit bats in Panama. *Journal of Mammalogy,* 61:20–29.

——— 1987. Roosting behavior. In *Demography and Natural History of the Common Fruit Bat,* Artibeus jamaicensis, *on Barro Colorado Island, Panama,* ed. C. O. Handley, pp. 131–136. Washington, DC: Smithsonian Institution Press.

Moss, C. J. & Poole, J. H. 1983. Relationships and social structure of African elephants. In *Primate Social Relationships: An Integrated Approach,* ed. R. A. Hinde, pp. 315–325. Oxford: Blackwell Scientific.

Nelson, J. E. 1964. Vocal communication in Australian flying-foxes (Pteropodidae; Megachiroptera). *Zeitschrift für Tierpsychologie,* 21:857–870.

Neuweiler, G. 1984. Foraging, echolocation and audition in bats. *Naturwissenschaften,* 71:446–455.

Nishida, T. 1990. *The Chimpanzees of the Mahale Mountains.* Tokyo: Tokyo University Press.

Norberg, U. M. & Fenton, M. B. 1988. Carnivorous bats. *Biological Journal of the Linnean Society,* 33:383–394.

Nowak, R. M. 1994. *Walker's Bats of the World.* Baltimore: Johns Hopkins University Press.

Obrist, M. 1995. Flexible bat echolocation: the influence of individual, habitat, and conspecifics on sonar signal design. *Behavioral Ecology and Sociobiology,* 36:207–219.

Ortega, J. & Arita, H. T. 1999. Structure and social dynamics of harem groups in *Artibeus jamaicensis* (Chiroptera: Phyllostomidae). *Journal of Mammalogy,* 80:1173–1185.

O'Shea, T. J. 1980. Roosting, social organization and the annual cycle in a Kenya population of the bat *Pipistrellus nanus. Zeitschrift für Tierpyschologie,* 53:171–195.

Owren, M. J. & Bernacki, R. 1988. The acoustic features of vervet monkey (*Cercopithecus aethiops*) alarm calls. *Journal of the Acoustical Society of America,* 83:1927–1935.

Pagel, M. D. & Harvey, P. H. 1989. Taxonomic differences in the scaling of brain on body size among mammals. *Science,* 244:1589–1593.

Petri, B., Pääbo, S., Von Haeseler, A. & Tautz, D. 1997. Paternity assessment and population subdivision in a natural population of the larger mouse-eared bat *Myotis myotis. Molecular Ecology,* 6:235–242.

Pinker, S. & Bloom, P. 1990. Natural language and natural selection. *Behavioral and Brain Sciences,* 13:707–784.

Porter, F. L. 1979. Social behavior in the leaf-nosed bat, *Carollia perspicillata. Zeitschrift für Tierpsychologie,* 50:1–8.

Racey, P. A. & Entwhistle, A. C. 2000. Life history and reproductive strategies of bats. In *Reproductive Biology of Bats,* ed. P. H. Krutzsch & E. G. Crichton, pp. 363–414. New York: Academic Press.

Ransome, R. D. 1991. Greater horseshoe bat. In *The Handbook of British Mammals,* ed. G. B. Corbet & S. Harris, pp. 88–94. Oxford: Blackwell Scientific Publications.

Richman, B. 1976. Some vocal distinctive features used by gelada monkeys. *Journal of the Acoustical Society of America,* 60:718–724.

——— 1987. Rhythm and melody in gelada vocal exchanges. *Primates,* 28:569–581.

Ridgway, S. H. & Brownson, R. H. 1984. Relative brain sizes and cortical surface areas of odontocetes. *Acta Zoologica Fennica,* 172:149–152.

Rossiter, S. J., Jones, G., Ransome, R. D. & Barratt, E. M. 2000. Parentage, reproductive success and breeding behaviour in the greater horseshoe bat (*Rhinolophus ferrumequinum*). *Proceedings of the Royal Society of London,* ser. B, 267:545–551.

Russ, J. M., Racey, P. A. & Jones, G. 1998. Intraspecific responses to distress calls of the pipistrelle bat, *Pipistrellus pipistrellus. Animal Behaviour,* 55:705–713.

Rydell, J. 1993. Variation in the sonar of an aerial-hawking bat (*Eptesicus nilssonii*). *Ethology,* 93:275–284.

Scherrer, J. A. & Wilkinson, G. S. 1993. Evening bat isolation calls provide evidence for heritable signatures. *Animal Behaviour,* 46:847–860.

Shane, S. H., Wells, R. S. & Wursig, B. 1986. Ecology, behavior and social-organization of the bottle-nosed dolphin: a review. *Marine Mammal Science,* 2:34–63.

Sluiter, J. W. & van Heedt, P. F. 1966. Seasonal habits of the noctule bat (*Nyctalus noctula*). *Archives Nierlandaise de Zoologie,* 16:432–439.

Smithers, R. H. N. 1992. *Land Mammals of Southern Africa.* Cape Town: Southern Book Publishers.

Smolker, R. & Pepper, J. W. 1999. Whistle convergence among allied male bottlenose dolphins (Delphinidae, Tursiops sp.). *Ethology,* 105:595–617.

Snowdon, C. T. 1990. Language capacities of nonhuman animals. *Yearbook of Physical Anthropology,* 33:215–243.

Snowdon, C. T. & Elowson, A. M. 1999. Pygmy marmosets modify call structure when paired. *Ethology,* 105:893–908.

Storz, J. F., Bhat, H. R. & Kunz, T. H. 2000. Social structure of a polygynous tent-making bat, *Cynopterus sphinx* (Megachiroptera). *Journal of Zoology,* 251:151–165.

Suga, N., Niwa, H., Taniguchi, I. & Margoliash, D. 1987. The personalized auditory cortex of the mustached bat: adaptation for echolocation. *Journal of Neurophysiology,* 58:643–654.

Suthers, R. A. 1988. The production of echolocation signals by bats and birds. In *Animal Sonar,* ed. P. E. Nachtigall & P. W. B. Moore, pp. 23–46. New York: Plenum Press.

Thomson, C. E., Fenton, M. B. & Barclay, R. M. R. 1985. The role of infant isolation calls in mother infant reunions in the little brown bat, *Myotis lucifugus* (Chiroptera, Vespertilionidae). *Canadian Journal of Zoology,* 63:1982–1988.

Tschapka, M. & Wilkinson, G. S. 1999. Free-ranging vampire bats (*Desmodus rotundus,* Phyllostomidae) survive 15 years in the wild. *Zeitschrift für Säugetierkunde,* 64:239–240.

Tuttle, M. D. & Stevenson, D. 1982. Growth and survival of bats. In *Ecology of Bats,* ed. T. H. Kunz, pp. 105–150. New York: Plenum Press.

Vaughan, T. A. & O'Shea, T. J. 1976. Roosting ecology of the pallid bat, *Antrozous pallidus. Journal of Mammalogy,* 57:19–42.

Wang, X. 2000. On cortical coding of vocal communication sounds in primates. *Proceedings of the National Academy of Sciences USA,* 97:11843–11849.

Watt, E. M. & Fenton, M. B. 1995. DNA fingerprinting provides evidence of discriminate suckling and non-random mating in little brown bats *Myotis lucifugus. Molecular Ecology,* 4:261–264.

Whiten, A. 1999. The Machiavellian intelligence hypothesis. In *MIT Encyclopedia of the Cognitive Sciences,* ed. R. A. Wilson & F. Keil, pp. 495–497. Cambridge, MA: MIT Press.

Whiten, A. & Byrne, R. W. 1988. Taking Machiavellian intelligence apart. In *Machiavellian Intelligence: Social Expertise and the Evolution of Intellect in Monkeys, Apes, and Humans,* ed. R. W. Byrne & A. Whiten, pp. 50–66. Oxford: Oxford University Press.

Wickler, W. & Seibt, U. 1976. Field studies on the African fruit bat *Epomophorus wahlbergi* (Sundevall), with special reference to male calling. *Zeitschrift für Tierpsychologie,* 40:345–376.

Wilkinson, G. S. 1984. Reciprocal food sharing in vampire bats. *Nature,* 309:181–184.

——— 1985a. The social organization of the common vampire bat. I. Pattern and cause of association. *Behavioral Ecology and Sociobiology,* 17:111–121.

——— 1985b. The social organization of the common vampire bat. II. Mating system, genetic structure, and relatedness. *Behavioral Ecology and Sociobiology,* 17:123–134.

——— 1986. Social grooming in the common vampire bat, *Desmodus rotundus. Animal Behaviour,* 34:1880–1889.

——— 1987. Altruism and cooperation in bats. In *Recent Advances in the Study of Bats,* ed. P. A. Racey, M. B. Fenton, & J. M. V. Rayner, pp. 299–323. Cambridge: Cambridge University Press.

——— 1992. Communal nursing in evening bats. *Behavioral Ecology and Sociobiology,* 31:225–235.

——— 1995. Information transfer in bats. *Symposium of the Zoological Society of London,* 67:345–360.

Wilkinson, G. S. & Boughman, J. W. 1998. Social calls coordinate foraging in greater spear-nosed bats. *Animal Behaviour,* 55:337–350.

Wilkinson, G. S. & McCracken, G. F. 2001. Bats and balls: sexual selection and sperm competition in the Chiroptera. In *Bat Ecology,* ed. T. H. Kunz. Chicago: Chicago University Press.

Zahn, A. & Dipple, B. 1997. Male roosting habits and mating behaviour of *Myotis myotis. Journal of Zoology* (London), 2243:659–674.

13. Dolphins Communicate about Individual-Specific Social Relationships

Armstrong, E. A. 1973. *A Study of Bird Song.* New York: Dover.

Axelrod, R. & Hamilton, W. D. 1984. *The Evolution of Cooperation.* New York: Basic Books.

Beecher, M. D., Campbell, S. E., Burt, J. M., Hill, C. E. & Nordby, J. C. 2000. Song-type matching between neighboring song sparrows. *Animal Behaviour,* 59:21–27.

Beer, C. G. 1970. Individual recognition of voice in the social behavior of birds. In *Advances in the Study of Behavior,* vol. 3, ed. J. S. Rosenblatt, C. G. Beer, & R. A. Hinde, pp. 27–74. New York: Academic Press.

Brown, E. D. & Farabaugh, S. M. 1997. What birds with complex social relationships can tell us about vocal learning: vocal sharing in avian groups. In *Social Influences on Vocal Development,* ed. C. Snowdon & M. Hausberger, pp. 98–127. Cambridge: Cambridge University Press.

Byrne, R. & Whiten, A. 1988. *Machiavellian Intelligence: Social Expertise and the Evolution of Intellect in Monkeys, Apes, and Humans.* Oxford: Clarendon Press.

Caldwell, M. C. & Caldwell, D. K. 1965. Individualized whistle contours in bottlenosed dolphins (*Tursiops truncatus*). *Science,* 207:434–435.

——— 1968. Vocalizations of naïve captive dolphins in small groups. *Science,* 159:1121–1123.

——— 1972. Vocal mimicry in the whistle mode by an Atlantic bottlenosed dolphin. *Cetology,* 9:1–8.

——— 1979. The whistle of the Atlantic bottlenosed dolphin (*Tursiops truncatus*): ontogeny. In *Behavior of Marine Animals,* vol. 3, *Cetaceans,* ed. H. E. Winn & B. L. Olla, pp. 369–401. New York: Plenum Press.

Caldwell, M. C., Caldwell, D. K. & Tyack, P. L. 1990. A review of the signature whistle hypothesis for the Atlantic bottlenose dolphin, *Tursiops truncatus.* In *The Bottlenose Dolphin: Recent Progress in Research,* ed. S. Leatherwood & R. Reeves, pp. 199–234. San Diego: Academic Press.

Connor, R. C., Smolker, R. A. & Richards, A. F. 1992. Aggressive herding of females by coalitions of male bottlenose dolphins (*Tursiops* sp.). In *Coalitions and Alliances in Humans and Other Animals,* ed. A. H. Harcourt & F. B. M. de Waal, pp. 415–444. Oxford: Oxford University Press.

Elowson, M. A. & Snowdon, C. T. 1994. Pygmy marmosets, *Cebuella pygmaea,* modify vocal structure in response to changed social environment. *Animal Behaviour,* 47:703–715.

Farabaugh, S. M., Linzenbold, A. & Dooling, R. J. 1994. Vocal plasticity in budgerigars (*Melopsittacus undulatus*): evidence for social factors in the learning of contact calls. *Journal of Comparative Psychology,* 108(1):81–92.

Giles, H. 1973. Accent mobility: a model and some data. *Anthropological Linguistics,* 15:87–109.

Hausberger, M. 1997. Social influences on song acquisition and sharing in the European starling (*Sturnus vulgaris*). In *Social Influences on Vocal Development,* ed. C. T. Snowdon & M. Hausberger, pp. 128–156. New York: Cambridge University Press.

Humphrey, N. K. 1976. The social function of intelligence. In *Growing Points in Ethology,* ed. P. P. G. Bateson and R. Hinde, pp. 303–317. Cambridge: Cambridge University Press.

Janik, V. M. 2000. Whistle matching in wild bottlenose dolphins (*Tursiops truncatus*). *Science,* 289:1355–1357.

Janik, V. M. & Slater, P. J. B. 1997. Vocal learning in mammals. In *Advances in the Study of Behavior,* 26:59–99.

——— 1998. Context-specific use suggests that bottlenose dolphin signature whistles are cohesion calls. *Animal Behaviour,* 56:829–838.

Janik, V. M., Denhardt, G. & Todt, D. 1994. Signature whistle variations in a bottlenosed dolphin, *Tursiops truncatus. Behavioral Ecology and Sociobiology,* 35:243–248.

Jones, G. & Ransome, R. D. 1993. Echolocation calls of bats are influenced by maternal effects and change over a lifetime. *Proceedings of the Royal Society of London,* ser. B, 252:125–128.

Kroodsma, D. E. & Parker, L. D. 1977. Vocal virtuosity in the brown thrasher. *Auk,* 94:783–785.

Loesche, P., Stoddard, P. K., Higgins, B. J. & Beecher, M. D. 1991. Signature versus perceptual adaptations for individual vocal recognition in swallows. *Behaviour,* 118:15–25.

Mann, J. & Whitehead, H. 2000. Female reproductive strategies of cetaceans. In *Cetacean Societies: Field Studies of Whales and Dolphins,* ed. J. Mann., R. Connor, P. L. Tyack, and H. Whitehead, pp. 219–246. University of Chicago Press, Chicago.

Mayr, E. 1976. Behavior programs and evolutionary strategies. In *Evolution and the Diversity of Life: Selected Essays,* pp. 694–711. Cambridge, MA: Harvard University Press.

McBride, A. F. & Kritzler, H. 1951. Observations on the pregnancy, parturition, and postnatal behavior in the bottlenose dolphin. *Journal of Mammalogy,* 32:251–266.

McCowan, B. & Reiss, D. 1995. Quantitative comparison of whistle repertoires from captive adult bottlenose dolphins (Delphinidae, *Tursiops truncatus*): a reevaluation of the signature whistle hypothesis. *Ethology,* 100:193–209.

——— 1997. Vocal learning in captive bottlenose dolphins: a comparison with humans and nonhuman animals. In *Social Influences on Vocal Development,* ed. C. T. Snowdon & M. Hausberger, pp. 178–207. Cambridge: Cambridge University Press.

Medvin, M. B. & Beecher, M. D. 1986. Parent-offspring recognition in the barn swallow (*Hirundo rustica*). *Animal Behaviour,* 34:1627–1639.

Miller, D. E. & Emlen, J. T. 1975. Individual chick recognition and family integrity in the ring-billed gull. *Behaviour,* 52:124–144.

Nowacek, D. P. 1999. Sound use, sequential behavior and foraging ecology of foraging bottlenose dolphins, *Tursiops truncatus.* Doctoral diss., Massachusetts Institute of Technology and Woods Hole Oceanographic Institution, MIT/WHOI 99–16.

Nowicki, S. 1989. Vocal plasticity in captive black-capped chickadees: the acoustic basis and rate of call convergence. *Animal Behaviour,* 37:64–73.

Payne, R. B. & Payne, L. L. 1997. Field observations, experimental design, and the time and place of learning bird songs. In *Social Influences on Vocal Development,* ed. C. T. Snowdon and M. Hausberger, pp. 57–84. Cambridge: Cambridge University Press.

Petrie, M. & Møller, A. P. 1991. Laying eggs in others' nests: intraspecific brood parasitism in birds. *Trends in Ecology and Evolution,* 6:315–320.

Richards, D. G., Wolz, J. P. & L. M. Herman. 1984. Vocal mimicry of computer-generated sounds and vocal labeling of objects by a bottlenosed dolphin, *Tursiops truncatus. Journal of Comparative Psychology,* 98:10–28.

Robert, M. & Sorci, C. 2001. The evolution of obligate interspecific brood parasitism in birds. *Behavioral Ecology,* 12:128–133.

Sayigh, L. S., Tyack, P. L., Wells, R. S. & Scott, M. D. 1990. Signature whistles of free-ranging bottlenose dolphins, *Tursiops truncatus:* stability and mother-offspring comparisons. *Behavioral Ecology and Sociobiology,* 26:247–260.

Sayigh, L. S., Tyack, P. L., Wells, R. S., Scott, M. D. & Irvine, A. B. 1995. Sex difference in whistle production of free-ranging bottlenose dolphins, *Tursiops truncatus. Behavioral Ecology and Sociobiology,* 36:171–177.

Sayigh, L. S., Tyack, P. L., Wells, R. S., Solow, A., Scott, M. D. & Irvine, A. B. 1999. Individual recognition in wild bottlenose dolphins: a field test using playback experiments. *Animal Behaviour,* 57:41–50.

Scherrer, J. A. & Wilkinson, G. S. 1993. Evening bat isolation calls provide evidence for heritable signatures. *Animal Behaviour,* 46:847–860.

Schusterman, R. J., Gisiner, R. & Hanggi, E. B. 1992. Imprinting and other aspects of pinniped-human interactions. In *The Inevitable Bond: Examining Scientist-Animal Interactions,* ed. H. B. Davis & A. Dianne, pp. 334–356. New York: Cambridge University Press.

Shepard, C. A., Giles, H. & Le Poire, B. A. 2001. Communication accommodation theory 25 years on. In *The New Handbook of Language and Social Psychology,* ed. W. P. Robinson & H. Giles. New York: Wiley.

Smolker, R. A. & Pepper, J. W. 1999. Whistle convergence among allied male bottlenose dolphins (Delphinidae, *Tursiops* sp.). *Ethology,* 105:595–617.

Smolker, R. A., Mann, J. & Smuts, B. B. 1993. Use of signature whistles during separation and reunions by wild bottlenose dolphin mothers and infants. *Behavioral Ecology and Sociobiology,* 33:393–402.

Stoddard, P. K. & Beecher, M. D. 1983. Parental recognition of offspring in the cliff swallow. *Auk,* 100:795–799.

Symmes, D., Newman, J. D., Talmadge-Riggs, G. & Lieblich, A. K. 1979. Individuality and stability of isolation peeps in squirrel monkeys. *Animal Behaviour,* 27:1142–1152.

Thorpe, W. H. & North, M. E. W. 1966. Vocal imitation in the tropical boubou shrike, *Laniarius aethiopicus major,* as a means of establishing and maintaining social bonds. *Ibis,* 108:432–435.

Tyack, P. 1986. Whistle repertoires of two bottlenosed dolphins, *Tursiops truncatus:* mimicry of signature whistles? *Behavioral Ecology and Sociobiology,* 18:251–257.

——— 1991. Use of a telemetry device to identify which dolphin produces a sound. In *Dolphin Societies: Discoveries and Puzzles,* ed. K. Pryor & K. S. Norris, pp. 319–344. Berkeley: University of California Press.

——— 1993. Why ethology is necessary for the comparative study of language and communication. In *Language and Communication: Comparative Perspectives,* ed. H. L. Roitblat, L. M. Herman, & P. Nachtigall, pp. 115–152. Hillsdale, NJ: Erlbaum.

——— 2000. Dolphins whistle a signature tune. *Science,* 289:1310–1311.

Tyack, P. L. & Sayigh, L. S. 1997. Vocal learning in cetaceans. In *Social Influences on Vocal Development,* ed. C. Snowdon & M. Hausberger, pp. 208–233. Cambridge: Cambridge University Press.

Vehrencamp, S. L. 2001. Is song-type matching a conventional signal of aggressive intentions? *Proceedings of the Royal Society of London, Series B: Biological Sciences,* 268:1637–1642.

Watwood, S., Wells, R. & Tyack, P. L. In revision. Whistle sharing in allied male bottlenose dolphins, *Tursiops truncatus. Behavioral Ecology and Sociobiology.*

West, M. J., King, A. P. & Freeberg, T. M. 1997. Building a social agenda for the study of bird song. In *Social Influences on Vocal Development,* ed. C. Snowdon & M. Hausberger, pp. 41–56. Cambridge: Cambridge University Press.

Case Study 13A. *Natural Semanticity in Wild Primates*

Cheney, D. L. & Seyfarth, R. M. 1988. Assessment of meaning and the detection of unreliable signals by vervet monkeys. *Animal Behaviour,* 36(2):477–486.

——— 1990. *How Monkeys See the World: Inside the Mind of Another Species.* Chicago: Chicago University Press.

Deacon, T. W. 1997. *The Symbolic Species.* New York: Norton.

Eimas, P. D., Siqueland, P., Jusczyk, P. & Vigorito, J. 1971. Speech perception in infants. *Science,* 171:303–306.

Premack, D. & Dasser, V. 1991. Perceptual origins and conceptual evidence for theory of mind in apes and children. In *Natural Theories of Mind,* ed. A. Whiten, pp. 253–266. Oxford: Blackwell.

Seyfarth, R. M., Cheney, D. L. & Marler, P. 1980. Vervet monkey alarm calls: semantic communication in a free-ranging primate. *Animal Behaviour,* 28(4):1070–1094.

Tomasello, M. & Call, J. 1997. *Primate Cognition.* New York: Oxford University Press.

Tomasello, M. & Zuberbühler, K. 2001. Primate vocal and gestural communication. In *The Cognitive Animal,* ed. M. Bekoff.

Zuberbühler, K. 2000a. Referential labeling in wild Diana monkeys. *Animal Behaviour,* 59:917–927.

——— 2000b. Interspecies semantic communication in two forest monkeys. *Proceedings of the Royal Society of London,* ser. B, 267:713–718.

——— 2001. Predator-specific alarm calls in Campbell's monkeys. *Behavioral Ecology and Sociobiology.*

Zuberbühler, K., Noë, R. & Seyfarth, R. M. 1997. Diana monkey long-distance calls: messages for conspecifics and predators. *Animal Behaviour,* 53:589–604.

Zuberbühler, K., Cheney, D. L. & Seyfarth, R. M. 1999. Conceptual semantics in a nonhuman primate. *Journal of Comparative Psychology,* 113:33–42.

14. Koshima Monkeys and Bossou Chimpanzees

Boesch, C. & Boesch-Achermann, H. 2000. *The Chimpanzees of the Taï Forest: Behavioural Ecology and Evolution.* Oxford: Oxford University Press.

de Waal, F. B. M. 2001. *The Ape and the Sushi Master: Cultural Reflections by a Primatologist.* New York: Basic Books.

Galef, B. G. 1990. The question of animal culture. *Human Nature,* 3:157–178.

Goodall, J. 1986. *The Chimpanzees of Gombe: Patterns of Behavior.* Cambridge, MA: The Belknap Press of Harvard University Press.

Hirata, S., Myowa, M. & Matsuzawa, T. 1998. Use of leaves as cushions to sit on wet ground by wild chimpanzees. *American Journal of Primatology,* 44:215–220.

Hirata, S., Yamakoshi, G., Fujita, S., Ohashi, G. & Matsuzawa, T. 2001a. Capturing and toying with hyraxes *(Dendrohyrax dorsalis)* by wild chimpanzees (*Pan troglodytes*) at Bossou, Guinea. *American Journal of Primatology,* 53:93–97.

Hirata, S., Watanabe, K. & Kawai, M. 2001b. "Sweet-potato washing" revisited. In *Primate Origins of Human Cognition and Behavior,* ed. T. Matsuzawa, pp. 487–508. Berlin: Springer.

Imanishi, K. 2001. *A Japanese View of Nature: The World of Living Things.* Richmond, UK: Curzon Press. Translation of *Seibutsuno Sekai* (1941).

Inoue-Nakamura, N. & Matsuzawa, T. 1996. Development of stone tool use by wild chimpanzees *(Pan troglodytes). Journal of Comparative Psychology,* 111:159–173.

Itani, J. 1963. Vocal communication of the wild Japanese monkey. *Primates,* 4(2):11–66.

Kawai, M. 1965. Newly acquired pre-cultural behavior of the natural troop of Japanese monkeys on Koshima islet. *Primates,* 6:1–30.

Kortlandt, A. 1962. Chimpanzees in the wild. *Scientific American,* 206:128–138.

——— 1986. The use of stone tools by wild-living chimpanzees and earliest hominids. *Journal of Human Evolution,* 15:77–132.

Mann, A. 1972. Hominid and cultural origins. *Man,* 7:379–386.

Matsuzawa, T. 1985. Use of numbers by a chimpanzee. *Nature,* 315:57–59.

——— 1994. Field experiments on the use of stone tools by chimpanzees in the wild. In *Chimpanzee Cultures,* ed. R. Wrangham, F. de Waal, & P. Heltne, pp. 169–209. Cambridge: Cambridge University Press.

——— 1996. Chimpanzee intelligence in nature and in captivity: isomorphism of symbol use and tool use. In *Great Ape Society,* ed. W. McGrew, L. Marchant, & T. Nishida, pp. 196–209. Cambridge: Cambridge University Press.

——— 1999. Communication and tool use in chimpanzees: cultural and social context. In *The Design of Animal Communication,* ed. M. Hauser & M. Konishi, pp. 645–671. Cambridge, MA: MIT Press.

——— 2001. *Primate Origins of Human Cognition and Behavior.* Berlin: Springer.

McGrew, W. C. 1992. *Chimpanzee Material Culture.* Cambridge: Cambridge University Press.

Nishida, T. 1990. *The Chimpanzees of the Mahale Mountains.* Tokyo: Tokyo University Press.

Sugiyama, Y. & Koman, J. 1979. Tool-using and making behavior in wild chimpanzees at Bossou, Guinea. *Primates,* 20:513–524.

Tomasello, M., Kruger, A. C. & Rander, H. H. 1993. Cultural learning. *Behavioral and Brain Sciences,* 16:495–552.

Watanabe, K. 2001. A review of 50 years of research on the Japanese monkeys of Koshima: Status and dominance. In *Primate Origins of Human Cognition and Behavior,* ed. T. Matsuzawa, pp. 405–417. Berlin: Springer.

Watanabe, K., Mori, A. & Kawai, M. 1992. Characteristic features of the reproduction in Koshima monkeys, *Macaca fuscata fuscata:* a 34-year summary. *Primates,* 33:1–32.
Whiten, A., Goodall, J., McGrew, W., Nishida, T., Reynolds, V., Sugiyama, Y., Tutin, C., Wrangham, R. & Boesch, C. 1999. Cultures in chimpanzees. *Nature,* 399:682–685.
Wrangham, R., McGrew, W., de Waal, F. & Heltne, P. 1994. *Chimpanzee Cultures.* Cambridge, MA: Harvard University Press.
Yamakoshi, G. 1998. Dietary responses to fruit scarcity of wild chimpanzees at Bossou, Guinea: possible implications for ecological importance of tool use. *American Journal of Physical Anthroplogy,* 106:283–295.
——— 2001. Ecology of tool use in wild chimpanzees: toward reconstruction of early hominid evolution. In *Primate Origins of Human Cognition and Behavior,* ed. T. Matsuzawa, pp. 537–556. Berlin: Springer.

Case Study 14A. *Movement Imitation in Monkeys*

Bugnyar, T. & Huber, L. 1997. Push or pull: an experimental study on imitation in marmosets *(Callithrix jacchus). Animal Behaviour,* 54:817–831.
Custance, D., Whiten, A. & Fredman, T. 1999. Social learning of an artificial fruit task in capuchin monkeys *(Cebus apella). Journal of Comparative Psychology,* 113:13–23.
Gibson, J. J. 1979. *The Ecological Approach to Visual Perception.* Boston: Houghton Mifflin.
Heyes, C. M. & Galef, B. G., Jr. 1996. *Social Learning in Animals: The Roots of Culture.* San Diego: Academic Press.
Heyes, C. M. & Ray, E. D. 2000. What is the significance of imitation in animals? *Advances in the Study of Behavior,* 29:215–245.
Spence, K. W. 1937. Experimental studies of learning and higher mental processes in infra-human primates. *Psychological Bulletin,* 34:806–850.
Tayler, C. K. & Saayman, G. S. 1973. Imitative behaviour by Indian Ocean bottle nose dolphins *(Tursiops adunctus)* in captivity. *Behaviour,* 44:286–298.
Thorpe, W. H. 1963. *Learning and Instinct in Animals,* 2d ed. London: Methuen.
Voelkl, B. & Huber, L. 2000. True imitation in marmosets. *Animal Behaviour,* 60:195–202.

15. Individuality and Flexibility of Cultural Behavior Patterns in Chimpanzees

Boesch, C. & Boesch, H. 1989. Hunting behavior of wild chimpanzees in the Taï National Park. *American Journal of Physical Anthropology,* 78:547–573.
Boesch, C. & Boesch-Achermann, H. 2000. *The Chimpanzees of the Taï Forest.* Oxford: Oxford University Press.

Busse, C. 1978. Do chimpanzees hunt cooperatively? *American Naturalist,* 112:767-770.

de Waal, F. B. M. 2001. *The Ape and the Sushi Master: Cultural Reflections by a Primatologist.* New York: Basic Books.

Goodall, J. 1971. *In the Shadow of Man.* Boston: Houghton Mifflin.

——— 1973. Cultural elements in a chimpanzee community. In *Precultural Primate Behavior,* ed. E. W. Menzel, pp. 144–184. Basel: Karger.

——— 1986. *The Chimpanzees of Gombe: Patterns of Behavior.* Cambridge, MA: The Belknap Press of Harvard University Press.

Hamilton, W. 1964. The genetical evolution of social behavior. *Journal of Theoretical Biology,* 7:1–52.

Heyes, C. M. 1993. Imitation, culture and cognition. *Animal Behaviour,* 46:999–1010.

Hosaka, K., Nishida, T., Hamai, M., Matsumoto-Oda, A. & Uehara, S. 2001. Predation of mammals by the chimpanzees of the Mahale Mountains, Tanzania. In *All Apes Great and Small,* vol. 1: *Chimpanzees, Bonobos, and Gorillas,* ed. B. M. F. Galdikas, N. Briggs, L. K. Sheeran, G. L. Shapiro, & J. Goodall. New York: Kluwer.

Huffman, M. A. 1997. Current evidence for self-medication in primates: a multidisciplinary perspective. *Yearbook of Physical Anthropology,* 40:171–200.

Hunt, K. D. 2000. Initiation of a new chimpanzee study site at Semliki-Toro Wildlife Reserve, Uganda. *Pan Africa News,* 7:14–16.

Imanishi, K. 1952. Evolution of humanity. In *Man,* ed. K. Imanishi, pp. 36–94. Tokyo: Mainichi-Shinbunsha.

——— 1957. Identification: a process of enculturation in the subhuman society of *Macaca fuscata. Primates,* 1:1–29.

——— 1961. The origin of human family: a primatological approach. *Japanese Journal of Ethnology,* 25:119–138.

——— 1974 [1941]. *The World of Living Things.* Tokyo: Kodansha.

Kawai, M. 1964. *The Ecology of Japanese Monkeys.* Tokyo: Kawadeshobo.

——— 1965. Newly acquired pre-cultural behavior of the natural troop of Japanese monkeys on Koshima Islet. *Primates,* 6:1–30.

Kawamura, S. 1959. The process of subculture propagation among Japanese macaques. *Primates,* 2:43–60.

——— 1965. Sub-culture among Japanese macaques. In *Monkeys and Apes: Sociological Studies,* ed. S. Kawamura & J. Itani, pp. 239–289. Tokyo: Chuokoronsha.

Kummer, H. & Goodall, J. 1985. Conditions of innovative behavior in primates. *Philosophical Transactions of the Royal Society of London,* ser. B, 308:203–214.

McGrew, W. C. 1992. *Chimpanzee Material Culture.* Cambridge: Cambridge University Press.

——— 2000. The chimpanzees of Mt. Assirik revisited. Presentation at the Conference of the Behavioral Diversity of Chimpanzees and Bonobos, Seewiesen, Germany, June 2000.

McGrew, W. C. & Tutin, C. E. G. 1978. Evidence for a social custom in wild chimpanzees? *Man* (n.s.), 13:234–251.

McGrew, W. C., Marchant, L. F., Scott, W. E. & Tutin, C. E. G. 2001. Intergroup differences in a social custom of wild chimpanzees: the grooming hand clasp of the Mahale chimpanzees. *Current Anthropology,* 42:148–153.

Nakamura, M. In press. Grooming-hand-clasp in Mahale chimpanzees: Comparison with branch-clasp grooming. In *The Behavioral Diversity of Chimpanzees and Bonobos,* ed. C. Boesch, G. Hohmann, & L. Marchant. Cambridge: Cambridge University Press.

Nakamura, M., McGrew, W. C., Marchant, L. & Nishida, T. 2000. Social scratch: another custom in wild chimpanzees? *Primates,* 41:237–248.

Nishida, T. 1980a. The leaf-clipping display: a newly-discovered expressive gesture in wild chimpanzees. *Journal of Human Evolution,* 9:117–128.

——— 1980b. Local differences in responses to water among wild chimpanzees. *Folia primatologica,* 33:189–209.

——— 1987. Local traditions and cultural transmission. In *Primate Societies,* ed. B. B. Smuts, D. L. Cheney, R. M. Seyfarth, R. W. Wrangham, & T. T. Struhsaker, pp. 462–474. Chicago: University of Chicago Press.

——— 1988. Development of social grooming between mother and offspring in wild chimpanzees. *Folia primatologica,* 50:109–123.

——— ed. 1990. *The Chimpanzees of the Mahale Mountains.* Tokyo: University of Tokyo Press.

——— 1994. Review of recent findings on Mahale chimpanzees: implications and future research directions. In *Chimpanzee Cultures* (R. W. Wrangham, W. C. McGrew, F. B. M. de Waal & P. Heltne, eds.), pp. 373–396. Cambridge, MA: Harvard University Press.

——— 1997. Sexual behavior of adult male chimpanzees of the Mahale Mountains National Park, Tanzania. *Primates,* 38:379–398.

——— 1999. *Where Has Human Nature Come From?* Kyoto: Kyoto University Press.

Nishida, T. & Nakamura, M., 1993. Chimpanzee tool use to clear a blocked nasal passage. *Folia primatologica,* 61:218–220.

Nishida, T., Wrangham, R. W., Goodall, J., and Uehara, S. 1983. Local differences in plant feeding habits of chimpanzees between the Mahale Mountains and Gombe National Park, Tanzania. *Journal of Human Evoution,* 12:467–480.

Nishida, T., Hasegawa, T., Hayaki, H., Takahata, Y. & Uehara, S. 1992. Meat-sharing as a coalition strategy by an alpha male chimpanzee? In *Topics in Primatology,* vol. 1: *Human Origins,* ed. T. Nishida, W. C. McGrew, P. Marler, M. Pickford, & F. B. M. de Waal, pp. 159–174. Kyoto: Kyoto University Press.

Nishida, T., Kano, T., Goodall, J., McGrew, W. C. & Nakamura, M. 1999. Ethogram and ethnography of Mahale chimpanzees. *Anthropological Science,* 107:141–188.

Premack, D. & Premack, A. J. 1994. Why animals have neither culture nor history. In *Companion's Encyclopedia of Anthropology,* ed. T. Ingold, pp. 350–365. New York: Routledge.

Sakamaki, T. 1998. First record of algae feeding by a female chimpanzee at Mahale. *Pan Africa News,* 5:1–3.

Stanford, C. B. 1998. *Chimpanzee and Red Colobus.* Cambridge, MA: Harvard University Press.

Takahata, Y., Hasegawa, T. & Nishida, T. 1984. Chimpanzee predation in the Mahale Mountains from August, 1979 to May, 1982. *International Journal of Primatology,* 5:213–233.

Takahata, Y., Hiraiwa-Hasegawa, M., Takasaki, H. & Nyundo, R. 1986. Newly acquired feeding habits among the chimpanzees of the Mahale Mountains National Park, Tanzania. *Human Evolution,* 1:277–284.

Tomasello, M. & Call, J. 1997. *Primate Cognition.* Oxford: Oxford University Press.

Whiten, A., Goodall, J., McGrew, W. C., Nishida, T., Reynolds, V., Sugiyama, Y., Tutin, C. E. G., Wrangham, R. W. & Boesch, C. 1999. Cultures in chimpanzees. *Nature,* 399:682–685.

Zamma, K. In press. Why do wild chimpanzees groom leaves? New evidence from Mahale. *Primates.*

CASE STUDY 15A. *Sex Differences in Termite Fishing among Gombe Chimpanzees*

Boesch, C. & Boesch, H. 1984. Possible causes of sex differences in the use of natural hammers by wild chimpanzees. *Journal of Human Evolution,* 13:415–440.

Clutton-Brock, T. H. 1977. Some aspects of intraspecific variation in feeding and ranging behaviour in primates. *Primate Ecology: Studies of Feeding and Ranging Behaviour in Lemurs, Monkeys and Apes,* pp. 539–556. New York: Academic Press.

Demment, M. 1983. Feeding ecology and the evolution of body size of baboons. *African Journal of Ecology,* 21:219–233.

Goodall, J. 1986. *The Chimpanzees of Gombe: Patterns of Behavior.* Cambridge, MA: The Belknap Press of Harvard University Press.

McGrew, W. C. 1979. Evolutionary implications of sex differences in chimpanzee predation and tool use. In *The Great Apes,* ed. D. A. Hamburg & E. R. McCown, pp. 441–463. Menlo Park, CA: Benjamin/Cummings.

——— 1992. *Chimpanzee Material Culture: Implications for Human Evolution.* Cambridge: Cambridge University Press.

——— 2001. The other faunivory: primate insectivory and early human diet. In *Meat-eating and Human Evolution,* ed. C. B. Stanford & H. T. Bunn, pp. 160–178. Oxford: Oxford University Press.

McGrew, W. C. & Marchant, L. F. 1999. Laterality of hand use pays off in foraging success for wild chimpanzees. *Primates,* 40(3):509–513.

Pandolfi, S. S., van Schaik, C. P. & Pusey, A. E. In preparation. Differential socioecological strategies affect sex differences in vegetable diet among Gombe chimpanzees.

Trivers, R. L. 1972. Parental investment and sexual selection. In *Sexual Selection and the Descent of Man, 1871–1971,* ed. B. Campbell, pp. 136–179. Chicago: Aldine.

van Schaik, C. P. 1999. The socioecology of fission-fusion sociality in orangutans. *Primates* 40(1):69–86.

Wrangham, R. W. 1996. Social ecology of Kanyawara chimpanzees: implications for understanding the costs of great ape groups. In *Great Ape Societies,* ed. W. C. McGrew, L. F. Marchant, & T. Nishida, pp. 45–57. Cambridge: Cambridge University Press.

16. Ten Dispatches from the Chimpanzee Culture Wars

Alp, R. 1997. "Stepping-sticks" and "seat-sticks": new types of tools used by wild chimpanzees (*Pan troglodytes*) in Sierra Leone. *American Journal of Primatology,* 51:45–52.

Benedict, R. 1935. *Patterns of Culture.* London: Routledge and Kegan Paul.

Bonner, J. T. 1980. *The Evolution of Culture in Animals.* Princeton, NJ: Princeton University Press.

Brown, D. E. 1991. *Human Universals.* New York: McGraw-Hill.

Byrne, R. W. & Russon, A. E. 1998. Learning by imitation: a hierarchical approach. *Behavioral and Brain Sciences,* 21:667–721.

Byrne, R. W. & Whiten, A., eds. 1988. *Machiavellian Intelligence: Social Expertise and the Evolution of Intellect in Monkeys, Apes and Humans.* Oxford: Clarendon Press.

Cronk, L. 1999. *That Complex Whole: Culture and the Evolution of Human Behavior.* Boulder, CO: Westview.

de Waal, F. 1996. *Good Natured: The Origins of Right and Wrong in Humans and Other Animals.* Cambridge, MA: Harvard University Press.

——— 2001. *The Ape and the Sushi Master: Cultural Reflections of a Primatologist.* New York: Basic Books.

Hewlett, B. S. & Cavalli-Sforza, L. L. 1986. Cultural transmission among Aka pygmies. *American Anthropologist,* 88:922–934.

Hirata, S., Watanabe, K. & Kawai, M. 2001. "Sweet-potato washing" revisited. In *Primate Origins of Human Cognition and Behavior,* T. Matsuzawa, pp. 487–508. Tokyo: Springer.

Hockett, C. F. 1960. The origin of speech. *Scientific American,* 203(9):89–96.

Imanishi, K. 1952. Evolution of humanity. In *Man,* ed. K. Imanishi. Tokyo: Mainichi-Shinbursha.

Kroeber, A. L. 1928. Sub-human cultural beginnings. *Quarterly Review of Biology,* 3:325–342.

Kummer, H. 1971. *Primate Societies.* Chicago: Aldine-Atherton.

——— 1995. *In Quest of the Sacred Baboon: A Scientist's Journey.* Princeton, NJ: Princeton University Press.

Kuper, A. 1999. *Culture: The Anthropologists' Account.* Cambridge, MA: Harvard University Press.

Lanjouw, A. 2002. Tool use in Tongo chimpanzees. In *Behavioral Diversity in Chimpanzees and Bonobos,* ed. C. Boesch, G. Hohmann, & L. F. Marchant. Cambridge: Cambridge University Press.

Marshall, A. J., Wrangham, R. W. & Arcadi, A. C. 1999. Does vocal learning affect the structure of vocalizations in chimpanzees? *Animal Behaviour,* 58:825–830.

McGrew, W. C. 1998. Culture in nonhuman primates? *Annual Review of Anthropology,* 27:301–328.

——— 2001. The nature of culture: prospects and pitfalls of cultural primatology. In *Tree of Origin: What Primate Behavior Can Tell Us about Human Social Evolution,* ed. F. B. M. de Waal, pp. 229–254. Cambridge, MA: Harvard University Press.

McGrew, W. C. & Tutin, C. E. G 1978. Evidence for a social custom in wild chimpanzees? *Man,* 13:234–251.

McGrew, W. C., Marchant, L. F., Scott, S. E. & Tutin, C. E. G 2001. Intergroup differences in a social custom of wild chimpanzees: the grooming-hand-clasp of the Mahale Mountains. *Current Anthropology,* 42:148–153.

Menzel, E. W., ed. 1973. *Precultural Primate Behavior.* Basel: S. Karger.

Morgan, L. H. 1868. *The American Beaver and His Works.* Philadelphia: Lippincott.

Mundinger, P. C. 1980. Animal cultures and a general theory of cultural evolution. *Ethology and Sociobiology,* 1:182–223.

Nakamichi, M., Kato, E., Kojima, Y. & Itoigawa, N. 1998. Carrying and washing of grass roots by free-ranging Japanese macaques at Katsuyama. *Folia primatologica,* 69:35–40.

Nakamura, M., McGrew, W. C., Marchant, L. F. & Nishida, T. 2000. Social scratch: another custom in wild chimpanzees? *Primates,* 41:237–248.

Nishida, T. 1987. Local traditions and cultural transmission. In *Primate Societies,* ed. B. B. Smuts, D. L. Cheney, R. M. Seyfarth, R. W. Wrangham, & T. T. Struhsaker, pp. 462–474. Chicago: University of Chicago Press.

Opie, I. & Opie, P. 1987. *The Language and Lore of School Children.* New York: Oxford University Press.

Rendall, L. & Whitehead, H. 2001. Culture in whales and dolphins. *Behavioral and Brain Sciences,* 24:309–324.

Savage-Rumbaugh, E. S. 1998. Scientific schizophrenia with regard to the language act. In *Piaget, Evolution, and Development,* ed. J. Langer & M. Killen, pp. 145–169. Mahwah, NJ: Lawrence Erlbaum Associates.

Stanford, C. B. 1998. The social behavior of chimpanzees and bonobos: empirical evidence and shifting assumptions. *Current Anthropology,* 39:399–420.

Strassmann, J. E., Zhu, Y. & Queller, D. C. 2000. Altruism and social cheating in the social amoeba *Dictyostelium discoideum. Nature,* 408:965–967.

Tomasello, M. 1999. The human adaptation for culture. *Annual Review of Anthropology,* 28:509–529.

Tylor, E. B. 1871. *Primitive Culture.* London: Murray.

Washburn, S. L. & Benedict, B. 1979. Non-human primate culture. *Man,* 14:163–164.

West, M. J. & King, A. P. 1996. Social learning: synergy and songbirds. In *Social Learning in Animals: The Roots of Culture,* ed. C. M. Heyes & B. G. Galef, pp. 155–178. San Diego: Academic Press.

Wheatley, B. P. 1999. *The Sacred Monkeys of Bali.* Prospect Heights, IL: Waveland Press.

Whiten, A. & Ham, R. M. 1993. On the nature and evolution of imitation in the animal kingdom: reappraisal of a century of research. *Advances in the Study of Behavior,* 21:239–283.

Whiten, A., Goodall, J., McGrew, W. C., Nishida, T., Reynolds, V., Sugiyama, Y., Tutin, C. E. G., Wrangham, R. W. & Boesch, C. 1999. Cultures in chimpanzees. *Nature,* 399:682–685.

Woolfenden, G. E. & Fitzpatrick, J. W. 1984. *The Florida Scrub Jay.* Princeton, NJ: Princeton University Press.

Wrangham, R. W. 1999. Evolution of coalitionary killing. *Yearbook of Physical Anthropology,* 42:1–30.

Yamakoshi, G. & Sugiyama, Y. 1995. Pestle-pounding behavior of wild chimpanzees at Bossou, Guinea: a newly observed tool-using behavior. *Primates,* 36:489–500.

Case Study 16A. *Spontaneous Use of Tools by Semifree-ranging Capuchin Monkeys*

Boesch, C. 1991. Teaching among wild chimpanzees. *Animal Behaviour,* 41:530–532.

Boesch, C. & Boesch, H. 1983. Optimization of nut-cracking with natural hammers by wild chimpanzees. *Behavior,* 83:265–286.

Fernandes, M. E. B. 1991. Tool use and predation of oysters (*Crassostrea rhizophorae*) by a tufted capuchin, *Cebus apella apella,* in brackish water mangrove swamp. *Primates,* 32(4):529–531.

Fragaszy, D. M. & Adams-Curtis, L. E. 1997. Developmental changes in manipulation in tufted capuchins (*Cebus apella*) from birth through 2 years and their relation to foraging and weaning. *Journal of Comparative Psychology,* 111(2):201–211.

Inoue-Nakamura, N. & Matzuzawa, T. 1997. Development of stone tool use by wild chimpanzees (*Pan troglodytes*). *Journal of Comparative Psychology,* 111(2):159–173.

Langguth, A. & Alonso, C. 1997. Capuchin monkeys in the Caatinga: tool use and food habits during drought. *Neotropical Primates,* 5(3):77–78.

Matsuzawa, T. 1994. Field experiments on use of stone tools by chimpanzees in the wild. In *Chimpanzee Cultures,* ed. R. W. Wrangham, W. C. McGrew, F. B. M. de Waal, & P. G. Heltne, pp. 351–370. Cambridge, MA: Harvard University Press.

Ottoni, E. B. & Mannu, M. 2001. Semifree-ranging tufted capuchin monkeys (*Cebus apella*) spontaneously use tools to crack open nuts. *International Journal of Primatology,* 22(3):347–358.

van Schaik, C. P., Deaner, R. O. & Merrill, M. Y. 1999. The conditions for tool use in primates: implications for the evolution of material culture. *Journal of Human Evolution,* 36:719–741.

Visalberghi, E. 1987. Acquisition of nut-cracking behaviour by 2 capuchin monkeys (*Cebus apella*). *Folia primatologica,* 49:168–181.

Westergaard, G. C. & Suomi, S. J. 1993. Use of a tool-set by capuchin monkeys. *Primates,* 34(4):459–462.

17. Society and Culture in the Deep and Open Ocean

Amos, W. 1999. Culture and genetic evolution in whales. *Science,* 284:2055a.

André, M. & Potter, J. R. 2000. Fast-ferry acoustic and direct physical impact on cetaceans: evidence, trends and potential mitigation. In *Proceedings of the Fifth European Conference on Underwater Acoustics, ECUA 2000,* ed. M. E. Zakharia, P. Chevret, & P. Dubail, pp. 491–496. Lyon, France: ECUA.

Arnbom, T., Papastavrou, V., Weilgart, L. S. & Whitehead, H. 1987. Sperm whales react to an attack by killer whales. *Journal of Mammalogy,* 68:450–453.

Avise, J. C., Neigel, J. E. & Arnold, J. 1984. Demographic influences on mitochondrial DNA lineage survivorship in animal populations. *Journal of Molecular Evolution,* 20:99–105.

Baird, R. W. 2000. The killer whale: foraging specializations and group hunting. In *Cetacean Societies,* ed. J. Mann, R. C. Connor, P. Tyack, & H. Whitehead, pp. 127–153. Chicago: University of Chicago Press.

Barrett-Lennard, L. G., Yurk, H. & Ford, J. K. B. 2001. A sound approach to the study of culture. *Behavioral and Brain Sciences,* 24:325–326.

Beran, M. J., Gibson, K. R. & Rumbaugh, D. M. 1999. Predicted hominid performance on the transfer index: body size and cranial capacity as predictors of transfer ability. In *The Descent of Mind,* ed. M. Corballis & S. E. G. Lea, pp. 87–97. Oxford: Oxford University Press.

Bérubé, M., Aguilar, A., Dendanto, D., Larsen, F., Notarbartolo di Sciara, G., Sears, R., Sigurjónsson, J., Urban, R. J. & Palsbøll, P. J. 1998. Population genetic structure of North Atlantic, Mediterranean Sea and Sea of Cortez fin whales, *Balaenoptera physalus* (Linnaeus 1758): analysis of mitochondrial and nuclear loci. *Molecular Ecology,* 7:585–599.

Best, P. B. 1979. Social organization in sperm whales, *Physeter macrocephalus.* In *Behavior of Marine Animals,* vol. 3, ed. H. E. Winn & B. L. Olla, pp. 227–289. New York: Plenum.

Best, P. B., Canham, P. A. S. & Macleod, N. 1984. Patterns of reproduction in sperm whales, *Physeter macrocephalus. Reports of the International Whaling Commission* (Special Issue), 6:51–79.

Boesch, C. 2001. Sacrileges are welcome in science! Opening a discussion about culture in animals. *Behavioral and Brain Sciences,* 24:327–328.

Boran, J. R. & Heimlich, S. L. 1999. Social learning in cetaceans: hunting, hearing and hierarchies. *Symposia of the Zoological Society* (London), 73:282–307.

Boughman, J. W. & Wilkinson, G. S. 1998. Greater spear-nosed bats discriminate group mates by vocalizations. *Animal Behaviour,* 55:1717–1732.

Boyd, R. & Richerson, P. 1985. *Culture and the Evolutionary Process.* Chicago: Chicago University Press.

——— 1996. Why culture is common, but cultural evolution is rare. *Proceedings of the British Academy,* 88:77–93.

Byrne, R. & Whiten, A. 1988. *Machiavellian Intelligence: Social Expertise and the Evolution of Intellect in Monkeys, Apes, and Humans.* Oxford: Clarendon.

Caro, T. M. & Hauser, M. D. 1992. Is there teaching in non-human animals? *Quarterly Review of Biology,* 67:151–174.

Cavalli-Sforza, L. L. & Feldman, M. W. 1981. *Cultural Transmission and Evolution: A Quantitative Approach.* Princeton, NJ: Princeton University Press.

Cavalli-Sforza, L. L., Feldman, M. W., Chen, K. H. & Dornbusch, S. M. 1982. Theory and observation in cultural transmission. *Science,* 218:19–27.

Christal, J. 1998. An analysis of sperm whale social structure: patterns of association and genetic relatedness. Ph.D. diss., Dalhousie University, Halifax, Nova Scotia.

Christal, J. & Whitehead, H. 2001. Social affiliations within sperm whale (*Physeter macrocephalus*) groups. *Ethology,* 107:323–340.

Christal, J., Whitehead, H. & Lettevall, E. 1998. Sperm whale social units: variation and change. *Canadian Journal of Zoology,* 76:1431–1440.

Clarke, M. R. 1977. Beaks, nets and numbers. *Symposia of the Zoological Society of London,* 38:89–126.

Connor, R. C. 2000. Group living in whales and dolphins. In *Cetacean Societies,* ed.

J. Mann, R. C. Connor, P. L. Tyack, & H. Whitehead, pp. 199–218. Chicago: University of Chicago Press.

——— 2001. Individual foraging specializations in marine mammals: culture and ecology. *Behavioral and Brain Sciences,* 24:329–330.

Connor, R. C., Mann, J., Tyack, P. L. & Whitehead, H. 1998. Social evolution in toothed whales. *Trends in Ecology and Evolution,* 13:228–232.

Dalebout, M. L., Hooker, S. K. & Christensen, I. 2001. Genetic diversity and population structure among northern bottlenose whales, *Hyperoodon ampullatus,* in the western North Atlantic. *Canadian Journal of Zoology,* 79:478–484.

Deecke, V. B., Ford, J. K. B. & Spong, P. 2000. Dialect change in resident killer whales: implications for vocal learning and cultural transmission. *Animal Behaviour,* 40:629–638.

de Waal, F. 2000. *The Ape and the Sushi Master: Cultural Reflections of a Primatologist.* New York: Basic Books.

Dufault, S. & Whitehead, H. 1998. Regional and group-level differences in fluke markings and notches of sperm whales. *Journal of Mammalogy,* 79:514–520.

Gordon, J. C. D. 1987. Sperm whale groups and social behaviour observed off Sri Lanka. *Reports of the International Whaling Commission,* 37:205–217.

Gordon, J. C. D., Moscrop, A., Carlson, C., Ingram, S., Leaper, R., Matthews, J. & Young, K. 1998. Distribution, movements and residency of sperm whales off the Commonwealth of Dominica, Eastern Caribbean: implications for the development and regulation of the local whale watching industry. *Reports of the International Whaling Commission,* 48:551–557.

Guinet, C. & Bouvier, J. 1995. Development of intentional stranding hunting techniques in killer whale (*Orcinus orca*) calves at Crozet Archipelago. *Canadian Journal of Zoology,* 73:27–33.

Haase, B. & Félix, F. 1994. A note on the incidental mortality of sperm whales (*Physeter macrocephalus*) in Ecuador. *Reports of the International Whaling Commission* (Special Issue), 15:481–483.

Herman, L. M. In press. Vocal, social, and self-imitation by bottlenosed dolphins. In *Imitation in Animals and Artifacts,* ed. C. Nehaniv & K. Dautenhahn. Cambridge, MA: MIT Press.

Hewlett, B. S. & Cavalli-Sforza, L. L. 1986. Cultural transmission among Aka pygmies. *American Anthropologist,* 88:922–934.

Hooker, S. K. & Baird, R. W. 1999. Deep-diving behaviour of the northern bottlenose whale, *Hyperoodon ampullatus* (Cetacea: Ziphiidae). *Proceedings of the Royal Society of London,* ser. B, 266:671–676.

Humphrey, N. K. 1976. The social function of intellect. In *Growing Points in Ethology,* ed. P. P. G. Bateson & R. A. Hinde, pp. 303–317. Cambridge: Cambridge University Press.

Jefferson, T. A., Stacey, P. J. & Baird, R. W. 1991. A review of killer whale interactions with other marine mammals: predation to co-existence. *Mammal Review,* 4:151–180.

Jefferson, T. A., Leatherwood, S. & Webber, M. A. 1993. *Marine Mammals of the World.* Rome: UNEP.

Kahn, B. 1991. The population biology and social organization of sperm whales *(Physeter macrocephalus)* off the Seychelles: indications of recent exploitation. M.Sc. thesis, Dalhousie University, Halifax, Nova Scotia.

Kato, H. 1984. Observation of tooth scars on the head of male sperm whale, as an indication of intra-sexual fightings. *Scientific Reports of the Whales Research Institute,* 35:39–46.

Laland, K. N. 1992. A theoretical investigation of the role of social transmission in evolution. *Ethology and Sociobiology,* 13:87–113.

Law, R. J., Stringer, R. L., Allchin, C. R. & Jones, B. R. 1996. Metals and organochlorines in sperm whales (*Physeter macrocephalus*) stranded around the North Sea during the 1994/1995 winter. *Marine Pollution Bulletin,* 32:72–77.

Lucas, Z. N. & Hooker, S. K. 2000. Cetacean strandings on Sable Island, Nova Scotia, 1970–1998. *Canadian Field Naturalist,* 114:45–61.

Lyrholm, T. & Gyllensten, U. 1998. Global matrilineal population structure in sperm whales as indicated by mitochondrial DNA sequences. *Proceedings of the Royal Society of London,* ser. B, 265:1679–1684.

Lyrholm, T., Leimar, O. & Gyllensten, U. 1996. Low diversity and biased substitution patterns in the mitochondrial DNA control region of sperm whales: implications for estimates of time since common ancestry. *Molecular Biology and Evolution,* 13:1318–1326.

Lyrholm, T., Leimar, O., Johanneson, B. & Gyllensten, U. 1999. Sex-biased dispersal in sperm whales: contrasting mitochondrial and nuclear genetic structure of global populations. *Proceedings of the Royal Society of London,* ser. B, 266:347–354.

Maestripieri, D. & Whitham, J. 2001. Teaching in marine mammals? Anecdotes vs. science. *Behavioral and Brain Sciences,* 24:342–343.

Marsh, H. & Kasuya, T. 1986. Evidence for reproductive senescence in female cetaceans. *Reports of the International Whaling Commission* (Special Issue), 8:57–74.

Maynard Smith, J. & Haigh, J. 1974. The hitch-hiking effect of a favourable gene. *Genetics Research,* 23:23–35.

McComb, K., Moss, C., Durant, S. M., Baker, L., and Sayialel, S. 2001. Matriarchs as repositories of social knowledge in African elephants. *Science,* 292:491–494.

McGrew, W. C. 1998. Behavioural diversity in populations of free-ranging chimpanzees in Africa: is it culture? *Human Evolution,* 13:209–220.

Mesnick, S. L. 2001. Genetic relatedness in sperm whales: evidence and cultural implications. *Behavioral and Brain Sciences,* 24:346–347.

Mesnick, S. L., Taylor, B. L., Le Duc, R. G., Treviño, S. E., O'Corry-Crowe, G. M.

& Dizon, A. E. 1999. Culture and genetic evolution in whales. *Science,* 284:2055a.

Møhl, B., Wahlberg, M., Madsen, P. T., Miller, L. A. & Surlykke, A. 2000. Sperm whales clicks: directionality and source level revisited. *Journal of the Acoustical Society of America,* 107:638–648.

Noad, M. J., Cato, D. H., Bryden, M. M., Jenner, M. N. & Jenner, K. C. S. 2000. Cultural revolution in whale songs. *Nature,* 408:537.

Norris, K. S. & Schilt, C. R. 1988. Cooperative societies in three-dimensional space: On the origins of aggregations, flocks and schools, with special reference to dolphins and fish. *Ethology and Sociobiology,* 9:149–179.

Olesiuk, P., Bigg, M. A. & Ellis, G. M. 1990. Life history and population dynamics of resident killer whales (*Orcinus orca*) in the coastal waters of British Columbia and Washington State. *Reports of the International Whaling Commission* (Special Issue), 12:209–243.

Papastavrou, V., Smith, S. C. & Whitehead, H. 1989. Diving behaviour of the sperm whale, *Physeter macrocephalus,* off the Galápagos Islands. *Canadian Journal of Zoology,* 67:839–846.

Payne, K. 1999. The progressively changing songs of humpback whales: a window on the creative process in a wild animal. In *The Origins of Music,* ed. N. L. Wallin, B. Merker, & S. Brown, pp. 135–150. Cambridge, MA: MIT Press.

Payne, K. & Payne, R. S. 1985. Large-scale changes over 17 years in songs of humpback whales in Bermuda. *Zeitschrift für Tierpsychologie,* 68:89–114.

Payne, R. & Guinee, L. N. 1983. Humpback whale, *Megaptera novaeangliae,* songs as an indicator of "stocks." In *Communication and Behavior of Whales,* ed. R. Payne, pp. 333–358. Boulder, CO: Westview Press.

Pitman, R. L., Ballance, L. T., Mesnick, S. I., and Chivers, S. J. 2001. Killer whale predation on sperm whales: observations and implications. *Marine Mammal Science,* 17:494–507.

Pryor, K., Lindbergh, J., Lindbergh, S. & Milano, R. 1990. A dolphin-human fishing cooperative in Brazil. *Marine Mammal Science,* 6:77–82.

Rendell, L. & Whitehead, H. 2001a. Culture in whales and dolphins. *Behavioral and Brain Sciences,* 24:309–324.

——— 2001b. Cetacean culture: still afloat after the first naval engagement of the culture wars. *Behavioral and Brain Sciences,* 24:360–373.

Rice, D. W. 1989. Sperm whale. *Physeter macrocephalus* Linnaeus, 1758. In *Handbook of Marine Mammals,* vol. 4, ed. S. H. Ridgway & R. Harrison, pp. 177–233. London: Academic Press.

Richard, K. R., Dillon, M. C., Whitehead, H. & Wright, J. M. 1996. Patterns of kinship in groups of free-living sperm whales (*Physeter macrocephalus*) revealed by multiple molecular genetic analyses. *Proceedings of the National Academy of Sciences USA,* 93:8792–8795.

Richerson, P. J. & Boyd, R. 1998. The evolution of human ultrasociality. In *Indoctrinability, Ideology and Warfare,* ed. I. Eibl-Eibesfeldt & F. K. Salter, pp. 71–95. London: Berghahn Books.

Rosel, P. E. & Rojas-Bracho, L. 1999. Mitochondrial DNA variation in the critically endangered vaquita *Phocoena sinus* Norris and Macfarland, 1958. *Marine Mammal Science,* 15:990–1003.

Secchi, E. R., Wang, J. Y., Murray, B. W., Rocha-Campos, C. C. & White, B. N. 1998. Population differentiation in the franciscana (*Pontoporia blainvillei*) from two geographic locations in Brazil as determined from mitochondrial DNA control region sequences. *Canadian Journal of Zoology,* 76:1622–1627.

Siemann, L. A. 1994. Mitochondrial DNA sequence variation in North Atlantic long-finned pilot whales, *Globicephala melas.* Ph.D. diss., Massachusetts Institute of Technology, Cambridge, MA.

Simmonds, M. P. 1997. The meaning of cetacean strandings. *Bulletin Institut Royal de Sciences Naturelles de Belgique Biologie,* 67(Suppl.):29–34.

Simöes-Lopes, P. C., Fabián, M. E. & Menegheti, J. O. 1998. Dolphin interactions with the mullet artisanal fishing on southern Brazil: a qualitative and quantitative approach. *Revista Brasileira de Zoologia,* 15:709–726.

Smolker, R. A., Richards, A. F., Connor, R. C., Mann, J. & Berggren, P. 1997. Sponge-carrying by Indian Ocean bottlenose dolphins: possible tool-use by a delphinid. *Ethology,* 103:454–465.

Steele, J. H. 1985. A comparison of terrestrial and marine ecological systems. *Nature,* 313:355–358.

Tiedemann, R. & Milinkovitch, M. 1999. Culture and genetic evolution in whales. *Science,* 284:2055a.

Watkins, W. A. & Schevill, W. E. 1977. Sperm whale codas. *Journal of the Acoustical Society of America,* 62:1486–1490.

Watkins, W. A., Moore, K. E. & Tyack, P. 1985. Sperm whale acoustic behaviors in the southeast Caribbean. *Cetology,* 49:1–15.

Weilgart, L. S. & Whitehead, H. 1988. Distinctive vocalizations from mature male sperm whales (*Physeter macrocephalus*). *Canadian Journal of Zoology,* 66:1931–1937.

——— 1997. Group-specific dialects and geographical variation in coda repertoire in South Pacific sperm whales. *Behavioural Ecology and Sociobiology,* 40:277–285.

Weilgart, L., Whitehead, H. & Payne, K. 1996. A colossal convergence. *American Scientist,* 84:278–287.

Whitehead, H. 1989. Formations of foraging sperm whales, *Physeter macrocephalus,* off the Galápagos Islands. *Canadian Journal of Zoology,* 67:2131–2139.

——— 1993. The behaviour of mature male sperm whales on the Galápagos breeding grounds. *Canadian Journal of Zoology,* 71:689–699.

——— 1995. Status of Pacific sperm whale stocks before modern whaling. *Reports of the International Whaling Commission,* 45:407–412.

——— 1996a. Babysitting, dive synchrony, and indications of alloparental care in sperm whales. *Behavioural Ecology and Sociobiology,* 38:237–244.

——— 1996b. Variation in the feeding success of sperm whales: temporal scale, spatial scale and relationship to migrations. *Journal of Animal Ecology,* 65:429–438.

——— 1998. Cultural selection and genetic diversity in matrilineal whales. *Science,* 282:1708–1711.

——— 1999a. Culture and genetic evolution in whales. *Science,* 284:2055a.

——— 1999b. Variation in the visually observable behavior of groups of Galápagos sperm whales. *Marine Mammal Science,* 15:1181–1197.

Whitehead, H. & Kahn, B. 1992. Temporal and geographical variation in the social structure of female sperm whales. *Canadian Journal of Zoology,* 70:2145–2149.

Whitehead, H. & Mann, J. 2000. Female reproductive strategies of cetaceans. In *Cetacean Societies,* ed. J. Mann, R. Connor, P. L. Tyack, & H. Whitehead, pp. 219–246. Chicago: University of Chicago Press.

Whitehead, H. & Weilgart, L. 2000. The sperm whale: social females and roving males. In *Cetacean Societies,* ed. J. Mann, R. C. Connor, P. Tyack, & H. Whitehead, pp. 154–172. Chicago: University of Chicago Press.

Whitehead, H., Waters, S. & Lyrholm, T. 1991. Social organization in female sperm whales and their offspring: constant companions and casual acquaintances. *Behavioural Ecology and Sociobiology,* 29:385–389.

Whitehead, H., Brennan, S. & Grover, D. 1992. Distribution and behaviour of male sperm whales on the Scotian Shelf, Canada. *Canadian Journal of Zoology,* 70:912–918.

Whitehead, H., Christal, J. & Dufault, S. 1997. Past and distant whaling and the rapid decline of sperm whales off the Galápagos Islands. *Conservation Biology,* 11:1387–1396.

Whitehead, H., Dillon, M., Dufault, S., Weilgart, L. & Wright, J. 1998. Non-geographically based population structure of South Pacific sperm whales: dialects, fluke-markings and genetics. *Journal of Animal Ecology,* 67:253–262.

Whitehead, H., Richerson, P., and Boyd, R. In press. *Selection.*

Whiten, A. 2000a. Primate culture and social learning. *Cognitive Science,* 24:477–508.

Whiten, A. 2000b. Social complexity and social intelligence. In *The Nature of Intelligence,* ed. G. R. Bock, J. A. Goode, & K. Webb, pp. 185–201. Chichester, UK: Wiley.

Whiten, A. & Ham, R. 1992. On the nature and evolution of imitation in the animal kingdom: reappraisal of a century of research. *Advances in the Study of Behavior,* 21:239–283.

Whiten, A., Goodall, J., McGrew, W. C., Nishida, T., Reynolds, V., Sugiyama, Y., Tutin, C. E. G., Wrangham, R. W. & Boesch, C. 1999. Cultures in chimpanzees. *Nature,* 399:682–685.

Williams, T. M., Friedl, W. A., Fong, M. L., Yamada, R. M., Dedivy, P. & Haun, J. E. 1992. Travel at low energetic cost by swimming and wave-riding bottlenose dolphins. *Nature,* 355:821–823.

Wrangham, R. W. & Rubenstein, D. I. 1986. Social evolution in birds and mammals. In *Ecological Aspects of Social Evolution,* ed. D. I. Rubenstein & R. W. Wrangham, pp. 452–470. Princeton, NJ: Princeton University Press.

Case Study 17A. *Do Killer Whales Have Culture?*

Bain, D. E. 1988. An evaluation of evolutionary processes: studies of natural selection, dispersal, and cultural evolution in killer whales *(Orcinus orca).* Ph.D. diss., University of California, Santa Cruz.

Baird, R. W. & Dill, R. W. 1995. Occurrence and behaviour of transient killer whales: seasonal and pod-specific variability, foraging behaviour, and prey handling. *Canadian Journal of Zoology,* 73:1300–1311.

Barrett-Lennard, L. G. 2000. Population structure and mating patterns of killer whales, *Orcinus orca,* as revealed by DNA analysis. Ph.D. diss., University of British Columbia, Vancouver, Canada.

Bigg, M. A. 1982. An assessment of killer whale (*Orcinus orca*) stocks off Vancouver Island, British Columbia. *Report of the International Whaling Commission,* 32:655–666.

Bigg, M. A., Olesiuk, P. F., Ellis, G. M., Ford, J. K. B. & Balcomb, K. C. 1990. Social organization and genealogy of resident killer whales *(Orcinus orca)* in the coastal waters of British Columbia and Washington State. *Report of the International Whaling Commission* (Special Issue), 12:383–405.

Caldwell, M. C. & Caldwell, D. K. 1972. Vocal mimicry in the whistle mode by an Atlantic bottlenosed dolphin. *Cetology,* 9:1–8.

Deecke, V. B., Ford, J. K. B. & Spong, P. 2000. Dialect change in resident killer whales: implications for vocal learning and cultural transmission. *Animal Behaviour,* 60:629–638.

Ford, J. K. B. 1989. Acoustic behaviour of resident killer whales *(Orcinus orca)* off Vancouver Island, British Columbia. *Canadian Journal of Zoology* 67:727–745.

——— 1991. Vocal traditions among resident killer whales *(Orcinus orca)* in coastal waters of British Columbia. *Canadian Journal of Zoology* 69:1454–1483.

Ford, J. K. B. 2002. Dialects. In *The Encyclopedia of Marine Mammals,* ed. W. F. Perrin, B. Wursig, & H. G. M. Thewissen, pp. 322–323. New York: Academic Press.

Ford, J. K. B. & Ellis, G. M. 1999. *Transients: Mammal-hunting Killer Whales of British Columbia, Washington, and Southeastern Alaska.* Vancouver: UBC Press.

Ford, J. K. B., Ellis, G. M., Barrett-Lennard, L. G., Morton, A. B., Palm, R. S. & Balcomb, K. C. 1998. Dietary specialization in two sympatric populations of killer whales (*Orcinus orca*) in coastal British Columbia and adjacent waters. *Canadian Journal of Zoology,* 76:1456–1471.

Ford, J. K. B., Ellis, G. M. & Balcomb, K. C. 2000. *Killer Whales,* 2d ed. Vancouver: UBC Press.

Guinet, C. 1991. Intentional stranding apprenticeship and social play in killer whales (*Orcinus orca). Canadian Journal of Zoology,* 69:2712–2716.

Guinet, C. & Bouvier, J. 1995. Development of intentional stranding hunting techniques in killer whale (*Orcinus orca*) calves at Crozet Archipelago. *Canadian Journal of Zoology,* 73:27–33.

Janik, V. M. & Slater, P. J. B. 1997. Vocal learning in mammals. *Advances in the Study of Behaviour,* 26:59–99.

Lopez, G. C. & Lopez, D. 1985. Killer whales *(Orcinus orca)* off Patagonia, and their behavior of intentional stranding while hunting shore. *Journal of Mammalogy,* 66:181–183.

Marler, P. & Tamura, M. 1962. Song "dialects" in three populations of white-crowned sparrows. *Condor,* 64:368–377.

Matkin, C. O. & Saulitis, E. L. 1997. Killer whale *(Orcinus orca)* NGOS Report. In *Restoration Notebook of the Exxon Valdez Oil Spill Trustee Council.* Anchorage, AK.

Matkin, C. O., Ellis, G. M., Olesiuk, P. & Saulitis, E. L. 1999. Association patterns and inferred genealogies of resident killer whales, *Orcinus orca,* in Prince William Sound. *Fisheries Bulletin,* 97(4):900–919.

Miller, P. J. O. & Bain, D. E. 2000. Within-pod variation in the sound production of a pod of killer whales, *Orcinus orca. Animal Behaviour,* 60:617–628.

Morton, A. B. 1990. A quantitative comparison of the behavior of resident and transient forms of the killer whale off the central British Columbia coast. *Report of the International Whaling Commission* (Special Issue), 12:245–248.

Osborne, R. W. 1986. A behavioral budget of Puget Sound killer whales. In *Behavioral Biology of Killer Whales,* ed. B. C. Kirkevold & J. S. Lockard, pp. 211–249. New York: A. R. Liss.

Saulitis, E., Matkin, C., Barrett-Lennard, L., Heise, K. & Ellis, G. 2000. Foraging strategies of sympatric killer whale (*Orcinus orca*) populations in Prince William Sound, Alaska. *Marine Mammal Science,* 16(1):94–109.

Simila, T. & Ugarte, F. 1993. Surface and underwater observations of cooperatively feeding killer whales in northern Norway. *Canadian Journal of Zoology,* 71:1494–1499.

Yurk, H., Barrett-Lennard, L. G., Ford, J. K. B. & Matkin, C. O. 2002. Cultural transmission within maternal lineages: vocal clans in resident killer whales in Southern Alaska. *Animal Behaviour,* 63:1103–1119.

18. Discovering Culture in Birds

Ball, G. F. & Hulse, S. H. Bird song. *American Psychologist,* 33:37–58.

Bird, D. M. 1999. *The Bird Almanac: The Ultimate Guide to Essential Facts and Figures of the World's Birds.* Toronto: Key Porter.

Cairns, R. B. 1979. *The Analysis of Social Interchanges.* San Francisco: W. H. Freeman.

Chaiken, M. L., Bohner, J. & Marler, P. 1993. Song acquisition in European starlings, *Sturnus vulgaris:* a comparison of the songs of live-tutored, untutored, and wild-caught males. *Animal Behaviour,* 46:1079–1090.

Chaiken, M. L., Gentner, T. Q. & Hulse, S. E. 1997. Effects of social interactions on the development of starling song and the perception of those effects by conspecifics. *Journal of Comparative Psychology,* 111:379–392.

Chapman, F. M. 1934. *Handbook of Birds of Eastern North America.* New York: D. Appleton and Co.

Engle, M. 2001. Social influences on song development and mimicry in the European starling *(Sturnus vulgaris).* Unpublished diss., Indiana University, Bloomington.

Freeberg, T. M. 1996. Assortative mating in captive cowbirds is predicted by social experience. *Animal Behaviour,* 51:1129–1142.

——— 1997. Cultural transmission of behaviors facilitating assortative courtship and mating in cowbirds *(Molothrus ater).* Unpublished diss., Indiana University, Bloomington.

——— 1998. The cultural transmission of courtship patterns in cowbirds, *Molothrus ater. Animal Behaviour,* 56:1063–1073.

——— 2000. Culture and courtship in vertebrates: a review of social learning and transmission of courtship systems and mating patterns. *Behavioural Processes,* 51:177–192.

Freeberg, T. M., King, A. P. & West, M. J. 1995. Social malleability in cowbirds (*Molothrus ater artemisiae*): species and mate recognition in the first 2 years of life. *Journal of Comparative Psychology,* 109:357–367.

Friedmann, H. 1929. *The Cowbirds: A Study in the Social Biology of Parasitism.* Springfield, IL: Charles C Thomas.

Friedmann, H., Kiff, L. F. & Rothstein, S. I. 1977. A further contribution to knowledge of the host relations of the parasitic cowbirds. *Smithsonian Contributions to Zoology,* 235:1–75.

Goldstein, M. H. 2001. Social mechanisms of vocal learning in human infants. Unpublished diss., Indiana University, Bloomington.

Goldstein, M. H. & West, M. J. 1999. Consistent responses by human mothers to prelinguistic infants: the effect of repertoire size. *Journal of Comparative Psychology,* 113:52–58.

Goldstein, M. H. & West, M. J. In preparation. Infant babbling is responsive to social contingencies: new evidence of the social modulation of vocal learning.

Greenewalt, C. 1968. *Bird Song: Acoustics and Physiology.* Washington, DC: Smithsonian Institution Press.

Hausberger, M. 1997. Social influences on song acquisition and sharing in the European starling (*Sturnus vulgaris*). In *Social Influences on Vocal Development,* ed. C. T. Snowdon & M. Hausberger, pp. 128–156. New York: Cambridge University Press.

King, A. P. & West, M. J. 1983a. Epigenesis of cowbird song: a joint endeavor of males and females. *Nature,* 305:704–706.

——— 1983b. Female perception of cowbird song: a closed developmental program. *Developmental Psychobiology,* 16:335–342.

——— 1987. Different outcomes of synergy between song production and song perception in the same subspecies (*Molothrus ater ater*). *Developmental Psychobiology,* 20:177–187.

——— 1988. Searching for the functional origins of cowbird song in eastern brown-headed cowbirds (*Molothrus ater ater*). *Animal Behaviour,* 36:1575–1588.

——— 1990. Variation in species-typical behavior: a contemporary theme for comparative psychology. In *Contemporary Issues in Comparative Psychology,* ed. D. A. Dewsbury, pp. 331–339. Sunderland, MA: Sinauer.

Kroodsma, D. E. & Pickert, R. 1980. Environmentally-dependent sensitive periods for avian vocal learning. *Nature,* 288:477–479.

Kroodsma, D. E. & Miller, E. H. 1996. Ecology of song development in passerine birds. In *Ecology and Evolution of Acoustic Communication in Birds,* ed. D. E. Kroodsma & E. H. Miller, pp. 3–19. Ithaca, NY: Comstock Publishing Associates.

Lehrman, D. S. 1970. Semantic and conceptual issues in the nature-nurture problem. In *Development and Evolution of Behavior: Essays in Memory of T. C. Schneirla,* ed. L. R. Aronson, E. Tobach, D. S. Lehrman, & J. S. Rosenblatt, pp. 17–52. San Francisco: W. H. Freeman.

——— 1974. Can psychiatrists use ethology? In *Ethology and Psychiatry,* ed. N. F. White, pp. 187–196. Toronto: University of Toronto Press.

Marler, P. & Nelson, D. A. 1993. Action-based learning: a new form of developmental plasticity in bird song. *Netherlands Journal of Zoology,* 43:91–103.

Mayr, E. 1974. Behavior programs and evolutionary strategies. *American Scientist,* 62:650–659.

Mountjoy, D. J. & Lemon, R. E. 1991. Song as an attractant for male and female starlings, and the influence of song complexity. *Behavioral Ecology and Sociobiology,* 28:97–100.

Nadler, R. D., Dahl, J. F., Collins, D. C. & Gould, K. E. 1994. Sexual behavior of chimpanzees (*Pan troglodytes*): male versus female regulation. *Journal of Comparative Psychology,* 108:58–67.

Nelson, D. A. & Marler, P. 1994. Selection-based learning in bird song development. *Proceedings of the National Academy of Sciences USA,* 91:10498–10501.

Nelson, D. A., Marler, P. & Palleroni, A. 1995. A comparative analysis of vocal learning: intraspecific variation in the learning process. *Animal Behaviour,* 30:83–97.

O'Loghlen, A. L. & Rothstein, S. I. 1995. Culturally correct song dialects are correlated with male age and female song preferences in wild populations of brown-headed cowbirds. *Behavioral Ecology and Sociobiology,* 36:251–259.

Pinker, S. 1994. *The Language Instinct.* New York: Harper Collins.

Schlichting, C. D. & Pigliucci, M. 1998. *Phenotypic Evolution: A Reaction Norm Perspective.* Sunderland, MA: Sinauer.

Smith, V. A., King, A. P. & West, M. J. 2000. A role of her own: female cowbirds (*Molothrus ater*) influence male song development and outcome of song learning. *Animal Behaviour,* 60:599–609.

Smith, V. A., King, A. P. & West, M. J. In press. Learning in a social context: affiliation patterns in a captive flock of brown-headed cowbirds *(Molothrus ater). Animal Behaviour.*

Taylor, S. E., Klein, L. C., Lewis, B. P., Gruenewald, L. L. Gorung, R. A. & Updegraf, J. A. 2000. Bio-behavioral response to stress in females: tend-and-befriend, not fight-or-flight. *Psychological Review,* 107:411–429.

West, M. J. & King, A. P. 1985. Social guidance of vocal learning by female cowbirds: validating its functional significance. *Ethology,* 70:225–235.

——— 1988. Female visual displays affect the development of male song in the cowbird. *Nature,* 334:244–246.

——— 1990. Mozart's starling. *American Scientist,* 78:106–114.

West, M. J., Stroud, A. N. & King, A. P. 1983. Mimicry of the human voice by European starlings: the role of social interaction. *Wilson Bulletin,* 95:635–640.

West, M. J., White, D. J. & King, A. P. In review. Female brown-headed cowbirds' (*Molothrus ater*) organization and behaviour reflects male social dynamics. *Animal Behaviour.*

Wexler, K. 1990. Innateness and maturation in language development. *Developmental Psychobiology,* 23:645–660.

White, D. J., King, A. P. & West, M. J. In press a. Facultative development of courtship and communication in juvenile male cowbirds (*Molothrus ater*). *Behavioral Ecology.*

White, D. J., King, A. P. & West, M. J. In press b. Social interaction with juvenile males improves adult males' courtship skills. *Behavior.*

White, D. J., King, A. P., Cole, A. & West, M. J. In press c. Opening the social gateway: early vocal and social sensitivities in brown-headed cowbirds (*Molothrus ater*). *Ethology.*

Whiten, A. 2000. *Social Complexity and Intelligence.* In *The Nature of Intelligence,* ed. G. R. Bock, J. A. Goode, and K. Webb, pp. 185–201. Chichester, UK: Wiley.

Whiten, A., Goodall, J., McGrew, W. C., Nishida, T., Reynolds, V., Sugiyama, Y., Tutin, C. E. G., Wrangham, R. W. & Boesch, C. 1999. Culture in chimpanzees. *Nature,* 389:682–685.

Williams, H., Kilander, K. & Sotanski, M. 1993. Untutored song, reproductive success and song learning. *Animal Behaviour,* 45:695–705.

Acknowledgments

The editors of this volume would like to express heartfelt thanks to the help received with organization of the 2000 conference in Chicago, especially from the Chicago Academy of Sciences. The planning of this conference occurred with the inspired help of Paul Heltne, who kept us focused on the project for years. Instrumental for its success were also the late Kenneth Norris and the staff of the two co-sponsoring institutions, the Living Links Center of Emory University and the Jane Goodall Institute. We thank Jane Goodall herself for offering us her wisdom during the final day of the conference. The conference could not have occurred without the program coordination of Darren Long of Living Links. Running of the conference and public relations were greatly facilitated by Betsy Altman, Jennifer Blitz, Steward Hudson, Betty Leslie, Lisa Noland, and Bryn Reese.

With regard to the present volume we are extremely grateful for the untiring coordination and editorial assistance of Darren Long, who kept all of the authors and reviewers moving along until the final deadline. Apart from the contributors to this volume, who mostly read and commented on each others' papers, we also had external reviewers to whom we are most grateful: Filippo Aureli, Marina Cords, Susan Farabaugh, Donald Griffin, Alexander Harcourt, Hans Kummer, David Watts, and Andrew Whiten. We are fortunate to have worked with such a distinguished group of scientists, all enthusiastic about placing their work in this broader framework, and to Harvard University Press and our editor, Ann Downer-Hazell, for helping us to convert the excitement generated by the conference into a permanent record of our converging views on the evolution of individualized societies.

1. Life History and Cognitive Evolution in Primates

We thank Rob Barton, Peter Holland, Valen Johnson, Chris Kirk, Charlie Nunn, and Signe Preuschoft for insightful discussion. We thank the following for making original comparative data available to us: Dr. Heiko Frahm for primate brain data available for analysis, Keyt Fischer for data on mammalian life histories and brain size, and Charles Nunn for data on primate terrestriality.

Case Study 1A. Sociality and Disease Risk

The ideas presented here benefited greatly from interaction and collaboration with Janis Antonovics, John Gittleman, and Carel van Schaik. I thank Kate Jones, Carel van Schaik, Peter Tyack, and Randy Wells for helpful comments on the manuscript. This material is based upon research supported by the National Science Foundation under a grant awarded in 2000 (Postdoctoral Fellowship in Biological Informatics).

2. Dolphin Social Complexity

Doug Nowacek provided much insight into the classification of feeding behaviors, and along with Denise Herzing and Richard Connor provided data for Table 2.1. Data from Sarasota dolphin genetic analyses have been provided by Debbie Duffield of Portland State University. Edward Owen produced the sociograms through Hal Whitehead's software, SOCPROG. Many of the ages of the resident Sarasota dolphins have been determined through the expertise of Aleta Hohn of the National Marine Fisheries Service. Sue Hofmann, Stephanie Nowacek, Kim Hull, and Kim Urian have collected much of the sighting data and developed and maintained the long-term sighting database for Sarasota dolphins. Peter Tyack, Frans de Waal, Toshisada Nishida, Katy Payne, and Edward Owen improved the manuscript greatly through their suggestions. Support for the research reported here has come from the Chicago Zoological Society, Earthwatch Institute, Disney Wildlife Conservation Fund, Dolphin Quest, and the National Marine Fisheries Service.

3. Sources of Social Complexity in the Three Elephant Species

For the new data presented in this chapter I am indebted to Andrea Turkalo, our host in the Dzanga field site in 2000, and to the sponsors of the Elephant Listening Project: the U.S. Fish and Wildlife Service, the International Fund for Animal Welfare, Wildlife Conservation Society, Conservation International, the Harry Frank Guggenheim Foundation, the Park Foundation, World Wildlife Fund, and the Cornell Laboratory of Ornithology. I thank Melissa Groo and Mya Thompson for a huge amount of help in preparing the data and manuscript, and I thank Donald Griffin, Peter Tyack, and Gerald Wilkinson for many useful suggestions.

Case Study 4B. Levels and Patterns in Dolphin Alliance Formation

The Shark Bay Dolphin Research project is a cooperative endeavor that at any one time includes up to seven projects and researchers from three continents. We thank the Monkey Mia Resort, the Department of Human Biology at the University of Western Australia, and Richard Holst, the Ceebie Trust, and the Getty Foundation for years of support. The male alliance project has received considerable support over the years from the National Geographic Society, and a fellowship from the National Institutes of Health.

6. Is Social Stress a Consequence of Subordination or a Cost of Dominance?

Our thanks to the people who made critical contributions to the original studies summarized here, especially Nancy Creel, Peter Waser, Jon Rood, and Steve Monfort. These studies were supported by the National Science Foundation, the Smithsonian Institution, and the Frankfurt Zoological Society. Jennifer Sands was supported by an EPA-STAR fellowship.

Case Study 6A. Sperm Whale Social Structure

Studies such as this rely on the generous assistance of many people in the field and laboratory; those who have shared their genetic data with us, helped with the analyses, and commented on the ideas presented here. We thank Hal Whitehead, Richard Connor, Luke Rendell, Margie Morrice, Mark

Hindell, Deborah Thiele, Jenny Christal, Joanna Bond, Robert Pitman, Amy Frye, Frank Cipriano, the Parks and Wildlife Service of Tasmania, and Environment Australia.

7. Equivalence Classification as an Approach to Social Knowledge

The preparation of this manuscript was supported by Office of Naval Research Grants N00014-99-1-0164 and N00014-00-1-0836 to R. J. Schusterman and a DoD AASERT Fellowship to C. Reichmuth Kastak. We thank those that organized the Animal Social Complexity conference and contributed to this volume for stimulating the development of our ideas about the cognitive basis of animal social complexity.

Case Study 10A. Emotional Recognition by Chimpanzees

This project was supported by NIH RR-00165 to the Yerkes Primate Center, and R01-RR09797 to F. B. M. de Waal, and NSF-IBN-9801464 to L. A. Parr. The author thanks F. B. M. de Waal, S. Fernàndez-Carriba, J. Flack, A. Lacreuse, T. Matsuzawa, S. Boysen, and P. Tyack for their comments on an earlier version of this study, and the Animal Care Staff at the Yerkes Primate Center. The Yerkes Primate Center is fully accredited by the American Association for Accreditation of Laboratory Animal Care.

11. Vocal Communication in Wild Parrots

The author wants to thank Sandra Vehrencamp, Kathryn Cortopassi, Jessica Eberhard, Tim Wright, Susan Farabaugh, and Peter Tyack for useful comments and suggestions on earlier drafts. Much of the research cited in this chapter was supported by NSF grants IBN 94-06217 and IBN 94-06217.

12. Social and Vocal Complexity in Bats

I am grateful to S. R. E. Steele, L. E. O. Braack, and N. Rautenbach for field assistance in South Africa, to B. Fenton for encouragement and to J. Bradbury, D. Griffin, and P. Tyack for helpful comments that improved the manuscript. J. Balcombe, G. McCracken, and J. A. Scherrer graciously provided recordings of infant bats. P. Ro, W.-W. Lin, C. Kang, and M.

Naganuma helped to digitize and score vocalizations. My work on bat social behavior and communication has been supported by the National Science Foundation, the American Philosophical Society, and the University of Maryland Graduate School.

13. Dolphins Communicate about Individual-Specific Social Relationships

I would like to thank Jan van Hooff, Meredith West, and Gerald Wilkinson for reviewing the initial draft of this chapter. Much of the research described in this chapter involved collaboration with Laela Sayigh and Stephanie Watwood while they were graduate students in my lab. The field research was supported by the U.S. Office of Naval Research Grant N0014-87-K-0236 and National Institutes of Health Grant 5 R29 NS25290. The preparation of this chapter was supported in part by NIH Grant R01 DC04191. This is contribution number 10695 from the Woods Hole Oceanographic Institution.

14. Koshima Monkeys and Bossou Chimpanzees

I would like to thank Jun'ichiro Itani and Masao Kawai for information about the historical background of primatological studies in Japan. The Chicago Academy of Science offered me three consecutive opportunities to participate in the international symposia on the intelligence of chimpanzees and other nonhuman animals. Thanks are also due to the editors, Frans de Waal and Peter Tyack, as well as Darren Long for their encouragement and patience. I am grateful to Satoshi Hirata for discussion and Dora Biro for corrections to the English in the original manuscript. The long-term studies were financially supported by various grants, and the preparation of the article was supported by a grant from the Japanese government (#12002009).

15. Individuality and Flexibility of Cultural Behavior Patterns in Chimpanzees

I thank Frans de Waal and Peter Tyack for their invitation to the stimulating conference in Chicago. I thank H. Hayaki, K. Kawanaka, K. Norikoshi, Y. Takahata, and K. Zamma for unpublished information. The fieldwork and data analyses on which this paper was based were supported financially by the LSB Leakey Foundation, a grant-in-aid for COE Research (#10CE2005 to O. Takenaka), and the Basic Research Fund A1 of the Japanese Ministry of Education, Culture, Sports, Science, and Technology (#12375003 to Toshisada Nishida). I also thank A. Harcourt, S. Pandolfi, and Frans de Waal for comments on the paper.

Case Study 15A. Sex Differences in Termite Fishing among Gombe Chimpanzees

Thank you to Ernestine Friedl and the Women's Studies Department of Duke University for funding; Ian Gilby, Jennifer Williams, and Elizabeth Vinson of the University of Minnesota for immeasurable help; Steve Churchill, Leslie Digby, Diane Brockman, Christine Drea, Kirstin Siex, and Robert Deaner of Duke University for their invaluable advice; Dave Pandolfi, Judi Jaffee, Jaime Jaffee, Robert Rich, Marilyn Scher, and Ernie Scher for moral support.

16. Ten Dispatches from the Chimpanzee Culture Wars

I thank Frans de Waal and Peter Tyack for organizing the conference that led to this chapter and the Chicago Academy of Sciences for hosting it; Hal Whitehead, Linda Marchant, Toshisada Nishida, and Frans de Waal for critical comments on the manuscript; Diana Deaton for manuscript preparation; and finally Toshisada Nishida, Caroline Tutin, and Richard Wrangham for enlightening discussions over decades.

17. Society and Culture in the Deep and Open Ocean

Many thanks to all those who helped study the sperm whales at sea, and to Tom Arnbom, Jenny Christal, Mary Dillon, Susan Dufault, Jonathan Gordon, Nathalie Jaquet, Benjamin Kahn, Luke Rendell, Kenny Richard, Sean Smith, Susan Waters, and Linda Weilgart for their substantial parts in the analysis and interpretation of the data. Our research on sperm whales has largely been funded by the Natural Sciences and Engineering Research Council of Canada, the Whale and Dolphin Conservation Society, the International Whaling Commission, the National Geographic Society, and the

Green Island Foundation. I am particularly grateful to Professor Eberhardt Gwinner for his hospitality at the Research Center for Ornithology of the Max Planck Foundation in Seewiesen, Germany, where this chapter was written. Comments from Frans de Waal, William McGrew, Sarah Mesnick, and Luke Rendell on the whole chapter, and Katy Payne on a part of it, were most helpful.

Contributors

CHRISTOPHE BOESCH

Department of Primatology, Max Planck Institute for Evolutionary Anthropology, Leipzig, Germany

SARAH T. BOYSEN

Department of Psychology, Ohio State University, Columbus, Ohio, United States

JACK W. BRADBURY

Laboratory of Ornithology, Cornell University, Ithaca, New York, United States

DOROTHY L. CHENEY

Department of Psychology, University of Pennsylvania, Philadelphia, Pennsylvania, United States

RICHARD C. CONNOR

Department of Biology, University of Massachusetts–Dartmouth, North Dartmouth, Massachusetts, United States

SCOTT CREEL

Department of Ecology, Montana State University, Bozeman, Montana, United States

ROBERT O. DEANER

Department of Biological Anthropology and Anatomy, Duke University, Durham, North Carolina, United States

FRANS B. M. DE WAAL

Living Links, Yerkes Primate Center, Emory University, Atlanta, Georgia, United States

ANDREW E. DIZON

Molecular Ecology Group, Southwest Fisheries Science Center, National Marine Fisheries Service, NOAA, La Jolla, California, United States

CHRISTINE M. DREA

Department of Biological Anthropology and Anatomy, Duke University, Durham, North Carolina, United States

ANNE ENGH

Department of Zoology, Michigan State University, East Lansing, Michigan, United States

SERGIO ESCORZA-TREVIÑO

Molecular Ecology Group, Southwest Fisheries Science Center, National Marine Fisheries Service, NOAA, La Jolla, California, United States

KAREN EVANS

Antarctic Wildlife Research Unit, School of Zoology, University of Tasmania, Hobart, Tasmania, Australia

LAURENCE G. FRANK

Department of Psychology, University of California–Berkeley, Berkeley, California, United States

KAREN I. HALLBERG

Department of Psychology, Ohio State University, Columbus, Ohio, United States

KAY E. HOLEKAMP

Department of Zoology, Michigan State University, East Lansing, Michigan, United States

LUDWIG HUBER

Institute of Zoology, Vienna, Austria

JOHN HYDE

Molecular Ecology Group, Southwest Fisheries Science Center, National Marine Fisheries Service, NOAA, La Jolla, California, United States

DAVID KASTAK

Institute of Marine Sciences, Long Marine Laboratory, University of California–Santa Cruz, Santa Cruz, California, United States

ANDREW P. KING

Department of Psychology, Indiana University, Bloomington, Indiana, United States

MICHAEL KRÜTZEN

Department of Biology, University of Massachusetts–Dartmouth, North Dartmouth, Massachusetts, United States

MASSIMO MANNU

Department of Experimental Psychology, University of São Paulo, São Paulo, Brazil

TETSURO MATSUZAWA

Department of Behavioral and Brain Sciences, Primate Research Institute, Kyoto University, Kyoto, Japan

W. C. MCGREW

Department of Zoology and Anthropology, Miami University, Oxford, Ohio, United States

SARAH L. MESNICK

Southwest Fisheries Science Center, National Marine Fisheries Service, La Jolla, California, United States

DOUGLAS A. NELSON

Department of Psychology, Ohio State University, Columbus, Ohio, United States

TOSHISADA NISHIDA

Department of Zoology, Kyoto University, Kyoto, Japan

CHARLES L. NUNN

Department of Biology, University of Virginia, Charlottesville, Virginia, United States

EDUARDO B. OTTONI

Department of Experimental Psychology, University of São Paulo, São Paulo, Brazil

STEPHANIE S. PANDOLFI

Department of Biological Anthropology and Anatomy, Duke University, Durham, North Carolina, United States

LISA A. PARR

Yerkes Primate Center, Emory University, Atlanta, Georgia, United States

KATY PAYNE

Cornell Bioacoustics Research Program, Laboratory of Ornithology, Cornell University, Ithaca, New York, United States

SUSAN PERRY

Department of Anthropology, University of California–Los Angeles, Los Angeles, California, United States

SIGNE PREUSCHOFT

Living Links, Yerkes Primate Center, Emory University, Atlanta, Georgia, United States

ANNE E. PUSEY

Department of Evolution, Ecology, and Behavior, University of Minnesota, St. Paul, Minnesota, United States

COLLEEN REICHMUTH KASTAK

Institute of Marine Sciences, Long Marine Laboratory, University of California–Santa Cruz, Santa Cruz, California, United States

JENNIFER L. SANDS

Department of Ecology, Montana State University, Bozeman, Montana, United States

RONALD J. SCHUSTERMAN

Institute of Marine Sciences, Long Marine Laboratory, University of California–Santa Cruz, Santa Cruz, California, United States

ROBERT M. SEYFARTH

Department of Psychology, University of Pennsylvania, Philadelphia, Pennsylvania, United States

BARBARA L. TAYLOR

Southwest Fisheries Science Center, National Marine Fisheries Service, La Jolla, California, United States

PETER L. TYACK

Biology Department, Woods Hole Oceanographic Institution, Woods Hole, Massachusetts, United States

JAN A. R. A. M. VAN HOOFF

Department of Ethology and Socioecology, University of Utrecht, Utrecht, the Netherlands

CAREL P. VAN SCHAIK

Department of Biological Anthropology and Anatomy, Duke University, Durham, North Carolina, United States

BERNHARD VOELKL

Institute of Zoology, University of Vienna, Vienna, Austria

SOFIA A. WAHAJ

Department of Zoology, Michigan State University, East Lansing, Michigan, United States

RANDALL S. WELLS

Center for Marine Mammal and Sea Turtle Research, Mote Marine Laboratory, Sarasota, Florida, United States

MEREDITH J. WEST

Department of Psychology, Indiana University, Bloomington, Indiana, United States

DAVID J. WHITE

Department of Psychology, Indiana University, Bloomington, Indiana, United States

HAL WHITEHEAD

Department of Biology, Dalhousie University, Halifax, Nova Scotia, Canada

GERALD S. WILKINSON

Department of Biology, University of Maryland, College Park, Maryland, United States

HARALD YURK

Department of Zoology, University of British Columbia, Vancouver, British Columbia, Canada

KLAUS ZUBERBÜHLER

Department of Primatology, Max Planck Institute for Evolutionary Anthropology, Leipzig, Germany

Index

Aardwolves, 123, 125. *See also* Hyenas
Accomodation, 359–360
Acoustic communication. *See* Auditory communication; Vocal communication
Adang, Otto, 283
Adaptation: social strategizing and, 5, 9–11, 22–23; hypothesis evaluation and, 6–9
Aerostat observation, 33, 39
African wild dogs, 164–166
Age at first reproduction (AFR), 11–12; variation in, 14–15; primates and, 16; dolphins and, 41; hyenas and, 128
Aggression: capuchin monkeys and, 111–114; rank and, 112; redirected, 112; separating interventions and, 113; dolphins and, 116; hyenas and, 126–127, 130–131, 134, 136–137, 146, 151; social stress and, 153–169; glucocorticoid levels and, 162–169; wolves and, 166–167; equivalence classification and, 192; sea lions and, 194, 196–199; chimpanzees and, 204–205; vervet monkeys and, 211, 215; problem solving and, 237–240 (*see also* Problem solving); conflict resolution and, 240–244, 249–253; silent bared-teeth display and, 264–265; laughter and, 286–287; emotional recognition study and, 289–292; squawking and, 303; intimidation displays and, 398–400. *See also* Predation
Alarm calls, 363; social knowledge and, 201–203; primates and, 255–256; parrots and, 303; prime-probe studies and, 364–367
Alliances, 115, 120; super-, 116–119; triadic interactions and, 117–119; reciprocal altruism and, 244–248; bats and, 327. *See also* Coalitions
Allomothers, 58–59, 65, 74, 343
Allonursing: elephant and, 59; sperm whale and, 451
Amboseli National Park, 57
American Sign Language, 255
Analysis of covariance (ANCOVA), 166–167
Ant fishing, and chimpanzees, 395
Anticipation, 99–101
Arboreality: primates and, 5–6, 15–16; bats and, 15–16; mortality and, 15–17; brain size and, 17–18; Taï chimpanzees and, 99
Association coefficients, 115
Atlantic spotted dolphins, 33
Attachment, hyenas and, 132–133
Auditory communication, 34, 102–104; elephants and, 73–79; hyenas and, 137–138; alarm and, 201–203, 255–256, 303, 363–367; vervet monkeys and, 210, 212–213; bats and, 328–341; semantics and, 362–367; prime-probe studies and, 364–367; drumming and, 398–399, 405, 408–410, 412; sperm whale codas, 452–453; humpback whale and, 455; bird songs and, 472–487. *See also* Laughter; Vocal communication

Baboons, 149, 209–212; social stress and, 156; glucocorticoid levels and, 157, 160; equivalence classification and, 193; juvenile, 220; laughter and, 268–269. *See also* Primates
Barrett-Lennard, L. G., 457
Bats, 258; maximum life span (MLS) and, 13; arboreality and, 15–16; group size and, 323–325; foraging and, 324–325, 328–329, 339–340; individuality and, 324–325, 345; body size and, 325; roost transfer and, 325–326;

Bats *(continued)*
group stability and, 325–327; female, 326–328; male, 326–328; mating and, 327–328, 331–333; echolocation and, 328, 330, 333–334; cognition and, 328–330, 339–341; vocal complexity and, 330; isolation calls and, 334–338; brain of, 339; grooming and, 339–340; primates and, 339–340; language and, 340–341; open communication systems and, 358–359
Bears, 271
Behavior: social strategizing and, 5, 9–11, 22–23; brain structure and, 8; disease risks and, 26–31; promiscuity and, 27–28; dolphins and, 32–35 (*see also* Dolphins); feeding, 53–55 (*see also* Foraging); elephants and, 59–62, 74–75, 79–84; mortality and, 81–83; attachment and, 132–133; hyenas and, 132–141, 149–152; social play and, 133–134, 266–279, 398 (*see also* Play); greeting ceremonies and, 134–135; reconciliation and, 134–135, 240–244, 249–253; mongooses and, 160–164; glucocorticoid levels and, 162–169; African wild dogs and, 164–166; wolves and, 166–167; sperm whales and, 170–174, 445–464; equivalence classification and, 179–181 (*see also* Equivalence classification); matching-to-sample reinforcement and, 180–187; contextual control and, 187–191; sea lions and, 193–200; vervet monkeys and, 207–220; chunking and, 224–226; problem solving and, 230–248; conciliatory, 241, 250–253; reciprocal altruism and, 244–248; laughter and, 260–287; silent bared-teeth display and, 262–265, 270, 276–279; emotional recognition and, 288–292; parrots and, 293–316; bats and, 323–341; grooming, 339–340, 402–405; food washing, 375–380; nut cracking, 380–387; chimpanzees and, 392–413 (*see also* Chimpanzees); troop-oriented, 393; ant fishing, 395; wood licking, 395–397; locality-specific, 395–405; leaf swallowing, 396, 398; rock licking, 396–397; drumming, 398–399, 405, 408–410, 412; intimidation, 398–400; rock throwing, 399–400; courtship, 400–402, 410–411; innovation and, 405–408; individuality in, 408–411 (*see also* Individuality); cultural identity issues and, 424–426; genetic diversity and, 452–453; killer whales and, 465–469, birds and, 470–492; synergy and, 471. *See also* Aggression; Cultural transmission; Predation
Bekoff, M., 266
Birds, 372–373, 423; parrots, 257, 293–316; play and, 266; hawks, 299; owls, 299; individuality and, 345; contact calls and, 359–360; songs of, 470, 472–487; cowbirds, 472–483; starlings, 472, 483–488
Bobby (chimpanzee), 317–321
Boesch, Christophe, 87–88, 93–110
Bond groups: elephants and, 64–66; hyenas and, 123; sea lions and, 193–200. *See also* Relationships
Bonner, John T., 423
Bossou chimpanzees, 380–387
Bottlenose dolphins. *See* Dolphins
Boysen, Sarah T., 255, 317–321
Bradbury, Jack W., 257, 293–316
Brain: comparative neuroanatomical approach and, 1, 6–11, 22–23; neocortex, 1, 8; tissue, 2; behavior and, 8; motor skills and, 8; scaling techniques for, 8–9; cerebral cortex, 93; matching-to-sample reinforcement and, 180–187; chunking and, 224–226; bats and, 339
Brain size, 1–4; measurement of, 8–9; sociality effects on, 10; life history and, 11–20, 25; MLS and, 12; arboreality and, 17–18; correlated evolution and, 18–20; elephants and, 57, 83, 85; primates and, 93; intelligence and, 122; sperm whales and, 448
Buffalo, 141
Bull groups, 70

Caldwell, David, 347, 349, 351, 353, 357
Caldwell, Melba, 347, 349, 351, 353, 357
Call, J., 192
Canaries, 476
Capuchin monkeys: foraging and, 111; aggressive gestures of, 111–112; subordinates and, 112–114; separating interventions and, 113; pseudo-coalitions and, 113–114; male-male coalitions and, 114; problem solving and, 235; cooperation and, 246–247; tool use and, 440–443; juvenile, 442–443. *See also* Monkeys
Cercopithecine monkeys, 130
Cerebral cortex, 93
Chaiken, M. L., 486–487
Cheney, Dorothy L., 176, 192–193, 207–229
Chimpanzees, 490; predation and, 87–88, 98–101; female, 95–96, 105–106, 108, 414–418; male, 95–96, 98–99, 101–107, 414–418; dependence and, 95–98; group hunt and, 98–101; territorial defense and, 102–107; frontal attack and, 104–105; back-and-forth attack and, 105–106; rearguard supported attack and, 106; lateral attack and, 106–107; commando tactics of, 107; foraging and, 107–109, 317; cognition and, 109–110; equivalence classifica-

tion and, 182; Goodall study on, 204–205; problem solving and, 234–236, 239–240; conflict resolution and, 240–244; laughter and, 272–273, 283–285; play and, 274, 283, 398; emotional recognition by, 288–292; vocal signaling and, 317–321; nut cracking and, 380–387; individuality and, 392–413; humans and, 394; ant fishing and, 395; wood licking and, 395–397; specific behavior of Mahale, 395–405; leaf swallowing and, 396, 398; rock licking and, 396–397; juveniles, 398; intimidation displays and, 398–400; courtship displays and, 400–402; grooming and, 402–405; innovation and, 405–408; flexible behavior of, 408–411; termite fishing and, 414–418; cultural identity issues and, 419–439; tradition and, 428–431; language and, 431–432. *See also* Primates

Chuckles, 266–267, 272–274

Chunking, 177, 224–226; structure value and, 227–229

Clans, 66, 90

Coalitions: defense and, 102–107; Taï chimpanzees and, 102–107; white-faced capuchins and, 111–114; alliances and, 115–120, 244–248, 327; dolphins and, 115–120, 343; hyenas and, 130–131, 149–152; sea lions and, 199–200; equivalence classification and, 204–205; problem solving and, 237–240

Coefficient of association. *See* Half-weight coefficient of association

Cognition, 175–178; arboreal lifestyle and, 2; Machiavellian intelligence and, 5, 7, 238, 322, 339, 444–445; social strategizing and, 5, 9–11; comparison of hypotheses, 6–9; brain structure and, 6–11; life history and, 11–20, 25; selective advantages of, 20–24; domain-general, 23–24; Taï chimpanzees and, 109–110; hyenas and, 141–146; sea lions and, 193–200; equivalence classification and, 206 (*see also* Equivalence classification); laughter and, 282–286; vocal signaling and, 317–321; phylogenetic effects and, 322; bats and, 328–330, 339–341; semanticity and, 362–367; infant babbling and, 490–491. *See also* Intelligence

Collaboration. *See* Cooperative strategies

Communication, 257–259; visual, 8, 103–104, 135–137; dolphins, 34, 37, 55–56, 342–361; elephants, 59–62, 64, 71, 73–79; hyenas, 135–140; body posture and, 136; vocal, 137–138 (*see also* Vocal communication); olfactory, 138–140, 346; sea lions, 194–195; alarm and, 201–203, 255–256, 303, 363–367; food calling and, 203; mothers and offspring, 203–204; vervet monkeys, 210, 212–213; laughter and, 260–287; silent bared-teeth display and, 262–265, 270, 276–279; ritualization and, 264–265; wild parrots, 293–316; loud contact calls, 301, 305–310; bats, 322–341; echolocation and, 328, 330, 333–334; mating and, 331–333; isolation calls and, 334–338; signature whistles and, 346–361; problem solving and, 360–361; primate semanticity and, 362–367; prime-probe studies and, 364–367; sperm whale codas, 452–453; humpback whale, 455; killer whales, 467–469; bird songs and, 470–487

Comparative neuroanatomical approach, 1; evaluation of hypotheses and, 6–9; scaling techniques and, 8–9; social strategizing and, 9–11, 22–23; cognition advantages and, 20–24; taxonomic scope and, 21

Conciliatory tendency (CT), 241, 250–253

Conflict resolution. *See* Reconciliation

Connor, Richard C., 115–120, 173, 189–190, 193

Contact call, parrots and, 301–302

Cooperative strategies, 89–92, 142–145; foraging and, 87–88; defined, 93–94; four levels of, 94; rarity of, 94–95; synchrony and, 94, 137, 143–144; group hunt and, 98–101; anticipation, 99–101; ambushers, 100; blockers, 100; chasers, 100; drivers, 100; coalitions, 102–107; territorial defense and, 102–107; limiting factors to, 109–110; capuchin monkeys and, 111–114, 246–247. *See also* Chimpanzees

Courtship displays, 410–411; leaf clipping, 400–401; shrub bending, 401–402; bipedal swagger, 402

Cowbirds: as parasites, 472; female, 473–483, 489–490; songs of, 473–483; male, 474–483, 489–490; captive conditions and, 475–478; juvenile, 476, 489; mating and, 476–477; flock conditions and, 478–483

Creel, Scott, 91, 153–169

Crematogaster ants, 395–396

Cultural transmission, 369, 373, 439; behavioral diversity and, 52–56, 424–426; cetaceans and, 52–56, 444–469; habitat lore, elephants and, 300–301; parrots and, 300–301; fidelity and, 370; mechanisms of, 371; hitchhiking and, 372, 458–461; definitions for, 374–375, 432–434, 453–454, 470–471; monkeys and, 374–387; food washing and, 375–380; nut cracking and, 380–387; movement imitation and, 388–391; chimpanzees and, 392–413, 419–439; development of, 405; flexibility in, 408–413;

Cultural transmission *(continued)*
termite fishing and, 414–418; identity issues in, 419–424; human facilitation and, 423–424; social learning and, 426–428; tradition and, 428–431; without language, 431–432; collectivity and, 434–436; anthropology and, 436–437; complexity and, 437–438; tool use and, 440–443; ocean environment and, 460–464; killer whales and, 465–469; birds and, 470–492; synergy and, 471

Dallal, N., 224–225
Darrell (chimpanzee), 317–321
Darwin, Charles, 261–264
Dasser, V., 211
Dawkins, M. Stamp, 271
Deaner, Robert O., 1–25
Defense strategies, 110; chimpanzees and, 102–107; sea lions and, 196–197; squawking, 303
de Waal, Frans B. M., 177, 230–248, 283
Diana monkeys, 201–203; alarm call study of, 364–367
Disease: fecal contamination and, 26; insects and, 26; sociality risks and, 26–31; promiscuity and, 27–28
Dizon, Andrew E., 170–174
Dogura (chimpanzee), 403
Dolphins, 2–3; long term studies in, 32–38; aerostat observation of, 33, 39; observation difficulties in, 33; range of, 33, 457; age determination of, 33–34; communication by, 34–37, 55–56, 342–361; reproduction and, 34–35, 38–41, 43–44, 51–52, 342–343; Sarasota Bay study of, 34–51, 54–55; capture of, 35, 37; tagging of, 35, 38; life history for, 39–41; mortality and, 41; social system of, 41–51; community definition of, 42; geographical effects and, 42–43; bands of, 43–45; female, 43–47, 115–116, 342, 346–347; intelligence and, 45, 453–454; variability in, 45; juvenile, 47–48, 342–343, 346–347; male, 48–51, 115–117, 343–344, 359; pairing and, 48–51; mating and, 51–52, 342; cultural transmission and, 52–56, 453–464; sponge use and, 53; foraging and, 53–55; association coefficients and, 115; alliance formation and, 115–120; triadic interactions and, 117–118; equivalence classification and, 182, 189–190; conflict resolution and, 243; individual-specific relationships and, 342–361; allomothers and, 343; coalitions and, 343–344; vision of, 346; signature whistles and, 346–361, 454–455; learning and, 349; social stability of, 456; innovation and, 457
Dominance. *See* Rank
Drea, Christine M., 88–89, 121–148
Drumming, 398–399, 405, 408–410, 412
Dunbar, Robin I. M., 324

Echolocation, 328, 330, 333–334
Elephants: brain size and, 57, 83, 85; habitat of, 57; allomothers and, 58, 65, 74; hierarchy of, 58, 64, 69, 71–73; life span of, 58; group size and, 58–59, 70; juveniles of, 58–59, 65; reproduction and, 58, 62–63, 70–73, 77–78; female, 58–69, 85; mortality and, 59, 80–83; individuality and, 59–60, 79–80, 83, 85; behavior and, 59–62, 74–75, 79–84; communication and, 59–62, 64, 71, 73–79; matriarchs and, 62–64, 69; memory of, 63; water sources and, 63; bond groups and, 64–66; cost of alliances and, 65–66; clans and, 66; range of, 66–69; male, 69–73; musth and, 71–73, 77–78; life history and, 83, 85
Emotion: smiling and, 256, 276, 279–282, 285–286; laughter and, 260–287; intuitive interpretation of, 263–264; joy, 284–286; wooing, 285–286; chimpanzees and, 288–292. *See also* Behavior; Social stress
Empathy, 288–292
Encephalization quotient, 13
Endocrine correlates. *See* Glucocorticoids (GC)
Engh, Anne, 90–92, 149–152
Engle, Marianne, 486–487
Environmental influence: primates and, 14–18; relationships and, 42–43; sperm whales and, 445–448, 460–464; cowbirds and, 478–483; starlings and, 484–486
Equivalence classification, 179, 206, 223; development of, 180; matching-to-sample context and, 180–187; mathematics and, 181, 187–188; reflexivity and, 181, 188; symmetry and, 181, 188; transitivity and, 181, 188, 220; dolphins and, 182, 189–190; functional classes and, 182, 184–187; sea lions and, 182–183, 193–200, 217–218; sequential organization and, 184; pigeons and, 185; contextual control and, 187–191; vervet monkeys and, 190, 201–204, 217–224; different approaches to, 191–193; primates and, 191–193; triadic relationships and, 191–193, 203; aggression and, 192; baboons and, 193; coalitions and, 204–205; multiple measurement and, 217–219; individuality and, 218–220; member substitution and, 219; training distortions and, 221–222
Escorza-Treviño, Sergio, 170–174
Evans, Karen, 170–174

Evolution: brain size and, 1–4; primate cognition and, 5–25; natural selection and, 18–20; *Homo* v. *Pan*, 21–22, 25; domain-general cognition and, 23–24; dominance and, 153, 155; laughter and, 261–264; open communication systems and, 358–359; cetacean culture and, 457–458; K-selection and, 462
Evolution of Culture in Animals, The (Bonner), 423
Extractive foraging, 6

Facial expression, 260–287; 288–292. *See also* Rombt display
"Family of Chimps, The" (Haanstra), 284
Fecal contamination, 26
Fixed action pattern, 261
Flanking, 51
Flight, 15–17
Food sharing: parrots and, 299; bats and, 324
Foraging, 6, 122; dolphins and, 53–55; elephants and, 66–69; humans and, 95, 98; chimpanzees and, 95–98, 317, 395–396; hyenas and, 123, 142–146; food calling and, 203; parrots and, 295, 300; bats and, 324–325, 328–329, 339–340; food washing and, 375–380; nut cracking and, 380–387, 441–443; ants and, 395–396; new foods and, 406; termite fishing and, 414–418; sperm whales and, 450–451; K-selection and, 462. *See also* Cooperative strategies
Frank, Laurence G., 88–89, 121–148, 250
Freeberg, Todd, 489
Free-recall, 227
Freud, Sigmund, 393
Functional classes, 182, 184–187
Fur seals, 194

Genetic exclusion testing, 34
Gisiner, R. C., 182, 184, 196–197, 199–200
Glucocorticoids (GC), 91, 168–169; baboons and, 157, 160; social stress and, 157–160; mongooses and, 162–163, 164; African wild dogs and, 165; wolves and, 165, 166–167
Goldstein, Michael, 490
Gombe National Park, 2, 99, 204, 370
Goodall, Jane, 204–205, 370
Greeting ceremonies, 134–135; elephants and, 60–61; hyenas and, 134–135
Grooming, 412; bats and, 339–340; leaf, 402; hand-clasp, 402–404; social scratch, 404–405
Group: bands, 43–45; troops, 376, 393
Group size, 7; parasites and, 27; elephants and, 58–59, 70; coalitions and, 102–107; hyenas and, 123; mongooses and, 160–161; sperm whales and, 170–171; language and, 322–323; bats and, 323–325; stability and, 325–327; troops and, 376

Haanstra, B., 284
Half-weight coefficient of association, 115
Hall, K. R. L., 238
Hallberg, Karen I., 255, 317–321
Hardy, J. W., 308
Hawks, 299
Hemelrijk, Charlotte, 283
Hi (dolphin), 115, 119
Hinde, Robert A., 231–232
Holekamp, Kay E., 90–92, 149–152, 177, 249–253
Hormones. *See* Glucocortoids (GC)
Huber, Ludwig, 370–371, 388–391
Humans: foraging and, 95, 98; imitation and, 109; labor division, 109; matching-to-sample reinforcement and, 180; group recognition and, 216–217; laughter and, 265–274, 283–284, 286; rombt display and, 271–272; language and, 322–323, 431–432; culture and, 394, 419–424; tradition and, 428–431; starling interaction and, 485–486; infant babbling and, 490–491
Humor. *See* Laughter
Humpback whales, 455, 458, 468
Humphrey, Nicholas, 444–445
Hunting. *See* Foraging; Predation
Hyde, John, 170–174
Hyenas, 121; predation and, 88–89, 123, 125, 140–142, 146–147; parental care and, 89; clans and, 90; female, 90, 124–128, 130, 132–133, 149–150, 203–204; reproduction and, 90, 124–129; intelligence and, 122, 141–152; bonding and, 123; group size and, 123; mating and, 123; foraging and, 123, 142–146; male, 123–124, 128–129; body size of, 125; life history of, 125–129; bite-shake attack, 126–127; aggression and, 126–127, 130–131, 134, 136–137, 146, 151; juveniles, 126–130, 150–151; rank and, 128–132, 143, 149–152; coalitions and, 130–131; communication and, 135–140, 203–204; olfactory cues and, 138–140; commuting and, 146; hostile encounters and, 146; individuality and, 146–147; conflict resolution and, 243, 249–253

Imanishi, Kinji, 369–370, 392–393; food washing study and, 375–380; nut cracking study and, 380–387

Imitation, 293, 360; human, 109; food washing and, 375–380; nut cracking and, 380–387, 441–443; of movement, 388–391; dolphins and, 453–455; sperm whales and, 453–455; goal emulation and, 455; killer whales and, 467–469; starlings and, 484–487
Immune system: leukocyte counts and, 26; STDs and, 27, 30; WBC types and, 28, 30
Imo (monkey), 376–379
Individuality, 231; elephants and, 59–60, 79–80, 83, 85; hyenas and, 146–147; dominance and, 153–157; social stress and, 153–169; sperm whales and, 172–173; equivalence classification and, 218–220; loud contact calls, 305–308; bats and, 324–325, 345; specific relationships and, 342–361; dolphins and, 344–361; birds and, 345; olfactory communication and, 346; signature whistles and, 346–361; monkeys and, 374–375; nut cracking and, 380–387; chimpanzees and, 392–413; drumming and, 398–399, 408–410. *See also* Behavior
Innovation, 405; new foods and, 406; leaf napkin and, 406–407; strange encounters and, 407–408; termite fishing and, 414–418; sperm whales and, 457
Insects, 26
Insley, S. J., 194–195
Intelligence: brain size and, 1–4, 18–20, 122; comparative neuroanatomical approach and, 1, 6–11, 22–23; social strategizing hypothesis and, 5; Machiavellian, 5, 7, 238, 322, 339, 444–445; primates and, 5–25, 89, 122; arboreality and, 17–18; dolphins and, 45, 349–358, 453–455; hyenas and, 121, 141–152; equivalence classification and, 179 (*see also* Equivalence classification); matching-to-sample reinforcement and, 180–187; without reinforcement, 190; sea lions and, 193–200; training distortion and, 221–222; chunking and, 224–226; structure value and, 227–229; problem solving and, 230–248; parrots and, 293, 300–301, 304–305, 316–317; vocal learning and, 304–305; phylogenetic effects and, 322; food washing and, 375–380; nut cracking and, 380–387, 441–443; tool use and, 380–387, 406–407, 414–418, 440–443; movement imitation and, 388–391; innovation and, 405–408, 414–418, 457; termite fishing and, 414–418; cultural identity issues and, 426–428; sperm whales and, 444–445, 453–455; killer whales and, 467–469; starlings and, 486–487, infant babbling and, 490–491. *See also* Cultural transmission
Intimidation displays, 398–400
Isolation calls, 334–338
Itani, Jun'ichiro, 392; food washing study and, 375–380; nut cracking study and, 380–387

Janik, V. M., 348–349, 356–359, 361
Joy, 284–286

Kahame community, 204–205
Kalunde (chimpanzee), 403
Kaplan, H., 95
Kasekela community, 204–205
Kastak, Colleen Reichmuth, 175, 179–206, 217–223
Kastak, David, 175, 179–206, 217–223
Kawai, Masao, 192, 237–238, 377, 392
Kawamura, Shunzo, 376–377, 392
Killer whales, 174; predation and, 447–448, 465–467; cultural transmission and, 465–469; female, 467; residents, grouping of, 467; matrilines of, 467–468; intelligence and, 467–469; vocal clans and, 467–469
King, Andrew P., 372–373, 470–492
Kinship, 64–66, 118; hyenas and, 131, 141–142; sperm whales and, 172; sea lions and, 195; social knowledge and, 208–212; vervet monkeys and, 210–212, 218–219; cetaceans and, 457–458
Köhler, Wolfgang, 177
Kohts, Nadia, 263
Koshima monkeys, 375–380, 387
Krützen, Michael, 115–120
K-selection, 16, 462
Kummer, Hans, 175, 231–232

Language, 201; evolution of, 230–248, 317–321, 362–367; equivalence classification and, 181–182; syntactical structure and, 233–237; group size and, 322–323; bats and, 340–341; cultural identity issues and, 431–432. *See also* Communication
Laughter, 256; musculature in, 260; development of, 261; fixed action pattern and, 261; universality of, 261; Darwin on, 261–264; silent bared-teeth display and, 262–265, 270, 276–279; intuitive interpretation of, 263–264; ritualization and, 264–265; human homologue for, 265–279; play and, 266–268; variants in, 268–275; rombt displays and, 271–272; motivational basis of, 275–276; smile origin and, 276; power asymmetry hypothesis and, 279–282; comment laugh and, 282–284; cognition and, 282–286; self-awareness and, 284–286; origins of, 286; aggression and, 286–287

Leaves: swallowing of, 396, 398; clipping of, 400–401; grooming and, 402; as napkins, 406–407
Leukocyte counts, 26–31
Life history: primates and, 5–20; correlation issues and, 11–12, 18–20; cognition and, 11–20, 25; encephalization quotient and, 13; variation in mammalian, 14–18; flying and, 15–17; birth intervals and, 22; dolphins and, 39–41; elephants and, 83, 85; hyenas and, 125–129
Lions, 98
Luit (chimpanzee), 239–240

Macaques, 210–211; laughter and, 262, 268–270; rombt display and, 271–272; play and, 274–275
McGrew, William, 371, 419–439
Machiavellian intelligence. *See* Social strategizing hypothesis
Macuda, T., 225–226
Mahale National Park, 99
Mama (chimpanzee), 239–240
Mangabeys, 137
Manning, A., 271
Mannu, Massimo, 371–372, 440–443
Matching-to-sample reinforcement: behavior and, 180–187; emotional recognition study and, 289
Mating: promiscuity and, 27–28; flanking and, 51; dolphins and, 51–52, 342; hyenas and, 123; bats and, 327–328; courtship displays and, 331–333, 400–402; cowbirds and, 476–477; starlings and, 483–484. *See also* Reproduction
Matriarchs, 62–64
Matrilineal system, 58–64, 118–120, 128–132, 149–152
Matsuzawa, Tetsuro, 234–235, 369, 374–387
Maximum life span (MLS), 11–12; encephalization quotient and, 13; body size and, 15
Meck, W., 224–225
Memory: chunking and, 224–226; elementary association and, 226–227; free-recall and, 227; training effects and, 227; structure value and, 227–229; bats and, 324–325
Menzel, Emil W., 436
Mesnick, Sarah L., 89, 170–174
Mills, M. G. L., 139
Mimicry. *See* Imitation
Mito, Satsue, 376
Mongooses, 148; dominance and, 153–155; group size and, 160–161; social stress and, 160–164
Monkeys, 363; alliances and, 118; mothers and offspring, 203; problem solving and, 237–240; reciprocal altruism and, 245–248; semanticity and, 362–367; prime-probe experiment and, 364–367; individuality and, 374–375; cultural transmission and, 374–387; troops of, 376; nut cracking and, 380–387; movement imitation and, 388–391; tool use and, 440–443. *See also individual species;* Primates
Mortality: body size and, 15; arboreality and, 15–17; natural selection and, 18–20; dolphins and, 41; elephants and, 59, 80–83
Moss, Cynthia, 57
Motor dexterity, 8
Mudumalai Wildlife Sanctuary, 57
Musth, 71–73, 77–78
Mutualism, 245–246

Natural selection, 18–20
Nelson, Douglas A., 255, 317–321
Neocortex, 1, 8
Networks, 3–4
Nikkie (chimpanzee), 239–240
Nishida, Toshisada, 369–371, 392–413
Nouabale-Ndoki National Park, 57
Nunn, Charles L., 2, 26–31
Nut cracking: chimpanzee tool use and, 234–235, 380–387; capuchin monkeys and, 441–443;

Olfactory communication: elephant and, 71, 79; hyenas and, 138–140; dolphins as without, 346
On the Expression of Emotions in Man and Animals (Darwin), 261
Ottoni, Eduardo B., 371–372, 440–443
Owls, 299

Pair duets, 302, 310–312, 314–315
Pandolfi, Stephanie S., 371, 414–418
Panting, 271
Parasites, 26; group size and, 27; WBC types and, 28; cowbirds as, 472, 483
Parr, Lisa A., 257, 288–292
Parrots, 257; intelligence and, 293; learning and, 293, 300–301, 304–305, 315–316; sound generation and, 294; various types of, 294; diet of, 295; foraging and, 295, 300; range of, 295–296; daily cycle of, 296–297; night roosts of, 297–298, 309; reproduction and, 298; predation and, 298–299; social structure of, 299–300; juveniles, 300–301; calls of, 301–314; warble songs and, 302, 312–314; dialects and, 308–309; vocal duets and, 310–312
Payne, Katy, 3–4, 57–85
PC-MC method, 241
Pepperberg, Irene, 293

Perry, Susan, 111–114
Personality. *See* Individuality
Pigeons, 184–185, 226
Play, 398; hyenas and, 133–134; social carnivores and, 169; chuckles and, 266–267, 272–274, 284; open-mouth display and, 267–272, 274–275; parrots and, 269; dance, 271; panting and, 271; rombt displays and, 271–272; silent bared-teeth display and, 276–279
Pongo-linguistics, 431–432
Pool-making, 379
Power asymmetry hypothesis, 279–282
Precultural Primate Behavior (Menzel), 436
Predation, 2; hyenas and, 88–89, 123, 125, 140–142, 146–147; dolphins and, 41; chimpanzees and, 87–88, 95–101; anticipation and, 99–101; sperm whales and, 174, 447–448, 461; parrots and, 298–299; alarm calls and, 363. *See also* Cooperative strategies
Preuschoft, Signe, 256–257, 260–287, 269–287
Primates: arboreal activities and, 5–6, 15–16; cognitive evolution of, 5–25; social strategizing and, 9–11, 22–23; life history and, 11–20, 25; age at first reproduction and, 16; analysis of hypotheses and, 20–22; *Homo* v. *Pan,* 21–22, 25; domain-general adaptations and, 23–24; leukocyte counts in, 26–31; intelligence and, 89; brain size and, 93; collaboration and, 94; similarities to hyenas, 125–129, 145, 149–150; equivalence classification and, 191–193; problem solving and, 230–233; laughter and, 262–263, 267–277, 279, 281–287; silent bared-teeth display and, 262–264, 270, 276–277, 279; play and, 266–279; rombt display and, 271–272; smiling and, 281–282; bats and, 339–340; natural semanticity in, 362–367; alarm calls and, 363; prime-probe experiment and, 364–367; cultural transmission and, 374–387; food washing and, 375–380; nut cracking and, 380–387; movement imitation and, 388–391. *See also* Alliances; Chimpanzees; Coalitions; Cooperative strategies; Monkeys; Reconciliation
Prime-probe studies, 364–367
Problem solving: primates and, 230–233; structure for, 233–237; coalitions and, 237–240; conflict resolution and, 240–244, 249–253; reciprocal exchange and, 244–248; communication and, 360–361; innovative behavior and, 405–408
Promiscuity, 27–28
Pusey, Anne E., 371, 414–418
Ranging patterns: dolphins and, 41–43; elephants and, 66–69; parrots and, 269; sperm whales and, 449–450
Rank, 90–91, 177–178; elephants and, 58, 64, 69, 71, 73; aggression and, 112; hyenas and, 128–132, 143, 149–152; social play and, 133–134; inheritance of, 149–152; dominance and, 153–156; social stress and, 153–169; cost of, 156; mongooses and, 160–164; glucocorticoid levels and, 162–169; African wild dogs and, 164–166; wolves and, 166–167; social knowledge and, 212–215; vervet monkeys and, 212–217; problem solving and, 237–240; troops and, 376
Rats, 224–227
Real Notch (dolphin), 115, 119
Reciprocal altruism, bats and, 244–248
Reconciliation, 240–244, 249–253; hyenas and, 134–135
Red colobus monkeys, 87–88, 99–101
Reflexivity, 181–182, 188
Reinforcement: matching-to-sample, 180–187; learning without, 190
Relationships: group size and, 7; primates and, 9–11, 22–23; disease risk and, 26–31; geography and, 42–43; dolphins and, 43–51, 115–120, 342–361; elephants and, 58–73; bond groups, 64–66, 123, 193–200; clans, 66, 90; family groups, 66–69; bull groups, 70; rank and, 90–91 (*see also* Rank); aggression and, 111–114; association coefficients, 115; alliances, 115–120, 244–248, 327; triadic, 117–119, 131–132 (*see also* Triadic relationships); kinship and, 118 (*see also* Kinship); hyenas and, 129–135; attachment and, 132–133; social play and, 133–134; greeting ceremonies and, 134–135; reconciliation and, 134–135; social stress and, 153–169; mongooses and, 160–164; African wild dogs and, 164–166; wolves and, 166–167; sperm whales and, 170–174, 448–453, 448–464; sea lions and, 193–200; dyadic, 222–223; problem solving in, 230–248; conflict resolution and, 240–244; reciprocal altruism and, 244–248; silent bared-teeth display and, 262–265, 270, 276–279; parrots and, 299–300; bats and, 323–328; individual-specific, 342–361. *See also* Cooperative strategies
Rendell, Luke, 453
Reproduction: sexually transmitted diseases and, 27, 30; sperm competition 28; dolphins and, 34–35, 38–41, 43–44, 51–52, 342–343; elephants and, 58, 62–63, 70–73, 77–78; hyenas and, 90, 124–129; mongooses and, 160–164;

African wild dogs and, 164–166; sea lions and, 194–197, 199–200; vervet monkeys and, 221; parrots and, 298; sperm whales and, 449–450. *See also* Mating
Rio (sea lion), 217–218, 222
Roberts, W. A., 225–226
Rock throwing, 399–400
Rombt (relaxed open-mouth bared-teeth) display, 271–272

Sandegren, F. E., 195–196
Sands, Jennifer L., 91, 153–169
Sarasota Bay, 34–36, 354
Sayigh, L. S., 350–352, 354, 356–357
Scaling techniques, 8–9
Schilder, Matthijs, 283
Schusterman, Ronald J., 175, 179–206, 217–218, 220–223
Sea lions, 175–176; equivalence classification and, 182–183, 193–200, 217–218; functional classes and, 185–187; kinship and, 193–196; aggression and, 194, 196–199; communication by, 194–195; reproduction and, 194–197, 199–200; female, 194–197, 199–200; male, 194–200; triadic relationships and, 197, 199; coalitions and, 199–200; training distortion and, 221–222
Separating interventions, 113
Sex. *See* Reproduction
Seyfarth, Robert M., 176, 192–193, 207–229
Sidman, M., 180–181, 185, 190
Signature whistle. *See* Whistles
Silk, J. B., 209, 215, 221
Similarity cooperation, 94
Slater, Peter J. B., 348–349, 357–359
Smiling, 256, 276; power asymmetry hypothesis and, 279–282; chimpanzees and, 285; wooing function of, 285–286; aggression and, 286–287. *See also* Laughter
Social complexity, 89–90; intelligence and, 1–4 (*see also* Intelligence); disease risk and, 26–31; cultural transmission and, 52–56 (*see also* Cultural transmission); cooperative strategies and, 93–110; problem solving in, 230–248; power asymmetry hypothesis and, 279–282. *See also* Rank
Social knowledge: complexity and, 207–208, 215–216; kinship and, 208–212; rank and, 212–215; humans and, 216–217; chunking and, 224–226; free-recall and, 227; training distortions and, 227; structure value and, 227–229. *See also* Equivalence classification
Social play, 133–134. *See also* Play
Social strategizing hypothesis, 5, 7, 9–11, 238; behavior and, 9–11; weaknesses of, 10; birth intervals and, 22; life history and, 22–23; bats and, 322, 339; sperm whales and, 444–445.
Social stress, 91, 168–169; dominance and, 153–160; fighting and, 156–157; endocrine correlates and, 157–160; subordination and, 156–160; mongooses and, 160–164; glucocorticoid levels and, 162–169; African wild dogs and, 164–166; wolves and, 166–167
Spatiotemporal mapping, 6–7
Sperm competition, 28
Sperm whales, 89, 170–173; predation and, 174, 447–448, 461; range of, 444, 448–449, 457; intelligence and, 444–445, 453–454; environment and, 445–448, 460–464; body size and, 448; brain size and, 448; male, 448–450; social structure of, 448–453; juveniles, 449–451; reproduction and, 449–450; group size and, 450; foraging and, 450–451; female, 450–453; communication by, 452–453; cultural transmission and, 453–464; imitation and, 454–455; social stability of, 456; innovation and, 457; genetic diversity in, 458–460; K-selection and, 462; destruction of, 463–464
Sponge use, by dolphins, 53
Squawking, 303
Starlings: songs of, 472, 483–488; mating and, 483–484; environmental effects and, 484–486; imitation and, 484–487; human interaction and, 485–486; social sonar of, 487–488
Stimuli: matching-to-sample reinforcement and, 180–187; functional classes and, 182, 184–187; sequential organization and, 184; contextual control and, 187–191; equivalence classification and, 216–220; chunking and, 224–226; elementary association and, 226–227; free-recall and, 227. *See also* Equivalence classification
Subordination. *See* Rank
Sugiyama, Yukimaru, 369–370
Sukumar, R., 57
Sweet potato washing, 375–380
Symmetry, 181, 188
Synchrony cooperation, 94, 137, 143–144
Syntactical structure. *See* Problem solving
Syntax, 230–248

Taï chimpanzees. *See* Chimpanzees
Taylor, Barbara L., 170–174
Termite fishing, 414–418
Tomasello, M., 192, 206
Tool use, 420, 423; nut cracking and, 380–387,

Tool use *(continued)*
441–443; ant fishing, 395; leaf napkins, 406–407; termite fishing and, 414–418; capuchin monkeys and, 440–443
Tradition, 428–431
Transitivity, 181, 188, 220
Triadic relationships, 222–223; hyenas and, 131–132; equivalence classification and, 191–193, 203; sea lions and, 197, 199; problem solving and, 237–240; conflict resolution and, 243–244
Turkalo, Andrea, 57, 59–60
Tyack, Peter L., 258–259, 342–361

van Hooff, Jan A. R. A. M., 256–257, 260–287
van Schaik, Carel P., 1–25, 371, 414–418
Vaughan, W. Jr., 184–185
Vervet monkeys: equivalence classification and, 190, 217–224; group size and, 208; male, 208–209, 215; female, 209–216, 219–220; grooming and, 209, 213–214; social recognition and, 209–210; communication and, 210, 212–213; aggression and, 211, 215; kinship and, 211–212, 218–219; rank and, 212–217; sequential training and, 214–215; individuality and, 218–220; juveniles, 219–220; reproduction and, 221; training distortion and, 221–222
Visual communication, 8, 103–104, 135–137
Vocal communication: parrots and, 293–316; loud contact call, 301, 305–310; soft contact call, 301–302; begging call, 302; distress, 302–303; pair duets, 302, 310–312, 314–315; preflight call, 302; warbles, 302, 312–314; alarms, 303; chimpanzees and, 317–321; acoustic analyses and, 319–321; echolocation and, 328, 330, 333–334; bats and, 328, 330–341; signature whistles and, 346–361; primate semanticity and, 362–367; prime-probe studies and, 364–367; sperm whale codas, 452–453; humpback whale, 455; killer whales, 467–469; bird songs and, 470–487; social sonar and, 487–488
Vocal learning: avian, 293–294; parrots and, 304–305; dolphins and, 349
Voelkl, Bernhard, 370–371, 388–391

Wahaj, Sofia A., 177, 249–253
Warble songs, 302, 312–314
Wells, Randall S., 2–3, 32–56, 343
West, Meredith J., 372–373, 470–492
Wheat placer mining, 378–379
Whistles, 346–348; learning of, 349; development of, 350–352; modification of, 352–357, 359–360; use of, 357–358
White, David J., 372–373, 470–492
White blood cells (WBCs), 2, 27–31
Whitehead, Hal, 3, 89, 173–174, 259, 372, 444–464
Whiten, A., 471, 487–488
Wildebeests, 141
Wilkinson, Gerald S., 257–258, 322–341
Wolves: social stress and, 160, 166–167; glucocorticoid levels and, 165, 166–167
Wooing, 285–286
World of Living Things, The (Imanishi), 392

Yoerg, Sonja, 136–137
Yurk, Harald, 465–469

Zebras, 141, 278
Zuberbühler, Klaus, 201, 255, 362–367